# PRINCIPLES AND PRACTICE OF CHROMATOGRAPHY

## ELLIS HORWOOD SERIES IN ANALYTICAL CHEMISTRY

*Series Editors:* Dr R. A. CHALMERS and Dr MARY MASSON, University of Aberdeen
*Consultant Editor:* Prof. J. N. MILLER, Loughborough University of Technology

S. Alegret — **Developments in Solvent Extraction**
S. Allenmark — **Chromatographic Enantioseparation—Methods and Applications**
G.E. Baiulescu, P. Dumitrescu & P.Gh. Zugravescu — **Sampling**
H. Barańska, A. Łabudzińska & J. Terpiński — **Laser Raman Spectrometry**
G.I. Bekov & V.S. Letokhov — **Laser Resonant Photoionization Spectroscopy for Trace Analysis**
K. Beyermann — **Organic Trace Analysis**
O. Budevsky — **Foundations of Chemical Analysis**
J. Buffle — **Complexation Reactions in Aquatic Systems: An Analytical Approach**
D.T. Burns, A. Townshend & A.G. Catchpole — **Inorganic Reaction Chemistry Volume 1: Systematic Chemical Separation**
D.T. Burns, A. Townshend & A.H. Carter — **Inorganic Reaction Chemistry: Volume 2: Reactions of the Elements and their Compounds: Part A: Alkali Metals to Nitrogen, Part B: Osmium to Zirconium**
J. Churáček — **New Trends in the Theory & Instrumentation of Selected Analytical Methods**
E. Constantin, A. Schnell & M. Mruzek — **Mass Spectrometry**
R. Czoch & A. Francik — **Instrumental Effects in Homodyne Electron Paramagnetic Resonance Spectrometers**
T.E. Edmonds — **Interfacing Analytical Instrumentation with Microcomputers**
Z. Galus — **Fundamentals of Electrochemical Analysis, Second Edition**
S. Görög — **Steroid Analysis in the Pharmaceutical Industry**
T. S. Harrison — **Handbook of Analytical Control of Iron and Steel Production**
J.P. Hart — **Electroanalysis of Biologically Important Compounds**
T.F. Hartley — **Computerized Quality Control: Programs for the Analytical Laboratory**
Saad S.M. Hassan — **Organic Analysis using Atomic Absorption Spectrometry**
M.H. Ho — **Analytical Methods in Forensic Chemistry**
Z. Holzbecher, L. Diviš, M. Král, L. Šůcha & F. Vláčil — **Handbook of Organic Reagents in Inorganic Chemistry**
A. Hulanicki — **Reactions of Acids and Bases in Analytical Chemistry**
David Huskins — **Electrical and Magnetic Methods in On-line Process Analysis**
David Huskins — **Optical Methods in On-line Process Analysis**
David Huskins — **Quality Measuring Instruments in On-line Process Analysis**
J. Inczédy — **Analytical Applications of Complex Equilibria**
M. Kaljurand & E. Küllik — **Computerized Multiple Input Chromatography**
S. Kotrlý & L. Šůcha — **Handbook of Chemical Equilibria in Analytical Chemistry**
J. Kragten — **Atlas of Metal-ligand Equilibria in Aqueous Solution**
A.M. Krstulović — **Quantitative Analysis of Catecholamines and Related Compounds**
F.J. Krug & E.A.G. Zagotto — **Flow Injection Analysis in Agriculture and Environmental Science**
V. Linek, V. Vacek, J. Sinkule & P. Beneš — **Measurement of Oxygen by Membrane-Covered Probes**
C. Liteanu, E. Hopîrtean & R. A. Chalmers — **Titrimetric Analytical Chemistry**
C. Liteanu & I. Rîcă — **Statistical Theory and Methodology of Trace Analysis**
Z. Marczenko — **Separation and Spectrophotometric Determination of Elements**
M. Meloun, J. Havel & E. Högfeldt — **Computation of Solution Equilibria**
M. Meloun, J. Militky & M. Forina — **Chemometrics in Instrumental Analysis: Solved Problems for IBM PC**
O. Mikeš — **Laboratory Handbook of Chromatographic and Allied Methods**
J.C. Miller & J.N. Miller — **Statistics for Analytical Chemistry, Second Edition**
J.N. Miller — **Fluorescence Spectroscopy**
J.N. Miller — **Modern Analytical Chemistry**
J. Minczewski, J. Chwastowska & R. Dybczyński — **Separation and Preconcentration Methods in Inorganic Trace Analysis**
T.T. Orlovsky — **Chromatographic Adsorption Analysis**
D. Pérez-Bendito & M. Silva — **Kinetic Methods in Analytical Chemistry**
B. Ravindranath — **Principles and Practice of Chromotography**
V. Sedivec & J. Flek — **Handbook of Analysis of Organic Solvents**
O. Shpigun & Yu. A. Zolotov — **Ion Chromatography in Water Analysis**
R.M. Smith — **Derivatization for High Pressure Liquid Chromatography**
R.M. Smith — **Handbook of Biopharmaceutic Analysis**
K.R. Spurny — **Physical and Chemical Characterization of Individual Airborne Particles**
K. Štulík & V. Pacáková — **Electroanalytical Measurements in Flowing Liquids**
J. Tölgyessy & E.H. Klehr — **Nuclear Environmental Chemical Analysis**
J. Tölgyessy & M. Kyrš — **Radioanalytical Chemistry, Volumes I & II**
J. Urbanski, *et al.* — **Handbook of Analysis of Synthetic Polymers and Plastics**
M. Valcárcel & M.D. Luque de Castro — **Flow-Injection Analysis: Principles and Applications**
C. Vandecasteele — **Activation Analysis with Charged Particles**
F. Vydra, K. Štulík & E. Juláková — **Electrochemical Stripping Analysis**
N. G. West — **Practical Environmental Analysis using X-ray Fluorescence Spectrometry**
J. Zupan — **Computer-supported Spectroscopic Databases**
J. Zýka — **Instrumentation in Analytical Chemistry**

# PRINCIPLES AND PRACTICE OF CHROMATOGRAPHY

B. RAVINDRANATH, Ph.D.
Head of Chemical Sciences
Vittal Mallya Scientific Research Foundation, Bangalore, India

**ELLIS HORWOOD LIMITED**
Publishers · Chichester

Halsted Press: a division of
**JOHN WILEY & SONS**
New York · Chichester · Brisbane · Toronto

First published in 1989 by
**ELLIS HORWOOD LIMITED**
Market Cross House, Cooper Street,
Chichester, West Sussex, PO19 1EB, England
*The publisher's colophon is reproduced from James Gillison's drawing of the ancient Market Cross, Chichester.*

**Distributors:**
*Australia and New Zealand:*
JACARANDA WILEY LIMITED
GPO Box 859, Brisbane, Queensland 4001, Australia
*Canada:*
JOHN WILEY & SONS CANADA LIMITED
22 Worcester Road, Rexdale, Ontario, Canada
*Europe and Africa:*
JOHN WILEY & SONS LIMITED
Baffins Lane, Chichester, West Sussex, England
*North and South America and the rest of the world:*
Halsted Press: a division of
JOHN WILEY & SONS
605 Third Avenue, New York, NY 10158, USA
*South-East Asia*
JOHN WILEY & SONS (SEA) PTE LIMITED
37 Jalan Pemimpin # 05–04
Block B, Union Industrial Building, Singapore 2057
*Indian Subcontinent*
WILEY EASTERN LIMITED
4835/24 Ansari Road
Daryaganj, New Delhi 110002, India





**British Library Cataloguing in Publication Data**
Ravindranath, B. *1947–*
Principles and practice of chromatography.
1. Chromatography
I. Title
543′.089

**Library of Congress Card No.** 88–13889

ISBN 0–7458–0296–6 (Ellis Horwood Limited)
ISBN 0–470–21328–0 (Halsted Press)

Typeset in Times by Ellis Horwood Limited
Printed in Great Britain by Hartnolls, Bodmin

# Table of contents

## PART II: GAS CHROMATOGRAPHY

# Preface

Over the past three decades, chromatography has come a long way from its early confines of the chemical laboratory as a separation method, to its present status as a very powerful analytical tool in experimental biology and chemistry. It not only revolutionized the research activity in these fields but also made considerable impact on several other user disciplines such as medicine, pharmacy, chemical technology, food science and many more. Recent applications of chromatography include environmental pollution control and therapeutic drug monitoring, in addition to the more traditional areas. Parallel with these developments are the advances in the subject of chromatography itself, with each of its different aspects, namely theory, techniques, instrumentation and applications, developing endlessly and at a stupendous rate. The chromatographic techniques range from the relatively simple paper chromatography to the more recent supercritical fluid chromatography. The tools employed vary from an inexpensive microslide to the most advanced microprocessor. The variety of chromatographic methods available, which include affinity chromatography and field-flow fractionation, is equally impressive. Almost all classes of compounds, from small inorganic ions to high molecular weight natural and synthetic polymers, have been analysed using chromatography. Countless reviews, monographs, treatises, serial publications and journals on chromatography have come into being and their number is increasing rapidly. Under these circumstances, embarking on writing an introductory general text on this multifaceted subject needs sufficient justification.

For one thing, much of the chromatographic literature caters to the needs of the specialist chromatographer, or to one who knows what he is looking for. However, the subject has now grown too wide, and the possible variations too bewildering, for a non-specialist chromatographer to know what to look for and where to look when there is a separation problem. I believe there are many of the latter type, since modern applications of chromatography encompass such diverse fields as industrial process control, molecular biology and clinical chemistry, and it is unlikely that most workers would have entered these fields with adequate exposure to the subject. It is also a fact that the subject has been growing too fast for even university curricula to adopt a comprehensive and unified approach. This is probably the reason why the subject is most often learnt by individuals outside the traditional curriculum, mainly through short-term courses offered by instrument companies and chromatography discussion groups; the main limitation of such courses is the inevitable narrowness of coverage, while the subject itself is extremely wide. This book is an effort to bring out, in a nutshell, the various aspects of the subject in a unified manner, with an emphasis on the principles.

This book is written with two categories of reader in mind. One is the student, who should recognize that though the different chromatographic methods appear to be so diverse in technique, they are all based on a common principle, namely, separation of a mixture of compounds on the basis of their differences in distribution between two phases, one of which is stationary. The theoretical principles, which are common to all the chromatographic methods, are included in the first part of the book to emphasize this point. The discussion on the impact of the chromatographic techniques on research and industry in the last part (Part IV) of the book could not be comprehensive; certainly not in a book of this size. However, representative examples of a number of different types of application are given so that the reader can perceive the copiousness of human ingenuity which led to the present state of development of the subject. The other two parts of the book, which belong to the two kinds of mobile phases, namely gas (Part II) and liquid (Part III), bring out the diversity in methodology. The organization followed in this book is an attempt to bring together all related topics and develop the subject systematically. Emphasis is on such practical aspects as the factors affecting the choice of the stationary phase, the mobile phase, the detection methods and their optimization. This should enable the other category of reader, namely the non-specialist chromatographer, to obtain a clear perspective of the techniques and options available before attempting a separation.

In pursuance of the above objectives, the state-of-the-art account of as many chromatographic techniques as possible is given. However, if one wants to include everything from Tswett's experiments to computer-assisted optimization of HPLC and still keep the size of the book within manageable limits, many compromises have to be made. Inevitably, older techniques have had to be de-emphasized; some of them have nevertheless been included, if only to keep the developments in perspective. No attempt is made to compile the chromatographic systems and the retention data in relation to different solute classes. Such information is too vast to be included in this volume and partial information is not very useful. References are kept to the minimum required to meet the demands of courtesy and information. Where possible, references are made to the most recent books and reviews from which the primary sources may be obtained.

I am indebted to several of my friends and colleagues who have reviewed the manuscript at different stages and made helpful suggestions. I shall, of course, take the full responsibility for all errors of judgement, of omission and of commission.

B. Ravindranath

*To*
*Viji and Savitr*

# List of abbreviations and acronyms

## ABBREVIATIONS

| | |
|---|---|
| Å | ångström(s) |
| atm | atmosphere(s) |
| ag | attogram(s) |
| cm | centimetre(s) |
| *d* | diameter |
| eq. | equation |
| eV | electronvolt(s) |
| fg | femtogram(s) |
| fmole | femtomole(s) |
| ft | feet |
| g | gram (s) |
| h | hour(s) |
| in. | inch |
| i.d. | inside diameter |
| km | kilometre(s) |
| kV | kilovolt(s) |
| l | litre(s) |
| m | metre(s) |
| m.p. | melting point |
| μl | microlitre(s) |
| μg | microgram(s) |
| μm | micrometre(s) |
| μsec | microsecond(s) |
| meq | milliequivalents |
| ml | millilitre(s) |
| mm | millimetre(s) |
| msec | millisecond(s) |
| mg | milligram(s) |
| mV | millivolt(s) |

| | |
|---|---|
| min | minute(s) |
| ng | nanogram(s) |
| nl | nanolitre(s) |
| nm | nanometre(s) |
| nmol | nanomole(s) |
| o.d. | outside diameter |
| ppm | parts per million ($10^{-6}$ parts) |
| ppb | parts per billion ($10^{-9}$ parts) |
| ppt | parts per trillion ($10^{-12}$ parts) |
| pg | picogram(s) |
| pmol | picomole(s) |
| $P$ | pressure |
| psi | pounds/inch$^2$ |
| rpm | revolutions per minute |
| sec | second(s) |
| sq. | square |
| wt. | weight |

## ACRONYMS

| | |
|---|---|
| AAS | Atomic absorption spectroscopy |
| AFID | Alkali flame ionization detector |
| API | Atmospheric pressure ionization |
| ARF | Absolute response factor |
| ASTM | American Society for Testing and Materials |
| AUFS | Absorbance units full-scale |
| BMD | Brunel mass detector |
| BPC | Bonded-phase chromatography |
| BSS | British Standards Specifications |
| C | Celsius (centigrade) |
| CC | Column chromatography |
| CCCC | Centrifugal countercurrent chromatography |
| CCD | Countercurrent distribution |
| CI | Chemical ionization |
| CLC | Column liquid chromatography |
| CLD | Chemiluminescence detector |
| CPI | Carbon-presence index |
| CRF | Chromatographic response factor |
| CRT | Cathode-ray tube |
| CSD | Cross-section detector |
| CSP | Chiral stationary phase |
| DACC | Donor–acceptor complex chromatography |
| DAD | Diode-array detector |
| DCCC | Droplet countercurrent chromatography |
| DEC | Donnan exclusion chromatography |
| DLI | Direct liquid introduction |
| DS | Data system |

| | |
|---|---|
| ECD | Electron-capture detector |
| ECL | Effective chain length |
| ECTEOLA | Epichlorohydrin triethanolamine |
| ED | Electrochemical detector |
| EFFF | Electrical field-flow fractionation |
| EI | Electron ionization |
| ELISA | Enzyme-linked immunosorbent assay |
| EMIT | Enzyme-multiplied immunoassay technique |
| EST | External standardization technique |
| FD | Field desorption |
| FED | Flame-emission detector |
| FFF | Field-flow fractionation |
| FFFF | Flow field-flow fractionation |
| FI | Field ionization |
| FID | Flame-ionization detector |
| FPD | Flame photometric detector |
| FPI | Fluorescent polarization immunoassay |
| FPLC | Fast (protein) liquid chromatography |
| FSOT | Fused-silica open-tubular columns |
| FTIR | Fourier transform infrared |
| GC | Gas chromatography |
| GFC | Gel filtration chromatography |
| GLC | Gas–liquid chromatography |
| GPC | Gel permeation chromatography |
| GSC | Gas–solid chromatography |
| HAFID | Hydrogen-atmosphere flame-ionization detector |
| HECD | Hall electrolytic conductivity detector |
| HEEP | Height equivalent to an effective theoretical plate |
| HETP | Height equivalent to a theoretical plate |
| HIC | Hydrophobic interaction chromatography |
| HPAC | High-performance affinity chromatography |
| HPIEC | High-performance ion-exchange chromatography |
| HPLC | High-performance liquid chromatography |
| HPSEC | High-performance size-exclusion chromatography |
| HPTLC | High-performance thin-layer chromatography |
| HR | High resolution |
| IAC | Immunoaffinity chromatography |
| IC | Ion chromatography |
| IEC | Ion-exchange chromatography |
| IMD | Ion-mobility detector |
| IPC | Ion-pair chromatography |
| IR | Infrared |
| IST | Internal standardization technique |
| IUPAC | International Union of Pure and Applied Chemistry |
| LEC | Ligand-exchange chromatography |
| LFER | Linear free energy relationship |
| LLC | Liquid–liquid chromatography |

| | |
|---|---|
| LR | Low resolution |
| MAGIC | Monodisperse aerosol generator for introduction of chromatographic effluents |
| MASS | Mass spectral search system |
| MCAC | Metal–chelate affinity chromatography |
| MCD | Microcoulometric detector |
| MDL | Minimum detection level |
| MFG | Mass fragmentography |
| MPD | Microwave plasma detector |
| MS | Mass spectrometry |
| MS–MS | Tandem mass spectrometry |
| MW | Molecular weight |
| MWD | Molecular weight distribution |
| NICI | Negative-ion chemical ionization |
| NMR | Nuclear magnetic resonance |
| NPD | Nitrogen–phosphorus detector |
| OPGV | Optimum practical gas velocity |
| OPTLC | Over-pressured TLC |
| OTC | Open-tubular columns |
| PBM | Probability-based matching |
| PGC | Pyrolysis gas chromatography |
| PIC | Paired-ion chromatography |
| PID | Photo-ionization detector |
| PLOT | Porous-layer open-tubular |
| PMD | Programmed multiple development |
| PS | Pore size |
| PSD | Piezoelectric sorption detector |
| PTGC | Programmed-temperature GC |
| PTLC | Programmed-temperature LC |
| RAD | Radioactivity detector |
| RF | Response factor |
| RI(U) | Refractive index (units) |
| RIA | Radio immuno assay |
| RID | Refractive index detector |
| RPLC | Reversed-phase liquid chromatography |
| RRF | Relative response factor |
| SCIC | Single-column ion chromatography |
| SCOT | Support-coated open-tubular |
| SEC | Size-exclusion chromatography |
| SFC | Supercritical fluid chromatography |
| SIC | Suppressor-ion chromatography |
| SFFF | Sedimentation field-flow fractionation |
| SIM | Selected-ion monitoring |
| SLFIA | Substrate-labelled fluorescent immuno assay |
| SN | Separation number |
| TAS | Thermomicro application–separation |
| TC | Template chromatography |

| | |
|---|---|
| TCD | Thermal conductivity detector |
| TDM | Therapeutic drug monitoring |
| TDS | Transport detector system |
| TFFF | Thermal field-flow fractionation |
| TFG | Thermo-fragmentography |
| TICC | Total ion-current chromatogram |
| TID | Thermionic detector |
| TIM | Total-ion monitoring |
| TLC | Thin-layer chromatography |
| TOF | Time of flight |
| TSI | Thermospray ionisation |
| USD | Ultrasonic detector |
| UV | Ultraviolet |
| WCOT | Wall-coated open-tubular |
| ZPC | Zero point of charge |

# *Part I*
# *Basic principles*

# 1

# Chromatography: an overview

## 1.1 INTRODUCTION

Isolation of extractable components of animal and plant tissues, their separation and their purification, are of interest to both biologists and chemists. During the entire period of modern scientific research, chemists have been relentlessly pursuing the goal of separation of materials into individual compounds. This is not without reason since homogeneity of the chemical substance is a prerequisite for any meaningful investigation to understand its structure, properties and reactivity; on the other hand, non-homogeneity of the materials often leads to erroneous results and conclusions. More recently, biologists have also come to appreciate the value of dealing with the individual components at the molecular level rather than the whole system. Today, there is hardly a branch of chemical or biological science, or industry, that does not use some form or other of separation process. In several areas (e.g., in the modern biotechnological processes), the efficiency of separation may be the limiting factor for the success of a process. Except when separation is not the end in itself (as in manufacturing activity) it is followed by qualitative analysis (identification) or quantitative determination with the aid of other techniques.

Every little difference in the chemical and physical properties of the molecules is sought to be exploited to effect the separation of a mixture into individual compounds. These include differences in chemical reactivity, polarity and molecular size. Chemical properties such as extractability of acids and bases into basic and acidic media, respectively, and complexation or derivatization reactions, have been used for a very long time. While these methods may be useful for group separations, they have very limited utility for the separation of individual compounds belonging to similar functionalities or groups. A majority of separation methods in chemistry are based on volatility (distillation) and solubility (crystallization, precipitation or partition between two immiscible liquids) [1]. These methods are very simple and convenient if the properties of the compounds to be separated are widely different.

Even when the properties are not appreciably different, separation is possible by fractional distillation and crystallization or countercurrent distribution, but the methods are tedious, time-consuming and seldom clean-cut; the problem is even more formidable if the material is a complex mixture of compounds and the quantities are very minute. However, these methods are still very useful for preparative and industrial separations. Also, the principles of distillation, partition and several other separation techniques have considerable bearing on the chromatographic methods.

### 1.1.1 Distillation

At a given temperature, a liquid is in equilibrium with the vapour above its surface. When the liquid is heated, the mobility of the molecules increases and they tend to leave the liquid and enter the vapour phase. When the vapour pressure equals the atmospheric pressure, the liquid boils, and when the vapour comes into contact with a cooler surface, it returns to the liquid phase. If the liquid is a mixture of two or more compounds, the vapour and the condensed liquid are richer in the more volatile compound and a separation based on the relative volatility is possible. To make the process more efficient, a fractionating column is introduced between the still pot and the condenser. The fractionating column may have several protrusions on the inner side (a Vigreux column) or be packed with small helices or rings; the purpose is to increase the area of contact between the liquid and the vapour. Some of the vapour condenses and runs down the column and contacts the vapour that is moving up. It is possible to envisage a fractionating column that has separate, individual stages or plates (bubble caps), wherein each plate corresponds to an evaporation–condensation step. A plot of the composition of the vapour and the liquid at equilibrium at different temperatures is called the phase diagram. Fig. 1.1 represents a phase diagram of an ideal two-component mixture, A and B, whose boiling point increases linearly with the proportion of B in the mixture. If the mixture at a starting composition C1 is heated to the boiling point of the mixture, T1, the composition of the vapour (and therefore that of the condensed liquid) is given by intersecting the vapour-composition curve at T1, that is, composition C2. Continuing the process produces liquids containing higher and higher proportions of A. Each stage of conversion of the liquid to vapour and back to liquid represents one plate. A distillate from a two-plate column will have a composition C3 and that from a three-plate column the composition C4, and so on. However, the composition of the distillate may vary from what is expected from the phase diagram since, in a real system, the plates may not be efficient enough for the vapour and the liquid to attain equilibrium. Also, most fractionating columns do not have individual plates and it is customary to refer to the *theoretical plate*, which is the length of the column yielding a concentration change corresponding to one equilibrium stage. The number of theoretical plates ($n$) may be calculated from the composition of the distillate and the phase diagram. The length of the column divided by the number of theoretical plates gives the height equivalent to a theoretical plate (HETP or $h$), that is, $h = L/n$. Thus, the lower the value of $h$, the more efficient is the column. While this is an oversimplified picture, the concept of the theoretical plate is very useful both in distillation and, as will be seen, in chromatography.

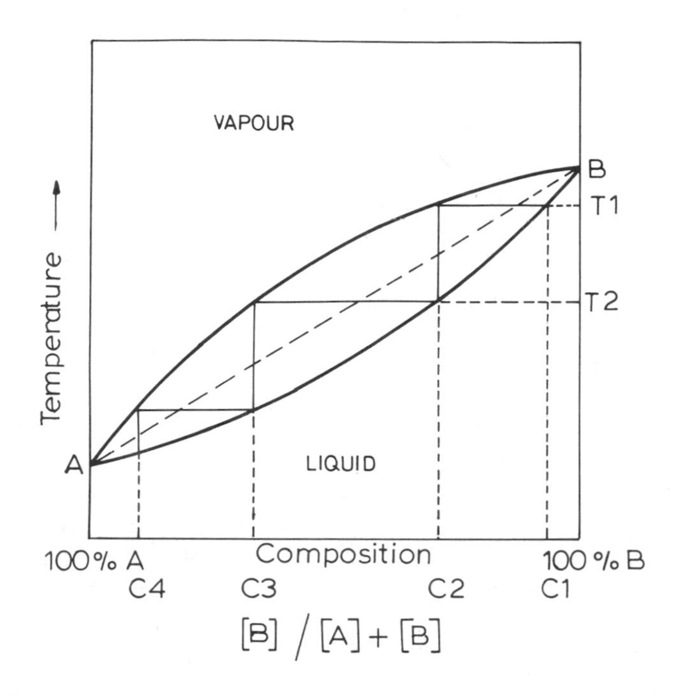


Fig. 1.1 — Phase diagram of an ideal two-component mixture.

### 1.1.2 Solubility and partition

It is well known that polar compounds like sodium chloride and sucrose dissolve in polar solvents, such as water, and non-polar substances in non-polar solvents (e.g. naphthalene in hexane); in other words, *like dissolves like.* The overall polarity of a substance depends on a number of factors such as the nature and geometry of the functional groups, the size of the molecule, and the environment. However, it is possible to both qualitatively and quantitatively determine the intermolecular interactions that influence solubility [2]. Briefly, there are four major, distinct and relevant intermolecular forces; namely, dipolar, induction, dispersion and hydrogen-bonding interactions. The interaction between two permanently dipolar molecules increases with increasing dipolar character of the molecules and is stronger at lower temperatures. The potential energy of the interaction is inversely proportional to the sixth power of the intermolecular distance. When a non-polar but polarizable molecule is close to a dipolar molecule, a redistribution of the electron density occurs in the former, resulting in an induced dipole. The more loosely the electrons are held, that is, the greater the number of double bonds (preferably conjugated) the molecule contains, the greater is the polarizability. Again, the induction energy is inversely proportional to the sixth power of the intermolecular distance, but is independent of the temperature. The dispersion forces, operating between every pair of adjacent molecules, are also independent of temperature. These forces mainly account for the interaction energy that holds molecules together in the liquid phase. The interaction of non-polar substances in water is termed hydrophobic interaction, the driving force being not the non-polar interactions between them but rather the effect of being squeezed out of the aqueous medium. When hydrogen is attached to an electronegative atom and is spatially close to another electronegative atom, there may exist a hydrogen-bond. The strength of the hydrogen-bonding is in the order of the electronegativities of the atoms involved: F···H—F > O···H—O > O···H—N > N···H—N > O···H—C, etc., and the bonding is stronger at lower temperatures. In addition to these are the familiar acid–base and complexation reactions which can greatly affect the solubility behaviour of a substance in a solvent. The differences in solubility of two substances in a given solvent may be attributed to the differences in one or more of the above interactions under the given conditions.

When a substance comes into contact with two immiscible phases it distributes itself between the phases. If one of the phases is a solid, the interaction of the substance with it is normally limited to its surface and the retention, if any, of the substance on the solid phase is called *adsorption*. When the phases are immiscible liquids or one of them is a gas and the other a liquid, and the substance in question has a finite solubility in the two phases, distribution or *partition* of the substance in the two phases would occur, depending on the relative solubility of the substance in the two phases. The ratio of the concentration of the substance in the two phases is called the *partition coefficient* or the *distribution constant*, $K$, and is a constant number for a given substance (solute), solvent pair and temperature. This statement is valid only for ideal or dilute solutions and in the absence of any interference from accompanying substances. The distribution constant is also independent of the relative volumes of the two solvents, being a function of the relative concentrations and not the absolute quantities. Thus, if 1 mole of a substance with a partition coefficient of 1 between two solvents **1** and **2** is dissolved in solvent **1** (called the raffinate) and equilibrated with an equal volume of solvent **2** (extractant) the amount of the solute remaining in the raffinate is 0.5 mole or 50%. If the raffinate is re-extracted with an equal volume of solvent **2**, the quantity of solute in the raffinate is again 50% of 0.5 mole or 25% of the original. After five such extractions, only about 3% of the solute is retained; that is, the quantity extracted is 97%. On the other hand, if the raffinate is extracted once with five volumes of the solvent **2**, the quantity extracted is only 83.3%. In more general terms, when a quantity $m$ of extractant (M) is used to extract a quantity $s$ of raffinate (S) containing $X_0$ concentration of the solute, the residual concentration $X_1$ in the raffinate is given by:

$$X_1 = \frac{X_0}{(1 + mK/s)} = \frac{X_0}{(1 + E)} \tag{1.1}$$

the quantity $mK/s$ being the extraction factor, $E$; the ratio, $m/s$, of the volumes of the two phases is called the phase ratio, $\beta$. Assuming that the two solvents are essentially immiscible, the amount of the solute ($X_n$) remaining in the raffinate after n successive extractions with the same quantity ($m$) of the extractant is given by:

$$X_n = \frac{X_0}{(1 + E)^n} \tag{1.2}$$

The above method of multiple extraction is very convenient for isolation of substances when $K$, defined as the ratio of the concentration of the solute in the raffinate (S) to that in the extractant (M), is small. However, in view of the large volumes of solvent required, it is rarely used for larger scale operations where economy is important and $K$ is close to unity. In the above example, let us assume that the solvent S, containing the solute, is divided into four equal portions and the first portion is successively extracted four times with an equal volume of the solvent M. The successive extracts are then used to extract the second portion of the original solution in the same sequence employed with the first portion and the process is repeated with the remaining two portions of the raffinate. It can be easily calculated that the combined extract would now contain about 73% of the solute compared to the 50% extraction if the total raffinate was equilibrated with an equal volume of the extractant in a single stage. The larger the number of portions of the raffinate and extractants used, the higher will be the efficiency of extraction and a ten-portion pattern would yield nearly 90% in the combined extract. This process of repeatedly contacting successive portions of both the phases is called countercurrent extraction since, in effect, the two phases move in opposite directions relative to each other even though one phase (S) is apparently stationary.

### 1.1.3 Countercurrent distribution

An important parameter for determining the separability of two compounds A and B by partition is the ratio of their distribution constants, $K_A/K_B$. This ratio is called the separation factor, $\alpha$, and is usually expressed as a value larger than unity so that $K_A > K_B$; when $\alpha = 1$, that is, $K_A = K_B$, there can be no separation by partition. When two substances in a mixture have widely different $K$ values in a pair of solvents, it seems reasonable to expect that they could be separated by partition. Let us assume that the compounds A and B have distribution constants 10 and 0.1, respectively, between two solvents S and M. Partition of the mixture containing equal amounts of A and B between equal volumes of the solvents S and M ($\beta = 1$) would leave 91% of A along with 9% of B in solvent S; correspondingly, solvent M would contain 9% of A and 91% of B. Further purification of A (and B) to 99% can be achieved simply by re-extraction of the respective fractions with M and S, but the yield is only 83%. Repetition of the process would produce A and B of 99.99% purity but the yield would drop to 75%. It must also be noted that, in this example, the separation factor is very large ($\alpha = 100$). Smaller separation factors would require a larger number of extractions and, as the number of extractions increase, the purity of the individual components increases but the yield of the solutes in different fractions drops much faster. Thus, the separation by partition becomes very tedious and often impractical unless a countercurrent procedure can be used [1].

Craig [3] developed a simple device that mimics tandem-arranged separatory funnels and demonstrated the practical utility of the countercurrent distribution (CCD) process. A six-stage CCD is represented diagrammatically in Fig. 1.2 using seven tubes numbered $n = 0$ to 6. The tubes are charged with equal portions of solvent S. The first tube is then charged with an equal volume of solvent M ($\beta = 1$) along with the solute; the quantity of the solute is taken as 64 units and its distribution constant between the two solvents is taken as unity ($K = 1$) for convenience of

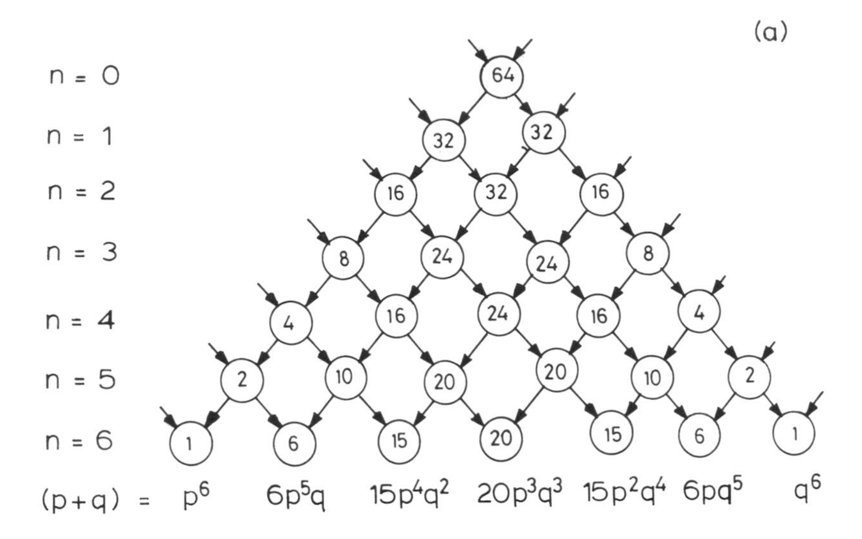


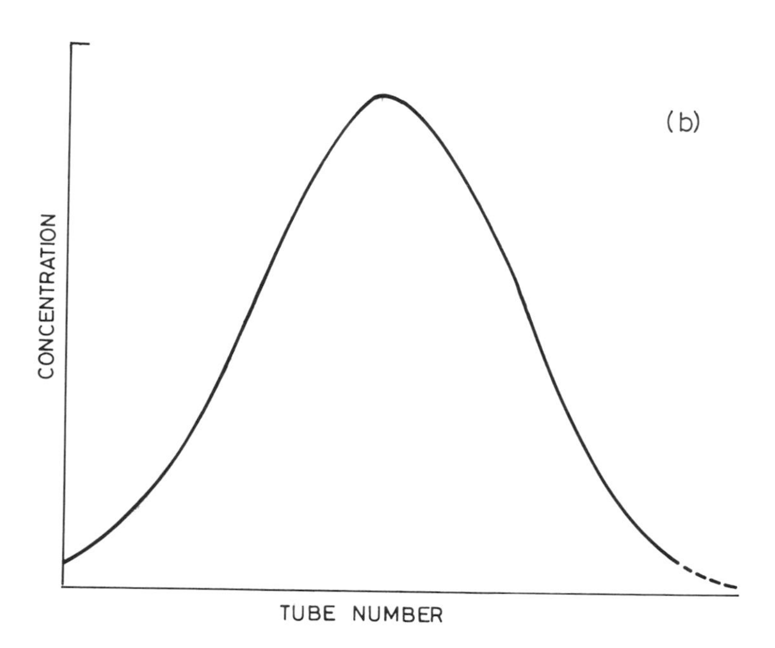


Fig. 1.2 — Diagrammatic representation of a six-stage countercurrent distribution. Note the relationship of (a) the concentration distribution and (b) the peak shape.

calculation. After equilibration, the solvent M, containing 50% of the solute is completely transferred to the second tube ($n = 1$) and a fresh portion of the same volume of M is added to the first tube. This completes the first stage ($r = 1$); the symbol $T_{r,n}$ is used to denote the fraction of the solute in each tube so that $T_{1,0} = T_{1,1} = 32/64$ at the end of the first stage. The process is repeated three more times to give a distribution of the solute as shown in Fig. 1.2. In general terms, let $p$ and $q$ be the mole fractions of the solute in the two phases M and S in the first tube, so that $p = E/(E+1)$ and $q = 1/(E+1)$. The distribution of the solute is then given by the binomial expansion of $(p+q)^n$ and its amount in a given tube is:

$$T_{r,n} = \frac{r!}{n!(r-n)!} \tag{1.3}$$

From the above equation it can be easily calculated that the ratio of the fractions in any two adjacent tubes is given by:

$$\frac{T_{(r+1),n}}{T_{r,n}} = \frac{r+1}{(r+1)-n}q \tag{1.4}$$

However, if the $n$th tube happens to contain the maximum amount of the solute, the above ratio will be nearly 1; that is, the fraction in the tube $n_{max}$ changes very little upon two successive transfers. If $r$ is sufficiently large, $r+1$ is approximately equal to $r$ and

$$q = \frac{r-n}{r} = 1 - \frac{n}{r} \tag{1.5}$$

Since $q = 1-p$, eq. (1.5) gives us: $n_{max} = rp$. Since $p$ and $q$ are functions of the distribution constant, it follows that the number of the tube containing the maximum amount of the solute is also a function of the distribution constant and if two compounds have different $K$ values their $n_{max}$ would also be different. In other words, the two compounds may be separated if $n$ is sufficiently large; the distribution of the solutes in a Craig apparatus is then similar to that in a familiar elution chromatogram (Fig. 1.3). Craig had originally developed a 25-stage CCD apparatus and progressively increased the number of stages; a 1000-stage automated apparatus was soon built. Large-scale CCD processes have been extensively used in metallurgy and in the production of antibiotics and certain other biologically active compounds. Though largely replaced in recent years by the chromatographic methods, the technique continues to attract attention in the form of high-performance countercurrent distribution [4] and its variants.

### 1.1.4 Biochemical separations

Many separation methods, which have been popular in biochemical research, take advantage of the large differences in the size of the molecules. Notable among these are ultracentrifugation [1] and dialysis [5]. In the common, density-gradient ultracentrifugation, a mixture of macromolecules is ultracentrifuged (rotated about an axis) at 40,000 to 80,000 rpm (compared to 1,000 to 5,000 rpm in the ordinary laboratory centrifugation) in a concentrated solution of a low molecular weight

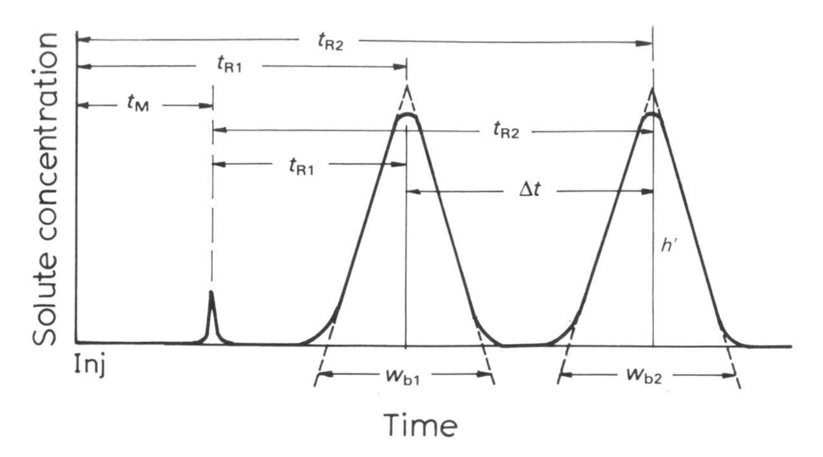


Fig. 1.3 — Elution profile of a typical two-component mixture by chromatography; for explanation of terms, see Table 1.1.

compound such as sucrose, urea or certain alkali-metal salts; more recently, ultracentrifugation media based on dextran and silica have become popular. A concentration gradient leading to a density gradient is maintained in the tube so that, after prolonged centrifugation, the macromolecules position themselves in order of their molecular density, which is a function of their molecular weight and shape. In favourable cases, the macromolecules are separated into discrete bands which can be carefully removed and separated from the smaller molecules by dialysis. Dialysis and several related techniques such as osmosis and ultrafiltration make use of the permeability of molecules of only certain size and shape across certain porous membranes; cellulose acetate and polyamide (nylon) are the most commonly used materials for the purpose. Semipermeable membranes permit the passage of only certain types of molecule, as occurs across the natural membranes in several biological transport phenomena. Use of such membranes in the dialysis of blood (i.e. artificial kidney) is well known. Electrodialysis involves migration of ions across a membrane on application of an electric field. Positive ions (cations) move towards the negatively charged (cathode) side and anions towards the anode; usually stacks of cation- and anion-selective membranes are used to facilitate the process.

Movement of particles under the influence of an electric field is extensively used in another separation technique; namely, electrophoresis [6]. In the simplest form, the technique involves the use of a filter paper strip or a gel, moistened with an electrolyte and supported on a glass plate. The sample to be separated is applied on to the plate at the centre, usually as a streak perpendicular to the length of the strip, and an electric current is applied at the ends. The mobility of the ions depends on their size and shape, the applied field strength and the viscosity of the medium. A high voltage (50 to 200 V/cm) is used for low molecular weight compounds and a low voltage (2 to 20 V/cm) is used for high molecular weight compounds. The net charge on the molecule, which determines its mobility, may be altered by changing the pH of the electrolyte. There are several variations of the technique of electrophoresis, which is widely used for the separation of amino acids, peptides, proteins, nucleosides, nucleotides and nucleic acids.

## 1.2 VISTAS OF CHROMATOGRAPHY

The separation methods in both organic and biological chemistry, as well as in other fields, have been largely replaced by the chromatographic methods, which are rapid, reliable and reproducible. The chromatographic process, by definition, involves differential distribution of substances between two phases, one of which is stationary and the other mobile. The stationary phase is a solid or a liquid supported as a thin film on a solid. In most cases, the selectivity of interaction of the stationary phase with the solute molecules determines the differential migration of the latter and thus contributes to the success of the chromatographic method. The selectivity may also be achieved by varying the mobile phase, which may either be a gas or a liquid. The mobile phase in gas chromatography (GC) is usually an inert gas such as helium or nitrogen, but may be replaced by (or admixed with) a vapour or by a supercritical fluid. The mobile phase in liquid chromatography (LC) may be water, an aqueous solution of an acid, base or salt, most of the organic solvents or mixtures thereof.

The stationary phase is usually packed into a long and narrow cylindrical column, made of glass or metal, with an arrangement to introduce the mobile phase and the sample at one end and collect or monitor the effluent at the other. In the case of LC, the stationary phase may also be in the form of a paper, or a thin layer, spread evenly on a plate (thin-layer, chromatography or TLC). The size of the solid particles is of the order of 150 to 180 $\mu$m for GC or traditional LC, and about 5 $\mu$m or less in TLC or modern, high-performance LC (HPLC). The sample to be separated is introduced into the stationary phase bed and the separation is achieved by allowing the mobile phase to flow through the bed. The success of the method depends on the ability of the mobile phase to selectively elute the components retained by the stationary phase as it flows through the column. A plot of the concentration of the solute in the effluent against the elution volume, or time, would typically appear as in Fig. 1.3, showing separation of the individual components or fractions. This technique of achieving the separation of the sample components is called the *elution development*. Alternatives to elution developments, are the frontal and the displacement developments, which are much less frequently used.

The *frontal analysis* or frontal development involves continuously percolating through the column, a solution of the sample to be separated in an appropriate solvent. If the mixture contains three components, A, B and C, and if compound A is less strongly retained than B and C, compound A appears first in the effluent from the column, followed by a mixture of A and B and then by a mixture of A, B and C. A plot of the solute concentration *vs*. effluent volume would appear as in Fig. 1.4a. This method of development obviously cannot be used for preparative separation of A, B and C, but is useful for the removal of small quantities of strongly retained impurities from the bulk of the material; for example, the purification of gases. Frontal development is also used in several non-analytical applications which have been reviewed [7].

*Displacement development* involves the use of a displacing agent, say D, which

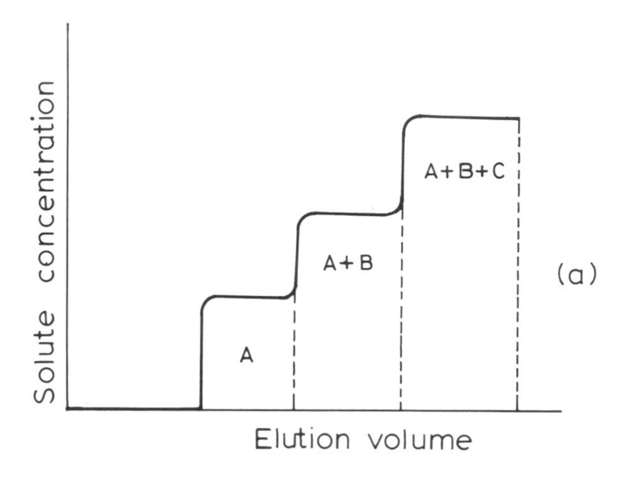


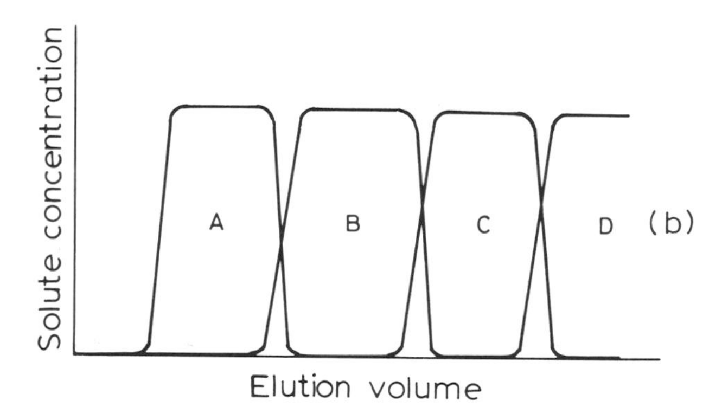


Fig. 1.4 — Solute concentration profile of the column effluent in (a) frontal development and (b) displacement development.

has a stronger affinity toward the stationary phase than any of the sample components, A, B and C. After the introduction of the sample into the stationary phase bed, a solution of D in a suitable solvent is allowed to pass through the column. As it moves down the column, compound D competes with the sample components for the active sites on the stationary phase and displaces them. Compound C, in turn, displaces the less strongly retained components A and B, and so on, resulting in a series of contiguous zones, one for each component (Fig. 1.4b). The zones of pure compounds are interspersed with those of the mixed fractions which may be collected and recycled. Displacement chromatography is primarily important for preparative separations as it uses the column capacity much better than in elution [8].

Chromatography is certainly the most efficient and valuable separation method known and, in this feature, it is unlikely to be surpassed by any other method in the foreseeable future. However, had it remained essentially a method of preparative separation, we would probably not have seen such a phenomenal growth in its development and applications. What gives it the distinctive features and enormous applicability is the feasibility of on-line detection and determination. A variety of detectors have been developed for different applications in both GC and LC. Some of these detectors are selective and at the same time highly sensitive to certain types of compounds. It is common knowledge that a chromatographic system is much more efficient in separating the components if the sample size is small. The development of high-sensitivity detectors has made it possible to use smaller and smaller samples making chromatography a powerful analytical tool. The main reason for the success of chromatography over the other methods of analysis is the possibility of interfacing with other analytical instrumentation. This feature greatly enhances its capability with respect to the three aspects of analytical chemistry; namely, separation, identification and quantitation. Chromatography has been conveniently combined with ultraviolet (UV), infrared (IR), nuclear magnetic resonance (NMR) and mass spectrometry (MS). All these techniques use very small quantities of the sample and the analysis time is very short. The spectral analysis is thus possible 'on the fly' without having to collect the sample after elution from the column. A few years ago, tandem mass spectrometry or MS–MS [9] appeared to threaten chromatography in qualitative and quantitative analysis. In this technique, a mixture of compounds (without prior separation) is subjected to mass-spectral ionization (bombardment of the molecules with high energy electrons in a high-vacuum chamber) and the resulting, characteristic ions are then led to a second mass spectrometer for further analysis. However, the possibility of MS–MS displacing chromatography is rather remote and, in fact, the two techniques should be considered complementary.

The utility of chromatography in qualitative analysis is twofold: (1) determination of the number of compounds in a given sample, and (2) their identification. The first of these is relatively simple if it is assumed that all the compounds are separated by the chromatographic bed and that all the compounds are detectable by the particular detection method used. Thus, a picture of the type shown in Fig. 1.3 indicates that the sample contains two compounds which elute at retention times $t_{R1}$ and $t_{R2}$ under the particular chromatographic conditions. It is possible that more than one compound may elute at the same time. It is also possible that a particular detector may not respond to a certain type of compound. Multiple separation and detection systems may have to be used to ascertain or rule out these possibilities. The retention times $t_{R1}$ and $t_{R2}$ are characteristic of the given compounds under the particular conditions of chromatography; namely, sample size, column length, diameter, nature and amount of stationary phase, nature and flow rate of the mobile phase, temperature, etc. The retention times can be, and are, used for identification of the compounds. However, in view of the several variables that affect the absolute retention times, direct comparison with the literature retention data may not be dependable and several alternatives have been suggested which include co-chromatography, comparison of relative retention times with respect to a standard, use of retention-index systems, etc. Microscale manipulations in chromatography (pre- and post-column derivatization techniques) are useful in several types of qualitative analysis that can be carried out in extremely small scale, and changes monitored by chromatography. These tests, the response to which may be characteristic of the nature of, or the functional groups in, the compounds under examination, may also be used for identification of the compounds. The response to certain types of selective detectors can also be characteristic of certain types of compounds.

The response of most chromatographic detectors is directly proportional, in a certain range (called the linear dynamic range of the detector), to the quantity or concentration of the component being detected. Thus, it is possible, to quantitatively determine several compounds at a time by measuring the area (or the height) of the peaks in a chromatogram. The sensitivity of many detectors is in the nanogram to picogram range and that of some selective detectors is even higher. The accuracy of quantitation by chromatography depends on several factors such as the method of sampling and sample introduction, efficiency of the column, linearity of the detector and the methods used to convert the detector response into the amount of the compound. More recently, scanning densitometers have been used to convert visualized thin-layer chromatograms into the more familiar elution chromatograms for easy quantitation. The methods of data handling of GC, LC and TLC-densitometry are essentially the same and are discussed in detail at appropriate places in the text.

### 1.2.1 Historical perspective

The history of chromatography has been discussed by several authors. The history of gas chromatography was briefly reviewed by Ettre, who was closely associated with its early developments [10]. More recently, he also discussed at length the history of liquid chromatography [11]. The personal recollections of fifty-six leading scientists on the historical aspects of chromatography have been published [12]. The purpose of the present account is not to give a chronological account of all the events in chromatographic history but to discuss the evolution of the subject in perspective so that it could also lead to an understanding of the development of the technique. Interestingly, the practice of many of the chromatographic methods remained essentially the same as that developed by the pioneers. However, the more recent advances in microprocessor technology of instrument control and data handling have made the technique of chromatography very different from what it was just a decade or two ago. From this point of view, the history of chromatography is probably still in the making.

An examination of the evolution of chromatography reveals certain interesting features which are probably true of other branches of science as well. For example, as happened on several occasions during the early stages of development of chromatography, any innovation made before the time is ripe is likely to go into oblivion, sometimes to be revived later. Thus, in spite of his student Goppelsroeder's efforts to popularise it, Schoenbein's 1861 discovery of *Kapillaranalyse* (capillary analysis, now recognizable as paper chromatography) remained largely unused for 80 years until Martin and Synge introduced paper partition chromatography in 1941 [13]. Similarly, Tswett's invention of column liquid chromatography (1903) was dormant for nearly three decades before its grand revival by Kuhn's group in essentially the same form [14]. On the other hand, Martin's second invention, gas-liquid chromatography (1952) [15], grew by leaps and bounds, due largely to the support of one of the potential users (the petroleum industry), thus, in a short span of ten years, it attained the technical sophistication that we know today.

The phenomenon of chromatography is probably as old as nature itself since many of the geological processes, such as the soil and ore formations, sedimentation and diagenesis, migration and accumulation of minerals and petroleum, etc., can be considered to be accompanied by, or resulting from, chromatographic processes [16]. Indeed, the earliest laboratory demonstration of the chromatographic process was recorded in a paper entitled 'A suggestion as to the origin of Pennsylvania Petroleum' by the famous American geochemist, David Talbot Day [17]. He allowed crude oil to diffuse upward through a column of fuller's earth and showed that unsaturated hydrocarbons as well as aromatics were preferentially retained in the lower portion of the tube; the first fractions were similar to light petroleum obtained by distillation, to be followed by heavier oils and finally petroleum jelly.

The pioneer of column liquid chromatography is undisputedly the Russian botanist, Mikhail S. Tswett [18]. (The official English transliteration of the Cyrillic spelling is Tsvet; however, Tswett is better known and was used by the inventor himself.) His life and work has been narrated by Sakodynskii in a series of articles, which make interesting reading [19]. Towards the end of the nineteenth century, many scientists were interested in the pigments of green leaves, namely, the chlorophylls. These pigments were very labile to chemical methods of separation and purification and one could not be sure of the authenticity of the isolated sample. Tswett was therefore looking for a physical method of separation and checked a large number of solvents for extracting the pigments and over 100 solid substances capable of selective retardation of individual pigments through adsorption. He mainly used calcium carbonate, tamped firmly into a glass tube. When a petroleum ether solution of the pigments was filtered through the column, they were resolved, according to the adsorption sequence, into various coloured zones. The separation was practically complete when, after the pigment solution had flown through, a stream of pure solvent was passed through the column. Tswett immediately recognized the potential of the technique for both qualitative and quantitative analysis and also demonstrated that the method was by no means limited to chlorophylls and coloured compounds. He was also the first to correctly interpret the theoretical basis of the separation phenomena and suggest acceptable nomenclature. It is a tribute to his genius that Tswett's original method of adsorption chromatography and elution development of the chromatogram are still widely used in many laboratories for both preparative and analytical purposes. It is for these reasons that Tswett is credited with the discovery of chromatography though there have been several earlier instances of the use of chromatographic methods of some sort. Subsequently, in 1922, a number of chromatographic experiments were reported in America by Palmer and, during the 1930's, adsorption chromatography was extensively used by Kuhn, Winterstein, Lederer, Zechmeister, Karrer, Reichstein and several other leading organic chemists. However, Tswett and his work remained largely unrecognised during his lifetime (1872–1919) for several reasons [20].

The introduction of partition chromatography in 1941 by Martin and Synge was indeed a turning point in the history of chromatography. The events that led to the development of the technique were recounted by Martin in his Nobel lecture of 1952 [21]. Martin had a great personal interest in distillation and, while watching Winterstein demonstrate the separation of carotenoids on a chalk column (1933), was fascinated to note the relationship between the chromatographic and distillation columns; there was a relative movement of two phases and it was their interaction at many points that gave rise to good separations. Apparently, this was the origin of the theoretical plate concept which Martin introduced in the chromatographic theory he later evolved. In 1938, Martin and Synge were working on the separation of amino acids of wool as their acetates. Finding the conventional countercurrent distribution between chloroform and water too tedious for separation of the acetylamino acids, they supported the water on a column of silica gel and allowed chloroform to percolate down the column, thus bringing two liquids into close contact. On addition of methyl orange to the liquid on the column, the acetylamino acids could be seen moving down the column as red bands. Encouraged by the success of this method, they, along with Consden and Gordon, extended it to the separation of free amino acids using filter paper as the support for the stationary phase. The first chromatograms were circles of paper in a petri dish containing water, and water saturated with butanol was fed by capillary to the centre by a tail on which a drop of amino acid solution was placed. When the solvent reached the edge, the paper was dried, sprayed with ninhydrin solution in butanol and heated to visualize the amino acids. Later, they used test tubes and more suitable containers (tanks) with troughs containing the mobile phase into which the tops of the paper strips could dip. An important development was running the chromatogram in two directions. The first solvent spread the amino acids in a line near one end of the paper from a spot near the corner; then, after drying, the paper was turned by 90° and, using a second solvent, the line of the spots was resolved in a two-dimensional pattern [22]. Paper chromatography was soon successfully applied to the separation of peptides, sugars, flower pigments and metals by different groups.

Martin and Synge, in their first paper on partition chromatography [13], had evolved a theory relating the mobility of the zones to the partition coefficient. Further, by introducing the concept of the theoretical plate, the shape and width of the zones could be predicted. Later, by assuming that the free energy of transfer of a compound from one phase to another was an additive function of the free energies of individual atoms or groups, it was possible to predict, with reasonable accuracy, the mobility of peptides and many other substances; they found a log relationship between the retention volumes and the number of carbon atoms in a molecule. Thus, within a short period, Martin and Synge developed partition column liquid chromatography, paper chromatography, methods of circular and two-dimensional irrigation, reversed-phase chromatography (in which the solid support is coated with a non-polar liquid and the mobile phase is an aqueous or other polar liquid) and techniques of visualizing chromatograms by use of spray reagents; and they proposed very successful theoretical concepts. The most fascinating aspect of the new technique was in connection with the quantity of the material needed for analysis. Accepted methods of amino acid analysis prior to the advent of partition chromatography required about half a kilogram of protein and about six months of work. The silica gel partition columns required a few milligrams and paper chromatography required only a few micrograms; in either case, the time required was a small fraction of that required earlier. It is thus not surprising that the techniques of partition chromatography would immediately find application in the separation of several classes of compounds. Synge, who shared the Nobel prize with Martin, described in his Nobel lecture the various applications of the new technique. He particularly elaborated the contributions of Sanger, Stein and Moore, who used the partition chromatographic techniques for the analysis of amino acids and peptides.

Moore and Stein are probably better known for their pioneering work on the use of ion-exchange chromatography (IEC) for automated analysis of amino acids [23]. Prior to its introduction in biochemistry, IEC was extensively used (1935–1945) by several leading inorganic chemists to separate inorganic ions, especially in connection with the development in the United States of the atomic bomb. Indeed, IEC is one area of chromatography which had its origin in inorganic analysis and, as mentioned earlier, soil chemists and geochemists were the first to recognize chromatographic phenomena [16]. The pioneer soil chemists Thomson and Way were probably the first to independently demonstrate (1850) the power of the soil to absorb ammonia from ammonium sulphate and simultaneously release an equivalent amount of calcium. Selective base preference was also demonstrated by Way who showed that whereas calcium in the soil could be displaced by ammonia, sodium was not. Investigations during the later part of the nineteenth century remained largely in the area of geochemistry, particularly soil chemistry and, based on the data accumulated, Gans (1907) was able to develop methods of softening of water using both natural and synthetic zeolites. Application of the ion-exchange processes to industrial processes during this period was hampered by the low exchange capacities and slow exchange rates of the zeolites. An important advance was made by Adams and Holmes (1935) who condensed various polyhydroxy and polyamino aromatics with formaldehyde to obtain weak cation and anion exchange resins. These synthetic resins have also now been replaced by cross-linked styrene-divinylbenzene polymers which can be sulphonated or chloromethylated and aminated to give cation and anion exchange resins, respectively.

The unfunctionalized styrene-divinylbenzene polymers also find extensive application in another area of chromatography, namely, size-exclusion chromatography (SEC). In this technique, the molecules are separated on the basis of their molecular size and shape, depending on their ability to enter the pore structure of the stationary phase. Molecules larger than the pores are unretained while those that could enter the pore structure are subsequently ejected (excluded) by the solvent molecules in the order of their molecular size, with the larger molecules emerging first. The technique of separation of molecules on the basis of their molecular size, incidentally, owes its origin to certain observations in connection with the ion-exchange phenomena. Subsequent to O. Samuelson's report (1944) that the anion exchanger, wofatit, is not capable of adsorbing lignin sulphonic acid, R. Kunin and R. J. Meyers (1949) studied the relationship between the degree of cross-linking, swelling properties and capacity for larger ions of a polystyrene anion exchanger and concluded that the pores were too small and prevented the molecules from entering the network. Subsequently, size discrimination by ion-exchange resins of not only ions but also neutral molecules and sieving effects of non-ionic porous materials such as charcoal, metal oxides, cellulose and starch were recorded by several authors. Many disadvantages, however, were associated with natural products such as starch; and the technique was placed on a firm practical foundation in 1959 when Porath and Flodin [24] synthesized cross-linked dextran gels using epichlorohydrin. In 1964, J. C. Moore prepared uncharged polystyrene gels of controlled porosity and demonstrated their outstanding utility for separation of high molecular weight synthetic polymers in organic solvents. Both the dextran and styrene-based gels are now extensively used for separation and analysis of natural and synthetic polymers.

Porath is also credited with the discovery of the cyanogen bromide method of activation of the dextran gels thus enabling covalent binding of enzyme substrates and antigens to dextran [25]. This led to a new technique of separation on enzymes and antibodies based on biospecific interaction of the bound molecules with the components in the mobile phase. The technique was also independently developed by Cuatracasas *et al.* [26], who, in addition to introducing the term affinity chromatography, demonstrated the use of specific adsorbents in enzyme purification. Actually, as early as 1910, E. Starkenstein, in a preparation of amylase on insoluble starch, illustrated the inherent advantages of such specific interactions with the stationary phase in chromatographic separation. The separation of anti-hapten antibodies was described in 1951 by L. S. Lerman, who also developed a system for purification of tyrosinase based on the same principle using resorcinol-cellulose conjugate [27].

The possibility of using chromatographic methods of separation outside the column was recognized very early. Apart from the interesting reference, dating back to the beginning of the Christian era, to the use of papyrus by Pliny the Elder (AD 23–79), there are several instances of systematic studies on the use of paper and cloth for separating dyes and other materials [28]. The explosive growth in the use of paper partition chromatography following the work of Martin and Synge has been discussed above. In recent years, paper chromatography has largely been replaced by thin-layer chromatography, mainly because of its speed and convenience. In fact, TLC was developed much earlier, before the work of Martin and Synge. As early as 1889, Beyerinck demonstrated the separation of hydrochloric and sulphuric acids by diffusion through a thin layer of gelatin. More interesting was the work of Wijsman (1898) who used the same technique to separate two enzymes in malt diastase, (only one of which split maltose obtained from soluble starch). He was also the first to use the fluorescence phenomenon to detect a zone on a thin layer at a level of 1/28,000,000 of a milligram (about 40 pg) of maltose [29]. In 1938, Izmailov and Shraiber devised what they called 'spot chromatography' mainly as an exploratory means to solvent selection and optimization of Tswett's column chromatography. Their innovation consisted of depositing a drop of the material being investigated on a flat layer of adsorbent and observing the separation into concentric circular zones. Amplification of the zones could be effected by application of several drops of the solvent, just as in ordinary chromatography. The flat layer was obtained by covering a microslide with a slurry of the adsorbent in water and allowing it to dry. Modern TLC is largely the outcome of the efforts of Kirchner (1945–1957) and Stahl (1956–1976). The former's contributions include introduction of $<150$ $\mu$m silica particles, use of an inorganic binder (gypsum), preparation of larger plates, development in a closed tank, two-dimensional TLC and development of specific spray reagents [29]. Stahl, who proposed the name thin-layer chromatography for the technique, was mainly responsible for the standardization of several TLC parameters and for the introduction of a number of gadgets such as spreaders for preparation of uniform layers, the sandwich chamber, the thermomicro application–separation (TAS) oven and the TASOMAT [30]. He also invented several useful techniques like the two-dimensional SRS (separation–reaction–separation), gradient development and thermo-fragmentography. Many of the routine TLC manipulations, such as preparation of the plates, application of the sample, development of the chromatogram and its visualization, can now be automated. An instrumental variation of the technique, using specially made plates with an adsorbent of a narrow particle size distribution (2 to 5 $\mu$m) and about 200 $\mu$m thickness, is sometimes called high-performance TLC or HPTLC in view of its superior results [31].

An important step that set the pace for the rapid progress of modern chromatography and instrumentation is the development of gas–liquid chromatography (GLC or, more commonly, GC) by James and Martin in 1951. Actually, even a decade earlier, Martin and Synge [13] noted that the mobile phase in chromatography could just as well be a vapour (gas) or a liquid. In fact, the practice of using a gas as the mobile phase existed much earlier, but only in connection with adsorption; i.e. gas–solid chromatography (GSC). Although selective adsorption for the separation of gases had its origin prior to the First World War, G. Hesse and co-workers (1942) were the first to correctly describe the gas chromatographic separation process for the separation of volatile organic acids. During 1945–1950, E. Cremmer (with F. Prior and R. Muller) developed a system consisting of a carrier-gas source, a sample introduction device, a thermostatted separation column and a thermal conductivity detector, and demonstrated its application to both the analysis of light gases and physico-chemical measurements.

In 1951, Martin and James set out to develop a better method than paper chromatography for the analysis of fatty acids. They used a quarter-inch diameter glass tube, about 15 inches long, packed with Celite. Nitrogen was introduced at one end of the column, the other end of which was provided with a capillary tube that dipped into a test tube containing indicator solution. Plotting the number of drops of titrant (required to neutralize the acid eluting from the column) against time yielded a series of steps; the height of the steps denoted the quantity of the acid emerging and their position on the time axis, the retention time. They first separated methylamines which could be run at room temperature. Later, using a steam-jacketed column (4 mm × 120 cm, packed with kieselguhr coated with DC-550 silicone oil containing 10% stearic acid), they separated the first members of the fatty acid series [15]. Next, they made an automatic titrating machine and, subsequently (1956), devised a gas-density balance with a detection limit of 1 $\mu$g. During the next four years, a series of very sensitive detectors of both general and specific applicability were introduced which used the principles of flame ionization, argon ionization and electron capture. The detection limit was now down to 1 pg which is not very different from what can be routinely achieved today. Around the same time, at the 1957 Lansing Symposium, Golay [32] demonstrated, using elaborate theoretical calculations and practical results, that the packed column could be replaced by a narrow open tube with a thin film of the stationary phase coated on the inner surface. Soon after, Desty invented the glass-drawing machine for the preparation of glass capillary columns [33]. However, the open-tubular columns did not immediately come into routine use because of certain practical problems such as the fragility of the glass capillaries; this problem was solved with the introduction of fused-silica columns in 1979 [34]. This apart, it is really astonishing how closely similar the first gas-chromatographic systems are to those still in use today. The 1960s saw rapid development of refined instrumentation, new stationary phases for different classes of compounds and techniques of derivatization to render non-volatile compounds volatile enough, and to decrease the response threshold for certain types of detectors. The coupling of gas chromatography and mass spectrometry by Ryhage [35] in 1964 was yet another breakthrough.

In addition to its application in various fields, the development of GC paid rich dividends in two areas. One is the theory of chromatography; the limited variability and inertness of the mobile phase made it possible to study the theoretical aspects such as mass transfer and diffusion phenomena more closely. Thus, in 1958, van Deemter derived the celebrated equation relating mobile phase velocity to the efficiency. The second area influenced by the progress in GC was liquid chromatography (LC), For almost two decades (1950s and 1960s), there was a virtual *status quo* in the technique of LC. Though it was recognized early that smaller stationary phase particles would greatly improve the efficiency of chromatographic separation, no progress could be made in that direction since the flow of the solvent through the column would then be too slow and there were no reliable pumps to deliver the solvent through such packed columns at a constant and reproducible flow rate. Though Tiselius had used an on-line differential refractive index detector for column liquid chromatography as early as 1948, the success of instrumentation in GC gave the needed impetus for developing sensitive detectors, solvent-delivery systems, bonded stationary phases, etc. More recently, open-tubular columns have been gaining popularity in LC, though success has been somewhat limited owing to permeability problems. Recent advances in supercritical fluid chromatography (SFC, where the mobile phase is a supercritical fluid such as carbon dioxide) appear to bridge the gap between the two techniques, taking advantage of the favourable aspects of both kinds of mobile phases and making the distinction artificial [36].

A collaborative effort involving a large number of people, organizations and instrument companies over the past two decades has resulted in an unprecedented growth of the subject of chromatography, encompassing theory, methodology, instrumentation and applications. With so many people belonging to different disciplines contributing to the development of chromatography, individual contributions become blurred, but it is possible to identify the current trends and envision the future prospects. Though originally developed as a method of preparative separation and isolation of individual components from a mixture, chromatography in recent years has become more of an analytical tool for quantitative and trace analysis. With regard to the separation capability, the earlier approach of enhancing the selectivity of the stationary phase to solute components has given way to increasing the efficiency of the column. The success of the open-tubular column, particularly in GC, has been phenomenal in this respect. However, there is still scope for speciality stationary phases, such as the chiral stationary phases, both in GC and LC. Also, widespread and routine use of chromatography in industrial quality control laboratories and process control (and in several other diverse fields, where the personnel may not be experts in chromatography) warrant tailor-made columns and instruments to suit specific applications.

Improvements in column technology, detectors and instrument design continue to be made for increased efficiency of separation, detection and quantitation. The objective is very rapid and accurate analysis of very small samples. Keeping with this trend, slower and less sensitive techniques have almost become obsolete, though some of them may remain historically important. Displacement of paper chromatography by TLC is one such example. Yet another trend in contemporary chromatographic practice is that commercial products are replacing what were normally being made in the laboratory. Thus, with the advent of pre-coated TLC plates, preparation of TLC plates in the laboratory has become a rarity. HPTLC plates, bonded-phase fused-silica columns and the microparticulate LC columns are generally obtained commercially. HPTLC, capillary GC and HPLC have also almost displaced the conventional TLC, packed column GC and the traditional, low-pressure column liquid chromatography. GC–MS which, until recently, was an expensive, sophisticated technique, requiring specialist training, has become a well-established, routine analytical tool; also LC–MS is fast becoming a reality, as is SFC and SFC–MS. Increased automation in sample introduction and data handling is the order of the day. Indeed, the faster (and largely automated) instrumental chromatographic techniques have become so fast that sample preparation for chromatography rather than the chromatography itself has become the determinant of the speed of the analysis. More recently, automated sample preparation and derivatization devices and laboratory robotics have been introduced. With the increasing use of automated sample processing methods, auto-injectors and high-efficiency separation systems there will be an enormous increase in the data output and, consequently, increased demands on the data-handling capability of the computers which form the modern chromatographic data processors.

In spite of the sophistication in instrumentation and considerable knowledge of the chromatographic behaviour of various compounds, the full potential of the technique has probably not been realized, due largely to the haphazard way in which the data had hitherto been recorded. There has been considerable discussion on the reproducibility of chromatographic data, which is very important in both qualitative and quantitative analysis. Availability of standardized pre-packed or pre-coated GC and LC columns and TLC plates (or the reusable sintered glass chromatorods), of guaranteed performance, well-defined mobile phase compositions and chromatographic conditions, and properly documented retention data (preferably relative to readily available standard compounds) will be very useful in qualitative analysis aimed at identification of compounds. The dominance of silica gel and octadecylsilyl-bonded reversed phases in LC is a welcome feature in view of the possible standardization. Even in GC, where there are about 500 stationary phases to choose from, use of only about ten of them is commonly found in recent literature. Also, among the commercially available, bonded-phase fused-silica columns, only a few are likely to remain popular. It is thus possible to compile and computerize solute, stationary phase and mobile phase combinations, typical or standardized procedures and the retention data for easy search and retrieval. It will also help projection, extrapolation and determination of the right system and conditions for a given set of compounds. Strangely, even today, the choice of a chromatographic system is largely empirical based on personal experience and literature information. Since chromatography is a triangular interaction between the solute, solvent and the sorbent, if the last two are held constant or varied systematically within limits, it should be possible to predict the chromatographic behaviour of the solute on the basis of its three-dimensional molecular structure or conversely, and more interestingly, determine the structure of the compound from its chromatographic behaviour. Earlier attempts to systematize structure–retention relationships have not been very successful owing mainly to the lack of standardization and the absence of a suitable mathematical model. Given the present day computer capabilities, this should not be very difficult. (See Chapter 9, section 9.2.3.5).

While chromatography may maintain its predominant position as a routine analytical tool and in trace analysis, there appears to be a revival of interest in its role as a technique for preparative separations, particularly in process-scale operations. Several commercial versions of process-scale chromatographs are already on the market. The great strides occurring in biotechnological developments will be a driving force for further refinements in separation technologies. Thus, several of the biologically active peptides, proteins, nucleotides and nucleic acids, which are very commonly handled in biotechnology research and which are very prone to denaturization, require very mild, selective and efficient systems for their separation and analysis. Simply stated, the future course of chromatography is likely to be dictated by the needs of the biologists.

### 1.2.2 Instrumentation in chromatography

Modern chromatography owes much of its convenience and success to the developments in instrumentation. In a typical chromatographic experiment, the sample is introduced at the head of the column, plate, or paper, by means of a syringe and through an appropriately designed injection port. The syringe is accurately calibrated and graduated so that the exact quantity of the sample transferred to the column is known. Human errors in repeated manual injections are eliminated by use of auto-injectors which can hold three to 100 sample vials and can deliver specific quantities of the sample to the column in a predetermined order and predetermined intervals. The injection port design should be such that the sample is transferred to the stationary phase as a plug; that is, without significant broadening of the band. Most gas chromatographs are equipped with an independently heated injection port which accepts the syringe through a septum; a septumless loop injector is more common in LC. The columns vary in length and diameter over a wide range. Most often, they are packed columns with the stationary phase in the form of solid particles of nearly uniform size, with or without a liquid coating. More recently, open-tubular columns, with the stationary phase supported by the column wall, have become popular in GC and, to an extent, explored in LC; these columns, of necessity, have much smaller diameters than the packed columns. The columns for GC and, occasionally, for LC are housed in a suitably thermostatted oven for maintaining the temperature accurately. The oven can be maintained at a certain temperature or programmed over a given range and at a given rate. While higher temperature operation or temperature programming is less common in LC, it often requires the facility to change the composition of the mobile phase during chromatography. The change of solvent is normally from low strength to high strength. The mobile phase enters the system near the injection port, carries the solute components along the column and delivers them to the detector, connected to the other end of the column. Usually the commercial gas chromatograph is an integrated unit consisting of the injection port, the column and the detector, along with temperature controls for independently maintaining the temperature of the three parts. Several variations are possible in each of these easily-interchangeable modules and these are discussed separately for GC and LC. The totally modular systems are more common in LC. More sophisticated systems may have multiple detectors, columns and switching valves for multidimensional chromatography. Multidimensional chromatography is the term used for a variety of operations other than the common unidimensional operation briefly described above. The operations include allowing the sample to pass through two or more detectors connected in parallel, back-flushing the column after chromatography to remove strongly retained components and heart-cutting which means allowing a portion of the effluent from one column through another column of a different selectivity for further separation.

The availability in recent years of inexpensive small computers (microprocessors) has revolutionized chromatographic instrumentation. Not only the data acquisition and manipulation but the operation of the chromatograph itself is often completely automated. Fully integrated, microprocessor-controlled chromatographs (e.g. Fig. 1.5) offer several advantages over the manual versions. Thus, the chromatographs

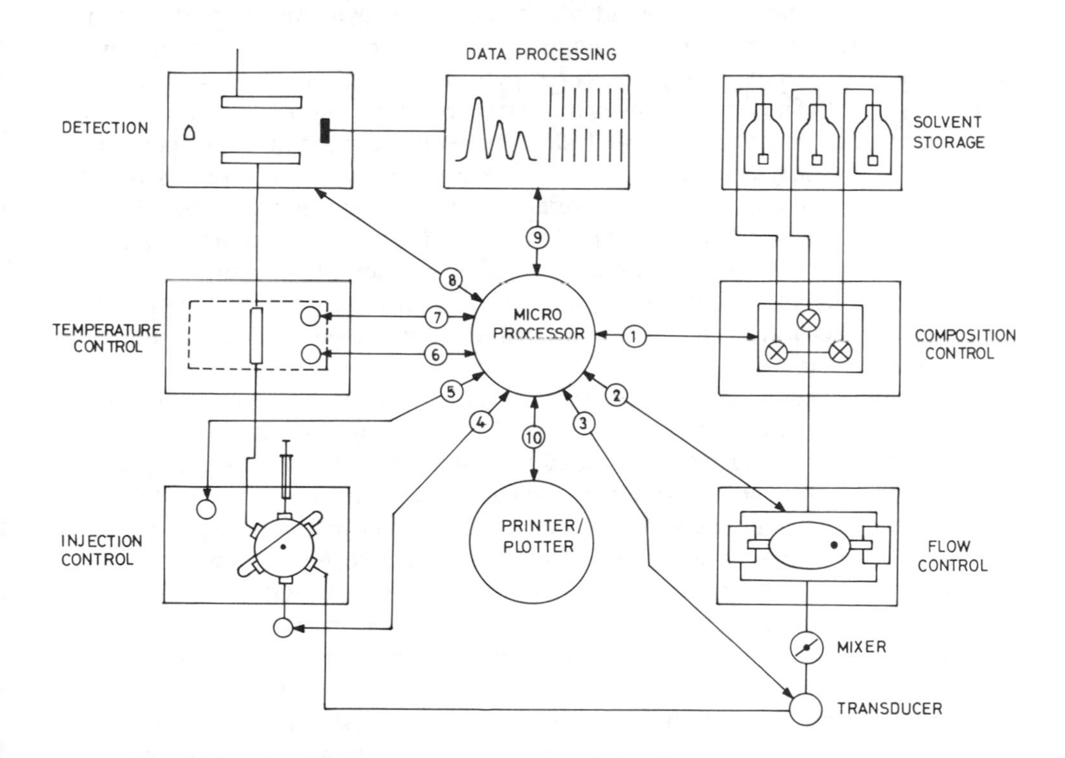


Fig. 1.5 — Schematic diagram of an integrated microprocessor-controlled liquid chromatograph. Illustration by courtesy of Spectra-Physics, Inc.

can automatically control, (1) mobile phase composition, (2) flow-control, (3) transducers, (4) injection valves, (5) auto-samplers, (6) column selection, (7) column temperature, (8) detectors, (9) data processors, (10) recorders, etc., in a predetermined manner. A number of combinations of operating parameters can be fed to the computer using the keyboard, stored and repeatedly used, ensuring accuracy and reproducibility. A precision of better than 0.1% in temperature, mobile-phase composition, flow rate and detector control is possible. Switching, and opening and closing of valves in multidimensional chromatography can be effected with an accuracy of milliseconds. Automatic control of injection systems and fraction collectors makes preparative chromatography less tedious. Much of the repetitive and routine analysis can be carried out unattended. The methods and programs, or the operating parameters, when recalled, are displayed on the cathode-ray tube (CRT or monitor). The methods can be changed as and when required, even during the run.

The output of the detector, which is in the form of an electrical signal, is fed to a potentiometric recorder which records the chromatogram on a strip chart. Fig. 1.3 is the most familiar form of a chromatogram recorded by such a device. From this, the data needed for both qualitative (retention times, etc.) and quantitative analysis (peak area or height) can be readily obtained. The dynamic properties of the recorder pen (i.e. the response of the recorder as a function of the signal) can be critical when the signal changes occur rapidly as in capillary or high-resolution chromatography. The larger number of very sharp peaks that are generated in this case also makes it difficult to accurately determine the retention times and peak areas manually. A number of reporting integrators are available which automatically plot the chromatogram and print out the retention times, and the peak areas, in both absolute units and as percentages of the total area. They can also be programmed to compute the quantitative data relative to either an internal or external standard. When a reporting integrator is used, the potentiometric recorder may be dispensed with and the electrical (analog) output is converted into the digital form, usable by the integrator systems.

Using the modern data processors, a large amount of the data output from chromatography can be conveniently stored, after analog-to-digital conversion, in the computer memory or disks, and retrieved on demand in a form suitable for the particular application. Presentation of the chromatogram by the data systems can either be in the real time (displayed on the CRT), or can be manipulated at the end of the run in many ways to obtain a standard form or a baseline compensated and normalized form. A hard copy of the raw or processed chromatogram , if needed, can be obtained by using a high-speed, microprocessor-based, printer-plotter.

In recent years, instrumentation and automation have also changed the practice of TLC [37]. Modern instrumental TLC uses pre-coated plates (made by automated machines), multiple sample applicators, solvent delivery systems, automated visualization equipment and scanning densitometers (for converting the thin-layer chromatograms into the traditional two-dimensional graphic pattern), coupled to data processors. Such systems are obviously invaluable in quality control applications involving a very large number of samples.

### 1.2.3 Nomenclature of chromatography

The term *chromatographic adsorption analysis*, coined by Tswett to describe his new technique of separation of compounds is frequently reduced to *chromatographic analysis* or *chromatography* (literally, colour writing). From this are derived other terms such as *chromatograph* and *chromatogram* to describe, respectively, the instrument used and the result of the analysis (either the paper, thin layer or column with the substances after separation, or a graphic presentation of the detector response *vs.* time, or effluent volume). It may be interesting to note that the term 'tsvet' in Russian means colour and there have been suggestions that, by calling it chromatography, Tswett was probably naming the technique after himself — something like *Tswettography*. However, 'chromatography', was defined as early as 1731 as 'treatise in colour' and there was also a book entitled *Chromatography*, published in 1835 [38].

Different authors defined the term chromatography in different ways and the one suggested by the International Union of Pure and Applied Chemistry is reproduced here [39]: 'A method, used primarily for separation of the components of a sample, in which the components are distributed between two phases, one of which is stationary while the other moves. The stationary phase may be a solid, or a liquid supported on a solid, or a gel. The stationary phase may be packed in a *column*, spread as a *layer*, or distributed as a film, etc.; in these definitions *chromatographic bed* is used as a general term to denote any of the different forms in which the stationary phase may be used. The mobile phase may be gaseous or liquid'.

It is obvious that a chromatographic system consists of three components, poetically describable as solute, solvent and sorbent, but more appropriately as the *sample*, *mobile phase* and *stationary phase*. Three more terms, used to describe the mobile phase and its derivatives, may be mentioned; they are: *eluent* (sometimes spelled *eluant*), *eluate* and *effluent* [40]. Eluent is the solvent (single or mixture) used for elution; that is, the mobile phase. Eluate is the mixture of the solute and the eluent exiting the column and effluent is the stream flowing out of the column with or without any separation taking place. Horvath (1980) had suggested a more precise term for the solute being chromatographed, namely, *eluite*, rhyming with analyte, but the term does not find wide acceptance.

Over the past few decades, a number of new chromatographic methods have been developed; these differ from one another in technique as well as the mechanism of separation. For example, in planar chromatography alone, one comes across terms such as thermomicro application–separation (TAS), thermo-fragmentography (TFG), high-performance thin-layer chromatography (HPTLC), etc., and development methods such as circular, anticircular, centrifugal, centripetal, temperature-programmed, vapour-programmed, programmed multiple development (PMD) and over-pressured, in addition to the more familiar ascending, descending, horizontal, continuous, stepwise, multiple, two-dimensional and gradient modes. Most of these have been developed by people with widely differing backgrounds and affiliations and the authors coined their own terminology, resulting in a bewildering array of terms and acronyms, which often represent techniques belonging, or closely related, to one of the general methods. Frequently, different terms are used to give the same meaning (e.g., solvent, wash liquid, developer, eluent, etc., for the mobile phase) or the same word is used to convey different meanings, e.g. carrier for the support in partition chromatography and for the mobile phase in gas chromatography). The terms elution and development are sometimes used interchangeably but the former is appropriate for column chromatography and the latter for planar chromatography.

Several national and international organizations have striven to develop a uniform nomenclature for chromatography, notable among them being the British Standards Institution (BSI), the International Union of Pure and Applied Chemistry (IUPAC) and the American Society for Testing and Materials (ASTM). The recommendations of these organizations have been published in several bulletins and have been excellently reviewed by Ettre in three parts [41]. There are considerable variations, anomalies and discrepancies in all the three systems but the ASTM recommendations, being the most recent (1979), are preferred here though some of the terms may be at variance from those in common usage. An encyclopaedic compilation of the terms and symbols used in chromatography has been made by Denney [42]. Some of the commonly used terms and their definitions are given in Table 1.1; a few general comments on their use may be in order.

A rather vexing aspect of chromatography is the seemingly endless proliferation of new terms and acronyms, ranging from the very familiar GC to the not-so-familiar donor–acceptor complex chromatography (DACC). Some of the more recent additions to the list may be interesting to note; for example, cartridge chromatography, metal-chelate chromatography, micellar chromatography, multi-ion chromatography, non-ideal chromatography, paramagnetic chromatography, reaction chromatography, redox chromatography, potential barrier chromatography and tandem chromatography. While many of these terms may not find general usage, a conscious effort seems warranted to combine related techniques to one representative term for easy access to the literature. Thus, the term planar chromatography is now often used to denote both paper and thin-layer chromatography. A few expressions (e.g. high-speed liquid chromatography (HSLC)) which were in common usage in the earlier literature have been replaced by more appropriate ones. There was also some controversy over the very familiar acronym HPLC, which is expanded as high-pressure liquid chromatography or high-performance liquid chromatography. The latter expression and its derivative, high-performance thin-layer chromatography (HPTLC) are generally accepted and widely in use, though both IUPAC and ASTM discourage use of such terms on the grounds that the distinction from the conventional gravity or capillary flow techniques is artificial. But since both traditional and instrumental techniques are still being used simultaneously in most laboratories, it is likely that the corresponding terms will also continue to be used. Recently, the seemingly unnecessary term high-resolution chromatography (to denote open-tubular chromatography and to distinguish it from the traditional and relatively low-resolution packed-column chromatography) has come to be used rather frequently. However, in the absence of a clear definition (e.g. in terms of solutes with the minimum separation factor that need to be resolved to qualify for the term) and the desirable general trend to constantly expect better resolution, the term is unlikely to be meaningful. Another artificial term is reversed-phase (not reverse phase) chromatography in which the mobile phase is more polar than the stationary phase; the alternative is considered *normal* only because it was in use earlier. Yet another terminology that has been the subject of some discussion is in relation to the dimensions of the columns. Columns in both GC and LC may be either packed or open tubular. In the latter case, the column wall itself acts as the support for the stationary phase, either coated or bonded. In order to ensure sufficient contact between the mobile phase and the stationary phase, the open tubular columns need to be of capillary dimensions and the terms open

**Table 1.1** — List of commonly used symbols in chromatography

| Symbol | Parameter | Formula |
|---|---|---|
| $a$ | Activity | $\gamma C$ |
| $\mathbf{a}$ | Mark–Houwink constant | |
| $a_M$ | Activity in the mobile phase | |
| $a_S$ | Activity in the stationary phase | |
| $A$ | Cross-sectional area of the tube | |
| $B^o$ | Permeability | |
| $C$ | Concentration | |
| $C_p$ | Peak capacity | |
| $d$ | Diameter | |
| $d_c$ | Column (inside) diameter | |
| $d_f$ | Stationary (liquid) phase thickness | |
| $d_p$ | Particle diameter of the packing in packed columns | |
| $D_M$ | Diffusion coefficient of the solute in the mobile phase | |
| $D_S$ | Diffusion coefficient of the solute in the stationary phase | |
| $D_T$ | Thermal diffusion coefficient | |
| $E$ | Efficiency | $t_R/w_b$ |
| $f$ | Fugacity | |
| $F$ | Mobile-phase flow-rate | |
| $F_c$ | Temperature-corrected flow-rate | |
| $G$ | Gravitational (or centrifugal) acceleration | |
| $h$ | Height equivalent to one theoretical plate (HETP) | |
| $H$ | Height equivalent to an effective theoretical plate (HEEP) | |
| $I$ | Kovats retention index | |
| $I_{TP}$ | Kovats index in PTGC | |
| $j$ | Mobile-phase compressibility factor | |
| $k$ | Capacity factor | $(V_R - V_M)/V_M$ or $(t_R - t_M)/t_M$ |
| $k'$ | Capacity factor (SEC) | $(V_R - V_0)/V_0$ or $(t_R - t_0)/t_0$ |
| $k_0$ | Ratio of the total to interstitial void volume | |
| $K$ | Distribution constant | $[C]_S/[C]_M$ |
| $\mathbf{K}$ | Mark–Houwink constant | |
| $K_0$ | Distribution constant (SEC) | $(V_R - V_0)/V_i$ |
| $L$ | Column length | |
| $M$ | Molar | |
| $M$ | Aspect ratio of ellipsoid | |
| $M_n$ | Number-average molecular weight | |
| $M_w$ | Weight-average molecular weight | |
| $n$ | Number of theoretical plates | $16(t_R/w_b)^2$ |
| $n$ | Number of moles of the solute | |
| $n_M$ | Number of moles of solute in mobile phase | |
| $n_S$ | Number of moles of solute in stationary phase | |
| $N$ | Number of effective theoretical plates | $16(t'_R/w_b)^2$ |
| $N$ | Avogadro number | |
| $P$ | Pressure | |
| $P_c$ | Critical pressure | |
| $r$ | Radial position in tube cross-section | |
| $R$ | Tube radius | |
| $R$ | Fraction of the solute in the mobile phase | $1/(1 + k)$ |
| $R$ $(R^*)$ | Gas constant | |
| $R_e$ | Radius of equivalent sphere | |
| $R_f$ | Mobility of solute relative to solvent front | $Z_s/Z_f$ |
| $R_G$ | Gyration radius | |
| $R_S$ | Resolution | $(t_{R2} - t_{R1})/w_b$ |

*Continued next page*

**Table 1.1** (cont.) — List of commonly used symbols in chromatography

| Symbol | Parameter | Formula |
|---|---|---|
| $t$ | Time | |
| $t_M$ | Mobile-phase hold-up time ($t^o$) | |
| $t_M^o$ | Corrected mobile-phase hold-up time | $jt_M$ |
| $t_N$ | Net retention time | $t_R^o - t_M^o$ |
| $t_R$ | Retention time | |
| $t'_R$ | Adjusted retention time | $t_R - t_M$ |
| $t_M^o$ | Corrected retention time | $jt_R$ |
| $t_w$ | Peak width in time scale | |
| $T$ | Temperature | |
| $T_c$ | Column temperature | |
| $T_c$ | Critical temperature | |
| $u$ | Mobile-phase linear velocity | $L/t^o$ |
| $u_e$ | Interstitial velocity | |
| $u_0$ | Mobile-phase velocity | |
| $u_{opt}$ | Optimum velocity | |
| $v$ | Reduced velocity | $ud_p/D_M$ |
| $V_M$ | Mobile-phase hold-up volume | $Ft_R$ |
| $V_M^o$ | Corrected mobile-phase hold-up volume | $jV_M$ |
| $V_N$ | Net retention volume | $V_R^o - V_M^o$ |
| $V_R$ | Retention volume | |
| $V'_R$ | Adjusted retention volume | $V_R - V_M$ |
| $V_R^o$ | Corrected retention volume | $jV_R$ |
| $V_S$ | Volume of stationary phase | |
| $w_b$ | Peak width at the base | |
| $w_h$ | Peak width at half height | |
| $w_i$ | Peak width at inflection points | |
| $x$ | Mole fraction | |
| $Z_f$ | Distance travelled by the solvent (TLC) | |
| $Z_0$ | Distance between spot and solvent level | |
| $Z_s$ | Distance travelled by the spot (TLC) | |
| $\alpha$ (alpha) | Separation factor | $t'_{R2}/t'_{R1}$ or $V'_{R2}/V'_{R1}$ |
| $\alpha^*$ | Relative retention time | $t_{R2}/t_{R1}$ or $V_{R2}/V_{R1}$ |
| $\beta$ (beta) | Phase ratio | $V_M/V_S$ |
| $\gamma$ (gamma) | Tortuosity factor | |
| $\gamma$ | Activity coefficient | |
| $\delta$ (delta) | Solubility parameter | |
| $\Delta$ (delta) | Difference between two values of the same parameter | |
| $\varepsilon$ (epsilon) | Molar extinction coefficient | |
| $\varepsilon$ | Dielectric constant (mobile phase) | |
| $\varepsilon^o$ | Solvent strength | |
| $\eta$ (eta) | Viscosity | |
| $[\eta]$ | Intrinsic viscosity | |
| $\lambda$ (lambda) | Packing irregularity factor | |
| $\mu$ (mu) | Chemical potential | |
| $\pi$ (pi) | 3.14159 . . . | |
| $\phi$ (phi) | Fluorescence yield | |
| $\rho$ (rho) | Density | |
| $\rho_S$ | Density of the stationary phase | |
| $\sigma$ (sigma) | Standard deviation | |
| $\sigma^2$ | Variance | |
| $\tau$ (tau) | Surface tension | |
| $\theta$ (theta) | Contact angle (TLC) | |
| $\xi$ (xi) | Proportionality constant relating gyration radius to spherical radius | |

tubular and capillary are often used interchangeably, but actually they are not synonymous. The expression capillary refers only to the diameter, usually less than 1 mm, either packed or open tubular.

Interestingly, there is no clear definition or expression for one of the most important parameters in chromatography, namely, $R$. It is used to denote the fraction of the solute in the mobile phase and simply called the *R value*. Terms such as retention or retardation factor are occasionally used to describe the $R$ value but obviously these are erroneous since solutes strongly retained or retarded by the stationary phase will have lower $R$ values; the term dragging has been suggested [43]. However, a word starting with the letter $R$, such as *releasability factor* may be more appropriate. The corresponding term in planar chromatography is denoted by $R_f$ (or $R_F$), defined as the ratio of the distance travelled by the compound to the distance travelled by the solvent front; in other words, mobility of the solute relative to the front. Both the symbols $R_f$ and $R_F$ are used by different people to describe the same parameter. Similarly, two symbols $k$ and $k'$ (more commonly $k'$) are used to represent the *capacity factor*, which is defined as the ratio of the amount of the solute in the stationary phase to that in the mobile phase. The term retention factor is probably more appropriate in this case. Since the other term may be mistaken to mean the capacity of the column, the IUPAC proposed the term *concentration distribution ratio*, $D$, which has not found much favour; the term capacity factor continues to be used almost universally. Ettre [41] has suggested that $k$ be used in all forms of chromatography except size-exclusion chromatography, where the retention time for the unretained molecules and that of the mobile phase molecules is not the same. Mobile phase molecules, being the smallest, elute last in SEC and therefore $k$, defined as given in Table 1.1, will have a negative value. More detailed explanations of the different symbols and their interrelationships are given in later chapters wherever appropriate.

### 1.2.4 Literature of chromatography

The literature of chromatography, though voluminous, is well organized and readily accessible. Journals, of course, are the primary source of information on new developments in any subject. Much of the early literature on chromatography was spread out in journals of different subjects; understandably so because many of the developments were made by scientists whose main activity was in fields other than chromatography and they chose to publish them in journals relevant to their area of application. With the introduction of the *Journal of Chromatography* in 1958 (Elsevier, Amsterdam), a nucleus was formed and a forum for chromatographers became available. Elsevier also published *Chromatographic Reviews* as a serial publication for some time, but later incorporated it into the *Journal of Chromatography*. (The practice of publishing *Chromatographic Reviews* as a separate volume of the journal has also been recently discontinued.) *Journal of Chromatography, Biomedical Applications* was started in 1977 under separate editorship but as a part of the main Journal. The periodicity and volume numbering of the Journal is rather unusual; for example, 40 volumes consisting of 61 issues were published in 1987. (The publication schedule appears on the inside back cover of the Journal.) Apart from reviews, full papers, notes and a majority of the important chromatographic symposia proceedings, the Journal has several important and useful features worthy of mention. The most important is the bibliography section which appears periodically. It is an exhaustive compilation of the literature on chromatography, divided into the main sections of gas chromatography, liquid chromatography, planar chromatography and electrophoretic techniques. Each section is divided into 37 subsections, based on the classes of compounds and their applications, which are appropriately further subdivided into smaller groups of compounds for easy location. In addition there is a subject index and an index of types of compounds chromatographed. The 200th bibliography in 25 years of publication of the Journal appeared in 1986 (Vol. 372); the total number of chromatographic references cited in these sections was a staggering one hundred and fifty eight thousand, eight hundred and seventy two!

As a technique of qualitative and quantitative analysis, chromatography is strictly a part of analytical chemistry and the journal, *Analytical Chemistry*, published by the American Chemical Society since 1929, covers a considerable portion of the literature on chromatography. Particularly useful are the two-yearly reviews; even years contain one issue of *Fundamental Reviews* of the developments in analytical techniques including chromatographic methods, and odd years cover *Application Reviews* pertaining to different areas of application. The August issue each year contains a *Lab Guide* listing the suppliers of the analytical instruments, chromatographic products and reagents.

From 1963 to 1968, Preston Technical Abstracts Co. (later Preston Publications Inc.), Niles, Illinois, published the *Journal of Gas Chromatography*, subsequently renamed the *Journal of Chromatographic Science.* In addition to research papers, one can find timely review articles on instrumentation and methodology in chromatography. An *International Chromatography Guide*, included in the journal, lists chromatographic products (from adsorbents to visualizing agents) in alphabetical order, their suppliers and the suppliers' addresses.

Marcel Dekker Inc., New York, publishes the *Journal of Liquid Chromatography* (monthly since 1978, 16 issues per volume) and *Separation Science and Technology* (10 issues per volume since 1979; formerly, from 1966 to 1978, *Separation Science*), as well as *Separation and Purification Methods* (2 issues per volume). Also from Marcel Dekker, are *Advances in Chromatography* and *Chromatographic Science Series*, both from 1965. The former is a collection of reviews on topics of general interest, and the individual numbers of the latter deal with selected topics. *Chromatographia* (monthly since 1968, Pergamon Press, Oxford) and the *Journal of High Resolution Chromatography and Chromatography Communications* (monthly since 1978, Dr Huethig Verlag GmbH, Heidelberg, FRG) are two other journals in the field. The latest additions to the list of chromatography journals are *Biomedical Chromatography* (bimonthly, beginning in 1987, Heyden, London) and *Journal of Planar Chromatography* (1988, Heuthig, Heidelberg). Apart from the original papers, most journals contain book reviews, news sections on new products, literature and meetings, and advertisements from the chromatographic products manufacturers and suppliers. The manufacturers of chromatographic products and instruments also publish useful literature in the form of newsletters, technical bulletins, notes and information brochures, which are available on request. Abstracts and bulletins are published by scientific bodies devoted to the subject of chromatography, e.g. *The Chromatographic Society Bulletin*, UK. Most of the literature is abstracted by a number of gas and liquid chromatography abstracting services and by the *Chemical Abstracts.* (*CA Selects* on gas chromatography, gel-permeation chromatography, high-performance liquid chromatography, ion chromatography, and ion-exchange, paper and thin-layer chromatography are extracts from the main volumes.)

There have been repeated attempts at compilation of the chromatographic literature in the form of bibliographies on individual methods; namely, gas, column, liquid, paper and thin-layer chromatography, etc., the most recent being *A Guide to HPLC Literature* [44]. The first three volumes this series include references covering the period 1966–1982; Volume 1 covers the period 1966–1979, Volume 2 covers 1980–1981 and Volume 3 covers 1982. The citations are mainly limited to the methods used most often, namely the articles on normal and reversed-phase LC. Articles on affinity, ion-exchange and size-exclusion chromatography are not included. Citations on planar chromatography are limited to articles that deal with the correlation between TLC and HPLC.

There have also been several attempts to compile the large amount of retention data on different compounds under different chromatographic conditions, the most extensive being the *Handbook of Chromatography* [45], published by the CRC Press in 1972. Volume I of this work contains tables of gas, liquid, paper and thin-layer chromatography. The compounds are divided according to the functional groupings (acids, alcohols, aldehydes, etc.), chemical class (carbohydrates, hydrocarbons, steroids, etc.) or according to the application (drugs, pesticides, plasticizers, etc.), depending on convenience. So far as is possible, complete information is given in each case, including the stationary phase, column dimensions, configuration, mobile phase and its velocity, temperature and other experimental conditions. LC tables contain information mainly on ion exchange and size separation,since work on the book commenced in 1969 when modern LC (HPLC) was still in its infancy. The tables are followed by an elaborate cross-index of over 12,000 compounds. Volume II contains a section on the principles and techniques with a brief introduction to the various chromatographic methods. The section on detection reagents contains 453 reagents, their methods of preparation, applications and results, again with an index. Information on selected methods of sample preparation and derivatization, sources of chromatographic materials and products, and a directory of chromatography books (published before 1970) completes the work. Though undoubtedly the most valuable compilation of chromatographic data at that time, it was inevitable that, considering the rate of growth of such data, a work of this type would soon be dated. Subsequently, the CRC Press continued publication of handbooks of chromatography on specific compound classes; namely, drugs [46], phenols and organic acids [47], carbohydrates [48], polymers [49], amino acids and amines [50], terpenoids [51], pesticides [52], inorganics [53], etc. With microcomputers becoming very common and the chromatographic data getting too unwieldy, it is more convenient to have the retention data on floppy discs for easy access. Computerized libraries of gas chromatographic retention data (for example, Kovats retention indices) on selected stationary phases are now available for several thousands of compounds.

## REFERENCES

[1] E. S. Perry and A. Weissberger (eds.), *Separation and Purification*, 3rd edn, Wiley, New York, 1978.
[2] M. R. J. Dack (ed.), *Solutions and Solubilities*, Wiley, New York, Parts I and II, 1975/1976.
[3] L. C. Craig, *J. Biol. Chem.*, **155**, 519 (1944).
[4] Y. Ito, *Adv. Chromatogr.*, **24**, 181 (1984).

[5] S.-T. Hwang and K. Kammermeyer, *Membranes in Separations*, Wiley, New York, 1975.
[6] A. T. Andrews, *Electrophoresis: Theory, Techniques and Biochemical and Clinical Applications*, 2nd edn, Clarendon, Oxford, 1986.
[7] J. F. Parcher, *Adv. Chromatogr.*, **16**, 151 (1978).
[8] C. Horvath, in *The Science of Chromatography*, F. Bruner (ed.), Elsevier, Amsterdam, 1985.
[9] F. W. McLafferty, *Tandem Mass Spectrometry*, Wiley, New York, 1983.
[10] L. S. Ettre, *J. Chromatogr.*, **112**, 1 (1975).
[11] L. S. Ettre, in *High Performance Liquid Chromatography: Advances and Perspectives*, C. Horvath (ed.), Academic Press, New York, Vol. 1, 1980.
[12] L. S. Ettre and A. Zlatkis (eds.), *75 Years of Chromatography: A Historical Dialogue*, Elsevier, Amsterdam, 1979.
[13] A. J. P. Martin and R. L. M. Synge, *Biochem. J.*, **35**, 91 (1941); **35**, 1358 (1941).
[14] R. Kuhn, A. Winterstein, and E. Lederer, *Hoppe-Seyler's Z. Physiol. Chem.*, **197**, 141 (1931).
[15] A. T. James and A. J. P. Martin, *Biochem. J.*, **50**, 679 (1952).
[16] A. S. Ritchie, *Adv. Chromatogr.*, **3**, 119 (1966).
[17] D. T. Day, *Proc. Am. Phil. Soc.*, **36**, 112 (1897).
[18] M. Tswett, *Ber. Deut. Bot. Ges.*, **XXIV**, 316 (1906); **XXIV**, 384 (1906); **XXV**, 71 (1907).
[19] K. I. Sakodynskii, *J. Chromatogr.*, **220**, 1 (1981); and citations therein.
[20] L. S. Ettre and C. Horvath, *Anal. Chem.*, **47**, 422A (1975).
[21] A. J. P. Martin, in *Nobel Lectures: Chemistry*, 1942–1962, Elsevier, Amsterdam, 1964.
[22] R. Consden, A. H. Gordon, and A. J. P. Martin, *Biochem. J.*, **38**, 224 (1944).
[23] S. Moore and W. H. Stein, *J. Biol. Chem.*, **192**, 663 (1951).
[24] J. Porath and P. Flodin, *Nature*, **183**, 1657 (1959).
[25] R. Axen, J. Porath, and S. Ernback, *Nature*, **214**, 1302. (1967).
[26] P. Cuatracasas, M. Wilchek and C. B. Anfinsen, *Proc. Natl. Acad. Sci. U.S.A.*, **61**, 636 (1968).
[27] M. Wilchek and W. B. Jakoby, *Meth. Enzymol.*, **34**, 3 (1974).
[28] J. Sherma and G. Zweig, *Paper Chromatography*, Academic Press, New York, 1971.
[29] J. G. Kirchner, *Thin-Layer Chromatography*, Wiley, New York, 2nd edn, 1978.
[30] E. Stahl, *J. Chromatogr.*, **165**, 59 (1979).
[31] A. Zlatkis and R. E. Kaiser (eds.), *HPTLC: High-Performance Thin-Layer Chromatography*, Elsevier, Amsterdam, 1977.
[32] M. J. E. Golay, in *Gas Chromatography (1957 Lansing Symposium*, V. J. Coates, H. J. Noebels, and I. S. Fagerson (eds.), Academic Press, New York, p. 1, 1958.
[33] D. H. Desty, J. N. Haresnape and B. H. F. Whyman, *Anal. Chem.*, **32**, 302 (1960).
[34] R. Dandeaneau and E. H. Zerenner, *J. High Resolut. Chromatogr. & Chromatogr. Commun.*, **2**, 351 (1979).
[35] R. Ryhage, *Anal. Chem.*, **36**, 759 (1964).
[36] P. J. Schoenmakers and F. C. C. J. G. Verhoeven, *Trends Anal. Chem.*, **6**, 10 (1967).
[37] W. Bertsch, S. Hara, R. E. Kaiser, and A. Zlatkis, (eds.), *Instrumental HPTLC*, Huethig, Heidelberg (1980).
[38] I. I. Williams and H. Weil, *Nature*, **170**, 503 (1952).
[39] IUPAC, *Pure and Appl. Chem.*, **37**, 447 (1974).
[40] S. Borman, *Anal. Chem.*, **59**, 99A (1987).
[41] L. S. Ettre, *J. Chromatogr.*, **165**, 235 (1979); **220**, 29 (1981) and **220**, 65 (1981).
[42] R. C. Denney, *A Dictionary of Chromatography*, Macmillan, London, 2nd edn, 1982.
[43] R. Delley, *Chromatographia*, **15**, 167 (1982).
[44] H. Colin, A. M. Krstulovic, J. Excoffier, and G. Guiochon, *A Guide to HPLC Literature*, Vols. 1–4, Wiley, New York, 1983–1986.
[45] G. Zweig and J. Sherma (eds.), *Handbook of Chromatography*, Vols. I and II, CRC Press, Cleveland, Ohio, 1972.
[46] R. N. Gupta (ed.), *Handbook of Chromatography: Drugs*, Vols. I and II, CRC Press, Boca Raton, Florida, 1981.
[47] T. Hanai (ed.), *Handbook of Chromotography: Phenols and Organic Acids*, CRC Press, Boca Raton, Florida, 1982.
[48] S. C. Churms (ed.), *Handbook of Chromatography: Carbohydrates*, CRC Press, Boca Raton, Florida, 1982.
[49] C. G. Smith, N. E. Skelly, C. C. D. Chow, and R. A. Solomon, *Handbook of Chromatography: Polymers*, CRC Press, Boca Raton, Florida, 1982.
[50] S. Blackburn, *Handbook of Chromatography: Amino Acids and Amines*, CRC Press, Boca Raton, Florida, 1983.
[51] C. J. Coscia, *CRC Handbook of Chromatography: Terpenoids*, Vol. 1, CRC Press, Boca Raton, Florida, 1984.

[52] J. M. Follweiler and J. Sherma, *CRC Handbook of Chromatography: Pesticides and Related Organic Chemicals*, Vol. 1, CRC Press, Boca Raton, Florida, 1984.
[53] M. Qureshi (ed.), *CRC Handbook of Chromatography: Inorganics*, CRC Press, Boca Raton, Florida, 1986.

# 2

# Theoretical concepts

## 2.1 INTRODUCTION

The theory of chromatography has been discussed by a number of authors from different points of view [1–8]. Only a brief and simplified account of the topic is presented here. The discussion, which is rather qualitative in character, is aimed at describing the factors that cause the observed peak shapes and separation, and at improving them where necessary. The treatment is directed towards improvement of the overall chromatographic efficiency which includes the peak shape (sharpness or narrowness and symmetry), resolution between two adjacent peaks, and the speed of the analysis. It may be assumed that, even in a complex mixture, the basic problem of resolution of the pair of compounds is the most difficult to resolve. The calculations and approaches to solving this problem can often be extended to the separation of the other components of the mixture. In some cases, however, several portions of the chromatogram may have to be independently optimized. It is common experience that in some complex samples the peaks may have a wide range of retention times, in which case, the separations under a given set of temperature, mobile phase and flow conditions may take an inordinately long time. Special strategies like programmed-temperature GC, solvent-programmed LC and multidimensional chromatography may then be needed for adequate separation in a reasonable time.

The factors affecting the chromatographic performance can be divided into two distinct categories; namely, physical and chemical. The present discussion is limited to the former; i.e. the effects of such parameters as the column dimensions and permeability, support-particle size and size distribution, stationary-phase thickness, mobile-phase viscosity and velocity, and solute diffusivity in the two phases, etc. The discussion is generally common to all chromatographic processes irrespective of the nature of the solute, the stationary phase and the mobile phase. The more specific interactions between the three components, which determine the selectivity or the relative mobility of different solutes during their passage through the chromatographic bed, are discussed at appropriate places in subsequent chapters. A majority of these interactions are again based on the physical properties of the molecules involved; e.g. partition, adsorption, size exclusion, etc. Some of the separations, however, may be considered to involve chemical interactions. Chromatographic separations based on ion exchange, charge transfer, covalent bonding, etc., may be included in this category. Affinity chromatography is considered biospecific but, at the molecular level, should belong to one of the above classes.

## 2.2 CHARACTERISTICS OF THE CHROMATOGRAM

Before embarking on a discussion of the theory of chromatography, a brief account of the characteristic features of the chromatogram, the different retention parameters and their interrelationships may be in order. Fig. 1.3 showed a typical graphic representation of the chromatographic separation of a two-component mixture. Some of the notations on the chromatogram are elaborated here.

### 2.2.1 The retention parameters

The rate of migration of a compound in any chromatographic system (e.g. liquid–liquid partition chromatography) is determined by its distribution constant, $K$, between the two phases:

$$K = \frac{[C]_S}{[C]_M} \tag{2.1}$$

where $[C]_S$ and $[C]_M$ are the concentrations of the solute in the stationary and the mobile phase, respectively. If the distribution constant is large, the compound at equilibrium is present mainly in the stationary phase, with only a small fraction of the solute in the mobile phase. Since the molecules in the stationary phase do not move to any noticeable extent the velocity ($u_1$) of the band 1 is determined by the number of molecules ($u_{1M}$) of the compound 1 in the mobile phase at any given time and this, in turn, is determined by its distribution constant, $K_1$. If the sample contains a second compound with a distribution constant ($K_2$) different from $K_1$, its velocity along the column ($u_2$) will be different from $u_1$. If $K_2 > K_1$, $u_1 > u_2$ and, hence, compound 1 moves faster; in other words, in elution chromatography, compound 1 elutes first. The ratio of the number of molecules ($n_M$) of a particular compound in the mobile phase to the total number of molecules ($n_M + n_S$, $n_S$ being the number of molecules in the stationary phase) is denoted by $R$. On the other hand, the capacity factor, $k$, is given by the ratio of the number of molecules in the stationary phase to that in the mobile phase:

$$k = \frac{n_S}{n_M} \tag{2.2}$$

and

$$k+1 = \frac{n_S}{n_M} + \frac{n_M}{n_M} = \frac{n_S + n_M}{n_M}$$

But,

$$R = \frac{n_M}{n_M + n_S} \tag{2.3}$$

Therefore,

$$R = \frac{1}{k+1} \tag{2.4}$$

The rate of migration ($u_1$) of the compound relative to that of the mobile phase ($u$) is given by:

$$u_1 = uR \quad \text{or} \quad R = \frac{u_1}{u} \tag{2.5}$$

The capacity factor can be related to the distribution constant, $K$. The number of moles of the solute in the two phases is given by the molar concentration in each phase multiplied by the volumes of the two phases, and if $V_M$ and $V_S$ are the volumes of the mobile and the stationary phases and β is the phase ratio:

$$k = \frac{n_S}{n_M} = \frac{C_S V_S}{C_M V_M} = K\frac{V_S}{V_M} \tag{2.6}$$

or,

$$K = k\frac{V_M}{V_S} = k\beta \tag{2.7}$$

Eq. (2.7) indicates that $k$ is directly proportional to the volume of the stationary phase, which means that retention of a compound can be increased or decreased by suitably varying the stationary phase loading.

If the length of the column (or the stationary-phase bed) is $L$, the time taken by the compound to travel from point of sample introduction to the end of the column (the retention time, $t_R$) is given by:

$$t_R = \frac{L}{u_1} \tag{2.8}$$

and that for the mobile phase molecules:

$$t_{M} = \frac{L}{u} \tag{2.9}$$

Therefore,

$$t_{R} = t_{M}\frac{u}{u_{1}} = \frac{t_{M}}{R} \tag{2.10}$$

Combining eqs. (2.4), (2.5) and (2.10),

$$t_{R} = t_{M}(1+k) \quad \text{or} \quad k = \frac{t_{R}-t_{M}}{t_{M}} = \frac{t_{R}'}{t_{M}} \tag{2.11}$$

The adjusted retention time ($t_{R}'$), or the retention time of the solute less the time taken by the mobile phase (or the unretained) molecules to elute, is a more accurate description of the retention of the solute by the stationary phase.

The above equations relate the capacity factor to the retention time. The $k$ value can also be expressed in terms of the retention volumes. If $F$ (ml/sec) is the flow rate:

$$V_{R} = t_{R}F \tag{2.12}$$

and

$$k = \frac{V_{R}-V_{M}}{V_{M}} = \frac{V_{R}'}{V_{M}} \tag{2.13}$$

The separation between two components is the relative retention or the separation factor, $\alpha$, so that:

$$\alpha = \frac{k_{2}}{k_{1}} = \frac{K_{2}}{K_{1}} \tag{2.14}$$

Since the separation of two compounds is determined by their distribution constants, any of the variables that alter the equilibrium distribution, i.e. the composition of either or both of the phases and/or the temperature, affects the separation. While the distribution constant is independent of the relative amounts of the two phases, the capacity factor can be changed by changing the phase ratio. The change in the phase ratio, however, does not affect the relative retention, $\alpha$, which is a function of the relative physicochemical interactions of the solutes with the two phases.

### 2.2.2 Peak shape and the theoretical plate

Ideally, a chromatographic peak has Gaussian shape, that is, describable by the Gaussian concentration distribution eq. (2.15) and by the curve shown in Fig. 2.1:

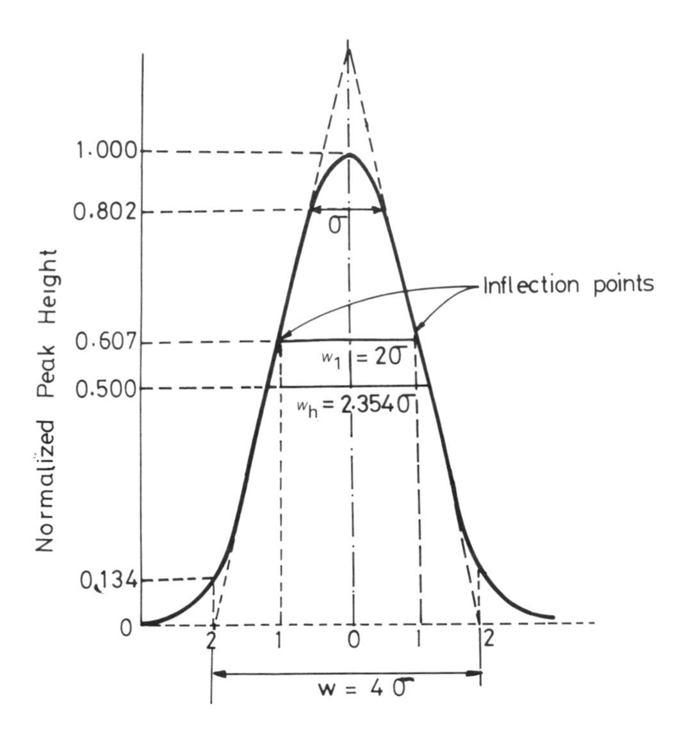


Fig. 2.1 — Characteristics of a Gaussian peak.

$$y = y_0 e^{-x^2/2\sigma^2} \tag{2.15}$$

where $y$ is the height at any point on the curve and $y_0$ is that at the peak maximum, $x$ is the distance between the point on the curve and the ordinate passing through the peak maximum and $\sigma$ is the standard deviation. The square of the standard deviation is the variance, $\sigma^2$. Since the Gaussian peak is symmetrical with respect to the ordinate, the peak width, $w$, at any point is:

$$w = 2x \tag{2.16}$$

The inflection points ($x = \sigma$) of the Gaussian peak occur at a height of 0.607 $y_0$ and the tangents to the peak drawn from the inflection points intercept the baseline at $x = 2\sigma$, so that the peak width at the inflection points ($w_i$), at half height ($w_h$) and at the base ($w_b$) are:

$$w_i = 2\sigma \tag{2.17}$$

$$w_h = 2\sigma\sqrt{2 \ln 2} = 2.335\sigma \tag{2.18}$$

and

$$w_b = 4\sigma = 2w_i \tag{2.19}$$

The width of the chromatographic peak is related to the efficiency of the chromatographic system, particularly the column. By analogy with distillation, the column efficiency is expressed by the number of theoretical plates, $n$, and the height equivalent to a theoretical plate, HETP or $h$ is given by:

$$h = \frac{L}{n} \tag{2.20}$$

where $L$ is the length of the column. A Gaussian peak is described by its variance, so that the variance divided by the length of the column is the plate height:

$$h = \frac{\sigma^2}{L} \tag{2.21}$$

Therefore, from eqs. (2.20) and (2.21),

$$h = \frac{L}{n} = \frac{\sigma^2}{L}$$

that is,

$$n = \frac{L^2}{\sigma^2} \tag{2.22}$$

Substituting $w_b$ for $\sigma$ from eq. (2.19),

$$n = \frac{L^2}{(w_b/4)^2} = 16\,\frac{L^2}{w_b^2} \tag{2.23}$$

Expressing the value of $L$, the distance travelled by a solute (i.e., through the length of the column) in the time scale (i.e., the retention time, $t_R$), we have:

$$n = 16\left(\frac{t_R}{w_b}\right)^2 \tag{2.24}$$

or,

$$n = 4\left(\frac{t_R}{w_i}\right)^2 = 5.54\left(\frac{t_R}{w_h}\right)^2 \tag{2.25}$$

The numbers in eq. (2.25) are derived from the inter-relationships expressed in eqs. (2.17) to (2.19). Thus, the number of theoretical plates can be calculated by measuring the peak width either at the base, at half height or at the inflection points. It is apparent that $n$ is related to the retention time and to the peak width and, in a given column, the value may be different for different solutes. For a given solute, the peak width decreases with increasing number of theoretical plates.

As noted earlier, the adjusted retention time ($t'_R$) rather than the retention time ($t_R$) is more representative of the actual retention of the solute by the stationary-phase bed and, hence, the number of effective plates ($N$) is defined as:

$$N = 16\left(\frac{t'_R}{w_b}\right)^2 = 16\left(\frac{t_R - t_M}{w_b}\right)^2 \tag{2.26}$$

In packed-column GC, where $t_R \gg t_M$, $N$ is approximately equal to $n$, but the difference between the two may be significant in open-tubular GC and in LC. They are related by:

$$N = n\left(\frac{k}{1+k}\right)^2 \tag{2.27}$$

As the value of $k$ increases, $(1+k)$ is approximately equal to $k$, and the effective plate number approaches the theoretical plate number. The height equivalent to an effective plate (HEEP, or $H$) is given by:

$$H = \frac{L}{N} \tag{2.28}$$

It is apparent from the above equations that the peak width ($w$) and the theoretical plate height ($h$) are functions of the standard deviation ($\sigma$) and the variance ($\sigma^2$). Since, for efficient chromatography, these values should be the lowest possible, optimization of the procedures can be guided by the contribution of different experimental factors to the variance. The advantage of considering the variance is that the total variance, $(\sigma_T)^2$, of a Gaussian curve is an additive function of the variances from different, independently variable causes:

$$\sigma_T^2 = \sigma_1^2 + \sigma_2^2 + \sigma_3^2 + \ldots \sigma_n^2 \tag{2.29}$$

where the subscripts 1, 2, 3, . ., $n$ qualify the variances from different sources. It is thus possible to consider the causes of band broadening independent of one another. The variance is proportional to the column length, the proportionality constant being the height equivalent to the theoretical platc, eq. (2.21), or the unit measure of band broadening, i.e. $h$:

$$h = \frac{\sigma_T^2}{L} = \frac{\sigma_1^2}{L} + \frac{\sigma_2^2}{L} + \frac{\sigma_3^2}{L} + \ldots \frac{\sigma_n^2}{L} \tag{2.30}$$

### 2.2.3 Resolution

Chromatography is basically a separation process and is considered to be more efficient if the peaks are well separated both at the apices and at the base. In quantitative terms, the separation is expressed by two parameters, i.e. the relative retention (or the separation factor, $\alpha$) and the resolution, ($R_S$). The resolution is defined as the difference in retention times of the two peaks divided by the average peak width at the base, expressed in time units:

$$R_S = \frac{t_{R2} - t_{R1}}{\frac{1}{2}(w_{b2} + w_{b1})} \tag{2.31}$$

where the subscripts refer to two adjacent peaks, 1 and 2. Since, normally, $w_{b1}$ is approximately equal to $w_{b2}$,

$$R_S = \frac{t_{R2} - t_{R1}}{w_b} \tag{2.32}$$

Thus, while the relative retention is indicative of the degree of separation at the apices ($\alpha = t'_{R1}/t'_{R2}$), the expression for resolution includes peak widths. The resolution, therefore, offers a more accurate description of the quality of separation. Indeed, as will be seen below, the expanded equation for resolution includes the relative retention ($\alpha$) also. The chromatographic separation can be optimized if the relationships between the various interdependent factors such as resolution, capacity factor, relative retention and the plate number are understood.

From eq. (2.11), we have:

$$k = \frac{t_R}{t_M} - 1 \quad \text{or} \quad t_R = t_M\,(1 + k) \tag{2.33}$$

i.e.

$$t_{R1} = t_M\,(1 + k_1) \quad \text{and} \quad t_{R2} = t_M\,(1 + k_2) \tag{2.34}$$

Substituting eq. (2.34) in eq. (2.32),

$$R_S = \frac{t_M (k_2 - k_1)}{w_b} \tag{2.35}$$

Eq. (2.24) can be written as:

$$w_b = \frac{4t_R}{\sqrt{n}} \tag{2.36}$$

Substituting the value of $t_R$ from eq. (2.34),

$$w_b = \frac{4t_M (1 + k_1)}{\sqrt{n}} \tag{2.37}$$

Inserting this expression for $w_b$ in eq. (2.35),

$$\begin{aligned} R_S &= \tfrac{1}{4} \sqrt{n} \frac{t_M (k_2 - k_1)}{t_M (1 + \mathrm{k}_1)} \\ &= \tfrac{1}{4}\sqrt{n} \frac{(k_2 - k_1)}{(1 + k_1)} \end{aligned} \tag{2.38}$$

Multiplying both the numerator and the denominator by $k_1$,

$$\begin{aligned} R_S &= \tfrac{1}{4}\sqrt{n} \frac{(k_2 - k_1)\, k_1}{(1 + k_1)\, k_1} \\ &= \tfrac{1}{4}\sqrt{n} \frac{(k_2 - k_1)}{k_1} \frac{k_1}{(1 + k_1)} \\ &= \tfrac{1}{4}\sqrt{n} \left(\frac{k_2}{k_1} - 1\right) \frac{k_1}{(1 + k_1)} \end{aligned} \tag{2.39}$$

The factor $k_2/k_1$ in the above equation can be substituted by $\alpha$ from eq. (2.14). Also, $k_1$ can be replaced by the more general $k$ to give:

$$R_S = \tfrac{1}{4} \sqrt{n}\, (\alpha - 1) \left(\frac{k}{1 + k}\right) \tag{2.40}$$

Eq. (2.40) is a fundamental relationship in chromatography and clearly indicates how changes in the values of $n$, $\alpha$ and $k$ affect the resolution. Though the three parameters, $n$, $\alpha$ and $k$ are inter-related, they can be optimized one at a time to improve the overall resolution.

The resolution between two peaks can be calculated using eqs. (2.31) or (2.32) but it is not always easy when two peaks are very close, or are in different relative proportions. This fact is apparent from Fig. 2.2 which shows the shape of the peaks at

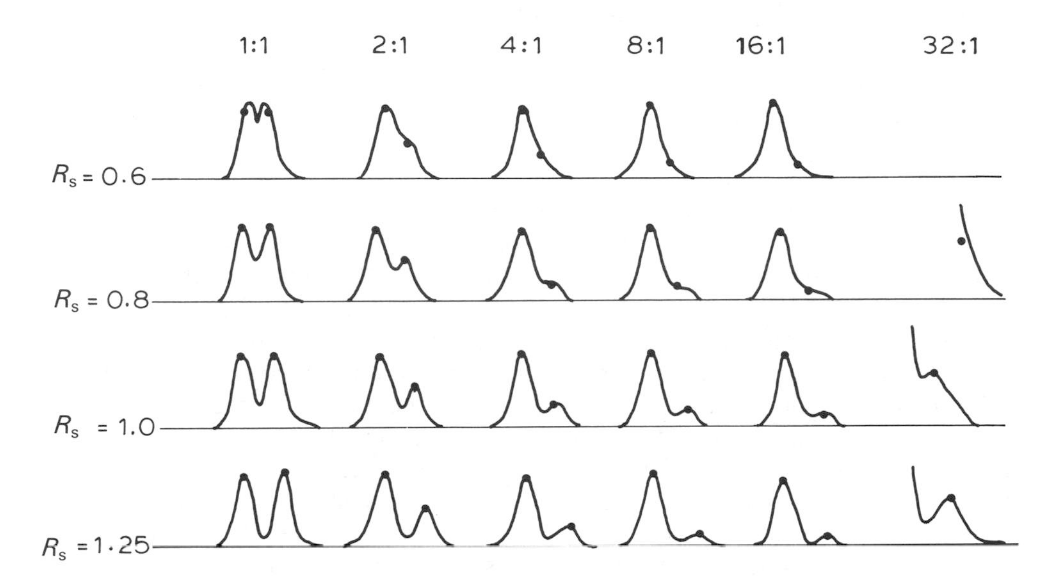


Fig. 2.2 — Band separation at different values of resolution and relative proportions. Reproduced from the *Journal of Chromatographic Science*, Ref. 9, by permission of Preston Publications, Inc.

different values of resolution and relative proportions [9]. The peaks are well separated at $R_s$ not less than 1. Normally, the higher the resolution, the better; but, improvement of resolution involves considerable time and effort and it is not always necessary to increase it to the maximum possible. The value of optimum or adequate resolution varies with the intended application. Thus, if the information sought from the chromatogram is limited to finding out whether a peak represents a single entity or if the tail is, in fact, another peak, an increase in the value of resolution to 0.8 is adequate in most cases unless the relative intensity of the peaks is larger than 16:1. Similarly, the minimum resolution is 0.8 if accurate retention data are needed. The dots at the apex of each peak in the figure indicate the true positions of the peak centres; at $R_S < 0.8$, they may not coincide with the apparent peak maxima. Also, the resolution is very important in preparative chromatography. At an $R_S$ of 0.6, cutting and collection of the fractions on either side of the centre of the peaks would furnish a fraction of about 90% purity, while an $R_S$ of 1.0 could yield materials of 98% purity provided, of course, the detector response for the two compounds is the same. A resolution of greater than 1.0 is also necessary for accurate determination of the peak heights and areas for quantitation. For comparable accuracy, resolution must be better for quantitation by peak area than for that by peak height. The curves in Fig. 2.2 would help determine, at a glance, the approximate resolution and decide if improvement is needed [9].

As indicated by eq. (2.40), the resolution can be improved by increasing the values of $n$, $k$ and $\alpha$ [10]. Once the stationary phase, mobile phase and the phase ratio are chosen (i.e. $\alpha$ and $k$ are held constant), resolution is directly proportional to the square root of $n$. For a given column length, $n$ is inversely proportional to the plate height, $h$. The strategies to decrease the plate height (which is a function of several physical parameters relating to the column, the stationary phase, the mobile phase and, most importantly, the mobile phase flow-rate) will be discussed in section 2.4 below. Assuming a constant value of $h$, $n$ is proportional to the column length, $L$, but, as will be seen later, this assumption is valid only within narrow limits. Also, since the resolution is proportional to the square root of $n$, very long columns are needed to achieve significant improvement of resolution by this method. While this is fairly easily achieved using open-tubular columns in GC, the practicability of this strategy in LC is limited.

An alternative to increasing $n$ is to increase the capacity factor, $k$. This may be effected by decreasing the phase ratio or by decreasing the column temperature (GC) or the mobile-phase strength (LC). Again, the resolution does not increase proportionally to $k$, but to $k/(1+k)$, which means that while an increase in $k$ from 0 to 1 would increase the resolution from 0 to 0.5, a further ten-fold increase of $k$ would increase it from 0.5 to 0.91, and so on. Since the analysis time increases linearly with $k$, the effectiveness of this strategy in terms of efficiency is limited. Nevertheless, it is the easiest to carry out and is usually the first strategy to be adopted, especially if the initial value of $k$ is low. Increasing the value of $\alpha$ is probably the most difficult and, to some extent, unpredictable. Since resolution is proportional to $\alpha - 1$, change in $\alpha$ can bring about dramatic improvement in resolution. But, that would require an understanding of the mechanisms of separation, which are discussed in the later chapters. Fig. 2.3 summarizes the effect of the different variables, $k$, $n$ and $\alpha$, on resolution.

## 2.3 BASIC PROCESSES IN CHROMATOGRAPHY

Fig. 2.4 shows the typical separation of a two-component mixture by thin-layer chromatography. The sample is applied at the bottom of the plate as a thin band or a small spot. As the mobile phase flows through the stationary-phase bed, the components also move along the plate. Depending on the relative mobilities of the two components, first a partial separation and then a complete separation of the bands could occur. Two characteristic features of the separation process can be recognized: (1) differential migration of the two components, and (2) spreading of the bands, as they move along the bed. By changing the form of the stationary-phase bed from a thin layer to that in a column and using an appropriate mobile phase, the same processes may be understood to occur in gas or column liquid chromatography as well. Differential migration, or the varying rate of mobility of the different components along the stationary phase, is the basic requirement of chromatographic separation since no separation occurs in the absence of it. On the other hand, the

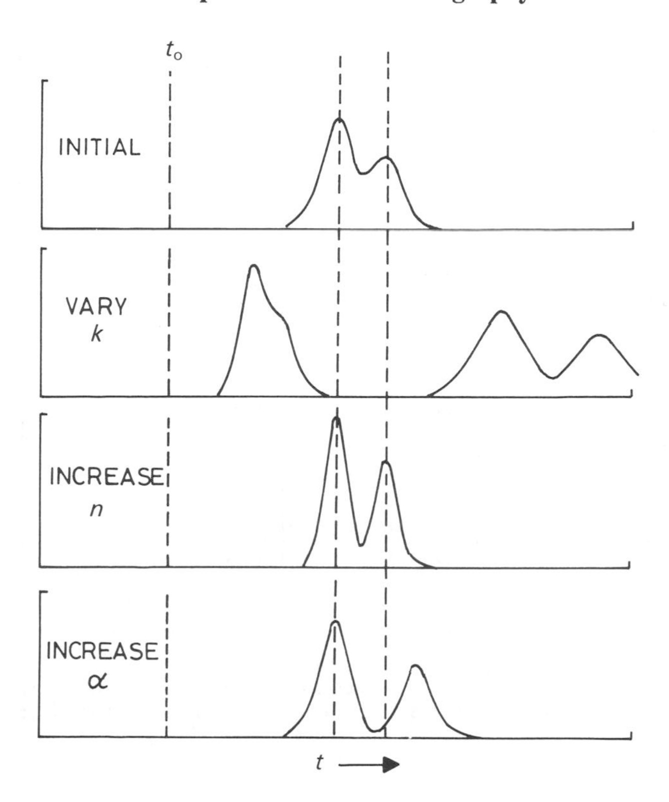


Fig. 2.3 — Effect of $k$, $n$ and $\alpha$ on resolution of two peaks. Reproduced from the *Journal of Chromatographic Science*, Ref. 10, by permission of Preston Publications, Inc.

approaches to improvement of the chromatographic efficiency invariably involve minimizing the band spreading.

The theory of chromatography was placed on a firm footing by Martin and Synge [11], who considered the process of partition chromatography as a series of transfers of the solutes between the stationary phase and the mobile phase. The system was considered as a miniaturized countercurrent distributor and, as in the case of the countercurrent extraction process or fractional distillation, the chromatographic system is visualized as being composed of a series of (theoretical) plates with the distribution of the solute between the two phases attaining equilibrium in each plate. Though inaccurate to the extent that the system in reality never attains equilibrium, the theoretical-plate model became very popular and was applied to systems where neither of the phases was a liquid (e.g. gas and adsorption chromatography). The success of the concept lay in its ability to easily predict the relative mobilities of different compounds on the basis of their partition coefficients and the relative amounts of the stationary and the mobile phases. As the distribution law (i.e. the ratio of the concentrations of the solute in the two phases being constant at a given temperature) is valid only for dilute solutions, reverse tailing (fronting) of peaks due to overloading and tailing due to adsorptive effects of the solid support, could be easily understood (see Fig. 2.6). But, a comprehensive chromatographic theory

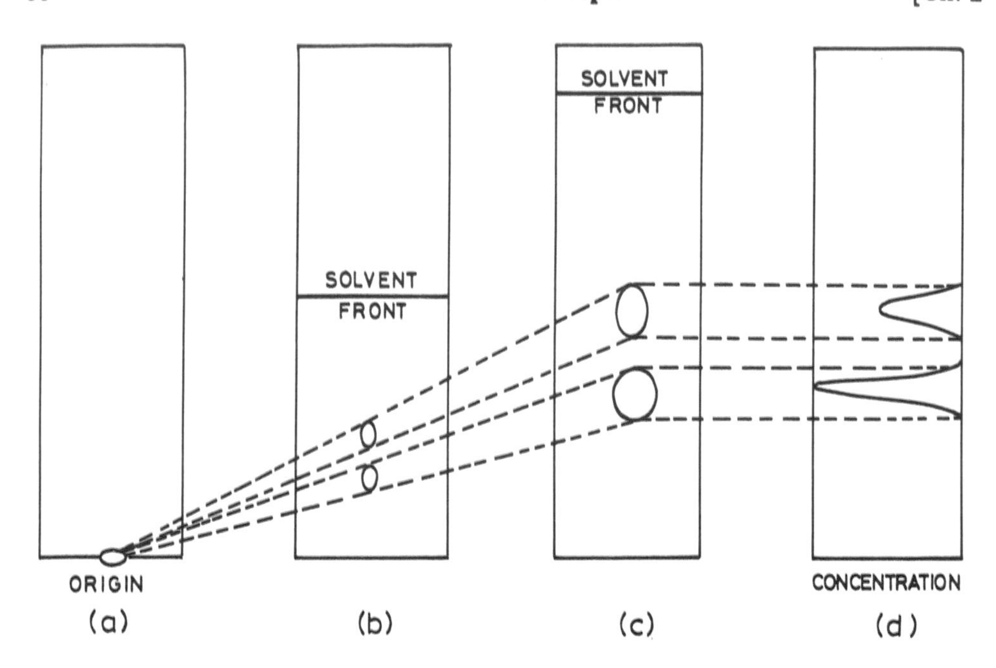


Fig. 2.4 — Separation of a hypothetical two-component mixture by thin-layer chromatography; illustration of band spreading with solute migration through stationary-phase bed and its relation to peak shape.

should satisfactorily account for the observed band broadening also. Subsequently, van Deemter *et al.* [12] developed the rate theory for gas chromatography, wherein the chromatographivc parameters are expressed in terms of retention times. They derived the now famous van Deemter equation for packed-column GC relating the efficiency of separation to various physical and operating parameters in chromatography. The rate theory was subsequently extended to open-tubular columns by Golay [13] and to LC by several others.

The rate theory invokes four basic processes that contribute to band broadening in a chromatographic column. Firstly, the path of the mobile phase through the interstitial space between the solid (support) particles in a packed column is not straight or uniform, but tortuous and random, as depicted pictorially in Fig. 2.5. Thus, some molecules of the mobile phase may have a shorter path than the others and move ahead of the average or the maximum concentration profile while those with longer paths fall behind. Band broadening due to this multipath process is understandably proportional to the support-particle diameter and to the non-uniformity of packing.

The second process that is responsible for the band broadening is the longitudinal diffusion. The solute molecules at any given point in a fluid do not remain stationary but diffuse in all directions until there is no concentration gradient. In practice, however, complete and uniform distribution of the solute throughout the mobile phase in the column does not occur because the mobile phase is continuously flowing in a certain direction and because the time spent by a solute molecule in the mobile

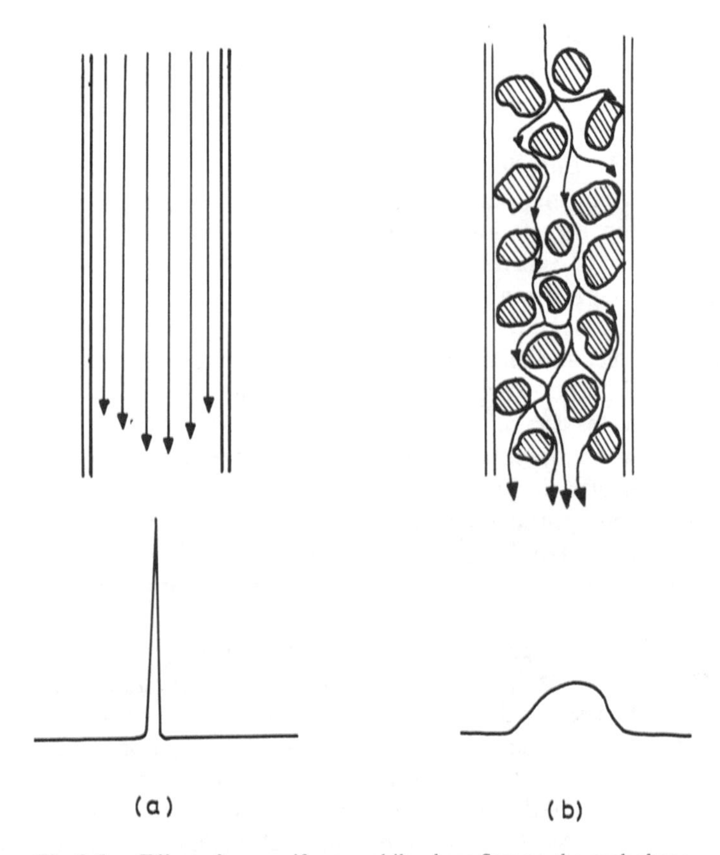


Fig. 2.5 — Effect of non-uniform mobile-phase flow on the peak shape.

phase is limited. Thus, the broadening due to this effect is directly proportional to the interstitial tortuosity and the diffusivity of the solute in the mobile phase and is inversely proportional to the mobile-phase linear velocity.

The two other major factors which contribute to band broadening relate to the resistance to mass transfer in the two phases. In a two-phase system, the transfer of a solute from one phase to another is not instantaneous because the solute molecules take a finite amount of time to traverse from the bulk of one phase to the interface and then enter the other phase. As one phase is continuously moving the solute molecules will also move a certain distance in the mobile phase during this time and equilibrium does not occur. The dispersion is further amplified by the parabolic flow profile of the mobile phase. Thus, the layer of the mobile phase adjoining the stationary phase surface is essentially static and the velocity of the different layers of the mobile phase increases with their distance from the stationary-phase surface. This results in some of the solute molecules in the bulk of the mobile phase moving ahead of those which are also in the mobile phase but closer to the interface. The resistance to mass transfer in the mobile phase is thus a function of the distribution constant, the phase ratio, the support-particle diameter, the diffusivity in the mobile phase, and the mobile-phase linear velocity. The effect of the resistance to mass transfer in the stationary phase is similar to that in the mobile phase; the thickness of the stationary-phase film is an important parameter that contributes to this effect.

### 2.3.1 Partition and adsorption

As already noted, chromatographic retention of a substance is effected by its distribution between the stationary and the mobile phases. In the absence of a chemical interaction, the distribution may be broadly classified as one of the two general mechanisms, partition and adsorption. The process is termed partition when the substance is transferred from the bulk of one phase into the bulk of the other, so that the solute molecules are completely surrounded by the molecules of the particulate phase; in other words, the solute molecules are in an isotropic environment in both the phases. On the other hand, when the solute molecules are adsorbed on the stationary phase, the interaction is limited to its surface; the solute molecules are thus in an anisotropic environment consisting of both the stationary-phase surface and the mobile-phase molecules. Depending on the affinity of the stationary phase to the components of the mobile phase, the composition of the liquid layer on the stationary-phase surface may be different from that of the bulk of the mobile phase.

In the absence of a solute, the surface of the stationary phase is covered by the mobile-phase molecules and, for adsorption to occur, a fraction of these molecules must be displaced by the solute. Three conditions may be envisaged: (1) The affinity of the solute molecules to the stationary-phase surface may not be strong enough to displace the mobile-phase component that is bound to the surface, in which case, either the solute is excluded or it partitions between the strongly bound component and the free mobile phase. (2) The solute molecules can more strongly bind to the stationary phase so that the role of the mobile-phase components in adsorptive retention is limited. (3) The affinity of the solute and the mobile-phase components may be comparable, in which case, situations intermediate between the above two conditions would operate. Thus, though conceptually distinct, the boundary between partition and adsorption under practical conditions is not clear-cut.

Retention in size-exclusion chromatography may be regarded as partitioning between two liquid phases, i.e. the interstitial bulk liquid and the pore fluid. Though the pore fluid is probably of the same composition, it may be considered to be different from the bulk mobile phase in view of the fact that the rotational and translational freedom of the solute molecules is reduced within the pores. Two extreme cases of solute–stationary phase interactions can be recognized. The strongest interactions probably occur in gas–solid chromatography, with the weakest interactions being those in the SEC, as just described.

The energy of interaction of the solute molecules with those of the gaseous mobile-phase molecules is much weaker than that with the liquid phases because the density of the former is about three orders of magnitude less. Thus, the retention in GC is largely determined by the interaction energy with the stationary phase and the free energy of vaporization of the solute. The role of the mobile phase in GC is thus limited to that of a carrier to transport the solute through the column. The interaction between the solute and the mobile-phase molecules, in contrast, could be very strong in LC and this would contribute significantly to the retention process. The distinction between GC and LC, however, gets blurred as the density of the gaseous mobile phase is increased by operating under supercritical conditions.

Retention of the solute by either adsorption or partition, and in either GC or LC is determined by the overall equilibrium distribution of the eluite between the stationary and the mobile phases. If the distribution constant remains unaltered by change in the concentration of the solute in the two phases, the sorption isotherm (i.e. the plot of the concentration distribution of the solute in the two phases at a given temperature) is said to be linear and the result is a symmetrical (Gaussian) distribution of the solute in the column effluent (Fig. 2.6). Linear isotherms may be

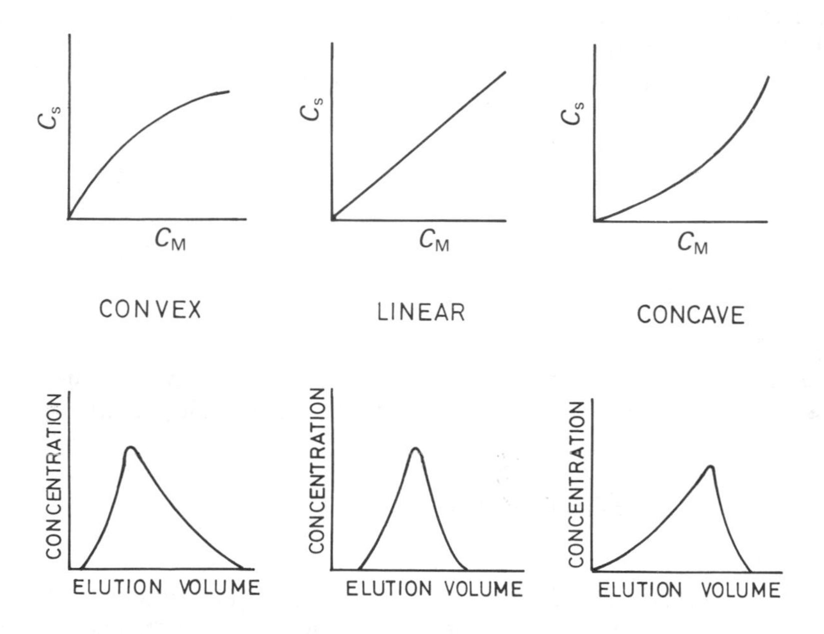


Fig. 2.6 — Effect of the shape of adsorption isotherms on the peak shape.

obtained over a limited concentration range and in the absence of selective interactions such as association, dissociation or complexation. When such interactions occur in either the mobile or the stationary phase, the sorption isotherm may be convex or concave, leading to tailing or fronting of the eluted peak shape as shown in the figure.

The linear isotherm is reflected in the distribution constant remaining unchanged. In thermodynamic terms, the equilibrium signifies that the chemical potential of the solute in the two phases is equal. The chemical potential, which is a function of the temperature, pressure and concentration of all the components of the system including that of the solute, is given by

$$\mu = \mu^{\circ} + RT \ln a \tag{2.41}$$

where $\mu^o$ is the chemical potential under standard conditions and $a$, the solute activity in the particular phase. The solute activity is a product of the activity coefficient ($\alpha$) and the concentration ($C$) so that

$$a = \gamma C \tag{2.42}$$

The activity may be defined as the ratio of the fugacity (i.e. tendency of the solute to escape from the condensed phase to the vapour phase) in the system to the fugacity in the standard state. The activity coefficient is the ratio of the activity of a substance to its mole fraction in solution. If $\mu_S$ and $\mu_M$ are the chemical potentials of the solute in the two phases, we have at equilibrium,

$$\mu_S = \mu_M \tag{2.43}$$

so that

$$\mu_S^o + RT \ln a_S = \mu_M^o + RT \ln a_M \tag{2.44}$$

The logarithm of the equilibrium (distribution) constant is given by the difference between the standard chemical potentials and the logarithmic activity coefficients:

$$\ln K = \frac{\mu_M^o - \mu_S^o}{RT} + (\gamma_S - \gamma_M) \tag{2.45}$$

If it is assumed that the standard chemical potentials of the solute in the two phases are approximately equal, the distribution constant and hence the chromatographic retention is determined by the ratio of the activity coefficients of the solute in the two phases. The chromatographic selectivity ($\alpha_{2,1}$) between two components, 1 and 2 can be obtained from the relation

$$\alpha_{2,1} = \frac{\gamma_{1,S}\ \gamma_{2,M}}{\gamma_{2,S}\ \gamma_{1,M}} \tag{2.46}$$

where the subscripts denote the identity of the corresponding components and the phases. The selectivity of the chromatographic system can thus be modulated by either changing one of the phases to obtain a more favourable activity coefficient ratio or by exploiting the temperature dependence of the coefficients.

### 2.3.2 Mobile-phase flow profile

While much of the discussion of the distribution of the solute between two phases is based on the solute attaining equilibrium, the requirement of continuous flow of the mobile phase through the bed at a finite velocity prevents solute equilibration and causes band broadening. The band broadening is enhanced by the non-uniformity of the flow profile or the spatial distribution of the flow velocities in the column. It is easy to imagine that the magnitude of axial dispersion increases with the mobile phase velocity. When a liquid or a gas flow through a tube, three types of flow profile are possible. In the ideal plug flow, the advancing boundary is perpendicular to the longitudinal axis of the tube and there is no band spreading. The near-ideal turbulent flow, whose profile is similar to that of the plug flow, occurs only at very high flow velocities and pressures. The more common laminar flow has a parabolic profile due to the viscous drag exerted by the tube wall on the fluid so that the particles at the centre of the tube move ahead of those near the wall (see Fig. 6.11b). The linear velocity of the mobile-phase molecules at a distance $x$ from the centre is given by

$$u_x = u_{\max}\left(1-\frac{x^2}{r^2}\right) \tag{2.47}$$

where $u_{\max}$ is the velocity at the centre of the tube of radius $r$. Given the viscosity of the mobile phase ($\eta$), tube length ($L$) and the pressure drop across the tube ($\Delta P$), the maximum linear velocity of the fluid at the centre is given by

$$u_{\max} = \Delta P\,\frac{r^2}{4\eta L} \tag{2.48}$$

The average linear velocity $u$ being half the maximum velocity,

$$u = \Delta P\,\frac{r^2}{8\eta L} \tag{2.49}$$

and, since the volumetric flow rate, $F = u\pi r^2$,

$$F = \Delta P\,\frac{\pi r^4}{8\eta L} \tag{2.50}$$

For calculating the linear velocity of a fluid in a packed column, the Darcy's equation gives:

$$u_{\mathrm{D}} = \Delta P\,\frac{B_1}{L} \tag{2.51}$$

where $B_1$ is the permeability coefficient and $u_{\mathrm{D}}$ is the velocity at the column outlet (Darcy velocity). $B_1$ multiplied by the viscosity, gives the specific permeability ($B_0$) so that

$$u_D = \Delta P \frac{B_0}{\eta L} \tag{2.52}$$

The linear velocity in the tube is related to the Darcy velocity by a factor determined by the fractional interparticle free space, $\varepsilon (0 < \varepsilon < 1)$ and is given by the equation:

$$u = \Delta P \frac{B_0}{\varepsilon \eta L} \tag{2.53}$$

The value of $\varepsilon$ is generally in the range of 0.38 to 0.42 in well-packed columns and thus is normally taken as 0.40.

The effective or the hydraulic radius ($r_H$) available for the fluid to flow in a packed column is related to the fractional interparticle free space and the specific surface area, $S_0$, according to the equation:

$$r_H = \frac{\varepsilon}{S_0 (1 - \varepsilon)} \tag{2.54}$$

Assuming the packing material is made up of spherical particles, the specific surface area is given by $6/d_p$ and substituting the value of $r$ in equation (2.49) and rearranging gives

$$u = \Delta P \frac{d_p^2 \varepsilon^2}{180 \eta (1 - \varepsilon)^2 L} \tag{2.55}$$

The above equation, known as the Kozeny–Carman equation, shows that the mobile-phase linear velocity in a packed column is directly related to the pressure drop and the square of the packing particle diameter and inversely to the column length. It also indicates that the linear velocity in LC can be increased by increasing the temperature which, in effect, decreases the viscosity. On the other hand, in GC (where the viscosity of the gaseous mobile phase increases with temperature), maintenance of constant flow rate at higher temperatures would require the use of higher inlet pressures. For a more detailed mathematical treatment of the flow profile, see [7,8].

As already noted, the solid adsorbent or the support particles cause irregular flow profile of the mobile phase; this would result in band broadening due to the different channels and flow velocities that the mobile phase molecules could take. In quantitative terms, the variance arising from any phenomenon is given by the product of the number of steps involved (denoted as $n'$ to distinguish from the plate number, $n$) and the square of the distance ($l$) over which the phenomenon occurs; i.e.

$$\sigma^2 = n' l^2 \tag{2.56}$$

Here, $n'$ is a measure of the number of particle layers in the column and can be substituted by $L/d_p$, $L$ and $d_p$ representing the column length and particle diameter respectively. On the other hand, $l$ is proportional to the particle diameter, giving:

$$l^2 = 2\lambda d_p^2 \qquad (2.57)$$

where $2\lambda$ is the geometrical constant and a measure of the packing non-uniformity in the column. The contribution to the peak variance from this source, $(\sigma_1)^2$, is thus,

$$\sigma_1^2 = 2\lambda d_p L \qquad (2.58)$$

The contribution of the packing non-uniformity to plate height is, therefore,

$$h_1 = \frac{\sigma_1^2}{L} = 2\lambda\ d_p \qquad (2.59)$$

Normally, the mobile phase in planar chromatography flows over the stationary-phase bed by capillary pressure difference on either side of the front. The pressure drop is proportional to the surface tension of the solvent and inversely proportional to the size of the particles. The linear velocity of the solvent front is not uniform, but advances faster along the narrower channels and slower in a layer of small particles. Solvents with low surface tension and high viscosity move slower and the velocity decreases as the solvent moves along the sorbent. A further complication arises from the fact that the solvent evaporates during the run, depending on the degree of saturation of the chamber. Also, in a multicomponent mobile-phase system, different solvents advance at different rates and the vapour components are adsorbed by the sorbent ahead of the solvent front. Rigorous mathematical treatment of the flow profile is, therefore, not available.

### 2.3.3 Diffusion and diffusivity

Diffusion is one of the most important transport properties of a fluid. As a contributor to the mass transport from one point or phase to another, the phenomenon of diffusion is of major significance in the chromatographic process, particularly in connection with band broadening. Two types of diffusion may be distinguished, namely, self-diffusion (i.e. diffusion of the same kind of atoms or molecules into one another) and interdiffusion or diffusivity of a solute in a medium (solvent), usually resulting in the relaxation of a concentration gradient. Fick's first law of diffusion states that the flux or flow of matter is equal to a negative concentration gradient multiplied by a proportionality constant, called the diffusion coefficient ($D$). The value of $D$ is approximately given by the Stokes–Einstein equation:

$$D = \frac{RT}{3\pi\eta d} \qquad (2.60)$$

where $T$ is the absolute temperature, $\eta$ is the viscosity of the medium and $d$ is the diameter of the diffusing particle. When the diffusion is restricted to one direction, Fick's second law relates the change of concentration as a function of time to that as a function of distance:

$$\frac{dc}{dt} = D\frac{d^2c}{dx^2} \tag{2.61}$$

Thus, if a solute is introduced into a system at a point along the longitudinal axis of the system, the increase in concentration at some distance $x$ from the point source will be related exponentially to the time and is given by the equation [1]:

$$\frac{C}{C_0} = \frac{1}{2\sqrt{\pi Dt}}e^{(-x^2/4Dt)} = \frac{e^{(-x^2/2\sigma^2)}}{\sigma\sqrt{2\pi}} \tag{2.62}$$

where $C$ is the concentration at distance $x$ after time $t$, and $C_0$ is the concentration of the solute at the source point. The similarity of the above equation, which also represents the Gaussian concentration distribution, to eq. (2.15) is quite apparent. The standard deviation ($\sigma$) and the variance ($\sigma^2$) are thus:

$$\sigma = \sqrt{2Dt} \tag{2.63}$$

and

$$\sigma^2 = 2Dt \tag{2.64}$$

The variance arising out of longitudinal diffusion of the solute along a unit length of the tube can be written as in eq. (2.65), by substituting $t$ in the above equation by $L/u$ and $D$ by $D_M$, the diffusivity of the solute in the mobile phase:

$$\sigma_2^2 = \frac{2D_ML}{u} \tag{2.65}$$

The above equation indicates that the band broadening arising out of longitudinal diffusion is smaller when the velocity of the mobile phase is higher; this is also understandable in qualitative terms since the higher the velocity, the shorter is the time available for the molecules to diffuse. However, the equation does not take into consideration the retarding effect of the packing material on the longitudinal diffusion and, for packed columns, the equation must be modified by including a labyrinth or tortuosity factor, $\gamma$, whose value may be between 0.5 and 1.0 depending on the packing [12]:

$$h_2 = \frac{\sigma_2^2}{L} = \frac{2\gamma D_M}{u} \tag{2.66}$$

### 2.3.4 Mass-transfer resistances

The transfer of the solute molecules from the stationary phase to the mobile phase, across the interface and vice versa, is not instantaneous and requires an energy barrier to be overcome. This resistance to mass transfer from one phase to another is different in the two phases and is determined by the solvation energy, fugacity and diffusivity of the solute in the two phases. A finite period of time elapses before a solute molecule moves from the bulk of one phase to the interface, overcomes the energy barrier and crosses over to the other phase. During this period, since the mobile phase is continuously flowing through the stationary-phase bed, the solute-concentration profiles in the two phases do not match. As shown diagrammatically in Fig. 2.7, the solute plug in the mobile phase tends to move ahead while that in the stationary phase is retarded.

Considering the stationary phase first; $n'$, the number of transfers (desorptions) from this phase into the mobile phase, is given by time ($t_R'$) spent by the solute in the stationary phase, divided by the unit time $\tau$, required by the molecule to migrate once from the stationary phase into the mobile phase.

$$n' = \frac{t_R'}{\tau} \tag{2.67}$$

If $d_f$, the thickness of the stationary phase, is the distance over which the molecule has to diffuse, at random, from the bulk of the phase to reach the interface (to permit the migration into the other phase) and $D_S$ is the diffusivity of the solute in the stationary phase, Einstein's diffusion law requires that:

$$\tau = C_1 \frac{d_f^2}{D_S} \tag{2.68}$$

where $C_1$ is a geometrical constant. From eqs. (2.9) and (2.11), we have,

$$t_R' = \frac{kL}{u} \tag{2.69}$$

and

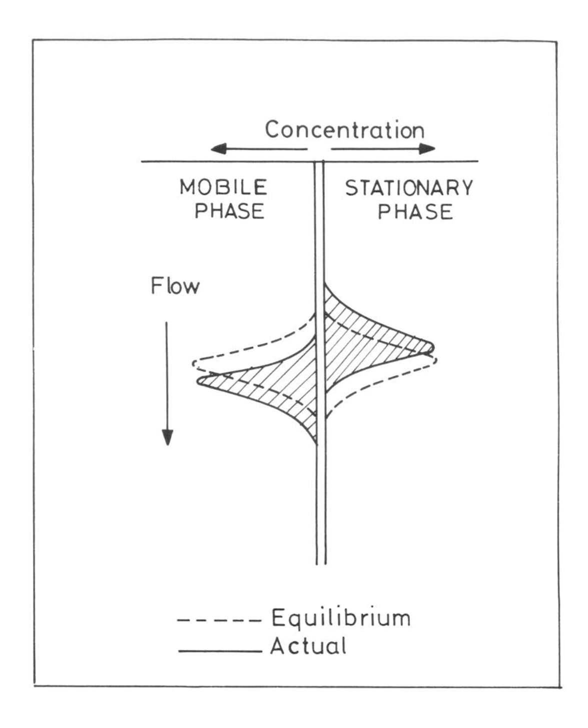


Fig. 2.7 — Illustration of the displacement of the solute concentration profile from that in hypothetical instantaneous equilibrium (shaded area) in the stationary and the mobile phases. Reproduced from Ref. 7 by courtesy of Elsevier Scientific Publishing Co.

$$n' = \frac{1}{C_1} \frac{D_S}{d_f^2} \frac{kL}{u} \tag{2.70}$$

The total time taken by the solute to pass from one end of the column to the other (i.e. the distance, $L$) is $t_M$ and the velocity of the band ($u_S$) in the stationary phase is therefore,

$$u_S = \frac{L}{t_R} = \frac{ut_M}{t_R} = \frac{1}{(1+k)} u \tag{2.71}$$

The distance travelled by the solute during its residence time is thus given by $t'_S u_S$, and during the period of one migration across the interface

$$l = \frac{t'_R u_S}{n'} \tag{2.72}$$

Substituting eq. (2.72) in eq. (2.56),

$$\sigma^2 = n'\left(\frac{t_R' u_S}{n'}\right)^2 \qquad (2.73)$$

Substituting the appropriate terms from eqs. (2.69), (2.70) and (2.71),

$$\sigma_3^2 = C_1 \frac{d_f^2}{D_S} \frac{u}{kL} \frac{u^2}{(1+k)^2} \frac{k^2 L}{u^2} \qquad (2.74)$$

The plate-height contribution of this phenomenon is obtained by cancelling the appropriate terms of the numerator and the denominator in the above equation and rearranging:

$$h_3 = \frac{\sigma_3^2}{L} = C_1 \frac{d_f^2}{D_S} \frac{k}{(1+k)^2} u \qquad (2.75)$$

or,

$$h_3 = C_S u \qquad (2.76)$$

where

$$C_S = C_1 \frac{k}{(1+k)^2} \frac{d_f^2}{D_S} \qquad (2.77)$$

The value of $C_1$ may be shown to be $8/\pi^2$ for packed columns and 2/3 for open-tubular columns [12].

The resistance to mass transfer in the mobile phase may be derived in a similar fashion with the values of $n'$ and $l$ as given in eqs. (2.78) and (2.79).

$$n' = \frac{D_M}{C_2 d_M^2} \frac{L}{u} \qquad (2.78)$$

$$l = \frac{C_2\, d_M^2}{D_M} \frac{\text{k}}{(1+\text{k})} \qquad (2.79)$$

where $d_M$ is the distance a solute molecule has to travel to reach the interface. The plate-height contribution of the mass-transfer resistance should then be,

$$h_4 = \frac{\sigma_4^2}{L} = C_2 \frac{k^2}{(1+k)^2} \frac{d_M^2}{D_M} u \qquad (2.80)$$

However, several other factors need to be considered in quantifying the resistance to mass transfer in the mobile phase, particularly in relation to the flow profile. Thus, the velocity distribution term ($h_5$) takes into account the variance arising out of the diffusion within mobile phase streams of different velocities (e.g. inside and outside the porous particles. This term, which is a function of the particle diameter and the mobile-phase diffusivity, is given by:

$$h_5 = \frac{\sigma_5^2}{L} = C_3 u \tag{2.81}$$

where

$$C_3 = C_3' \frac{d_p^2}{D_M} \tag{2.82}$$

$C_3'$ being a constant. Though nominally independent, the variances $(\sigma_4)^2$ and $(\sigma_5)^2$ do interact, and the total variance due to the resistance to mass transfer in the mobile phase should include a $C_4$ term, which is a function of the $C_2$ and the $C_3$ terms and whose plate-height contribution is given by:

$$h_6 = \frac{\sigma_6^2}{L} = C_4 \frac{k}{(1+k)} \frac{d_p\, d_M}{D_M} u \tag{2.83}$$

In open-tubular columns, we have

$$d_p = d_M = r \tag{2.84}$$

where $r$ is the column radius; the total plate-height contribution of the resistance to mass transfer in the mobile phase may be written as:

$$h_4 + h_5 + h_6 = \frac{C' + C''k + C'''k^2}{(1+k)^2} \frac{r^2}{D_M} u \tag{2.85}$$

where $C'$, $C''$ and $C'''$ are constants derivable from the values of $C_2$, $C_3$ and $C_4$ [14]. Assuming the reasonable parabolic profile for the mobile-phase flow in open-tubular columns, Golay [13] derived the following expression for the mass-transfer resistance in this phase:

$$h_4 + h_5 + h_6 = \frac{1}{24} \frac{(1 + 6k + 11k^2)}{(1+k)^2} \frac{r^2}{D_M} u \tag{2.86}$$

or simply,

$$h_4 + h_5 + h_6 = C_M u \tag{2.87}$$

where $C_M$, the total resistance to mass transfer in the mobile phase, is:

$$C_M = \frac{1}{24} \frac{(1 + 6k + 11k^2)}{(1 + k)^2} \tag{2.88}$$

The value of the second term is essentially constant at $k > 10$ and thus the mass transfer term is independent of the value of $k$. At $k = 10$, eq. 2.86 becomes:

$$h_4 + h_5 + h_6 = \frac{0.4\, r^2}{D_M} \tag{2.89}$$

The above equation indicates that for faster analysis at high flow rates a mobile phase of low viscosity and high diffusivity is desirable. It also indicates that the plate height is reduced rapidly by decreasing the column radius.

## 2.4 THE VAN DEEMTER EQUATION

The van Deemter equation [12] represents a landmark in the development of the theory of chromatography. It inter-relates the physical parameters within the column that contribute to band broadening and thus to the efficiency of chromatography. Thus, it relates the HETP ($h$) to the size of the support particles, the diffusivity of the solute in the two phases, the thickness of the stationary phase, the column diameter in open-tubular columns, and the velocity (flow-rate) of the mobile phase. In its original form, the van Deemter equation related the HETP ($h$) to the mobile-phase linear velocity ($u$) using three constants representing the contribution to peak broadening from eddy diffusion ($A$), molecular diffusion ($B$), and resistance to mass transfer ($C$):

$$h = A + \frac{B}{u} + Cu \tag{2.90}$$

Though commonly referred to as eddy diffusion, the first term of the equation (i.e. $A$) does not actually represent any diffusional phenomena but arises out of flow non-uniformity in packed columns. The plate height contribution of this term has been derived earlier (eq. (2.59)) and is given by $2\lambda d_p$. The $B$ term in the above equation represents the contribution to plate height of the solute diffusivity in the mobile phase and is given by eq. (2.66) as $2\gamma D_M$. The original $C$ term of the van Deemter equation (2.90), which represented resistance to mass transfer (that is, the tendency of the solute molecules to remain in a particular phase), was split into four independent variables, $h_3$, $h_4$, $h_5$ and $h_6$, as described in the preceding section. The complete plate-height equation can thus be written as:

$$h = h_1 + h_2 + h_3 + h_4 + h_5 + h_6$$
$$= A + \frac{B}{u} + (C_S + C_M)\, u \tag{2.91}$$

Also, though the above equation suggests that the $A$ term is independent of the mobile-phase velocity, there is a slight positive dependence and, on the basis of experimental evidence, the first term is occasionally substituted by $Au^{0.33}$, so that:

$$h = A\, u^{0.33} + \frac{B}{u} + (C_S + C_M)\, u \tag{2.92}$$

However, this modification does not significantly affect the following discussion on the relative contribution of the different parameters and, as such, will not be considered further.

The significance of eq. (2.91) or its simpler form, eq. (2.90), lies in the fact that it relates the different, independently variable physical parameters that contribute to the efficiency of a chromatographic system. For a given length of the column, the number of theoretical plates and thus the efficiency are inversely related to the HETP ($h$). The HETP can be reduced by reducing the values of the terms $A$, $B$ and $C$, one at a time. For a given set of values of $A$, $B$ and $C$, the plate height can be minimized by choosing an appropriate value of the mobile-phase velocity, $u$. The change of $h$ with $u$ is given by the van Deemter plot (Fig. 2.8). The shape of the plot is

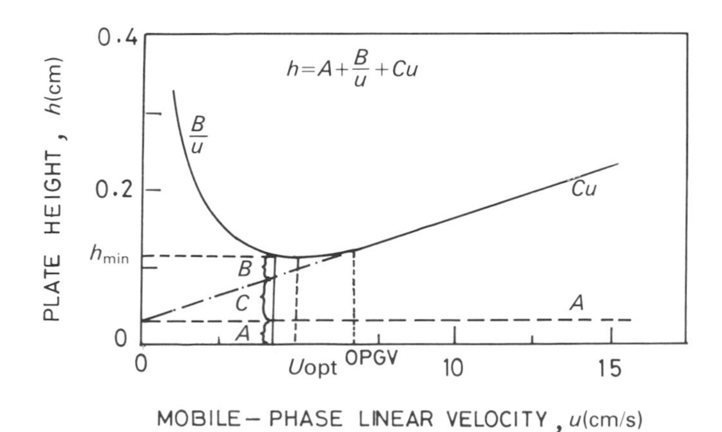


Fig. 2.8 — The van Deemter plot showing the change of plate height with mobile-phase linear velocity.

essentially the same for all chromatographic systems though the position of the minimum on the ordinate and abcissa (denoting the optimum mobile-phase velocity for obtaining the minimum plate height) and the slopes of the right- and the left-hand branches of the plot may vary significantly depending on the conditions. Even a qualitative examination of the basic features of the plot is rewarding. There is a minimum value for $u$ ($u_{opt}$) below which the value of $h$ increases steeply; in other words, the chromatography becomes very inefficient. The shape of the $h$ vs. $u$ plot, according to eq. (2.90), is determined by the values of $B$ and $C$, with $B$ controlling the left-hand branch ($u < u_{opt}$) and $C$ controlling the right-hand branch ($u > u_{opt}$). It pays to keep $u > u_{opt}$ in two ways: (1) since the slope of the right branch is less steep, the efficiency of the system does not fall so drastically as when $u < u_{opt}$; and (2) a higher value of $u$ means faster elution and shorter analysis time. The effect of the $B$ term can therefore be ignored for all practical purposes by selecting a mobile-phase velocity higher than or equal to $u_{opt}$. The van Deemter plot also shows why high pressure drops across the column should be avoided. The value of $h$ is at a minimum and the efficiency of the column at a maximum only over a small range of values of $u$. In other words, large pressure drops, which cause large differences in the mobile-phase velocity along the column, cause the system to operate much less efficiently along the major portion of the column. A closer examination of the terms $A$, $B$, $C_S$, and $C_M$, as defined by the eqs. (2.59), (2.66), (2.77) and (2.88) would indicate the approaches to improvement of column efficiency.

Two of the above terms, i.e. the eddy diffusion term $A$ and the velocity distribution term $C_3$ (eq. (2.72)), are functions of the particle diameter, $d_p$. The $A$ term is given by $2\lambda d_p$, where $\lambda$ is the packing constant. While the expression indicates that thc particle diameter should be minimal, practical considerations of the requirement of high head pressure limit the particle size to about 100–120 mesh (approximately 125–150 $\mu$m) or at best to 120–140 mesh (100–125 $\mu$m) in GC. Thus, the $A$ term determines the minimum value of $h$ in packed columns. In open-tubular columns, however, the $A$ term completely vanishes (and so does the $C_3$ term for films less than $\sim$0.5 $\mu$m thick) as there is no packing material. The particle size, which determines the interstitial velocity of the mobile phase influences the velocity distribution term $C_3$, which, for a given mobile phase, is proportional to the square of the particle diameter. The flow inequality in a packed column is due to the multichannel profile of the mobile-phase flow. Thus, the velocity of the mobile phase is higher between the particles than at the column wall and in irregularly placed and loosely packed regions; also, the mobile phase within the particle may be considered to be essentially still while that in the interparticle region moves. The flow inequality and thus the $A$ and the $C_3$ as well as the tortuosity factor ($\gamma$) in the $B$ term (eq. (2.66)) can be effectively reduced by using uniformly packed columns of small particles of uniform diameter. In a column of given radius ($r$), the $B$ term is directly proportional to the mobile-phase difffusivity, $D_M$. Thus, in order to decrease the $B$ term, a dense mobile phase of low diffusivity (e.g. nitrogen or argon in GC) is called for. However, at the normal mobile-phase velocity ($u > u_{opt}$), the $B$ term is insignificant. Also, $D_M$ appears in the denominator of two of the $C$ terms, i.e, $C_2$ and $C_3$, which become more important at $u > u_{opt}$. Thus, a mobile phase of high diffusivity (e.g. hydrogen or helium in GC) is to be generally preferred.

Eq. (2.77) indicates that the resistance to mass transfer term, $C_S$, increases exponentially with the thickness of the stationary phase, $d_f$, which can be decreased by decreasing the stationary-phase loading. Adsorption effects of the support particles may increase at lower loading, but by careful deactivation, it is possible to decrease the stationary-phase loading to as low as 0.1%; about 0.5 to 3.0% loading is now common. Though the resistance to mass transfer is smaller with a stationary phase of low diffusivity, polymeric stationary phases of relatively high viscosity are often preferred in GC mainly because of their stability and the consequent longer column life. Also, the notion that diffusivity is inversely proportional to the molecular weight and viscosity appears to be inaccurate, as shown by the high diffusivities of the longer-chain methylpolysiloxanes.

The third factor which is directly related to the $C_S$ term is the function $k/(1 + k)^2$ (eq. (2.77)). It can be easily calculated that the function has the maximum value of 0.25 when $k = 1$ and slowly levels off and eventually approaches zero as $k$ tends to infinity. Above $k = 8$ and for all practical values of $k$, the value of this function may be taken as approximately 0.1. However, between $k = 0$ and $k = 1$, the value of the term increases rapidly from 0 to 0.25. While such low values of $k$ are uncommon in GC, it is interesting to note that adjacent peaks in this region can have large $\alpha$ (relative retention) values. Where necessary, the value of $k$ can be reduced significantly by changing the phase ratio ($\beta$) or the distribution coefficient ($K$), according to eq. (2.7). While the phase ratio in a given system can be changed only within limits, the value of $K$, being a function of the chemical nature and the temperature, can be varied from zero to infinity.

The resistance to mass transfer in the mobile phase, $C_M$ (more precisely, $h_4$), is directly proportional to the function $(k/(1 + k))^2$ (eq. 2.80), unlike the $C_S$ term which varies with $k/(1 + k)^2$. The significance is that $C_2$ varies continuously but asymptotically from zero at $k = 0$ to nearly unity at $k = 100$. Eq. (2.80) also shows that $h_4$ is inversely related to the mobile-phase diffusivity, $D_M$, and exponentially to the depth of the mobile phase. It has already been pointed out that mobile phases with high diffusivity are generally preferred. The effective depth of the mobile phase in a column is best reduced by decreasing the support-particle size and the size distribution. Both these requirements have also been pointed out earlier.

On the basis of the above discussion, it may be concluded that in order to obtain a sharp, narrow chromatographic peak, the system should be optimized by using a small support-particle size and size distribution, stationary and mobile phases of high diffusivities, minimum stationary-phase thickness and mobile-phase velocity near or slightly above the $u_{opt}$ value determined by the van Deemter plot for the particular system.

Though the above generalizations were derived in connection with optimization of GC, they are also generally valid for LC; some important differences may, however, be mentioned. Diffusivity of a solute in a liquid phase, for example, is about four orders of magnitude lower than that in a gaseous mobile phase, which makes the $B$ term insignificant in LC. But, at the same time, the values of the $h_4$ and $C_3$ terms, whose expressions (eqs. (2.80) and 2.82)) have $D_M$ in the denominator, become very large. This would correspondingly require a $u_{opt}$ value that would be too small for LC to be practical, since, in that case, the analysis time would be impossibly large. However, three aspects of modern LC compensate for the very low diffusivity in the mobile phase. Firstly, the particle diameter is about 5 to 10 $\mu$m (or even 3 to 5 $\mu$m) compared to over 100 $\mu$m in GC; this considerably reduces the $A$ term of the van Deemter equation. The mobile-phase flow rate is also an order of magnitude less than that in packed column GC. Thirdly, as the mobile phase moves round the smaller particles and continuously flows through the interstitial space between the particles, convective mixing of the mobile phase occurs, thus decreasing the resistance to mass transfer in the mobile phase. However, the mixing of the mobile phase does not cause turbulence at the low linear velocities normally used in LC.

For several reasons, quantitative treatment of the van Deemter-type of relationships in LC is extremely complex [4]. Thus, in liquid–liquid (partition) chromatography, the static mobile phase contained in the support particles contributes significantly to $h$, but not to selectivity. This effect can be reduced by ensuring that the pores of the particles are completely filled with the stationary phase. It has been shown that the resistance to mass transfer in the static phase (i.e. the stationary phase plus the static mobile phase) has a predominant effect on $h$ at high mobile-phase velocity. At very low values of $u$, the effect of the longitudinal diffusion is important. As noted above, at medium flow rates, convective mixing and resistance to mass transfer in the mobile phase contribute significantly to $h$. It is therefore desirable to operate at relatively high flow-rates to reduce resistance to mass transfer in the mobile phase. This may, however, cause shearing of the stationary phase.

A somewhat different situation exists in liquid–solid (adsorption) chromatography. A molecule at the surface of the adsorbent has a potential energy ($E_p$) due to the intermolecular forces which retain the molecule on the stationary phase and a kinetic energy ($E_k$) due to the vibrational movement, which releases the molecule from the adsorbent into the mobile phase. Therefore, the molecule is adsorbed when $E_p > E_k$ and conversely, is desorbed when $E_p < E_k$. The solute molecules at different points in the chromatographic system have a range of energies depending on their mass, shape, temperature and environment. At any given point in space and time, some molecules have $E_k > E_p$ and are, therefore, released into the mobile phase while others must wait for their statistical chance to achieve sufficient $E_k$. The reverse is true for the adsorption process. The finite time interval between the adsorption or desorption of the first and the last molecule causes a finite amount of band broadening. The situation is further complicated by the fact that the different adsorption sites have different energies, determined by the surface activities, which, in turn, determine the linearity of the adsorption isotherm. Thus, the conditions for adsorption are different from that for desorption because they depend not only on the molecules having different energies but also on the chance of their striking the adsorbent surface at the site of a certain energy.

Although equations to describe the relations of $h$ to $u$ have been derived taking into account the peculiar situations existing in LC, they contain many empirical constants which vary from one stationary phase to another, depending upon the surface structure. The absolute values of $h$ derived from these equations may, therefore, be somewhat inaccurate. Nevertheless, the generalizations are essentially valid and the shape of the $h$ vs. $u$ plot for LC is similar to that shown in Fig. 2.8. Thus, in both GC and LC, the van Deemter equation points the way to reducing the plate height. The reduced plate height leads to a corresponding increase in the efficiency of the column and is thus the most important parameter that determines the peak width and thus the resolution.

## 2.5 EXTRA-COLUMN BAND BROADENING

Though the column is the most important part of the chromatographic system and the performance or the separation efficiency of the system is largely limited by its capabilities, modern chromatography invariably uses a considerable amount of instrumentation for convenience and for enhancing the speed of the analysis. While the various instrumental parts offer significant improvements in the output of the chromatograph, they also have a finite dead volume, whose relative contribution to the band broadening increases with increasing efficiency of the column. As noted earlier, the total variance of the Gaussian peak is an additive function of the various contributing variances and, analogous to eq. (2.29),

$$\sigma_T^2 = \sigma_{col}^2 + \sigma_{inj}^2 + \sigma_{conn}^2 + \sigma_{det}^2 \tag{2.93}$$

where the subscripts of the different terms on the right-hand side refer to the column, injection port, connecting tubes and the detector, respectively.

The effect of the various column parameters on $(\sigma_{col})^2$ and the strategies to minimize it have been discussed earlier. A detailed theoretical treatment of the subject of extracolumn band broadening is given by Sternberg [15] with reference to GC and by Martin *et al.* [16] with reference to LC. When the contribution to peak variance from other sources is insignificant,

$$\sigma_T^2 = \sigma_{col}^2 \tag{2.94}$$

While the relationship in eq. (2.94) should be the goal of the instrument designers, it is seldom achieved. Obviously, once an instrument or a component is chosen, very little can be done to decrease the dead volume and its extracolumn effects. Nevertheless, a brief discussion of the principles is in order.

The band broadening due to the injection is a function of three variables; namely, the injector design, $(\sigma_{V_o}^2)$, sample volume, $V_{inj}$, and the sampling mode:

$$\sigma_{inj}^2 = \sigma_{V_o}^2 + f\,V_{inj} \tag{2.95}$$

$(\sigma_{V_o}^2)$ is probably the lowest in on-column injection but this method of sample introduction may eventually cause peak distortion owing to the deformation of the column packing after repeated injections. On-column injection also minimizes $f$, the proportionality factor, which depends on the injection profile or the sampling mode. The contribution of $V_{inj}$ to band broadening is estimated to be 5 to 10% if the sample volume is 25 to 40% of the peak of interest. The band broadens rapidly as the sample volume exceeds 40% of the peak volume. As a rule of thumb, the sample volume should not exceed 1/3 of the volume of the first peak of interest.

The mobile phase is conducted from the injection port to the column, and from the column to the detector, *via* narrow connecting tubes. The diameter and uniformity of the tubing is very critical since radial dispersion of the mobile phase along the cross-section of the tube causes significant band broadening. The peak variance is given by:

$$\sigma^2_{\text{conn}} = \frac{ur^4L}{24D_M} \tag{2.96}$$

where $r$ and $L$ are the radius and the length of the tube and $D_M$ and $u$ are the diffusion coefficient and the linear velocity, respectively, of the mobile phase. The connecting tubes should, therefore, be as short and narrow as possible; the radius, however, cannot be reduced beyond a certain limit because of the pressure drop across the capillary which is also proportional to $r^4$. The tubes are connected to the injection port, column ends and the detector by means of ferrules and nuts of very low dead volume (Swagelok fittings). It is important to ensure that there are no crevices and unswept areas in the whole system from the injector to the detector.

The flow properties of the detector cell (resulting from the shape of the flow cell), its volume and response time can all contribute to band broadening. The major contribution arises from the flow mixing in the detector cell, an extreme case being complete turbulent mixing, where the variance is equal to the cell volume. At low flow-rates complete turbulent mixing is unlikely and the peak variance from this source may be less. If the cell volume is less than 10% of the peak volume, the variance arising from this source is insignificant.

In addition to the above, operations such as column switching, on-line, pre-column or post-column derivatization, etc., all cause band broadening; these are also predictable on the same concepts and can be minimized by suitably designing the systems.

## REFERENCES

[1] H. Purnell, *Gas Chromatography*, Wiley, New York, 1962.
[2] J. C. Giddings, *Dynamics of Chromatography*, Marcel Dekker, New York, 1965.
[3] F. Helfferich and G. Klein, *Multicomponent Chromatography. Theory of Interference*, Marcel Dekker, New York, 1970.
[4] R. P. W. Scott, *Contemporary Liquid Chromatography*, Wiley, New York, 1976.
[5] J. R. Conder and C. L. Young, *Physicochemical Measurement by Gas Chromatography*, Wiley, New York, 1979.
[6] A. S. Said, *Theory and Mathematics of Chromatography*, Heuthig, Heidelberg, 1981.
[7] C. Horvath and W. R. Melander, in *Chromatography* E. Heftmann (ed.), Part A, Elsevier, Amsterdam, 1983.
[8] R. J. Laub, in *Inorganic Chromatographic Analysis* J. C. MacDonald (ed.), Wiley, New York, 1985, Ch. 2.
[9] L. R. Snyder, *J. Chromatogr. Sci.*, **10**, 200 (1972).
[10] L. R. Snyder, *J. Chromatogr. Sci.*, **10**, 369 (1972).
[11] A. J. P. Martin and R. L. M. Synge, *Biochem. J.*, **35**, 1358 (1941).
[12] J. J. van Deemter, F. J. Zweiderweg, and A. Klinkenberg, *Chem. Eng. Sci.*, **5**, 271 (1956).
[13] M. J. S. Golay, in *Gas Chromatography* 1958, D. H. Desty (ed.), Butterworths, London, 1958.
[14] W. L. Jones, *Anal. Chem.*, **33**, 829 (1961).
[15] J. C. Sternberg, *Adv. Chromatogr.*, **2**, 205 (1966).
[16] M. Martin, C. Eon, and G. Guiochon, *J. Chromatogr.*, **108**, 229 (1975).

# *Part II*
# *Gas chromatography*

# 3

# Gas chromatography I: general features and inlet systems

## 3.1 INTRODUCTION

The use of a gas or a vapour as the mobile phase in chromatography was suggested as early as 1944 by Martin (and Synge) and was realized by the same inventor (in collaboration with James) eight years later. The subsequent decades have seen continuous development and application of the technique and, though it can now be considered a very refined method, further developments are not ruled out. Gas chromatography (GC) has remained the most used analytical technique for over two decades now. The immense popularity of GC as an analytical tool is due to several of its inherent advantages. Its high resolving power for solutes of even closely related structures such as isomers and homologues remains unmatched. Nanogram-level ($10^{-9}$ g) detection is routine while picogram ($10^{-12}$ g) sensitivity is not difficult to attain. Using certain selective and highly sensitive detectors, it is possible to detect certain compounds at femtogram levels ($10^{-15}$ g) as well. The accuracy and reproducibility are very high and short analysis times and feasibility of automation are added advantages. The availability of inexpensive, reliable and durable instrumentation has also contributed to its widespread application in various fields. The technique of GC has in fact become so common that its principles and basic instrumentation hardly need elaboration. Several basic texts [1–4] provide excellent coverage of the subject. None the less, the present discussion is included to provide a unified approach to the subject of chromatography. Also, GC happens to be the forerunner of many of the principles, techniques and instrumentation of modern liquid chromatography. Indeed, the current practices of GC and LC have so many features in common (sample preparation and data handling, for instance) that distinction based on the mobile phase appears artificial. The recent developments in supercritical fluid chromatography (SFC), wherein the mobile phase has the fluidity of a gas with the density and solvating properties of a liquid, further blurs the distinction.

The most important difference in the principles of GC and liquid chromatography (LC) is the role of the mobile phase. The most commonly used mobile phase in GC is an inert gas, whose function is to carry the solute from injection port through the column to the detector. The mobility of the solute is, thus, essentially dependent on its solubility in the gas and this is directly related to the vapour pressure of the solute at a given temperature. On the other hand, a number of intermolecular interactions between the solute and the mobile phase may be exploited in LC.

## 3.2 BASIC INSTRUMENTATION

The design of the gas chromatograph is fairly well standardized and most commercial instruments offer the capabilities generally required. The chromatograph (Fig. 3.1)

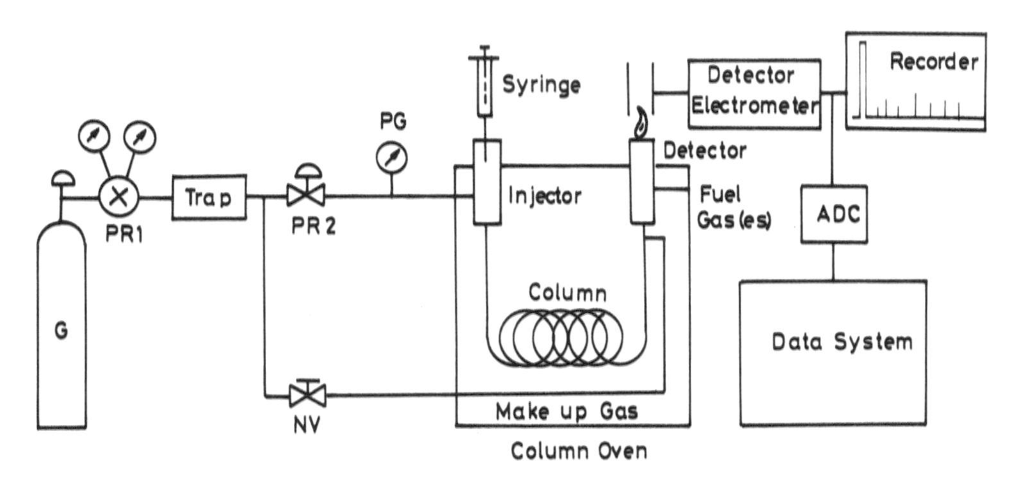


Fig. 3.1 — Basic configuration of a gas chromatograph; G, gas cylinder; PR1 and PR2, two-stage regulators; PG, pressure gauge; NV, needle valve for adjusting the make-up gas flow-rate. Reproduced with permission from Ref. 3; copyright (1984) John Wiley & Sons, Inc.

consists of an injection port, a column and one or more detectors, connected in series, in that order. The detectors may also be connected in parallel, after splitting the column effluent stream using T or Y joints. The interconnections are made of low-volume, leakproof Swagelok fittings, usually consisting of front and back ferrules and a nut. The ferrules may be made of graphite, metal, polyimide (DuPont's Vespel) or mixtures thereof. They are thermally stable (up to 350–400°C), chemically inert and expand slightly on heating to form a leak-tight seal. Capillary columns and effluent splitters are conveniently coupled by means of shrinkable Teflon tubing or fittings such as Supelco's capillary column butt connector, which is a double-tapered ferrule in a stainless-steel compression housing.

Most often, the gas chromatograph is offered as a single integrated unit, though each of the these parts can be conveniently and independently changed to suit the particular application. The mobile phase is generally delivered to the system from a commercial pressurized (2500 psi or 150–160 atm) gas cylinder, equipped with a two-stage regulator for coarse and fine control of the exit pressure (40–100 psi) of the gas and, thereby, its flow through the column. The gas is usually passed through a trap or a series of traps for removing the moisture, organic matter and oxygen, and then through 5–10 $\mu$m frits to filter off any particulate matter. Several simple devices for in-line gas purification are available from the supply houses. Unless the carrier gas is over 99.995% pure, contamination of the connecting tubes, flow controllers and columns may occur, causing additional detector noise and drift; this would adversely affect the sensitivity of the GC experiment.

The chromatograph may have additional, thermostatically controlled pneumatics to prevent drift and to maintain a constant flow of the gas at the given cylinder-head pressure. The flow control is particularly difficult and critical in open-tubular GC where the flow-rates are an order of magnitude lower than those in packed-column GC. A pressure regulator allows a wide range of operating flow-rates, and maintaining a constant pressure gradient across the column ensures a constant carrier gas flow. The pressure-regulated systems are very sensitive to changes in ambient pressure which causes a change in the pressure drop across the column and, consequently, in the mobile-phase flow-rate. Flow-rate changes are also inevitable in temperature-programmed GC since the gas viscosity increases with temperature and this would cause a decrease in the flow rate; for example, a 40 ml/min flow at 50°C (at a constant head pressure of around 14 psi) would be reduced to 20.4 ml/min at 300°C [4]. In principle, flow control is preferable over pressure regulation since several GC detectors are sensitive to changes in the flow-rate. However, the low flow-rates used in open-tubular column GC are difficult to control with the flow controllers available at present, and pressure regulation is often employed in this case. Automatic, microprocessor-based flow controllers with feedback capability and usable in the 1–500 ml/min, or wider, range, have become available and are very convenient. The flow control is very important in both qualitative and quantitative analysis by GC. A quantitative accuracy of 1% would require a flow stability within 0.2% variation and the extent of retention-time variability is equal to that of the flow-rate (see Chapter 9).

The gas flow can be either measured using a soap-film flow-meter at the outlet of the column, or read electronically on the instrument panel. The soap-bubble meter is simple, inexpensive and reasonably accurate (within 1% error). Thermal mass flow-meters have a flow-rate range of 0–100 ml/min, but are not very suitable in the very low flow-rates used in open-tubular GC. If the hold-up volume of the capillary column is known, the carrier-gas velocity is accurately and conveniently determined by measuring the time taken by an unretained sample (e.g, methane) to elute from the column.

The column is housed in an oven (typically of a size, 30 x 30 x 15 cm) that can normally accommodate 1 to 4 columns and is equipped with fans to ensure uniform and rapid distribution of the temperature. The oven walls should be well insulated so that the column temperature is not affected by changes in the injector, detector and the ambient temperatures. Most commercial instruments offer a temperature range of −50° to +450°C; the subambient-temperature operation would normally require an auxiliary coolant such as liquid nitrogen or carbon dioxide. The oven temperature is precisely settable and programmable from 0 to greater than 30°/min. The accuracy and reproducibility of temperature control of the oven depends on the type of controller. In the order of sophistication, these may be based on: (a) constant voltage or power supply, (b) full supply and cut off, (c) proportional power supply and (d) microprocessor-controlled feedback loop. The microprocessors can combine standard heating, controlled ambient air mixing and cryogenic cooling modes to achieve precise and accurate oven-temperature control over the whole range of available temperature.

Sample introduction into the gas chromatograph is most often carried out using a hypodermic syringe whose needle can pierce through an elastic septum retained in the injection port by means of a nut. The most commonly used silicone-rubber septa may contain monomeric and oligomeric impurities which may bleed into the column above a certain temperature (200°C), resulting in unsteady baseline and ghost peaks. Also, septum particles, generated by repeated piercing by the injection needles, may get lodged in the injection port and slowly decompose into volatile products that pass into the column. Recently, a variety of low-bleed, Teflon-faced silicone and Teflon-type elastomeric septa have become available which can be used at very high temperatures, some of them (e.g. Pyrosep S-1, from Supelco, U.S.A.) up to 400°C. Some of the modern injection port designs include a septum-purge system which diverts part of the carrier-gas stream to the vent and prevents column contamination by the products of septum bleed.

Occasionally, the liquid stationary phase, or its decomposition products, also bleed into the detector yielding spurious signals. In some GC applications, the column may have to be heated over a range after injecting the sample at a certain temperature (programmed-temperature GC or PTGC, see section 3.4). The effect of column bleed (which increases with the column temperature) results in an unacceptable drift in the baseline of the chromatogram.

The detector noise as well as the baseline drift in temperature-programmed GC can be suppressed by using a dual-column–dual-detector system. In these systems, the carrier-gas flow through the two identical columns is maintained at the same rate and, since both the columns are in the same oven, they are also at the same temperature. The detector is operated in a differential mode so that the bleed profiles from the two columns are nullified. Microprocessor-manipulated gas chromatographs can store the bleed profile of a single column and subtract it from the chromatogram and thus eliminate the need for using two columns and detectors. Many instruments, however, have provision for accommodating two each of the injection ports, columns and detectors (and amplifiers) so that they can be conveniently used either in the compensation mode or as two independent systems.

Columns and detectors in the above-described multicolumn-multidetector chromatographs may either be similar or different. By use of a splitter either after the injection port or the column, several interesting possibilities emerge. Allowing the injected sample to be split into two streams, feeding to two dissimilar columns, often permits qualitative analysis by ratioing of the retention times. The ratio of response to two dissimilar detectors for the split effluent from the same column makes it possible to distinguish between two compounds not separated by the column. It may be advantageous to use a highly specific detector in one of the channels (the other being a universal detector like the flame-ionization detector (FID)) so that trace components of interest are not missed. A portion of the effluent from the column can also be diverted to a second column of a different selectivity for further separation. Two- or multidimensional gas chromatography is thus possible [5]. The two columns may also be similar, in which case shorter analysis times are possible by diverting only the early-eluting compounds to the second column; the late-eluting compounds, which are often well resolved need not be subjected to further chromatography.

The injection-port and detector temperatures are independently settable and are generally maintained about 50°C above the maximum temperature of the column to prevent condensation of the sample components in these parts. Certain detectors such as the thermal conductivity and electron-capture detectors are required to be maintained at a very steady temperature. The thermal conductivity detector (TCD) is normally kept at as low a temperature as possible but at 15–50°C above the column temperature. The electron-capture detector (ECD) is also temperature-sensitive and the optimum temperature is determined by the solute and the radioactive source. The FID response is not so sensitive to temperature changes and is normally operated at 300 to 400°C.

The output of most GC detectors is in the form of an extremely small electrical current (e.g. 1 nanoampere to 1 picoampere for FID). A suitable amplifier or electrometer is required for this small output to be converted into a signal readable by the recorder or the integrator. Modern digital electrometers can routinely suppress the detector noise and amplify genuine signals without operator intervention.

The instrumental requirements for open-tubular GC are generally the same as those for the packed columns, but, in view of the much lower sample capacity, lower mobile-phase flow-rates and faster detector response required in the former, certain design modifications are required. The main differences are in the injection-port design (discussed in some detail later in this chapter) and in the addition of a 'make-up' gas line for the detector using suitable pneumatics. Very few detectors are specially designed for use with the restricted flow-rates of open-tubular GC. The most frequently used flame-ionization detector, for example, requires a 1–1.5 : 1 : 10 ratio of the carrier gas, hydrogen and air, with the hydrogen flow of around 30 ml/min to sustain the flame. The maximum possible carrier-gas flow-rate in open-tubular GC being only about 10% of this figure, the remainder is made-up by introducing an auxiliary gas (e.g. nitrogen for flame-ionization and argon–methane for electron-capture detectors) just before the column outlet. When 'low-volatile' compounds are suspected to be present in the analyte, it is advisable to pre-heat the make-up gas to prevent condensation of these components in the column. Make-up gas adaptor kits are commercially available from most supply houses, but can also be home-made with minimal facilities. Most modern chromatographs are equipped with make-up gas plumbing.

The performance of the gas chromatographic system is dependent on the different parts of the instrument; namely, the injection port, connecting tubes, column, and the detector; and on parameters such as the method of injection, flow-rate, etc. If the chromatograph is well designed, the contributions to band broadening from the injection port, detector and the connecting tube are minimal and constant. This may be ascertained by injecting methane which should give a needle-sharp peak. If the methane peak is not sharp, leaks, cracks or dead pockets are to be suspected. Once a sharp peak is obtained on methane injection, the column performance may be evaluated using test mixtures containing *n*-alkanes, an aromatic hydrocarbon (benzene or naphthalene), an olefin (e.g. 1-octene), polar solutes such as an alcohol (*n*-butanol or 1-octanol), an ester (methyl butyrate) and an ether (di-*n*-butyl ether) [3] (see also reference [40] of Chapter 4). Excessive retention, broader peaks and tailing of the polar solutes is indicative of adsorptive sites within the column. A mixture containing 2,6-dimethylaniline and dimethylphenol may be used to determine the neutrality of the column; strong retention of either of the solutes indicates acidic or basic character, respectively.

The number of theoretical plates ($n$), or the number of effective plates ($N$), or the corresponding plate heights ($h$ or $H$), is generally taken as the measure of the efficiency. Since the plate number (total or per unit length) does not take the time factor into account, the rate of production of the plates (i.e. the plate number divided by the retention time) may be more appropriate. The separation number (Trennzahl, $TZ$) is probably a more realistic measure of the separation efficiency of a column. In practice, the separation between two homologous compounds, A and B is measured as:

$$TZ = \frac{t_{R2} - t_{R1}}{w_{h2} + w_{h1}} - 1 \tag{3.1}$$

where $t_R$ and $w_h$ are the retention time and the peak width at half height of the compound denoted by the subscript. The separation number varies with the capacity factor and depends on the affinity of the stationary phase to the given type of solutes; e.g. alcohols have much higher $TZ$ values on a polar phase like Carbowax 20M than do, say, hydrocarbons. The separation number ($SP$) of the system is given by the sum of the separation numbers of the *n*-alkanes over the whole analytical range divided by the adjusted retention time of the last peak, so that $SP = \Sigma TZ/t'_R(\text{final})$.

The several available performance parameters and the column evaluation methods notwithstanding, the plate number ($n$) remains the simplest and the most popular measure of efficiency of the column. If used routinely and periodically, the column evaluation methods [4] can be very useful in alerting the chromatographer to any column deterioration and instrumental defects that may affect the optimization of the gas chromatogaphic experiment.

## 3.3 OPTIMIZATION OF GAS CHROMATOGRAPHY

The optimization of GC is aimed at achieving the maximum resolution, speed and sensitivity of analysis. The parameters to be optimized include every of aspect of chromatography; i.e. (1) sample preparation; (2) derivatization, if necessary; (3) sample size; (4) method of sample introduction; (5) injection-port design; (6) choice of the mobile phase and its flow-rate; (7) column dimensions and the material; (8) nature of the solid support; (9) the particle diameter and the size distribution; (10) choice of the stationary phase and its thickness; (11) method of coating the stationary phase on the support; (12) method of packing the stationary-phase-coated material into the column; (13) column conditioning; (14) column temperature and its variation; (15) multi-dimensional methods; (16) choice of the detector and its performance characteristics such as sensitivity, linear dynamic range and selectivity; and, finally, (17) data handling. Not all of these, however, are relevant in open-tubular GC which is fast replacing the traditional packed-column GC in most critical as well as routine applications. The first five aspects, dealing with the sample introduction into the gas chromatograph, are discussed in this chapter, while the next ten, pertaining to the separation system, form the subject matter of the next chapter. Detection in GC forms Chapter 5 and data handling methods, which are essentially the same for both GC and LC, are discussed in Chapter 9.

The theoretical concepts discussed in Chapter 2 should be useful in understanding the factors affecting the performance of GC. In general, optimal output is dependent on the smallest detectable sample size, plug injection, a mobile phase of high diffusivity, optimum flow-rate, longest column with smallest support-particle size (consistent with low pressure drop), minimum of stationary phase uniformly coated on the support, lowest column temperature and an appropriate detector in its optimum sensitivity range. It may be noted that most of these features can be realized only in open-tubular GC. The 'openness' of the column leads to high specific gas permeability with minimal pressure drop across the column and results in the elimination of multiple flow paths contributed by the column packing in the packed-column GC. The stationary phase, which is in the form of a thin film (most often 0.1 to 0.4 μm), also contributes to the high efficiency and low sample requirement of the open-tubular GC. The increased efficiency of the open-tubular columns is reflected in the large number of theoretical plates (about 10,000/metre compared to about 1000 in packed columns) that can be generated. The low pressure drop across the column also offers a very practical way to increase the plate numbers several fold simply by increasing the column length from about 2 m in packed-column GC to 25 m (or even up to 100 m) in open-tubular GC; thus, a million theoretical plates can be generated. Where such high efficiencies are not needed (as in some routine quality control applications), part of the efficiency can be sacrificed to increase the speed of the analysis by increasing the carrier-gas flow. This would result in the elution of sharper and more concentrated bands, thus increasing the peak height. Several other practical aspects of the operation also significantly enhance the signal-to-noise ratio, leading to much higher sensitivity in open-tubular GC. For example, the use of make-up gas renders the detector less sensitive to change in carrier-gas flow through the column. The flow-rate being a small fraction of that in packed-columns, the impurities, if any, in the carrier gas do not significantly contribute to the noise. Thus, while the peak area from a given quantity of the solute may be the same, the signal-to-noise ratio is enhanced by a factor of 100 resulting in a much lower detection limit. Fig. 3.2 compares the results of an analysis on a packed column and an open-tubular column [6].

The nature of the column (that is, the dimensions, packed or open-tubular, GSC or GLC, the nature of the stationary phase, its amount, etc.) obviously depends on the application. Selection of these parameters is quite straightforward in the sense that for preparative separations (that is, where the separated compounds may be collected using cold traps for further studies), a packed column of large diameter (8–200 mm) is necessary; open-tubular columns of diameter around 0.25 mm have a

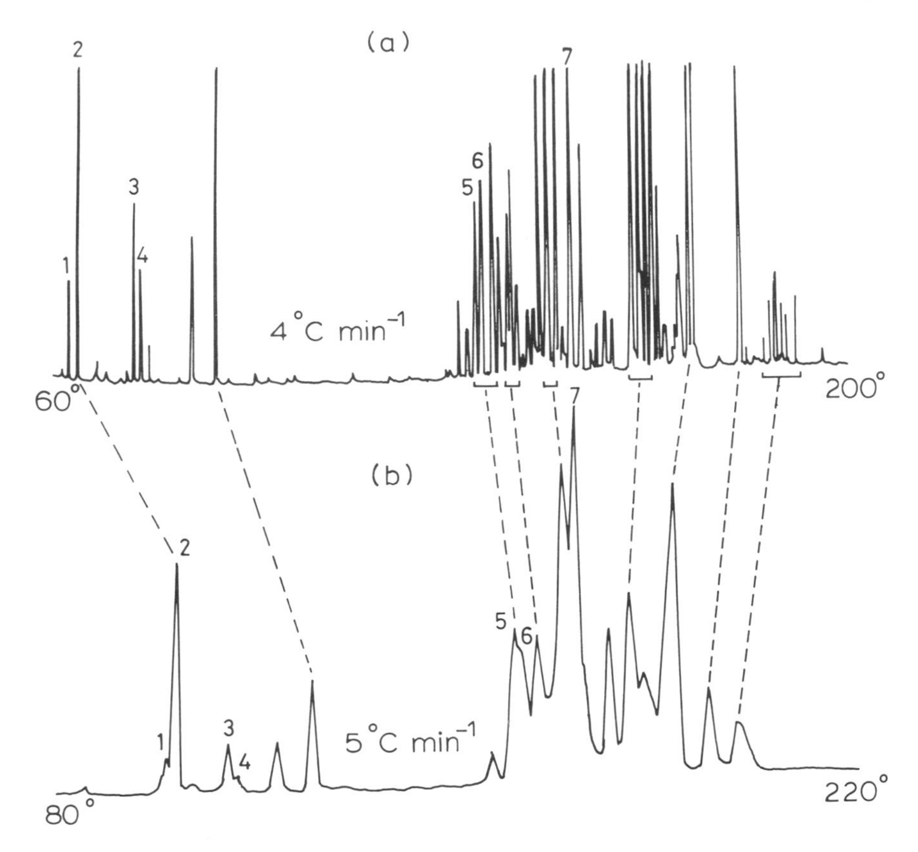


Fig. 3.2 — Comparison of gas chromatograms of calmus oil on (a) a 50 m × 0.3 mm i.d. OV-1 glass open-tubular column and (b) a 4 m × 3 mm i.d. packed column (5% OV-1 on 60–80 mesh Gaschrom Q). Reproduced from Ref. 6 by courtesy of Dr Alfred Huethig Verlag.

distinct advantage in analytical applications. The intermediate range packed (2 to 4 mm i.d.) and wide-bore, open-tubular (0.6 to 1 mm i.d.) columns may be used for routine analytical and quality control applications and trace analysis where easy disposability and higher sample capacity may be advantageous. The choice between GSC and GLC is also easy. GSC on molecular sieves or porous polymers is used for the analysis of permanent gases, low molecular weight hydrocarbons and other non-polar highly-volatile compounds, while GLC using a coated or bonded liquid phase is the obvious choice for all other types of substances. The choice of the liquid phase for general purpose or speciality applications is wide open (Chapter 4). Generally, a stationary phase of similar polarity (i.e. non-polar liquid phases for non-polar analytes and polar phases for polar solutes) may be preferred. In the absence of indications to the contrary, a cross-linked polydimethylsiloxane fused-silica column could be the first choice.

On a given column, the stationary-phase load, or the film thickness, determines the phase ratio, the volume of the mobile phase within the column being nearly constant. As can be expected, the more the stationary phase (i.e., the thicker the stationary phase), the greater is the retention. Heavier loading may thus be used for analysis of highly volatile compounds. On the other hand, increased film thickness increases the contribution of the 'resistance to mass transfer in the stationary phase', $h_3$ (eq. (2.75)), to the plate height and thus lowers the efficiency. Fig. 3.3 illustrates the role of the stationary-phase thickness in retention and resolution [6].

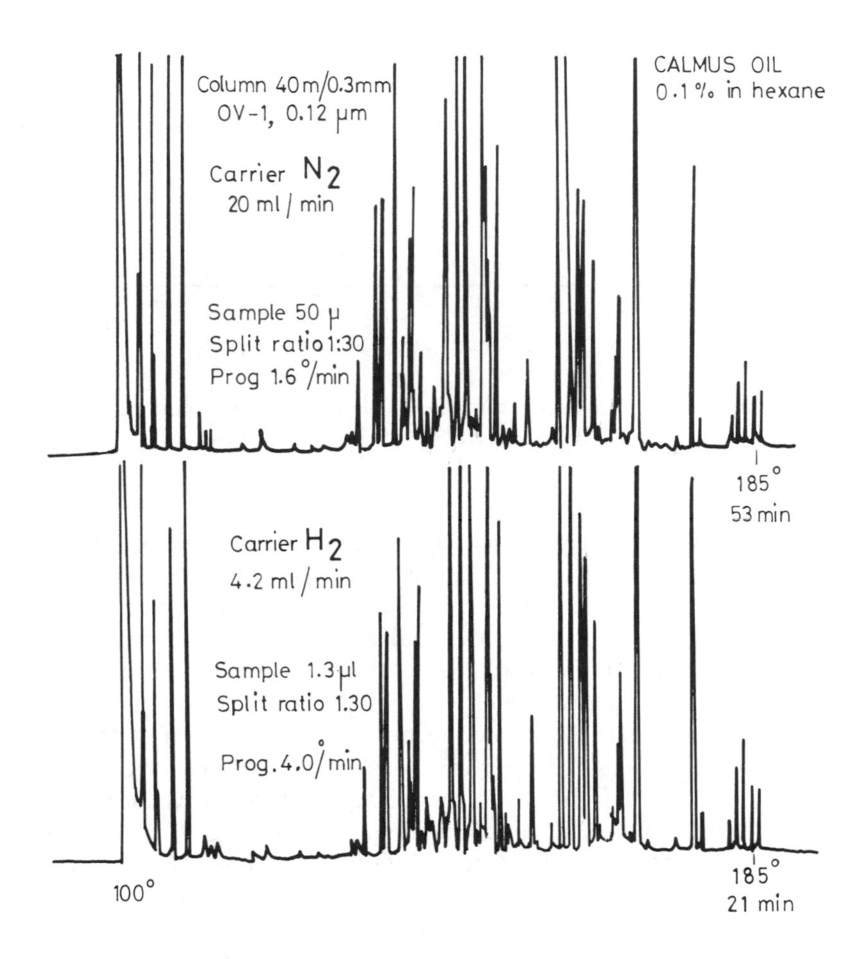


Fig. 3.3 — Comparison of nitrogen and hydrogen as mobile phases. Reproduced from Ref. 6 by courtesy of Dr. Alfred Huethig Verlag.

Among the usable mobile phases (see Chapter 4), the choice between the most commonly used gases; namely, helium and nitrogen, is often based on cost and availability. Fig. 3.4 shows the van Deemter plots for the common mobile phases in

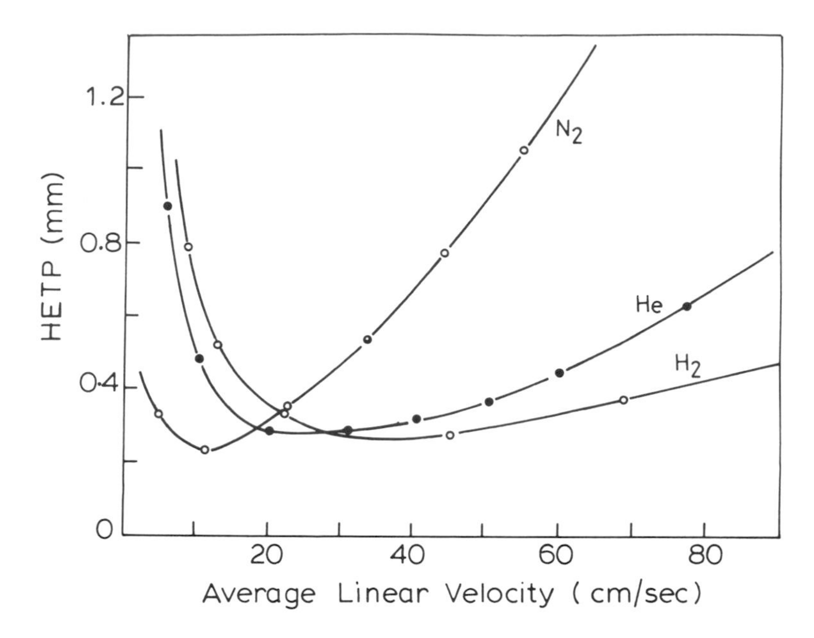


Fig. 3.4 — van Deemter plots for the common mobile phases in gas chromatography; by courtesy of Hewlett-Packard, Inc.

GC which indicate that while nitrogen (or argon) can yield high plate numbers at low flow-rates, helium (or hydrogen) can maintain reasonably good resolution over a wider range of mobile phase velocities. Higher mobile-phase flow-rates decrease the analysis time; faster elution improves the peak sharpness and the signal height. As Fig. 3.5 shows, for comparable resolution and signal intensity, use of hydrogen permits completion of the analysis in less than 40% of the time and with 25% of the sample relative to nitrogen as the mobile phase [6]. Operation at the optimum mobile-phase velocity ($u_{opt}$) permits use of shorter columns and hence faster analysis. In open-tubular GC, however, a much higher flow rate can be used without significant loss of resolution. The optimum practical gas velocity (OPGV, defined as the velocity at which the van Deemter plot becomes linear, Fig. 2.8) ensures maximum efficiency per unit time. The flow rates and several other optimization parameters such as the sample size, phase ratio, column length and resolution are dictated by the column diameter. These are summarized in Table 3.1.

Once the above parameters are optimized, the speed and selectivity of gas chromatography is determined by the column (oven) temperature. As noted earlier, the mobility of a solute in GC depends on its volatility, and the elution of the compound from the column is faster at higher temperatures. The retention time is approximately halved for every 30°C rise in the column temperature. But the selectivity of interaction between the solutes and the stationary phase is better at lower temperatures. The selected temperature of chromatography is often a compromise between the two opposing requirements and the choice is rather arbitrary.

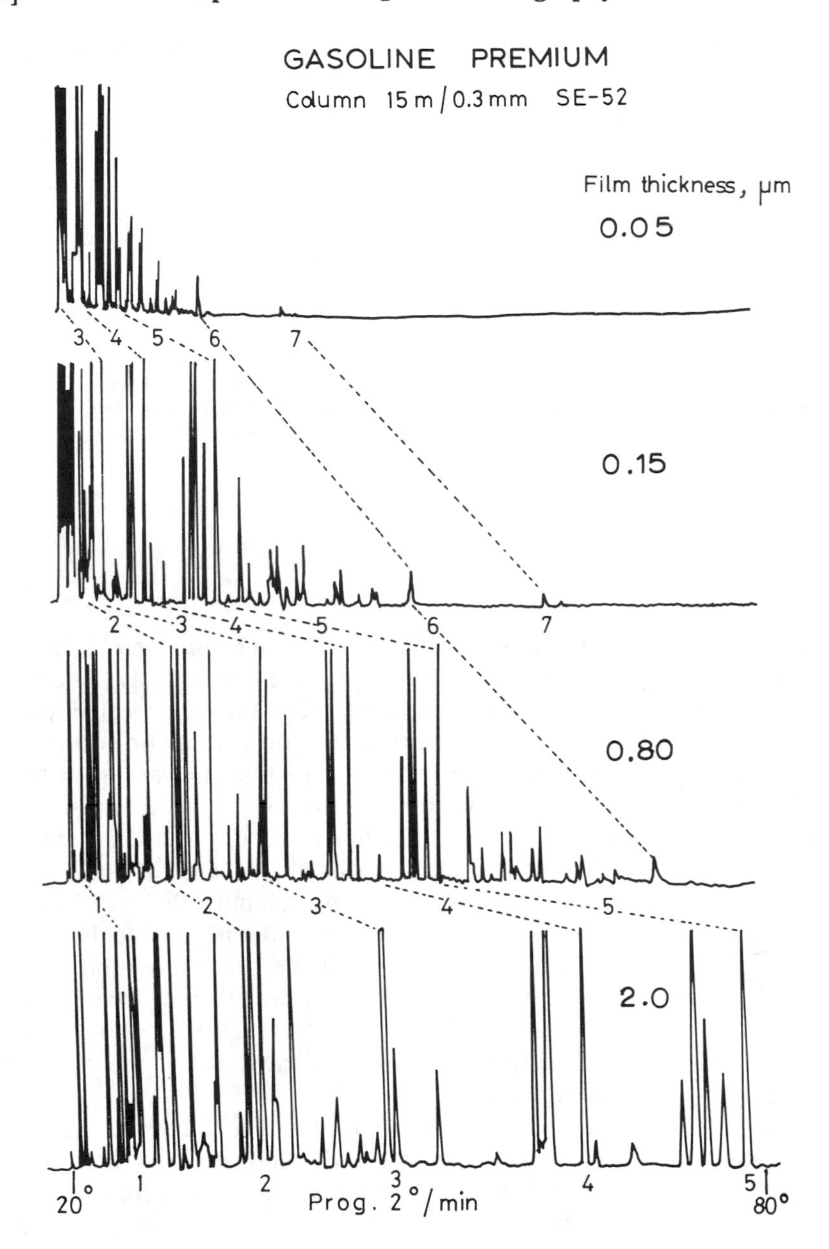


Fig. 3.5 — Effect of stationary-phase-film thickness on solute retention: (1, *n*-pentane; 2, benzene; 3, toluene; 4, *o*-xylene; 5, 1,2,4-trimethylbenzene; 6, naphthalene; 7, 2-methylnaphthalene). Reproduced from Ref. 6 by courtesy of Dr Alfred Huethig Verlag.

The column efficiency or the plate height ($h$) is related to the temperature by an expression similar to the van Deemter equation and shows a minimum as a function of temperature:

**Table 3.1** — Characteristics of gas chromatographic columns†

| Property‡ | Packed columns | | Open-tubular columns | | |
|---|---|---|---|---|---|
| | Regular | Micropacked | Wide-bore | Regular | Microbore |
| 1. i.d. (mm) | 2 | 1 | 0.53 | 0.25 | 0.1 |
| 2. Sample size | 0.2 mg | 2 mg | 2 mg | 100 ng | 10 ng |
| 3. $L$ (m) | 2 | 15 | 15 | 30 | 10 |
| 4. $n$/m | 1000 | 2500 | 2500 | 5000 | 12,500 |
| 5. $\Delta P$ | 4 | 30 | 0.1 | 1 | 65 |
| 6. $\beta$ | 12.5 | 50 | 175 | 250 | 250 |
| 7. $F_c$ (ml/min) | 30 | 20 | 10 | 2 | 0.4 |

†The properties given are only representative; the actual values may vary over a wide range.
‡i.d. = internal diameter; $L$ = length of the column; $n$/m = number of theoretical plates per metre; $\Delta P$ = pressure drop across the column in atm.; $\beta$ = phase ratio; $F_c$ = mobile-phase flow rate.

$$h = A + BT + \frac{C}{T} \tag{3.2}$$

Thus, the plate height varies with the temperature and there is an optimum temperature similar to the optimum mobile-phase linear velocity. The log $n$ ($n$ = number of theoretical plates) is also linearly related to the reciprocal of the absolute temperature. The van Deemter plots for different hydrocarbons on non-polar phases show that the slope of the right-hand branch of the plot (i.e. at higher flow rates) tends to be nearly horizontal at higher temperatures [7]. In certain cases the $h$ may even decrease with increasing mobile-phase linear velocity at higher temperatures. In other words, the flow-rate can be substantially higher (and the analysis faster) at higher temperatures without significant loss of column efficiency. For a given column, each solute has an optimum temperature and it is possible that two solutes may have the same optimum temperature and the same retention time at that temperature. Separation may be possible in such cases at a temperature away from the optimum. If two solutes are well separated, the highest possible temperature permitted by the thermal stability of the solutes and the stationary phase may be used so that the analysis time is minimized.

Change in temperature occasionally brings about a change in the stationary-phase selectivity. As the temperature is raised, there is a tendency towards an increase in the polarity of the stationary phase, resulting in an increased retention of the more polar solutes relative to non-polar solutes. Under these circumstances it is possible that the elution order of certain peaks may also change.

There are several computer-assisted approaches to optimization of the various gas chromatographic parameters like the flow-rate and the column temperature. The Simplex routine [8] involves measuring the separation efficiency at three different temperatures (flow-rates or programs) and marking them on a plot with the flow rate and the temperature forming the two axes and the three points forming a triangle (Fig. 3.6). The least effective point (point 3) is rejected and the triangle is reflected about the plane of the other two points. If the new point (point 4) is not better than the earlier one, the size of the triangle (Simplex) is maintained at the same level and

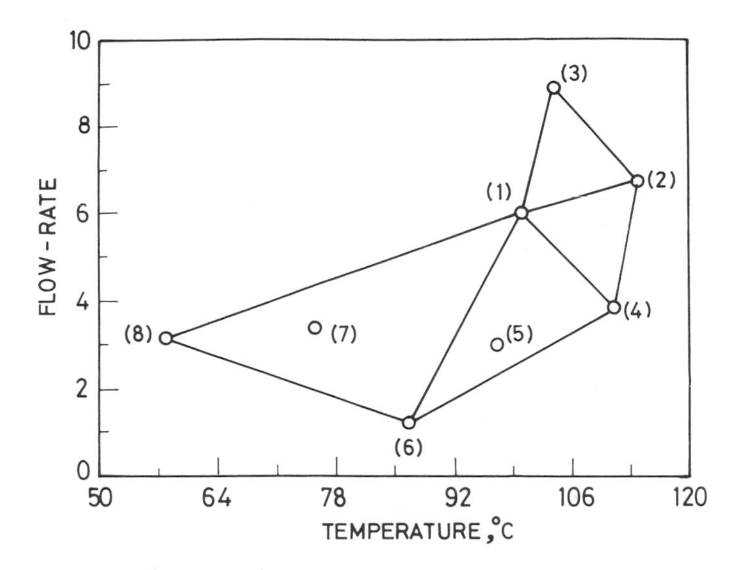


Fig. 3.6 — Illustration of the Simplex optimization procedure to improve separation efficiency. Reproduced from Ref. 8 by courtesy of Elsevier Scientific Publishing Co.

is reflected along the best axis giving a new point, 5. If the new point is better than the previous ones, the Simplex is expanded in the same direction taking care not to exceed the limits set in terms of the analysis time, etc. This procedure is repeated until the change in parameters does not produce any significant improvement in the separation efficiency. Optimization of programmed-temperature GC is also possible using similar protocols [9].

## 3.4 PROGRAMMED-TEMPERATURE GAS CHROMATOGRAPHY

It is often observed in GC that all components of a sample may not elute in a reasonably short time and with adequate resolution at a given temperature. For example, Fig. 3.7 shows the separation of some hydrocarbons and halogenated compounds [7]. At a column temperature of 45°C, only five components are eluted in about 30 min, at 120°C, eight compounds are seen. But, at the higher temperature, the first three components elute very fast and the separation is inadequate; in both cases, the later-eluting peaks are rather broad. On the other hand, increasing the temperature continuously from 30–180°C over the same period of time results in the elution of nine compounds, all of them appearing as sharp and well-separated peaks. At constant temperature or isothermal GC, the peak widths increase linearly with the retention time while the latter increase exponentially with the carbon number (in a homologous series). On the other hand, if the temperature is raised at an appropriate rate during chromatography (programmed-temperature GC, or PTGC), the peak widths remain constant while the retention times increase linearly with the carbon number. The above features; namely, faster analysis of samples

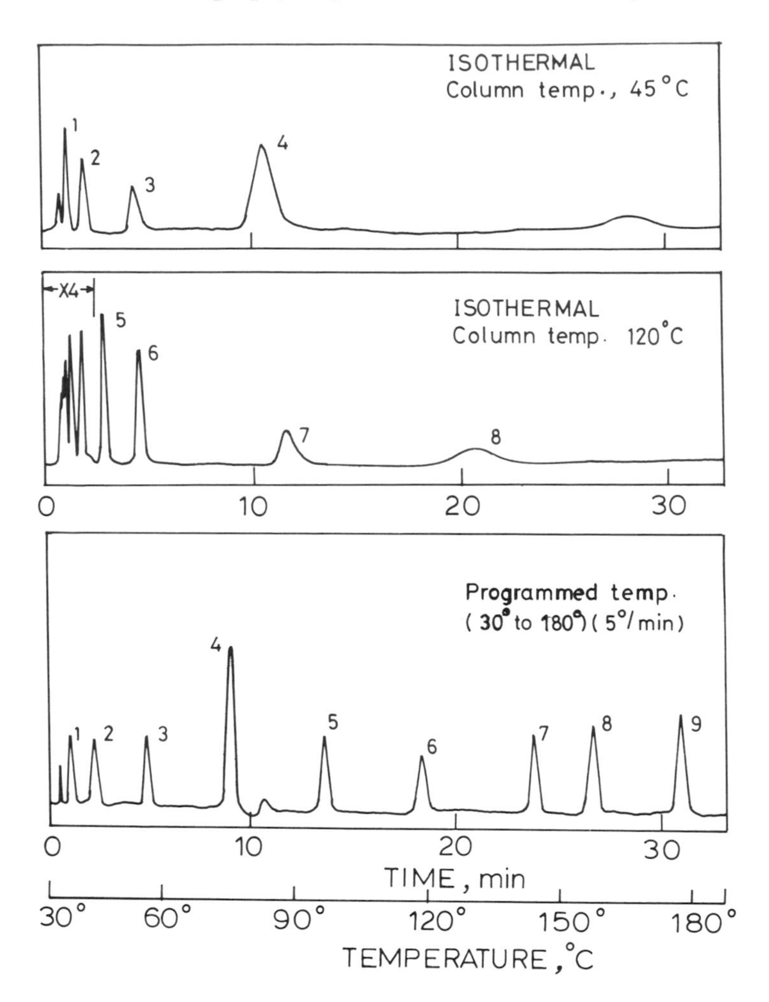


Fig. 3.7 — Comparison of isothemal and programmed-temperature GC. Peak identity: 1–6. *n*-pentane to *n*-octane, 7. bromoform, 8. *m*-chlorotoluene, 9. *m*-bromotoluene. Reproduced with permission from Ref. 7; copyright (1966) John Wiley & Sons, Inc.

containing compounds of a wide range of boiling points and improved detectability of late-eluting peaks owing to sharper peaks, are characteristic of PTGC.

In addition to the reduction of analysis time, the major advantage of PTGC is the possibility of using large sample volumes. If the initial temperature is considerably lower than the retention temperature ($T_R$, see below for definition), all the solute molecules may be retained by the first few plates in the column, just as in plug injection, with the result that the peaks are sharper even when large volumes are injected. The large sample volumes, sharp peaks and, thence, enhanced detector sensitivity are especially useful in trace analysis.

The convenience of handling compounds over a wide range of volatility without pre-fractionation has increased the popularity of PTGC and brought in some fundamental and lasting changes in the desirable characteristics of the stationary phases as well as in the design of gas chromatographs. The stationary phase, naturally, has to remain in the liquid state over a wider range of temperatures and the viscosity changes with temperature should be minimal. The changes in instrumentation include: (1) independent heating systems for the injection port, column oven and the detector; (2) built-in carrier-gas flow controllers in addition to the pressure regulators on the gas cylinders; (3) dual-column systems for detector reference-cell compensation where required; and (4) column-oven design and insulation to permit rapid equilibration of temperature. The PTGC capability is now a standard feature in almost all commercially available gas chromatographs. The presettable parameters are: the initial temperature, initial 'hold' time (i.e. the period of time after injection during which the temperature remains constant), the rate of increase of the temperature (number of degrees per minute), the final or maximum temperature, and the final hold time for the remaining compounds to be eluted.

There are a number of ways of temperature programming. Ballistic programming (increasing the oven temperature very rapidly) is convenient for removing the strongly retained and unwanted compounds after the compounds of interest are eluted. Stepwise increase of temperature is useful when analysing solutes which may appear as well-separated groups of peaks in isothermal GC. Linear temperature programming (that is, continuously increasing the temperature at a constant rate) is probably the most frequently used technique. Multiple ramping (linear programming at different rates and interspersed by isothermal runs of varying intervals) is feasible in most of the modern gas chromatographs.

The retention temperature ($T_R$) in PTGC is defined as the temperature at which a particular compound is eluted. The retention temperature varies with both the flow-rate ($F$) and the rate of increase of temperature ($r$) and is thus a function of the ratio $r/F$ (the program); the retention time ($t_R$) is given by $(T_R - T_0)/r$, where $T_0$ is the initial temperature in the column. At small $r/F$ ratios, increasing the heating rate does not significantly reduce the analysis time, but in open-tubular GC where the flow-rates are low and the $r/F$ ratios are consequently high, the analysis time is proportionally reduced with increase in the heating rate. Since the flow-rates are small and since smaller $r/F$ ratios yield better resolution, the heating rates in open-tubular GC are normally below 5°C per minute. Reproducibility of flow rates in PTGC is very difficult owing to the temperature dependence of viscosity and, therefore, the use of retention indices or retention temperatures for identification purposes may be inaccurate. PTGC may be considered equivalent to isothermal GC, provided the column temperature in the latter case is at the so-called significant temperature ($T'$). The significant temperature is usually about 40°C lower than the retention temperature. Thus, a separation system that is otherwise optimized for isothermal GC is also optimized for PTGC. For a given column and flow-rate, the speed and resolution is a function of the rate of increase of the temperature. Too fast a rise in temperature understandably impairs the resolution and it is generally recommended that the rate of increase of temperature should not exceed $12/t_M$, $t_M$ being the time taken by the unretained component to elute from the column. The initial temperature should also be as low as possible but too low a starting temperature would unnecessarily prolong the analysis time. The lowest possible initial temperature is also determined by the viscosity and the freezing point of the liquid phase. The initial temperature and the program may be such that the first components elute at a $k$ value around 5. The final temperature, naturally, is set to elute the least volatile compound in a reasonable period, subject to the thermal stability of the stationary phase.

Contrary to the popular notion, PTGC does not improve separability. Since PTGC resolution increases with decreasing heating rate, isothermal GC may be viewed as a limiting case of PTGC for maximized resolution. In other words, the resolution in PTGC may approach but does not equal that obtainable in isothermal GC. PTGC is thus a trade-off between higher resolution and faster analysis. Also, while a single PTGC run may speed up the analysis, it does not necessarily lead to shorter analysis time in multiple and routine analysis of a large number of samples since, after each run, the column requires to be cooled and equilibrated before the next analysis can begin. Isothermal GC may thus be preferred if the analysis could be completed in about 10–20 min. PTGC, on the other hand, is the most useful approach for the initial trial separations of unknown samples and method development. A single run over the entire permissible temperature range of the stationary phase and at a high heating rate would give a clear picture of the sample complexity and permit quick optimization of the column temperature.

## 3.5 PRESSURE (FLOW) PROGRAMMING

The mobile-phase flow programming by progressively increasing the inlet pressure is complementary to temperature programming and helps reduction in the analysis time. It is particularly useful in handling thermally labile compounds, where high temperatures cannot be used in the analysis. Operation at lower temperatures also reduces column bleed, improves the baseline stability and prolongs the column life. Unlike in PTGC, where considerable time may be needed for cool-down of the oven and restabilization, the initial flow-rate can be instantaneously reduced, permitting a larger number of analyses in a given time. On the other hand, higher than optimal flow-rates decrease resolution and cause peak distortion. Also, excessive pressure build-up could be hazardous. However, both these problems are minimized in open-tubular GC, where the van Deemter curves are relatively flat and where high flow-rates can be achieved with much less pressure drop across the column, particularly when short columns are used. When flow-sensitive detectors such as the flame-ionization detector are used, flow compensation using the make-up gas or the fuel gas is possible.

The flow or pressure programming can be stepwise, linear, logarithmic or exponential. Stepwise programming may be convenient for analysing the volatile compounds at low flow-rates and then raising the flow-rate to remove the more strongly retained ones. Exponential 'flow-rise' is not normally employed owing to the requirement of high inlet pressures. Linear or logarithmic programmes, which can be easily realized using simple pneumatic systems, are commonly employed.

As noted above, it is advantageous to use short columns in flow-programmed GC since a wide range of flow-rates are possible without excessive build-up of inlet pressure; sample size and peak capacities, however, would then be relatively limited.

The results are readily correlated to the standard conditions since the retention volume does not appreciably change in flow-programmed GC. Flow programming can also be conveniently combined with temperature programming [3].

## 3.6 SAMPLING METHODS

The methods of sample preparation that are common to both GC and LC are discussed in Chapter 9. Some of the special techniques, which are particularly useful in GC, are discussed here.

### 3.6.1 Headspace sampling

Compounds that are volatile enough for GC (for example, the flavour components of fruit juices) may be expected to have a finite vapour pressure at ambient temperature. When placed in a closed container, these compounds equilibrate between the liquid (aqueous solution) and the space above its surface (headspace). The composition of the headspace with respect to the volatiles may, perceptibly, be more representative of the flavour (odour) profile of the food product; in other words, functionally reflective. There are several other instances where an analysis of the headspace could reveal the composition of the non-volatile portion; e.g. volatile metabolites of a micro-organism can reveal the identity of the micro-organism (see section 9.3.3). The environment in a laboratory, industry or even a city may be considered larger headspaces whose composition directly reflects the exposure hazards. This is analogous to the natural sensory perception by which animals obtain information required for food, self-protection and association. Apart from the obvious importance, headspace analysis can also be convenient since the sample, being in a vapour phase and free from undesirable matrix, can be directly injected into a gas chromatograph. Frequently, however, the concentration of the volatiles in the headspace is too low for direct analysis even by GC using the more sensitive detectors and a concentration step may be necessary [10].

The sampling methods for headspace analysis may be divided into two main categories. The static method involves withdrawal of a certain amount of the vapour from the space above the matrix after the material has attained equilibrium with the space above its surface in a closed container. The composition of the phases at equilibrium is altered by temperature and by the withdrawal of a portion of one of the phases. Therefore, rigorous control on temperature, the method of withdrawal, and the volume of sample, is required for reproducible results. The major drawback of this method is that the concentration of the trace components and compounds with low vapour pressure may be too low for detection. In the dynamic procedure, an inert gas is passed over or through the material to be analysed and the effluent is collected by a suitable method. Passing the gas over the matrix may take longer for removing sufficient material from the headspace but is convenient when the material is a solid that cannot be solubilized, or when bubbling through the solution or the fluid sample causes foaming. The stripper gas, after passing through the trap, may either be vented or recycled until the least sorbed component in the trap does not break through (conservation procedure) or until the most sorbed component is in equilibrium with the gaseous phase (equilibration procedure). The volatiles in the gaseous effluent may be collected using cryogenic or cold traps, solid adsorbents (usually porous polymers), solid supports coated with a liquid stationary phase, or a chemical reactant that can temporarily retain the organics (chemisorption). A number of ancillary devices are available from several suppliers for automated and multiple headspace sampling.

Both the static and the dynamic headspace techniques have their own individual characteristics and fields of application. The static procedure is chosen whenever the concentration of the analytes in the headspace is sufficiently large and the amount of water in the sample is low. In addition to the compounds obtainable from the static procedures, the dynamic or the strip-trap techniques often yield compounds of lower volatility or concentration; these procedures are thus ideal for sample enrichment of trace volatiles in headspace for analysis by high resolution GC. The sample enrichment procedure for the determination of trace organic volatiles should be well defined and reproducible. The sources of error in sampling include non-homogeneity of the material, sample handling methods, chemical reactivity of the analytes, incomplete adsorption and desorption by the solid trap, non-linearity of the detector in the trace range and inaccuracies in the calculation and extrapolation methods.

In the direct sampling of headspace from a closed vessel, withdrawal of the analyte using a gas-tight syringe can often yield a representative sample, but the method is often employed only for preliminary and qualitative experiments since the results are unsatisfactory for quantitative purposes. Direct connection between the headspace vessel and the chromatographic column is possible *via* the syringe and the technique can be easily automated so that the headspace volume transferred to the GC is exactly the same. It is essential that the syringe is at a higher temperature than the sample and the back-pressure in the GC injection port is less than that of the sample vessel.

Cryogenic trapping is a common method of headspace concentration. Here again, a gas-tight syringe is used to withdraw a certain volume of the headspace from the thermostatted sample; the contents in the syringe are then transferred to a cold trap where the volatiles are condensed and later evaporated into the GC column by heating the trap to a suitable temperature, either rapidly or over a period in a programmed manner. This method, even when about 100 ml of the headspace is condensed, does not result in sufficient enrichment of the trace constituents and is rarely used. Alternatively, the sample, if solid, is packed in a long, cylindrical column with an inlet and an outlet and, if liquid, is charged into a wash bottle type arrangement so that the inlet either dips into the liquid or is just above its surface. A purified inert gas (helium or nitrogen) is passed through the sample and the effluent is condensed in a U-tube or a coil, cooled under liquid nitrogen or solid carbon dioxide–acetone mixture. This method of trapping is often limited to a few hundred millilitres of headspace because of the large amount of water which also gets trapped, seriously hampering subsequent analysis. Interpolation of a desiccant may sometimes be helpful but often leads to poor recovery of the desired components. When large quantities of headspace are to be concentrated, the effluent from the above arrangement may be led into an extractor containing water and a small quantity of a volatile solvent, like ether or Freon 11; the organics are recovered by evaporation of the volatile solvent. Entrapment of the organics with the exclusion of moisture can also be achieved by using a porous polymer of the styrene-divinylbenzene type

(Porapak Q or Chromosorb 105) or poly(2,6-diphenyl-*p*-phenylene oxide) (Tenax), packed in a narrow tube. About 500 mg of the polymer can trap volatiles from up to 250 l of headspace, pumped at a rate of 50–300 ml/min. Closed circuit methods, in which the purge gas is recycled through the sample and the trap several times using a pump, have received considerable attention in recent years.

The organics can be recovered from the adsorbents by solvent extraction using an organic solvent such as carbon disulphide or acetone. Solvent extraction, however, is not generally favoured for the analysis of trace volatiles owing to the possibility of introducing artefacts unless the solvents are of very high purity. Also, owing to the large excess of the solvent, the solvent peak may mask several of the peaks due to the volatile trace constituents. Some of the volatiles may be lost during the pre-concentration step which is often required and which is a time-consuming step requiring careful handling. More commonly, the trapped volatiles are recovered by purging with an inert gas in the reverse direction at up to 100–150°C and collecting the effluent in a cooled trap. The trap tube is often conveniently inserted before the GC column and is rapidly heated to 200–300°C so that the analytes pass into the column. Tenax GC, owing to its high thermal stability, is the most suitable solid adsorbent when thermal desorption methods are used. The thermal desorption methods often require expensive and rather complicated hardware, consisting of heated switching valves, desorption devices and transfer lines. Introduction of the trap tube directly into the injection port is, therefore, widely used. When moisture is not a problem, the cooled column head itself may serve as a trap and this technique is particularly convenient when capillary columns are used. The volatiles may also be trapped on a small quantity (about 2 mg) of activated charcoal; desorption of the trapped volatiles is achieved by placing the charcoal directly in the injection port of the gas chromatograph and heating the injection block to above 250°C. Alternatively, the adsorbed compounds are extracted with a few microlitres of an organic solvent. In fact, activated carbon and graphitized adsorbents were used in several of the early studies on trapping of headspace volatiles directly. However, their general utility for the purpose is limited owing to their high affinity for water, high surface activity and the high temperature of desorption where most compounds might decompose.

There are several other materials and techniques of headspace collection which are essentially variations of those discussed above. Commercial headspace samplers are available for direct injection into capillary columns. In a majority of these devices, the sample is taken in a glass vial, equipped with a PTFE or aluminium-lined rubber septum. Up to about 100 vials can be accomodated on a cylindrical magazine and thermostatted. A hypodermic needle penetrates the septum and allows the carrier gas to enter the vial so that the pressure in the vial is more than or equal to that in the column. The carrier-gas flow to the column is momentarily switched off to permit the headspace sample to be released into the column (Fig. 3.8). While these devices are convenient and the injection volume is reproducible, the sample size is obviously limited but may be sufficient for capillary GC. In certain cases, the sample size can be increased by adjusting the pH (in the case of volatile acids and amines), salting out (of relatively non-polar solutes from dilute aqueous solutions) or by increasing the temperature. Under a given set of conditions the concentration of a compound in the headspace is directly proportional to its concentration in the sample, and headspace techniques are suitable for quantitative analysis. However,

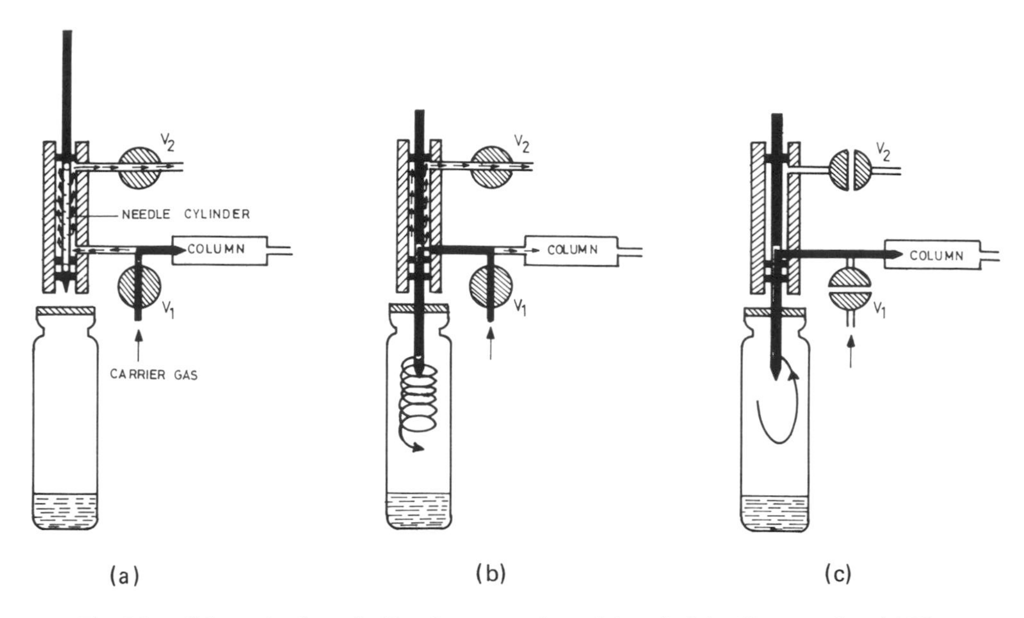


Fig. 3.8 — Schematic of a typical headspace sampler and the principle of its operation. (a) The carrier gas flows through the column via the valve $V_1$ and simultaneously flushes the needle cylinder, while the sample is thermostatted. (b) The needle is lowered to allow the carrier gas to mix with the sample headspace. (c) The valve $V_1$ is momentarily closed permitting the sample, now under slight positive pressure, to enter the column. (Illustration courtesy of Perkin Elmer & Co. GmbH, FRG.)

the partition coefficients are matrix dependent and, since the matrix composition is not easily reproduced in a model system, it is necessary to calibrate the sampling system using the standard addition or multiple extraction methods [11].

The headspace-sampling technique is mainly used for the determination of trace organic volatiles in samples where interference from the matrix components can be problematic; e.g. samples containing a large excess of late-eluting compounds and requiring long analysis times, polymeric compounds that may damage the column, or compounds that can obscure the analytes of interest. Certain non-homogeneous samples like blood, sewage, colloids and emulsions may require tedious sample clean-up protocols prior to analysis unless headspace techniques are used. This technique can also be used for functional group analysis of volatile compounds by subtractive methods; that is, forming a non-volatile derivative using the functional group and headspace analysis before and after derivatization. Simple examples of this technique involve varying the pH to retain acids and bases in the matrix, and treatment with concentrated sulphuric acid to remove all compounds except hydrocarbons. An interesting application is the determination of impurities in gases using the so-called reverse headspace analysis in which the gas phase is brought into thermodynamic equilibrium with a liquid, which is then analysed. Headspace analysis has been used for several physicochemical measurements, like ionization constants of acids and bases, stability constants of organometallics, and molecular weight determination of volatile compounds. The applications of headspace analysis in environmental control, industrial hygiene, flavour analysis of spices, beverages and dairy products, polymers, microbiology, medicine and forensic science have been reviewed [11,12].

### 3.6.2 Derivatization

Derivatization (that is, chemical modification of certain groups in a molecule) in chromatography has two main purposes: (1) to change the solubility properties and thus the chromatographic behaviour; and (2) to improve the detectability of a compound by certain detectors. Most frequently, derivatization in GC is to serve the first purpose, and in LC, the second. The solubility of a substance in a gas and therefore its partition coefficient between the gas and the stationary phase in GC depends on its volatility; in other words, the mobility of a substance under a given set of conditions depends on its volatility. Since a vast majority of compounds are not volatile enough for elution from a GC column in a reasonable period of time and a reasonable temperature range, early applications of GC were limited to a very few classes of compounds. Derivatization to render a non-volatile compound volatile has made it possible to extend the applicability of GC to a large variety of compounds including carbohydrates, proteins, nucleic acids and even inorganic compounds. In addition to volatility, derivatives containing nitro or halogenated groups can increase their sensitivity to certain detectors such as those based on the principle of electron capture; the signal enhancement in this case is proportional to the number of electron-hungry groups. In certain cases, derivatization may enhance the thermal stability of otherwise unstable compounds and, in certain others, improve the resolution of closely eluting or overlapping peaks by suitably altering their volatility. Derivatization also frequently improves the peak shapes by minimizing column interactions that cause peak tailing or irreversible adsorption [13].

The derivatization reagents, ideally, should be capable of reacting with several functional groups so that multi-step procedures may be avoided. The reactions should be mild, fast and reproducibly quantitative. Most of the commonly used derivatization reactions in GC satisfy these requirements. These reactions are generally simple and so is the equipment to prepare them. Derivatization for chromatography is usually a microchemical manipulation, requiring miniaturized apparatus and glassware. Tapered glass vials of 0.1–2.0 ml capacity, with Teflon-lined plastic screw caps, are most suitable. (It is advisable to rinse them with trimethylsilyl chloride solution in benzene to prevent any interaction of the analyte with glass.) Soluble samples and reagents are transferred using microsyringes. Mixing of the reactants may be by simple hand shaking, vortex mixing or by using small, Teflon-coated magnetic paddles and a magnetic stirrer. Some sluggish reactions may require heating to moderate temperatures (60 to 100°C). Long, screw-capped culture tubes are convenient for reactions at reflux temperature, as the air-cooled upper portion can act as the condenser when small volumes of solvent are used.

Trimethylsilyl (TMS) derivatives, which are readily formed by a variety of proton-active functional groups such as those containing hydroxyl, mercapto, amino and carboxylic functions, are the most favoured in GC. The ease of formation of these derivatives decreases in the order: alcohols > phenols > carboxylic acids >

amines > amides. Primary alcohols and amines yield the derivatives more readily than the secondary ones, and so on. TMS derivatives of thiols, sulphenic acids and phosphonic acids have also been prepared. A variety of reagents ranging from the simplest trimethylchlorosilane (TMCS) to *N*-trimethylsilylimidazole (TMSIM) have been used; the latter is probably the most effective and rapid for polyalcohol derivatives such as carbohydrates but does not react with amino groups. TMCS, on the other hand, is a poor silylating agent unless a base (e.g. pyridine or diethylamine) is used; it is most often employed as a catalyst for other silylation reactions. It also causes complications with unprotected ketones by forming enol ethers which are thermally unstable. Other common silylating reagents (in the order of their reactivity are: *N,O*-bis(trimethylsilyl)-trifluoroacetamide (BSTFA) > *N,O*-bis(trimethylsilyl)acetamide (BSA) > *N*-methyl-*N*-trimethylsilyltrifluoroacetamide (MSTFA) > trimethylsilyldiethylamine (TMSDEA) > *N*-methyl-*N*-(trimethylsilyl)acetamide (MSTA) > hexamethyldisilazane (HMDS). BSA and BSTFA are most commonly used for *N*-trimethylsilylation reactions with BSTFA enjoying some preference. Most silylation reactions are instantaneous and occur at room temperature. The rate may be affected by steric factors, catalysis and the solvent. Polar solvents such as pyridine, dimethylformamide, acetonitrile or THF, favour the derivative formation and are generally preferred. However, the silylating agents themselves are frequently good solvents and additional solvent is normally unnecessary.

In addition to the TMS derivatives, a number of trialkyl-, alkyldimethyl-, haloalkyl-, and haloaryl-substituted silyl derivatives have been introduced to improve their hydrolytic and thermal stability and to enhance the sensitivity and selectivity of detection. Isopropyldimethylsilyl and *t*-butyldimethylsilyl derivatives, for example, are two to four orders of magnitude more stable to solvolysis than the TMS analogues. The response to electron-capture detection of halogenated compounds decreases in the order I>Br>Cl≫F, and similarly for the boiling points. The relatively poor electron affinity of the fluoro derivatives is more than compensated by the enhanced volatility of the fluoro compounds and both volatility and sensitivity to electron-capture detection are increased by multiple substitution. Trifluoroacetyl (TFA) derivatives using trifluoroacetic anhydride (TFAA), trifluoroacetylimidazole (TFAI) or *N*-methylbis(trifluoroacetamide) (MBTFA), and pentafluoropropyl (PFP) or heptafluorobutyl (HFB) derivatives using similar reagents are favoured for electron-capture detection. Particularly sensitive (to electron-capture detection), diagnostic (in mass spectrometry) and suitable for steroids are the pentafluorophenyldimethylsilyl (FLOPHEMESYL) derivatives. The reactivity of 'flophemesyl' agents in pyridine is in the order: flophemesylamine > flophemesyl chloride > flophemesyldiethylamine>flophemesyldisilazane>flophemesylimidazole. Trialkylsilyl derivatives containing heterocyclic nitrogen, phosphorus and sulphur and sensitive to thermo-ionic and flame photometric detection, have also been reported. Most of these reagents are commercially available and are usually supplied with the derivatization procedure suitable for direct introduction into the gas chromatograph.

Alcohols, thiols, amines, imines, hydroxylamines, amides, etc., are readily convertible to their acetates. The corresponding perfluoroalkylacyl derivatives (e.g. trifluoroacetates, pentafluoropropionates and heptafluorobutyrates) are stable volatile derivatives, detectable by both flame-ionization and electron-capture detectors. The appropriate anhydride, acid chloride or acylimidazole may be used for the preparation of the derivatives. Methyl esters for carboxylic acids are readily prepared by treatment with the alcohol in presence of an acid catalyst (e.g. boron trifluoride-etherate) or by treatment with diazomethane. Other esters, particularly those useful for electron-capture detection (e.g. perfluorinated alcohols containing two to four carbon atoms), are obtained from the methyl esters by refluxing with the appropriate alcohol in the presence of an acid catalyst. Pentafluorobenzyl bromide (for carboxylic acids) and pentafluorobenzoyl chloride (for alcohols) are also used for preparing the corresponding esters. Other commonly used pentafluorophenyl-containing reagents are pentafluorobenzaldehyde for primary amines, pentafluorophenyl chloroformate for tertiary amines, pentafluorophenylhydrazine for carbonyl compounds, pentafluorophenacyl (or phenoxyacyl) chloride for alcohols, phenols and amines, etc. The pentafluorophenyl derivatives are very easy to form; they are volatile with good GC peak shapes and are very sensitive to electron-capture detection.

Apart from the silylation reactions of most proton-active groups, esterification of carboxylic acids, acylation of alcohols, phenols and amines, hydrazone formation of carbonyl groups and derivatization of different functionalities for electrochemical detection, etc., there are a number of innovative and unique derivatization reactions (particularly in the case of certain polyfunctional compounds), but they may not be of general utility. A number of comprehensive compilations on derivatization reagents, procedures and their applications are available [14–17]. The most dramatic changes in chromatographic properties probably occur in the case of inorganic ions through complexation reaction and preparation of organometallic compounds (Section 9.2.2.2). Another unique method of derivatization, useful in the analysis of the otherwise non-volatile polymers, is pyrolysis, which merits a special discussion and which is elaborated below.

### 3.6.3 Pyrolysis

Certain types of compounds, particularly polymers, which cannot be rendered volatile by derivatization, may be degraded by pyrolysis to smaller fragments to facilitate GC analysis. Pyrolysis, also known as thermal decomposition, degradation or cracking, may be defined as transformation of a compound by heat. These transformations can be specific and predictable and, under favourable circumstances, analysis of the fragmentation products can lead to identification of the parent structure as in mass spectrometry. Predictability of the pyrolytic transformations understandably depends on the precision and reproducibility of the pyrolytic conditions; i.e. the temperature (usually around 700°C), duration of exposure of the sample to the temperature (5 to 10 seconds), and control or, preferably, prevention of secondary reactions. A number of devices have been introduced to achieve these goals and several of them are commercially available. The pyrolysis products may be collected cryogenically or in a solvent (as in the headspace sampling methods described earlier) and injected into a GC column. Modern pyrolysers, with both the pyrolysis temperature and the period of exposure electronically controlled, may be directly attached to the gas chromatograph in place of the conventional injection port.

Pyrolysis for GC is normally effected in either continuous or pulsed modes. The continuous mode, in which a few milligrams of the sample is passed through a continuously heated zone, is particularly suited for volatile samples. The apparatus for this purpose consists of a small tube furnace or microreactor, operated either on-line or off-line. The sample is placed in a microboat or dropped into the heated region; alternatively, it is taken in a quartz capsule from which the products are released by breaking it after the pyrolysis. The main disadvantage of the continuous pyrolysers is that the furnace has to be maintained at a much higher temperature than that required for the pyrolysis and, heat transfer being slow, the products remain in the pyrolyser for a while to complete the operation, thus increasing the probability of secondary reactions. A pulse or filament heater (Fig. 3.9a), which requires only a few

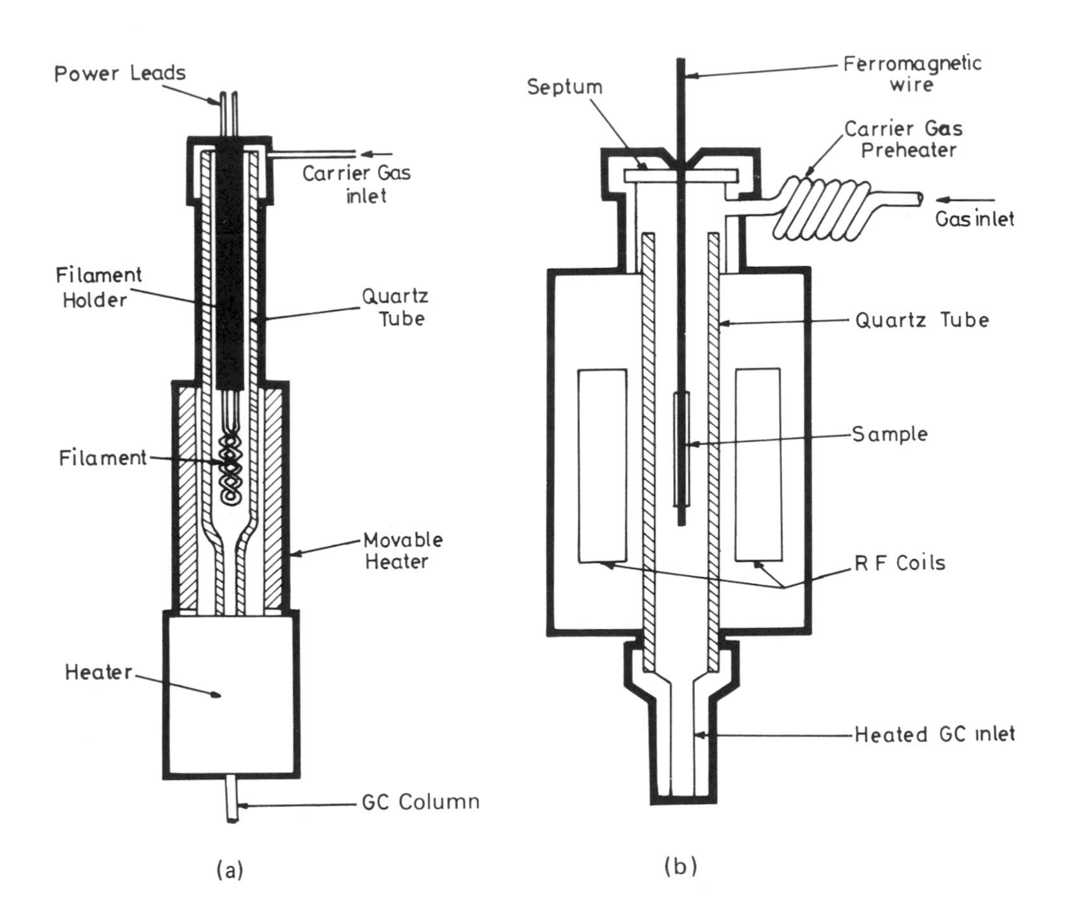


Fig. 3.9 — Schematic diagrams of (a) filament- or ribbon-type pyrolyser and (b) Curie-point pyrolyser. Reprinted with permission from Ref. 18; copyright (1980) American Chemical Society.

micrograms of the sample is more frequently used for polymeric samples. It consists of a coil (filament) or ribbon made of platinum or platinum–rhodium alloy, connected to a suitable power supply and a timer. The sample is coated on to the filament by simply dipping the filament into a solution of the sample in a volatile solvent. Insoluble samples are taken in a quartz boat around which the coil is wound. The temperature of the pyrolysis compartment, which is connected in series between the carrier-gas cylinder and the column and as close to the column as possible, can be raised very rapidly (1000°C in 20 msec). The temperature may also be raised in a programmed mode, say, about 300°C per minute. In the Curie-point pyrolyser (Fig. 3.9b), the sample is coated on a ferromagnetic filament and is heated inductively in a radiofrequency field to the temperature at which the alloy becomes paramagnetic (Curie point) and ceases to absorb the radiation. Since the Curie point of an alloy depends only on its composition, the pyrolysis temperature is highly reproducible.

An alternative to the above modes of pyrolysis is laser pyrolysis which is yet to come into wide use but which permits very controlled pyrolysis. Lasers can be used to direct very-high-energy radiation on a small area of the sample in the form of a series of short pulses of about 100 microseconds duration. It must be noted that the products of pyrolysis in all the three modes may be quite different. Even in a given type of pyrolyser, the composition of the products depends on many factors such as the pyrolysis temperature, the rate of heating, the sample size and homogeneity, the design of the pyrolyser–GC interface. Though many studies have been made on optimization of pyrolysis conditions, reproducibility of results from different laboratories has been poor; optimization of pyrolysers using reference standards may be useful.

By the very nature of the technique, pyrolysis GC is uniquely suited for studying the mechanisms of thermal degradation and structural analysis of compounds on the basis of the identity of the products of thermal degradation. In this sense, pyrolysis, where small structural differences may lead to strikingly different products, may sometimes be very informative. GC, pyrolysis and GC of selected peaks from the first GC, followed by mass spectrometric examination of the effluents, all in tandem, can be an extremely powerful combination. Even when identification of all the products of pyrolysis is not possible, the pyrogram (that is, the chromatogram of the pyrolysis products) can be a fingerprint for positive identification of the parent compound. Pyrolysis GC is used extensively in polymer characterization and analysis. The polymer may revert essentially to the monomers by a free radical mechanism and the identity and quantity of the monomers can reveal the nature of the polymer or the copolymer; the data can also be used for quantitation. The decomposition of the linear polymer of styrene and isoprene is a function of temperature; in other words, the ratio of the monomers formed reflects the temperature of pyrolysis. This concept of molecular thermometer has been used for standardization of pyrolysers. In addition to the above pyrolysis GC finds application in organic geochemistry, biochemical analysis, microbial taxonomy and trace analysis. Recent advances in pyrolysis methodology and commercially available instrumentation resulted in improved precision and reproducibility with even such diverse samples as powders, films, pellets, fibres, etc. This, coupled with high-resolution capillary GC and computerized data handling, provided solutions to many problems; microstructural information on synthetic and biological polymers and its application to forensic science may be given as examples [18–20].

## 3.7 SAMPLE SIZE

The size or the quantity of sample introduced into the chromatographic column has considerable influence on the resolution [21]. The maximum allowable sample size is one that does not decrease the efficiency of the column, as reflected by a decrease in the retention time and/or by an increase in the peak width. There is a range of sample sizes over which the column efficiency remains constant under a given set of other variables. The column may be said to be overloaded if the number of theoretical plates ($n$) is decreased by 10% of the maximum value. The maximum allowable sample size ($V_{max}$) is directly proportional to the corrected retention volume ($jV_R$), and inversely proportional to the square root of the number of theoretical plates; the proportionality constant has an approximate value of 0.02. If the mobile-phase compressibility is neglected ($j=1$), and if the number of theoretical plates in a packed column of about 2 mm i.d. is assumed to be a reasonable figure of 3000, the sample size can be calculated as 1/3000 of the retention volume. Thus, if the retention volume is 150 ml, the sample size should not exceed 150/3000 = 0.05 ml of the vapour or, assuming that the liquid expands 500-fold on vaporization, 0.05/500 ml or 0.1 $\mu$l of the liquid. In open-tubular columns of the normal internal diameter of 0.25 to 0.30 mm, the amount of stationary phase may be about 2–5% of that in packed columns and the phase ratio (β) may be over 20 times. Since $K=\beta k$, the capacity factor (and, therefore, the retention volume) is only a fraction of that in the packed column. Because of the high plate numbers (typically much over 100,000) and low retention volumes, the sample size in open-tubular columns is about 1/100 to 1/1000 of that tolerated by packed columns. The wide-bore capillary columns of 0.5 to 0.9 mm i.d. have sample capacities intermediate between the two types. Overloading or introduction of either a concentrated sample or a large volume of a dilute sample solution would often lead to decreased column efficiency and distorted peaks.

In quantitative analysis, it is also important that the sample size (more specifically, the concentration of the component of interest) does not exceed the linear dynamic range of the detector used. And, in addition to the column efficiency, if it is desired to take advantage of the sensitivity of the detector also, the sample size could be impracticably low. Thus, the common flame-ionization detector has a dynamic range of 1 pg to 1 $\mu$g, which means that even the 0.1 $\mu$l sample size (approximately 0.1 mg) is at least 10 times above the range and 100 million times more than what is needed. About 0.1 $\mu$l of a 0.01% solution is probably practical though even that is about 1000 times too large to exploit the full detector capability [2]. Thus, only the open-tubular columns are capable of realizing the full potential of GC, that is, the high resolving power, low sample size and the high sensitivity of the detectors.

## 3.8 INJECTION TECHNIQUES

It must be apparent from the foregoing discussion that the limiting factor in attaining the smallest needed sample size may be the difficulty in introducing such small quantities into the GC column; the problem is particularly formidable in the case of the open-tubular columns. While it may be impracticable to attain the smallest detectable sample size, it is important to keep the size to the minimum that can be introduced accurately and reproducibly. The samples are most often introduced by means of a hypodermic syringe; with the commercially available microsyringes, about 0.3 μl of a liquid can be injected with reasonable precision. The sample should be introduced into the column in the form of a gas or vapour so that it is uniformly distributed in the mobile phase before chromatography is initiated. However, a majority of samples for analysis are liquids, solids, or solutions thereof. The common practice is to dissolve such samples in a volatile solvent to facilitate its distribution between the two phases.

Gaseous samples are conveniently introduced into the column using either a sampling valve or a gas-tight syringe. Gas sampling valves generally consist of a rotating polymeric core, encased in a stainless steel body. The sample loop is filled with a known volume of gas in one position and, when the valve is rotated, the sample loop is swept by the carrier gas into the column. Reproducibility of injection volume with these valves is better than 0.5%. Syringe injection is more convenient and is less expensive but requires skill to achieve a high degree of precision; reproducibility may be around 1% at best. Several types of syringe and sampling valve, with injection volumes ranging from 0.001 to 50 ml, are available from the supply houses (e.g. Hamilton, U.S.A. and Rheodyne, U.S.A.).

Liquid samples of 0.05 to 10 μl can be injected using a microsyringe. Two types of syringe may be mentioned: the 'plunger-in-the-needle' type of syringe, which is very precise, and is recommended when very small samples are handled; and the 'plunger-in-the-barrel' type, which is more common for routine use albeit somewhat less accurate. In this type the injection volume is read on the barrel while the sample is displaced from the needle. Inaccuracies arise from different sources, depending on the technique used for transferring the sample. In the filled-needle technique, the sample does not enter the barrel; when the needle is inserted into a heated injection port, the sample evaporates and passes into the injection zone and movement of the plunger is not necessary. Alternatively, the sample is drawn into the barrel, leaving the needle empty. The sample is injected by depressing the plunger without allowing the needle to heat up. In these techniques, a major problem is sample discrimination due to uneven vaporization of the sample components; more volatile components quickly pass into the injection area leaving the part with the less volatile components in the needle. The problem may be partly overcome in the 'sample-in-the-barrel' approach by allowing the needle to heat up for about 3 to 5 seconds before depressing the plunger. Alternatively, a plug of solvent or air is drawn up by the plunger ahead of the sample so that the entire sample is pushed into the injection zone. Sample losses in the cold parts or crevices in the injection port may also be the source of inaccuracy in the injection volumes as read on the syringe barrel. However, with larger sample sizes, the proportion of the sample lost to the total sample injected is so small that it is hardly noticed in the normal packed-column GC. Also, it is a common practice to inject the sample as a solution in a volatile solvent which further reduces the losses due to the above factors.

Solid samples are usually introduced as solutions in a volatile solvent that is non-reactive with the sample components or the stationary phase and does not co-elute with the components of interest; more than one solvent may also be used, if necessary. Notched hypodermic needles or syringes for transfer of solid samples are also available. The various designs of hypodermic syringes for use in GC and the techniques of their handling may be well known, but precise sampling requires skill and observance of several precautions [22]. Alternatively, the samples may be encapsulated in glass capillaries and mechanically pushed into the heated injection block and crushed. It may be simpler to dissolve the sample in a volatile solvent, dip a filter paper, metal, glass or quartz strip into the solution and allow the solvent to evaporate, leaving the sample on the strip; the strip is then introduced into the flash heater and the vaporized sample swept into the column by the carrier gas. The advantages of this type of solid-sampling technique are that there is no interference from the solvent peaks, no contamination of the column owing to non-volatile components (these compounds remaining on the strips), and that dilute samples can be concentrated on the strips. A low-melting metal (e.g. Woods metal, m.p. 60.5°C) capsule, which melts in the heated injection port, may also be used. This technique is particularly useful when the presence of volatile compounds, which may be masked by the large solvent peak (or be lost during concentration), is suspected.

For repetitive or periodic injection of a large number of the same or different samples, autosamplers may be used. Apart from the feasibility of unattended operation, they are particularly advantageous in view of the consistently more precise results that can be obtained in comparison to the manual techniques. The single-syringe type is convenient for liquid samples, whereas the capsule samplers are more satisfactory for solid samples. Gas-selector valves and stream-selector valves (e.g. in process control) or headspace samplers, are generally used for gases. Automatic liquid samplers are quite popular and several versions of them are commercially available. They are mostly of the single-syringe type in which the syringe is repetitively filled and flushed, first with a suitable solvent and then with the sample solution, to eliminate sample memories (remnants of the previous sample) and to ensure precision in sampling. Electronic controls are used to inject samples, in predetermined volumes, frequencies and order, from a sample magazine that can accommodate up to 100 vials. A variable number of replicate sampling from predetermined vials is also possible. Sampling valves of the type generally employed in gas sampling are more convenient in process applications, where repetitive sampling from a process stream is needed; they could also be used in batch systems but for the sample carry-over problems.

For introduction of solids, emulsions or highly viscous liquids, aluminium or gold capsules and automatic filling and sealing devices may be used. The capsules from the holding magazine are mechanically drawn into the heated injection port and pierced with a metal needle, following which the contents of the capsule are flushed by the carrier gas. The main disadvantage of these injectors is that the sample vaporization may be incomplete and the exact amount of the sample entering the column may not be known. Automatic gas sampling valves are extremely useful in process monitoring, analysis of refinery gases, natural gas and ambient air and several other applications. Stream-selection valves are used to monitor multiple streams. The valves need a positive pressure to operate adequately, but if the pressure in the stream line is too high (500 psi), pressure reducers may be necessary. On the other hand, if the pressure is low or the sample size is not too large, manifolds or syringe systems are preferable. Automatic headspace samplers (discussed earlier) may be considered as a variation in gas-sampling systems.

Whatever the device used for measuring and transferring the sample, its introduction into the GC column is normally achieved *via* an injection port. The role of the injection port is to facilitate the plug injection (with minimum initial band width) by flash vaporization of the sample or otherwise. Its design should not cause extensive mixing-up of the sample vapour with the carrier gas, and it should be free of any deadspace or cold spots on which the sample may condense. The injection ports are different for packed columns and open-tubular columns. Wide-bore open-tubular columns, with an internal diameter of 0.5 to 0.9 mm are intermediate between the two types in both sample capacity and resolving power and can use either of the injection ports, depending on the carrier-gas flow-rates and the sample size. Use of packed-column injectors for the wide-bore open-tubular columns may require slight modification by way of appropriate glass inserts for both the injector and the detector. All the injection ports for the narrow-bore capillary columns can be used without modification and the injection process is much less critical; on-column injection into wide-bore capillary columns can be performed with the normal, rigid needles (0.47 mm o.d.).

## 3.9 PACKED-COLUMN INJECTORS

The typical injection port (Fig. 3.10a) for packed columns consists of a vaporizer tube (5 to 10 cm long) of minimum diameter (0.5 to 0.75 mm), usually glass-lined, removable, cleanable and replaceable. The glass liner prevents the sample from coming into contact with the heated metal surface. Encasing the vaporizer tube is another cylindrical tube of slightly larger diameter into which the carrier gas enters through a narrow side arm. At the outer end of the port is a self-sealing silicone septum, retained by a nut with an aperture which accepts the hypodermic syringe needle. The sample is introduced into the middle of the vaporizer tube and is swept by the heated carrier gas onto the column. The heated block maintains the temperature of the port high enough for the whole sample to be flash vaporized.

Heating of the injection port to a temperature higher than the boiling point of the highest boiling component of the sample effects flash vaporization of the sample before it is conveyed to the column. As this figure is generally not known, the injection-port temperature is generally chosen to be about 50 degrees higher than the maximum column temperature used, subject to the thermal stability of the compounds being analysed. Since band broadening is inevitable during the passage of the sample through the column and since narrow bands are desirable for better resolution, it is essential that the sample band at the head of the column be as narrow as possible. To achieve this, the sample is introduced as a plug, all at once and in the shortest possible time, not exceeding 1/10 of the duration of the narrowest (or the first) peak to be eluted. Ideally, the sample should be uniformly distributed through the first theoretical plate.

As already noted, a serious problem with the septum injectors that are commonly in use is the possibility of septum bleed; that is, monomer- and short-chain-polymer contaminants in the septum getting volatilized and eventually appearing as chromatographic signals or getting adsorbed on the stationary phase. Though specially made low-bleed septa are now available, alternative approaches may also be necessary. Adsorption of the organic solvents, particularly the chlorinated solvents, is yet another problem. The possibility of leakage of gases through the septum also cannot be ruled out. Septumless injectors which completely eliminate these problems are available but suffer from certain disadvantages such as large dead volume and

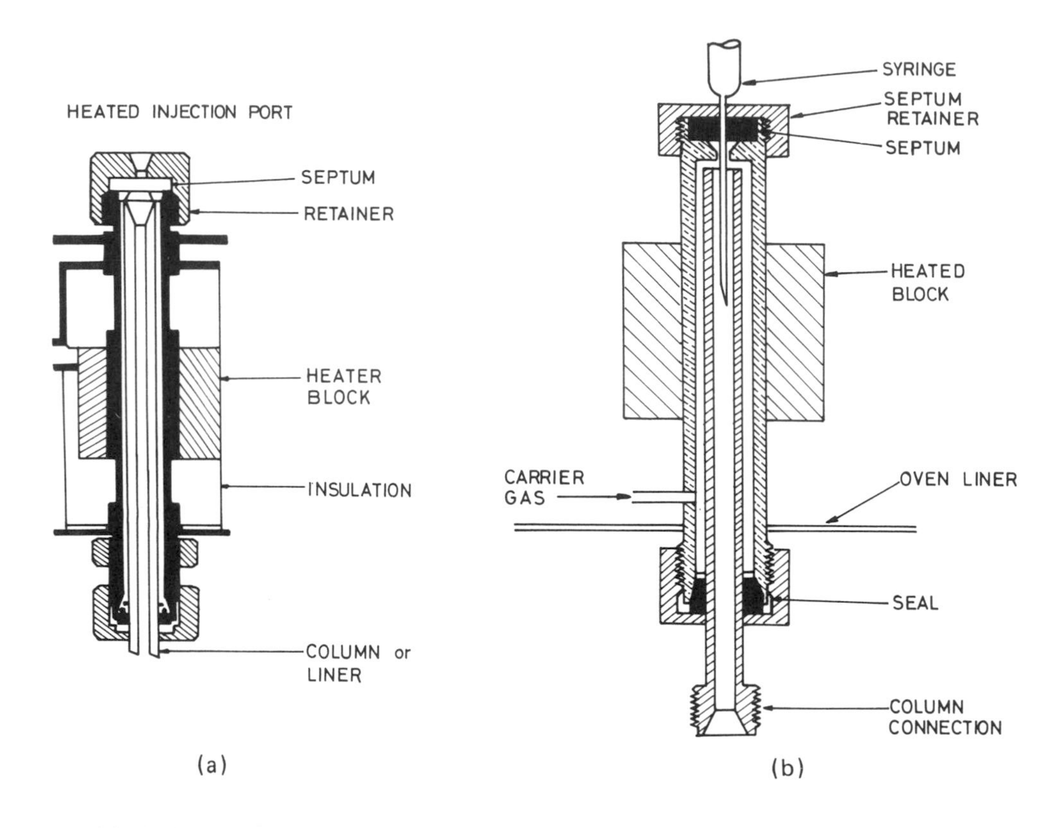


Fig. 3.10 — Schematic diagrams of (a) a typical heated injection port, and (b) on-column inlet for packed columns. Reproduced with permission from Ref. 4; copyright (1985) John Wiley & Sons, Inc.

are rather cumbersome. Some of the modern designs are useful in solid sampling and can be automated. Septum-cooled injectors and septum-bypass injectors have been developed but are not in common use. A more practical injector modification is the purged-septum injector in which the carrier-gas stream is split so that one portion enters the column and the other continuously purges the septum area and exits from the system. Thus, the gas that had come into contact with the septum is prevented from entering the column.

Chemical changes in the sample components, which may occur when the sample is injected into a heated port, are a source of error in both qualitative and quantitative analysis. The on-column injection technique may be employed if the sample is thermally too labile to withstand the high temperature of the flash vaporizer. On-column injection is also the method of choice when large volumes of dilute samples are to be injected, or when the samples contain solutes of wide boiling-point ranges and where quantitative accuracy using the flash-vaporization technique and peak-area measurement is inadequate. For on-column injection, the vaporizer tube may be replaced by a portion of the glass column so that the tip is within about 3 mm from the septum (Fig. 3.10b). In this case, even when the column temperature is high, the heat of vaporization of the sample causes it to cool and the sample does not fully vaporize. Therefore, the requirement of sample introduction as a plug is not achieved; in such cases, the gas flow may be interrupted for a while to allow sufficient time for sample vaporization. Also, when the column is inserted into the heating block, the temperature of the injection port should be below the temperature limit of the liquid phase to prevent its decomposition or stripping. In general, on-column injection is more precise for samples containing solutes with narrow boiling point differences, whereas the flash vaporization technique is preferable in other cases and for quantitation by peak height ratios. The best compromise probably is to use the on-column injection with additional heat from the vaporizer.

## 3.10 INJECTORS FOR OPEN-TUBULAR COLUMNS

Open-tubular columns are characterized by their very low sample capacities (about 0.1 to 1% of that of packed columns) and small internal diameter (about 0.25 mm), often smaller than the diameter of the needle of a microsyringe. Sample introduction into these columns therefore requires special devices, the design and performance of which is vital to the chromatography in terms of the peak shape, resolution and quantitation. Three recent monographs [23–25], dedicated to sample introduction in open-tubular GC, are indicative of the importance of the subject. A majority of injectors for open-tubular columns fall into one of the four types, i.e. inlet splitters, splitless or direct injectors, cold on-column injectors and programmed-temperature vaporizers. The headspace and the pyrolysis samplers discussed earlier are readily adaptable to open-tubular GC and may be considered for appropriate applications.

The aim of any sample introduction method is to introduce into the column a certain amount of the analyte as a narrow band whose composition reflects that of the original sample. Except in the case of the split injection, where only a tiny fraction of the sample enters the column, any practical volume of the sample would occupy a long segment of the narrow column unless some type of band concentration near the column head is effected. In other words, the conditions should be so adjusted as to promote strong retention of the sample components in the beginning of chromatography. This is generally achieved by temporarily increasing the phase ratio (e.g., by solvent effects, see below) and by lowering the temperature.

A number of factors determine the choice of the appropriate injection mode in open-tubular columns. The scope and limitations of the various options are discussed below. In general, split injection is the method of choice when the solutes have reasonable volatility and thermal stability and their concentrations fall within the range of 0.001 to 10% of the sample. The injection temperature, gas flow, split ratio, etc., are adjusted according to the volatilities of the sample components. When the analyte is presented as a dilute solution and the components are relatively high boiling, splitless sampling may be preferred. On-column injection, being a non-vaporizing technique, is especially suitable for thermolabile solutes which may decompose or rearrange at higher temperatures and for non-volatile analytes which cannot be volatilized at the normal operating temperatures of the heated injectors. With a sampling capacity of as much as 500 $\mu$l and no sample discrimination, it is ideal for trace and quantitative analysis.

### 3.10.1 Splitter injectors

As the name indicates, these injectors split the flash vaporized sample into two parts, allowing only the smaller portion of it into the column and venting the rest. Other design requirements of splitter injectors include minimization of extra-column band broadening of the injector, chemical inertness of the materials of construction and a wide operating range of temperature and split ratios. The split ratio in most commercial injectors may vary from 1:10 to 1:1000, but the more commonly used range is between 1:20 and 1:100. Successful split injection should ensure that the fraction that enters the column is representative of the total sample and that the split ratio is reproducible. The device (Fig. 3.11) typically consists of a long and narrow

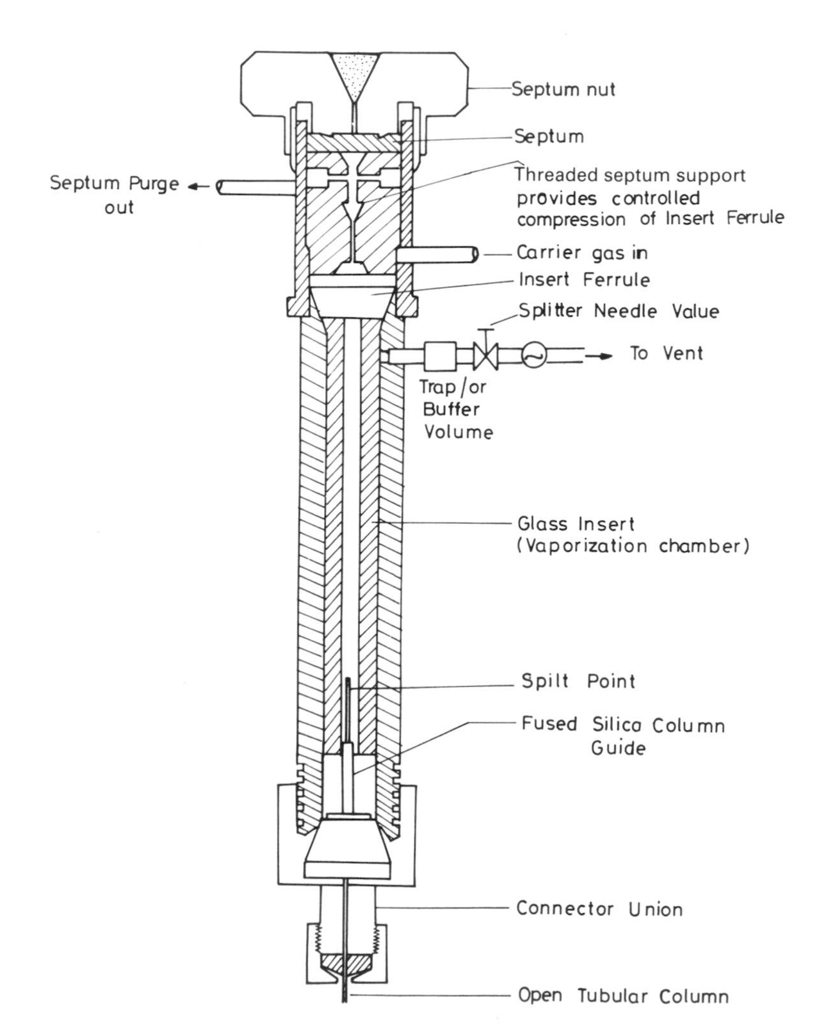


Fig. 3.11 — Schematic diagram of a typical split/splitless injector. Reproduced with permission from Ref. 3; copyright (1984) John Wiley & Sons, Inc.

vaporization chamber, a flow-mixer (e.g., a glass-wool packing) and a splitter valve. The buffer volume or filter trap between the vaporizer and the needle valve helps to retain the sample components so that viscosity changes due to the sample do not change the split ratio. The carrier gas, which enters the injector near the septum, is divided into two streams, one to purge the septum and the other, larger stream, to flow into the vaporizer and mix with the sample. The column is connected to the other end of the injector and is located within the heated zone to prevent recondensation of the sample components. The split ratio is determined by the ratio of the flow through the open-tubular column and the vent, and the flow at the vent is controlled by the interchangeable needle valve. Rapid injection, high vaporization temperature and high carrier-gas flow-rate through the vaporizer help minimize the residence time of the sample in the injector and, thus, the band broadening.

The split injection has been one of the most widely used techniques almost from the beginning of open-column GC. The main advantages are that the initial zone spreading is minimal at high split ratios (which can be easily adjusted for the sample volume and concentration used), and its compatibility with both isothermal and temperature-programmed operation. It is particularly useful for mixtures containing uniformly distributed components in high concentration, for method development and optimization and for routine analysis. Its operation is very simple, being similar to that of a packed column, once all the parameters are controlled. Proper selection of the injector temperature, speed and the split ratio,which depend on the sample, should ensure the linearity of sampling. The split ratio obviously depends on the capacity of the column and the sensitivity of the detector. A high injection temperature is recommended when the split ratio, the samplc volume and the carrier-gas flow-rate are large; the maximum temperature is also required when the volatility of the sample and the heat capacity of the carrier gas are low. A loosely packed glass-wool plug in the vaporization chamber increases the heat capacity as well as the mixing efficiency and improves the linearity and reproducibility of the injection. A major drawback of split sampling is the discrimination against less volatile components, making quantitative analysis somewhat unreliable when the sample contains compounds with widely different boiling points. However, if the composition and the split ratio are well-controlled, quantitative analysis is feasible using the internal standard method [4]. Use of an autosampler, which allows reproducible injections (precision of 1%), greatly improves the quantitative accuracy of the split sampling.

### 3.10.2 Splitless injectors

Often the same injector design is useful for both split and splitless injection, the only difference being the addition, in the latter case, of a solenoid valve downstream from the vent. In the splitless mode, the sample is injected along with a large amount (about 5 $\mu$l) of a more volatile solvent, with the splitter closed. The solvent boiling point is normally about 25°C below that of the first eluting component of the sample and thc column temperature is 15 to 30°C below the boiling point of the solvent. After 1 minute, the vent is opened and the injector is quickly purged of thc remnants of the sample by a temporary, large flow of the purge (carrier) gas, *via* the solenoid valve. If the sample is a gas, first a volatile solvent is injected, followed, after 15 to 30 sec, by about 1 ml of the gas. Interestingly, though a large volume is injected, a careful analysis of the events would show that splitless injection produces narrower bands due to the so-called solvent effect [23]. In the first stage of separation, the solvent shifts ahead of the sample and forms a condensed liquid layer (about 20 to 50 times thicker than the stationary phase), the concentration of the solvent increasing rapidly in the direction of the flow. As the sample components proceed, the front of the band is exposed to a denser stationary phase than the rear and hence is more strongly retained. The result, under properly chosen conditions, is a sharp reduction in the band width. Complete transfer of the sample in the splitless mode ensures quantitative and representative sample introduction. Also, since the sample is transferred slowly (40–60 sec) into the column, rapid vaporization is not required and the method may be used for thermally unstable compounds. It is also the method of choice for extremely sensitive determinations (trace analysis), for separation at the highest resolution of complex mixtures such as food flavours and for large volumes of dilute gases such as environmental air, etc.

For effective use of the solvent effect, larger volumes of the sample ($>1$ $\mu$l) in a medium volatile solvent (e.g. a C-7 hydrocarbon) is preferable; the sample volume, however, is limited by the capacity of the column. In general, a sample volume of 1 to 2 $\mu$l with a sampling rate of 0.1 $\mu$l/sec and a purge delay time of longer than 40 sec is recommended. For best results, a high dilution (1:100,000) and an initial column temperature of about 20 degrees below the boiling point of the solvent should be used. Optimal performance is thus obtainable only in temperature-programmed operation since focusing of the sample band requires a low initial column temperature and a much higher temperature is needed for completion of chromatography in a reasonable time. Among the drawbacks of the method are the facts that (a) quantitation is dependent on the technique employed and precise sample volumes, sampling time and optimization is necessary for reproducible results, (b) the solvent choice is an important factor in view of the requirement that the column temperature be below its boiling point for operation of the solvent effect, (c) volatile compounds, eluting before the solvent, are not effectively separated owing to band-shape distortions and the displacement chromatographic effect, unless there is a large difference in the $k$ values of the solutes and the solvent, and (d) high injection-port temperatures are still necessary to ensure complete transfer of high boiling components.

Flash vaporization and splitless injection is also possible in open-tubular GC using a miniaturized version of the packed-column injector (Fig. 3.10a). The 0.7 mm i.d. vaporization chamber accepts a standard 5 $\mu$l syringe of the 'plunger-in-the-needle' type with a needle diameter of 0.66 mm. The small-volume vaporization chamber generates a positive pressure gradient across the injector to permit all the injected sample to enter the column. Rapid vaporization of the sample is not necessary and the injector temperature is relatively low. It may be used in both isothermal and temperature-programmed operation. The usable sample range is an order of magnitude wider than in the splitless injection. About 1 $\mu$l samples can be easily injected in this system but, for isothermal operation, less than 0.5 $\mu$l sample size is recommended to minimize the initial bandwidth. Where larger samples (up to 10 $\mu$l) need to be injected (as in trace analysis), band spreading due to the large sample volume may be minimized by use of either a cold trap, temperature programming or the solvent effect [3].

### 3.10.3 Cold on-column injectors

As the name signifies, the cold on-column injector permits sample introduction directly on to the column at low temperature. Obviously, complete transfer of all the sample components, with no discrimination or loss is achieved. Excellent quantitative precision and accuracy for samples of wide volatility ranges and for those which are too thermally labile for vaporizing injectors are the obvious advantages of this technique [26]. The injector design is also relatively simple requiring, essentially, a syringe guide and a stop valve (Fig. 3.12). A cool-air circulator at the bottom of the

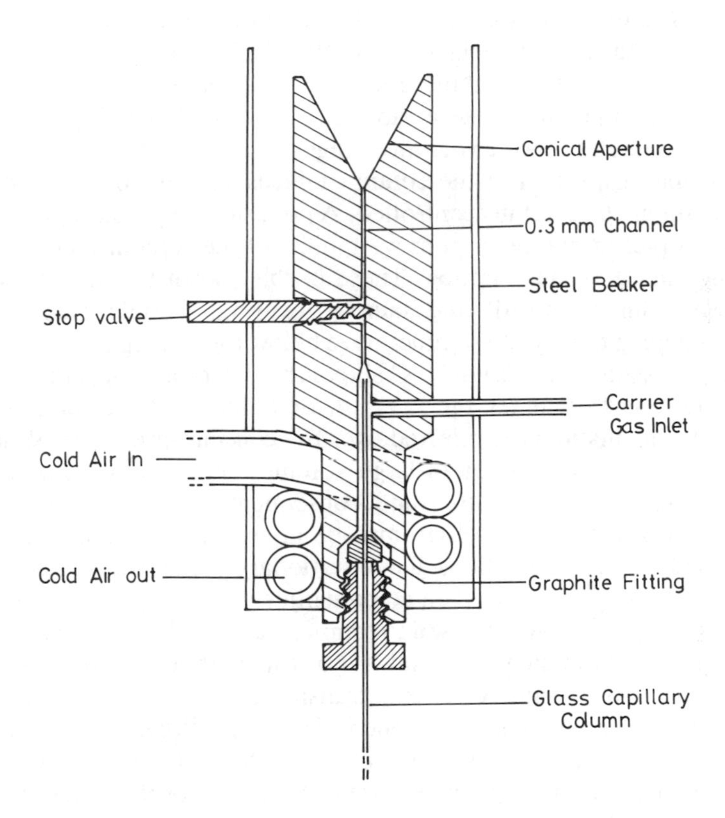


Fig. 3.12 — Cross-sectional view of an on-column injector. Reproduced from Ref. 26 by courtesy of Elsevier Scientific Publishing Co.

injector prevents loss of high volatiles during injection and allows sample injection in the liquid state so that there is no sample discrimination. The diameter of the syringe channel should match the outer diameter of the needle to minimize the pressure drop and carrier-gas leakage when the valve is opened for syringe insertion. Since the syringe is required to enter the capillary column, the normal syringes with 0.47 mm o.d. needles cannot be used. Syringes with stainless needles of 0.23 mm o.d., 0.1 mm i.d. and 85 mm length have been used for the common 0.32 mm open-tubular columns. Capillary columns of 0.2 mm i.d. require fused silica needles. Quantitative precision is satisfactory with sample volumes of 0.3 μl for syringes with the stainless steel needles and 0.2 μl for those with a fused silica needle. With large injection volumes of over 1 μl, peak splitting of the less volatile components is observed owing to droplet formation, and improper distribution of the sample in the initial portion of the column.

Introduction of the sample using the on-column technique is followed by raising the temperature of the column oven to start the chromatographic process. No auxiliary heating of the injection port is generally necessary since the column (oven) temperature is normally chosen to suit the boiling point of the solvent. However, in order to minimize the band spreading and peak tailing, independently heated on-column injectors have been designed; see Ref. 3. The temperature can be programmed from subambient to about 350°C. A low heating rate may be used for thermally labile compounds while a rapid rise (up to 180°C/min) is required for vaporization of samples of wide volatility ranges.

Initial sample dispersion at the column inlet is very important in on-column injection and should be carefully controlled. A poor sample profile at the inlet owing to flooding, droplet formation or poor wettability of the column wall would lead to peak splitting and poor quantitation. These problems can be minimized by using small sample volumes (<1 μl) and sample focusing. Sample focusing may be achieved by keeping the injector temperature below the boiling point of the solvent and the column temperature 10 to 20°C above that, to rapidly separate the solvent from the solute as they enter the column, and thus prevent column flooding. Constant-flow pneumatics are preferred over the constant-pressure systems for on-column sample focusing injection of large volumes because, in the latter, solvent vaporization causes pressure build-up and, consequently, flow reversal. Using a slow injection rate (0.1 μl/sec), up to 10 μl may be injected; if immobilized, stationary phases are used, no peak distortions due to stationary phase stripping are observed.

The main advantage of the cold on-column injectors is that, unlike the vaporizing injectors, the sample is never exposed to temperatures higher than that attained by the column during chromatography and is thus the method of choice for thermally unstable solutes. Also, since complete transfer of the sample is achieved, no discrimination among the sample components is noticed. Provided that the solvent is properly selected (that is, it is not too volatile at the column inlet temperature), highly precise quantitation is possible both in terms of absolute and relative peak areas. On the other hand, since vaporization occurs within the column (that is, in a region of relatively low volume), the sudden expansion of sample volume may cause selective sample losses owing to back ejection. Samples volumes larger than 1 μl will result in a long liquid inlet plug and irregular band shapes at the beginning of the separation process. Flooding of the column often results in peak spreading and peak splitting. Sample introduction into narrow-bore columns requires special devices that cannot be easily and routinely automated.

On-column injection using a length of uncoated capillary column (10–20 m) between the injection port and the normal, coated column as a *retention gap* permits injection of up to 100 μl of sample without peak distortion [27]. Injection of such large volumes is particularly convenient in trace analysis and requires much less sample preparation since repeated reconcentration steps can be avoided. It is well known that the loss of volatile sample components during such reconcentrations could be considerable and the retention-gap technique is suitable for both volatile and non-volatile samples. The effect of the retention gap, essentially, is to reconcentrate the sample on-column (at the head of the coated column), so that the volatile solutes are not lost and the low-volatile sample bands remain narrow.

On injection of a large volume of sample into the column, it forms several metres of a layer on the column wall; the length of the sample zone is roughly 25 cm per μl of the sample. If there is a coated stationary phase in this area, the sample components are unequally retained and a sharp initial band cannot be formed. On the other hand, the minimal retention by the uncoated column wall allows the sample components to be driven to the end of the flooded zone as the solvent evaporates. By appropriate choice of the solvent, the sample components are retained by the liquid and are released only at the end of the solvent evaporation. It is advantageous if the solvent boiling point is not less than about 30 degrees above that of the solute and the solvent polarity is such that it effectively retains the solute. It is obvious that the sharpness of the sample band depends on the relative retention powers of the coated and uncoated columns.

A prerequisite for the success of the sample reconcentration in the retention gap techniques is the wettability of the column wall by the solvent so that a uniform sample film is formed. The trimethylsilyl deactivation permits wetting by non-polar aliphatic hydrocarbon solvents; phenyldimethylsilyl-deactivated surfaces, on the other hand, are wetted by almost all solvents except water. It is also important, for the reasons noted above, that the length of the uncoated column should be long enough to prevent the sample flow into the separation column. The length of the sample plug depends not only on the sample volume but also on the rate of flow, the pressure drop across the sample plug, the internal diameter (more precisely, the internal surface area) of the uncoated column and the temperature relative to the boiling point of the solvent. Up to 150 μl of a sample could be injected on a 0.3 mm column of 20 m length at an inlet pressure of 1.5 atm (flow-rate, 7 ml/min) and column temperature about 20 degrees above the boiling point of the solvent. The allowable sample volume is only 100 μl near the boiling point and 70 μl if the temperature is 20 degrees lower.

In practice, it is convenient to couple a 15–20 m glass or fused silica uncoated column of 0.3–0.35 mm i.d. to a regular open-tubular column of 10–60 m and 0.3 mm i.d. *via* a PTFE shrinkable tubing or a butt connector. Any of the common stationary phases may be used provided the film thickness is at least 0.4 μm. The mobile phase of choice is hydrogen with a head pressure of 0.8 atm if the column length is 10 m and 2.5 atm if it is longer. With a 10 m retention gap, up to 50 μl of a sample (30 μl if the solvent is methanol) may be injected at 0.8 atm inlet pressure and at a temperature 5°C below the boiling point. Up to 80 μl may be injected if the temperature is raised to about 10°C above the boiling point and at a head pressure of 1.5 to 2.5 atm. Larger volumes can be injected with longer retention columns. The injection speed may be reduced from 20 μl/sec to 5 μl/sec during injection and, after the injection, the oven temperature may be raised by 10°C.

### 3.10.4 Programmed-temperature vaporizers

A new type of nearly universal injector that is gaining popularity is the programmed-temperature vaporizer, several versions of which have been described in the literature [24]. In principle, the technique involves injection of the sample into the cold chamber using an ordinary microsyringe and, after withdrawal of the syringe, heating the chamber ballistically to the vaporization temperature. The most interesting feature of this system is that it can be operated in either hot or cold, split or splitless, total sample introduction or solvent elimination modes. It can also be adapted to sample introduction in multiple parallel capillary columns and for pre-column sample concentration. Automated injection is possible using an ordinary liquid sampler and regular syringes.

For cold splitless injection, the sample is dissolved in a suitable low-boiling solvent and injected into the vaporizer kept at low temperature (i.e. the boiling point of the solvent), with the split valve closed. About 2–3 seconds after the injection, the injector block is heated rapidly to the temperature required for sample vaporization. After allowing sufficient time for the sample vapours to enter the column (30–90 sec), the split valve is opened to purge the injector of any remnants; the column temperature programme is simultaneously started. As soon as the purge valve is opened, the solvent peak drops to the baseline, allowing detection of volatile substances. 0.2 to 2 $\mu$l of the sample may be injected without adversely affecting the performance of the system. High molecular weight compounds are eluted normally, subject to column performance. In an automated mode, retention time reproducibility is claimed to be within a standard deviation range of 0.1 to 0.5%. Quantitative accuracy is comparable to the cold on-column injection described above.

When dilute samples are to be injected, the injector may be operated in the solvent-elimination mode. For this purpose, the split flow is suitably adjusted (to vent most of the solvent) prior to the injection and the valve is kept open during the injection with the injector temperature suitably low. After a few seconds, during which time the solvent leaves the system through the vent, the valve is closed and the injector block heated. After allowing sufficient time for sample vaporization and transfer to the column, the valve is again opened to purge the injector. This technique permits repetitive injection of very dilute samples without the danger of flooding the column or the detector. The possibility of highly volatile components being lost in the process can be minimized by keeping the injector initially at a low temperature, say 0°C, using a cryostat.

When required, split sampling is accomplished either by the usual way of injecting into the hot injector, or into the cold injector with rapid heating after removal of the syringe; either way, the split valve is kept open throughout. The limitations of flash vaporization and split injection, particularly in relation to quantitation, have been discussed earlier. Nevertheless, it remains one of the most convenient methods and the programmed-temperature technique eliminates most of the drawbacks. Thus, injection into the cold vaporizer eliminates the problems accompanying the insertion of a sample-filled syringe into the high-temperature zone. In fact, it is possible to introduce reproducibly small quantities of sample exactly into the same region of the injector, and the amount of the sample entering the column is proportional to the split ratio. Sudden flow-rate variations or split-ratio changes are not observed; neither is fog, aerosol, droplet formation, etc. There is also much less molecular-weight related sample discrimination. It has been demonstrated that a wide range of solvents, split ratios, sample concentrations, sample quantities and initial column temperatures could be used.

The programmed-temperature vaporizers score in yet another area where conventional injectors are not usable; that is, the microbore columns with less than 0.1 mm i.d., which are too narrow for cold on-column injection. While no instrumental modification is necessary, the programmed-temperature injection into microbore columns would require adaptation to the very low carrier-gas flow-rate; the transfer time should be more than a minute (80–90 sec) and a liner of much smaller diameter should be used. Since many of the problems associated with sample introduction are overcome by this technique, it appears very promising for ultra-trace analysis and for analysis of solutes which are very sensitive to heat.

## REFERENCES

[1] H. M. McNair and E. J. Bonelli, *Basic Gas Chromatography*, Varian, Walnut Creek, California, 1969.
[2] J. A. Perry, *Introduction to Analytical Gas Chromatography*, Marcel Dekker, New York, 1981.
[3] M. L. Lee, F. J. Yang, and K. D. Bartle, *Open Tubular Column Gas Chromatography*, Wiley, New York, 1984.
[4] R. L. Grob, *Modern Practices of Gas Chromatography*, 2nd edn., Wiley, New York, 1985.
[5] W. Bertsch, in *Recent Advances inCapillary Gas Chromatography*, W. Bertsch, W. G. Jennings, and R. E. Kaiser (eds.), Vol. 1, Huethig, Heidelberg, 1981.
[6] K. Grob and G. Grob, *J. High Resolut. Chromatogr. Chromatogr. Commun. (JHRCCC)*, **2**, 109 (1979).
[7] W. E. Harris and H. W. Habgood, *Programmed Temperature Gas Chromatography*, Wiley, New York, 1966.
[8] S. L. Morgan and S. N. Deming, *J. Chromatogr.*, **112**, 267 (1975).
[9] E. V. Dose, *Anal. Chem.*, **59**, 2420 (1987).
[10] A. J. Nunez, L. F. Gonzalez, and J. Janak, *J. Chromatogr.*, **300**, 127 (1984).
[11] B. V. Ioffe and A. G. Vittenberg, *Head-Space Analysis and Related Methods in Gas Chromatography*, Wiley, New York, 1984.
[12] B. Kolb (ed.), *Applied Headspace Gas Chromatography*, Heyden, London, 1980.
[13] J. F. Lawrence, *J. Chromatogr. Sci.*, **17**, 113 (1979).
[14] K. Blau and G. King (eds.), *Handbook of Derivatives for Chromatography*, Heyden, London, 1977.
[15] D. R. Knapp, *Handbook of Analytical Derivatization Reactions*, Wiley, New York, 1979.
[16] J. Drozd, *Chemical Derivatives in Gas Chromatography*, Elsevier, Amsterdam, 1981.
[17] R. W. Frei and J. F. Lawrence (eds.), *Chemical Derivatization in Analytical Chemistry*, Plenum, New York, 1981.
[18] C. J. Wolf, M. A. Grayson, and D. L. Fanter, *Anal. Chem.*, **52**, 348A (1980).
[19] W. J. Irwin, *Analytical Pyrolysis: A Comprehensive Guide*, Marcel Dekker, New York, 1982.
[20] S. A. Liebman and E. J. Levy (eds.), *Pyrolysis and GC in Polymer Analysis*, Marcel Dekker, New York, 1985.
[21] J. C. Sternberg, *Adv. Chromatogr.*, **2**, 205 (1966).
[22] C. H. Hamilton, in *Instrumentation in Gas Chromatography*, J. Krugers (ed.), Centrex, Eindhoven, 1968.
[23] K. Grob, *Classical Split and Splitless Injection in Capillary Gas Chromatography*, Huethig, Heidelberg, 1986.
[24] P. Sandra (ed.), *Sample Introduction in Capillary Gas Chromatography*, Vol. 1, Huethig, Heidelberg, 1986.
[25] K. Grob, *On-column Injection in Capillary Gas Chromatography. Basic Techniques, Retention Gaps, Solvent Effects,* Heuthig, Heidelberg, 1987.
[26] K. Grob and K. Grob, Jr., *J. Chromatogr.*, **151**, 311 (1978).
[27] K. Grob, Jr., G. Karrer and M.-L. Riekkola, *J. Chromatogr.*, **334**, 129 (1985).

# 4

# Gas chromatography II: separation systems

## 4.1 MOBILE PHASES

### 4.1.1 Inert gases

The mobile phase in GC is often referred to as the carrier gas, implying that its role is essentially to transport the sample components from the injection port through the column to the detector. Unlike in LC, the role of the mobile phase in achieving selectivity is very limited. However, as discussed in Chapter 2, the mobile-phase diffusivity and linear velocity through the column contribute significantly to the resolution. In addition, the sensitivity and performance of several detectors depend on the nature and purity of the mobile phase. Impurities in the mobile phase can also play havoc with the performance and life of the column; thus, in gas–solid chromatography, the impurities may occupy active sites on the solid surface and deactivate them. Also, as most organic compounds can react with oxygen, particularly on the active surface of the stationary phase, even traces of oxygen should be removed. Purification of gases for GC, therefore, received much attention in the early literature, but high-purity gases for GC have been commercially available for quite some time. In-line traps (gas purifiers)to remove particulate matter, oxygen, and moisture are also available from several supply houses. The main considerations in the choice of the mobile phase are thus determined essentially by the inertness of the gas, the cost, safety, and compatibility of the detector. The gases that are most often used in GC are hydrogen, helium, nitrogen and argon.

Among the different gases that may be used for GC, hydrogen and helium may be preferred for their high diffusivity and sensitivity in thermal conductivity detection; the heavier gases have poor thermal conductivity difference, relative to organic solutes, and are not well-suited for the conductivity detectors. The main drawback of using hydrogen as the mobile phase is the potential (but rarely experienced) explosion hazard when it comes into contact with air. Also, it adversely affects the stability and sensitivity of the flame detectors which use hydrogen to support the flame. Both the problems are, however, minimized in open-tubular GC in light of the very small flow-rate used and the possibility of using an auxiliary make-up gas other than hydrogen. Hydrogen can also react with a number of functional groups and the possibility is enhanced by the presence of any active (metallic) sites in the injection port, column or the detector. Further, the life of certain detectors which use thermistors may be decreased by hydrogen which reacts with the metal oxides present in them.

Both nitrogen and argon are completely inert and, because of their high density, have low diffusivity. In the van Deemter equation (Chapter 2), the mobile-phase diffusivity appears in the numerator in the formula for the $B$ term and, hence, the lower the gas diffusivity the better. However, at the high (practical) flow-rates, the contribution of the $B$ term to the plate height is negligible, and in the $C$ term, which now takes over, the mobile-phase diffusivity appears in the denominator and low mobile-phase diffusivity may be detrimental to high resolution. Thus, while low plate height (or high efficiency) may be attained with nitrogen or argon at low flow-rates, the slope of the linear velocity *vs* plate height rises very sharply and high efficiency can be maintained only over a very narrow range of flow-rates (Fig. 3.4). On the other hand, the minimum in the plots for hydrogen and helium, though not as low as nitrogen, is broader and does not rise as sharply at higher flow-rates. It may be noted that even the minimum of the van Deemter plot is also at a higher linear velocity (i.e. the optimum flow-rate is higher) for these gases. Therefore, the high-density gases such as nitrogen, argon and carbon dioxide may be chosen for high-efficiency GC at lower flow-rates and the low-density, high-diffusivity gases (hydrogen or helium), for faster analysis at higher flow-rates. As noted earlier, under a given set of conditions and resolution, use of hydrogen could shorten the analysis time to two-fifths of that required when nitrogen was used as the carrier-gas [1] (Fig. 3.3). It is common knowledge that the fast elution leads to narrower and taller peaks and, thus, the detectability (sensitivity) is also improved when hydrogen or helium is used as the mobile phase. The optimum flow-rate for a given column and carrier gas has to be determined experimentally. It is common practice to use higher than optimum flow-rate in order to hasten the analysis if the compounds are well separated.

### 4.1.2 Steam

The use of inert gases as mobile phases stems from the general perception that, in GC, interaction between the mobile phase and the solute (other than solubility) is unnecessary and undesirable. The limited role assigned to the mobile phase in GC requires that the solute–stationary phase interaction be weak enough for the solutes to be carried by the mobile phase. Application of gas–solid chromatography is thus largely limited to the analysis of permanent gases; polar solutes are too strongly retained on active solid surfaces and, even when the solids are coated with a non-volatile liquid phase (as in GLC), they tend to tail owing to the strong retention by the residual active sites. Several attempts have been made to improve the retention and peak shapes by addition of certain additives (organic vapours) to the carrier-gas. However, owing to the problems of detector incompatibility and several inconveniences, these approaches have not been practicable. On the other hand, deliberate mixing of water vapour with the carrier-gas sometimes resulted in greatly improved peak shapes without many of the problems associated with the organic additives. As a logical extension, steam itself could be used as the mobile phase. In view of its high polarity and other solvent effects, water vapour is expected to, and does, elute polar solutes rapidly.

Certain characteristic features of the 'steam-carrier GC' may be mentioned. Both high-boiling and highly polar solutes can be eluted very fast with insignificant tailing. Aqueous samples, including emulsions and suspensions, can be analysed without pretreatment. The chromatography, of necessity, has to be carried out at temperatures much above the boiling point of water in order to maintain the mobile-phase in a vapour phase at the relatively high head pressure. Thermal stability of the stationary phase is not a problem since the common adsorbents used are stable inorganic substances such as silica, alumina, magnesia or any of the common GC adsorbents and supports, including diatomaceous earth and charcoal. The mobile-phase selectivity is easily modified by addition of 10–20% of formic acid, hydrazine hydrate or ammonia to the water vapour; the formic acid helps elution of acidic solutes and the amines, the basic solutes, which may otherwise be strongly adsorbed on the active stationary phase.

Some modification of the equipment is obviously necessary. A steam generator (100 to 200 ml capacity, delivering steam at 2 to 3.5 atm. pressure) can be the source of the steam. Alternatively, water may be delivered to a column-head vaporizer at a flow-rate of 2 to 100 $\mu$l/min. using a syringe pump of the type used in LC (see Chapter 7). The vaporizer is of 5–10 ml capacity and is packed with granular silica (about 30 mesh) and heated to 150–180°C to generate a continuous flow of the vapour. The high-sensitivity flame-ionization detector can be readily used. Owing to the extremely fine and deeply porous nature of the common adsorbents, the peaks are somewhat broader. As the components are normally eluted very fast, peak broadening during injection directly affects the peak width of the eluted band. Occasional splitting of the single component peaks is ascribed to irregular adsorption near the injection port. The various aspects of the technique, namely, mechanism of separation, instrumentation, practice and applications have been reviewed [2].

A number of hydrocarbons (both low and high boiling), steroids, alcohols, aldehydes, ketones, ethers, esters, phenols, carboxylic acids, and amines have been analysed using this technique. Short analysis times, ease of sample handling, high sensitivity of the detection and low cost of the mobile phase should have made this technique especially suitable for routine analysis. However, the recent popularity of reversed-phase LC has been a set back to further development of this technique.

### 4.1.3 Supercritical fluids

It has been mentioned in the preceding section that steam can be used as the mobile phase in GC provided the temperature of the injection port, column and the detector is maintained well above 100°C. Though water completely vaporizes above 100°C at the atmospheric pressure, if the temperature of the chromatographic system is only slightly higher than 100°C, the operating pressure required for chromatography would readily condense it back to the liquid state. In fact, at any temperature below 374.4°C, water can be liquified by applying sufficient pressure (226.8 atm. at 374.4°C). The importance of the figure 374.4°C (the critical temperature, $T_c$) for water is that above this temperature, the water vapour cannot be condensed to a liquid at whatever pressure; the pressure (226.8 atm) required to condense water vapour at its critical temperature is called the critical pressure ($P_c$) and the density of the material at its critical temperature and pressure (critical point) is the critical density ($d_c$), which is 0.344 g/ml for water. At the critical point, the properties of the liquid and the gaseous phases converge and the phases become indistinguishable. Above the critical temperature and pressure, the substance is said to be in the supercritical state or phase. The properties of substances in their supercritical state are of immense importance to physical chemists and chemical engineers [3]. When a substance has an appropriate critical temperature (that is, above ambient temperature but not too much above), it may be used, in its supercritical state, as the mobile phase in chromatography. The properties which are of relevance to their use in supercritical fluid chromatography (SFC) are given in Table 4.1.

**Table 4.1** — Properties of supercritical fluids

| Compound | B.P. (°C at 760 mmHg) | *Critical properties* Temperature (°C) | Pressure (atm) | Density (g/ml) |
|---|---|---|---|---|
| Ammonia | −33.4 | 132.4 | 111.3 | 0.235 |
| Benzene | 80.1 | 288.9 | 48.3 | 0.302 |
| *n*-Butane | −0.5 | 152.0 | 37.5 | 0.228 |
| Isobutane | 11.7 | 134.9 | 36.0 | 0.221 |
| Carbon dioxide | −78.5* | 31.3 | 72.9 | 0.448 |
| Dichlorofluoromethane | 8.9 | 178.5 | 51.0 | 0.522 |
| Diethyl ether | 34.6 | 193.6 | 36.3 | 0.267 |
| Ethane | −88.0 | 32.4 | 48.3 | 0.203 |
| Ethanol | 78.4 | 243.4 | 63.0 | 0.276 |
| Isopropanol | 82.5 | 235.3 | 47.0 | 0.273 |
| Methanol | 64.7 | 240.5 | 78.9 | 0.272 |
| Nitrous oxide | −89.0 | 36.5 | 71.4 | 0.457 |
| *n*-Pentane | 36.3 | 196.5 | 33.3 | 0.237 |
| *n*-Propane | −44.5 | 96.8 | 42.0 | 0.220 |
| Sulphur dioxide | −10.0 | 157.5 | 77.6 | 0.524 |
| Sulphur hexafluoride | −63.8 | 45.6 | 37.1 | 0.752 |
| Water | 100.0 | 374.4 | 226.8 | 0.344 |
| Xenon | — | 16.5 | 57.6 | 1.113 |

*Sublimation temperature

A serious limitation in the applicability of traditional GC to the analysis of a vast majority of compounds has been the very low solvent power of the mobile phase for compounds of low volatility. The inferior solvent power of gases towards non-volatile substances is attributed to the very weak intermolecular interactions consequent to their low density and large intermolecular distances. The density of supercritical fluids is about 200 to 500 times that of the gas at the normal temperature and pressure (N.T.P.). The consequent high solvent power causes a dramatic decrease in the retention times and permits chromatography of thermally labile compounds at low temperatures. Since the solvent power can be increased by increasing the pressure, pressure programming can be substituted for the more common temperature programming of the traditional GC. While the density of a supercritical fluid is closer to that of the liquid, its viscosity, being about one hundredth of the liquid, is more akin to that of the gas. High flow-rates at low pressure drops are, therefore, possible, resulting in superior transport properties and much faster analyses.

Supercritical carbon dioxide has been the most popular mobile phase in SFC. In addition to its very convenient critical properties (Table 4.1), it is non-toxic, non-flammable, non-polluting, inexpensive and readily available. Addition of a modifier (e.g. methanol) to the supercritical fluid often decreases the analysis time by increasing the solubility of the solutes. Nitrous oxide has physical properties very similar to that of carbon dioxide. Mobile phases with higher critical constants (e.g. pentane) may also be used for separating high molecular weight compounds, but may require instrument modifications to attain supercritical conditions of the mobile phase, and safety systems to prevent any explosions or fire. Use of supercritical ammonia and such other mobile phases may need instruments made from corrosion-resistant components.

Retention in SFC is controlled by a combination of extraction and evaporation mechanisms, and selectivity can be achieved in a number of ways. As in normal-phase LC, use of a non-polar mobile phase and a polar stationary phase favours group separation. Chromatography in the reversed-phase mode causes separation based on molecular weights. At low pressures or high temperatures, elution follows the order of volatility; high pressures or low temperatures, on the other hand, permit separation based on adsorptive effects of the stationary phase towards the solutes.

Both fluid–liquid and fluid–solid modes may be used in SFC and the stationary phases normally used in LC are generally suitable for packed-column SFC. As in partition LC, the possibility of the liquid stationary phase dissolving in the mobile phase and migrating along the column does exist and bonded-phase adsorbents are preferable. Much of the early work on SFC was conducted using packed columns and the column dimensions were also similar to those in LC or somewhat longer (3 to 4 m). More recently, however, open-tubular columns, similar in column dimensions and chemical nature of the column materials to those commonly used in capillary GC, have come to be used almost universally. Open-tubular columns for SFC are intermediate between GC and LC in internal diameter; i.e. narrower than open-tubular GC columns and wider than open-tubular LC columns. Fused-silica columns of 50 $\mu$m i.d. and above, with polysiloxane stationary phases bonded to methyl, octyl, phenyl, cyanopropyl and liquid-crystalline functionalities, are commercially available and offer a wide range of solute–stationary phase interactions (see section 4.2.4 below). For comparison, the 50 $\mu$m column can generate between 5000 to 7000 plates/metre, similar to that obtained by a 300 $\mu$m i.d. column in GC. Since it is desirable to have minimal pressure drop across the column, the optimal length of the 50 $\mu$m column is around 40 m and that of a 25 $\mu$m i.d. column is only 5 m. Using a 34 m, 50 $\mu$m i.d. column, around 100,000 theoretical plates could be generated in 2 h (for a compound of $k$ value around 5).

Four different types of gradient elution are theoretically possible in SFC [4]. Pressure or density gradients may be used to shorten the analysis time and improve the resolution. The solvent power increases with increase in the density of the mobile phase. By gradual increase of the density of the mobile phase by pressure programming, compounds of a wide range of molecular weights may be eluted; as in the other modes of chromatography, slow programming rates give better results. It must be noted, however, that though the density increases directly with the pressure, the relationship is not linear unless the temperature is very high; therefore, linear pressure programming may often lead to inferior results. Solvent programming of the type used in LC and flow programming is also possible but less frequently employed. Temperature programming is not very useful in SFC since, while higher temperature is expected to increase the solubility of the solute in the mobile phase and thus help elution, it also causes reduction in the density of the supercritical fluid and decreases the eluting power of the mobile phase; most SFC separations are performed isothermally at a few degrees above (or around) the critical temperature.

The instrumentation for SFC is essentially a modified version of a gas or liquid chromatograph, the main difference being that the entire separation system rather than just the inlet section has to be maintained at constant high pressure. If the mobile phase is a liquid at ambient temperature, the pressure may be released soon after elution from the column so that the detection and fraction collection can be carried out at atmospheric pressure. The cooled and liquified mobile phase is delivered to the column using a pressure-controlled (rather than flow-controlled) LC pump; typical flow-rates being 50–80 μl per min (0.1 μl and above for capillary SFC), a pulse-free syringe pump is suitable. Alternatively, if the mobile phase is a gas at ambient pressure, it may be delivered from a high-pressure cylinder, electrically heated to provide suitable pressure. The whole SFC system, including the injection port, column and the detector, is maintained above the critical temperature; the common GC oven with a temperature range of 0 to 400°C is adequate for the purpose. The injection port for SFC is often a 4 or 6-port rotating valve, with an internal sample loop of 60 to 200 nl, using which up to about 75 nl may be conveniently injected. On-column injectors or valve injectors with flow splitters (for larger injection volumes) may also be used for capillary columns; however, on-column injectors may be problematic with the very narrow open-tubular columns currently popular in SFC. It is apparent that it is easier to adapt a liquid chromatograph for packed-column SFC work and a gas chromatograph for capillary SFC. Though the latter may be somewhat more difficult to realize [5], commercial systems based on this principle are available.

One of the major advantages of SFC is the versatility of the possible detection modes, which are generally based on several of the detection principles typical of both GC and LC, though adaptation to SFC could be occasionally problematic. Light absorption or fluorescence detectors may be used for SFC if appropriately modified for the high-pressure operation. When a mobile phase to which the detector does not respond is used, flame-based gas chromatographic detectors can be conveniently used after decompression. Also, as in GC and LC, the 'hyphenated techniques', namely, SFC–MS and SFC–FTIR, have become a reality with wide applicability [6]. Indeed, coupling of SFC with Fourier-transform infrared spectrometry or mass spectrometry is feasible with fewer practical difficulties than LC. The most attractive feature of SFC–MS is the possibility of using the mobile phase as the reagent gas in chemical-ionization mass spectrometry (see section 5.9.4)

SFC is considered a hybrid of GC and LC in that it combines several of the desirable characteristics of both gaseous and liquid mobile phases. SFC should, however, not be considered a replacement of either of the techniques, but as complementary to both of them. It can be used for the analysis of thermally labile and high-molecular-weight compounds that cannot normally be analysed by GC, but, for compounds that are stable under GC conditions, SFC is no match at present to the latter in separating power and efficiency. While the analysis is much faster than by LC with wider detection possibilities, it presently cannot offer the many different retention mechanisms available in LC. While most compounds analysed to date by SFC may also be conveniently analysed by either GC or LC, SFC is the answer for certain important types of compounds which are thermally too labile to be analysed by GC and which do not have a chromophore for UV or fluorescence detection. Though commercial instruments have become available only recently, considerable research effort has already been invested in the technique of SFC and its scope and limitations, as well as the factors effecting its performance, are well understood [6,7].

## 4.2 STATIONARY PHASES

### 4.2.1 Adsorbents for gas–solid chromatography

Gas–solid chromatography (GSC), in which the stationary phase is a solid adsorbent with no liquid phase coated on it, preceded gas–liquid chromatography (GLC) by several years, but was subsequently over-shadowed by the versatility of the latter. The main drawback of GSC is the tendency of the less volatile and more polar solutes to get strongly retained on the active adsorbents; the retention times and elution temperatures are consequently too high to be practical. Other limiting factors are the catalytic transformations of solutes on the active surfaces and the chemical and geometrical heterogeneity of most adsorbents, leading to non-linear adsorption isotherms, variable retention times, unsymmetrical peak shapes and excessive tailing, even with very small sample sizes. Nevertheless, GSC is important, albeit in a limited sphere. The main advantages are the long column life owing to the high stability of the adsorbents to heat and oxygen and the virtual absence of column bleed, permitting the use of high-sensitivity detectors like the helium-ionization detector. The selectivity of retention by the adsorbents is generally very high, permitting analysis of geometrical isomers and isotopes, inorganic (permanent) gases and volatile hydrocarbons, which are normally unresolved on the liquid stationary phases. Alumina, silica, molecular sieves and graphitized carbon blacks are the most frequently used adsorbents for GSC.

Considerable effort has been invested in the preparation of different adsorbents and, where necessary, modification of the stationary or the mobile phase to overcome the drawbacks noted above; as already noted, the simplest strategy has been to introduce water vapour into the mobile phase to reduce tailing. Macroporous silica, suitable for GSC is prepared from ordinary silica gel either by heating to 700–950°C or by hydrothermal treatment in an autoclave. Silica adsorbents with a wide range of specific surface areas and pore-sizes have been prepared [8]. Alumina and silica with similar surface areas show similar retention properties but different selectivities. Alumina and silica have been used for the separation of low molecular weight saturated and unsaturated hydrocarbons, their halogenated derivatives, and permanent gases. Short columns packed with microparticulate silica (7–10 or 25–35 $\mu$m average particle size) are used for high-speed GSC. Improved peak shape and efficiency is achieved by modification of these adsorbents. The modifiers generally used are inactive salts (alkali metal halides), salts that form donor–acceptor complexes (silver nitrate–dodecene and metal stearates) and organic substances such as urea and deoxycholic acid. Peak symmetry is obtained when the amount of the modifier is sufficient for at least a monolayer coverage of the adsorbent. Presumably, these modifiers selectively interact with the very active adsorption sites and thus decrease the heterogenecity of the solid surface. Faster analysis and improved column efficiency is also possible by mixing the adsorbents with inert solids such as diatomaceous earth or glass beads; the amount of the adsorbent relative to the total mixture in these columns is normally so small that the columns are called 'dusted columns'. The solute boiling point range and efficiencies of these columns are similar to those of the GLC columns. Mass transfer is very fast on these columns and high mobile-phase flow-rates can be used without significant loss in column efficiency, which could be of the order of 2400 plates/sec.

Molecular sieves are natural or synthetic zeolites (alumino-silicates of alkali or alkaline earth metals, commonly, sodium or calcium) having a well-ordered porous structure [9]. The passage to the inner pores may be imagined to be in the form of a tunnel, partially blocked by the cation, so that only molecules smaller than a certain size can pass through and be adsorbed by a variety of interactions with the charge field of the zeolite structure. They are generally designated by a number that reflects the size of the molecule that can enter the pores; e.g. 3Å, 4Å, 5Å, 10Å, 13Å, etc. (Å=ångström=$10^{-7}$ mm). In view of their capacity for efficient adsorption of small molecules like water, they are commonly used as in-line traps and can be easily regenerated by heating at 350°C for about an hour under a slow current of nitrogen. The 5Å-type is the most commonly used stationary phase for the separation of permanent gases, particularly nitrogen and oxygen. The elution order at room temperature is helium, neon, hydrogen, argon/oxygen, nitrogen, methane, carbon monoxide, water, and carbon dioxide.

Activated charcoal and several other forms of carbon have found limited use in view of the significant batch-to-batch variation. Unlike the zeolites, the carbon sieves are not highly crystalline, but show very strong affinity towards organic compounds. They may be used for purification of inorganic gases or for the separation of small polar molecules such as water, formaldehyde and hydrogen sulphide. Graphitized carbon blacks, useful for separation of polar compounds, are prepared by heating the charcoal to 1000°C to remove volatile tarry substances and then to about 3000°C. They are very brittle and great care must be taken during preparation, packing and use. Because of the residual polar active sites and crevices, they have to be coated with a liquid phase. Occasionally, the added liquids may impart a certain degree of selectivity to the stationary phase. Since usually a very low percentage of the liquid phase is added and since both the liquid and the solid may contribute to the retention and separation, it is also sometimes called gas–liquid–solid chromatography [10].

Several other materials, including bentonite, clathrate complexes, oxides of iron and chromium, barium sulphate, etc., have been suggested as adsorbents for GSC but have not found general application [11].

### 4.2.2 Porous polymers

Porous polymers are akin to the solid adsorbents discussed above in the sense that there is no added liquid phase as in the stationary phases for GLC. However, separations effected by these materials are independent of their pore size or the total pore volume and are probably based on partition, with the polymer behaving like an extended liquid. At low temperatures, however, adsorptive effects predominate and, in actual practice, both adsorption and partition may be operative. They are prepared by suspension co-polymerization of a large excess of a monofunctional monomer (e.g. styrene, acrylic acid derivatives or vinylpyrrolidone) with a difunctional monomer (e.g. divinylbenzene) in a suitable solvent. The cross-linking imparts the required rigidity to the beaded polymer structure and, on drying, the solvent evaporates, leaving a porous framework. The choice of the reactants, the solvent, catalyst and the reaction conditions like the temperature and the stirring speed determine the bead (or the particle) size and the porosity.

The commercially available porous polymers belong mainly to two series; namely, the Chromosorb century (Chromosorb 101 to 108) and the Porapak series. The Chromosorb 101, Chromosorb 102 and Porapak P are based on styrene–divinylbenzene and Porapak Q on ethylvinylbenzene–divinylbenzene. Chromosorb 103 and 106 are polystyrenes and Chromosorb 105, 107 and 108 are acrylic-ester based; Chromosorb 104 is a co-polymer of acrylonitrile and divinylbenzene. Porapak N, R and S are polyvinylpyrrolidones and Porapak P is an ethylene glycol–dimethacrylate copolymer. They all have a rather low usable temperature of around 250°C. Tenax, a linear polymer of *p*-2,6-diphenylphenylene oxide, has the advantage of high thermal stability and a usable temperature limit of up to 400°C in temperature-programmed chromatography. The chromatographic properties of these materials depend on a number of factors like the chemical nature, bead size, surface area and the pore size of the polymers (Table 4.2); each of the porous polymers has its own selectivity [12].

The porous polymers are widely used as stationary phases for the separation of permanent gases and certain short-chain polar compounds which yield poor peak shapes on the traditional GSC and GLC phases. The polymers with pore sizes of less than 0.01 μm are used for the separation of gases and those with pore diameters larger than 0.01 μm for the higher boiling compounds like alcohols, glycols, acids and amines. Generally, the larger the pore size, the faster is the analysis. Two features must be mentioned in relation to the use of these polymers. Firstly, the retention characteristics of the polymers continuously change when used for chromatography at higher temperatures and, secondly, some of the polymers may react with certain types of samples. Thus, Porapak Q and Chromosorb 102 react with nitrogen dioxide and Porapak S and Chromosorb 103 with nitro-alkanes.

### 4.2.3 Liquid stationary phases

The stationary phase for gas–liquid chromatography (GLC) is normally a non-volatile liquid, coated as a thin, uniform film on a solid support. The role of the stationary phase in chromatography is to selectively retain the sample components and that of the mobile phase is to carry them forward; as already noted, the selectivity and separation in gas chromatography is determined essentially by the former. The importance of the stationary phase in GLC is so well recognized that a major part of the first two decades of evolution of the technique has been devoted to the development of new and selective stationary phases in an effort to improve the separation. Over 400 liquid phases have been described in the literature and most of these are commercially available [13]. However, only a few of these are in common

**Table 4.2** — Properties of porous polymers*

| Polymer (Trade name) | Surface area ($m^2/g$) | Pore diameter ($\mu m$) | Suitable for separation of: | Not suitable for: |
|---|---|---|---|---|
| Chromosorb 101 | 50 | 0.3–0.4 | Alc,Ald,Es,Et, FA,Gly,Hcb,Kt | Amn,An |
| Chromosorb 102 | 300–500 | 0.0085 | Alc,Es,FA,Gly, Hcb,Kt,Ntr,NC,PG | Amn,An |
| Chromosorb 103 | 15–25 | 0.3–0.4 | Alc,Ald,Amd,Amn Hdz,Kt | FA,Gly, NC,Ntr |
| Chromosorb 104 | 100–200 | 0.06–0.08 | Amm,NC,NO,Ntr SG,Xy | Amn,Gly |
| Chromosorb 105 | 600–700 | 0.04–0.06 | Hcb,PG | Amn,Gly |
| Chromosorb 106 | 700–800 | 0.5 | Alc,FA,SG | Amn,Gly |
| Chromosorb 107 | 400–500 | 0.8 | Hcb | Amn,Gly |
| Chromosorb 108 | 100–200 | 0.25 | Alc,Ald,PG | |
| Porapak N | 225–350 | | Hcb,PG | Amn,Gly |
| Porapak P | 100–200 | | As for Chromosorb 101 | |
| Porapak Q | 500–700 | 0.0075 | As for Chromosorb 102 | |
| Porapak R | 450–600 | 0.0076 | Es,Et,NC,Ntr | Amn,Gly |
| Porapak S | 300–450 | 0.0076 | Alc,Hcb,Kt | Amn,FA |
| Porapak T | 250–300 | 0.009 | As for Chromosorb 107 | |
| Tenax–GC | 18.6 | | Most polar compounds | |

Alc=alcohols, Ald=aldehydes, Amd=amides, Amm=ammonia, Amn=amines, An=anilines, Es=esters, Et=ethers, FA=fatty acids (volatile acids), Gly=glycols, Hcb=hydrocarbons, Hdz=hydrazines, Kt=ketones, NC=nitro compounds, NO=nitrogen oxides, Ntr=nitriles, PG=permanent gases, SG=sulphur gases, Xy=xylenes.
*Data from Ref. 12.

use and the recent trend has been to discourage their proliferation except for speciality applications.

The selective retention of the sample components by the stationary liquid phase in GLC is controlled by their solubility in the liquid and this is determined by a number of intermolecular interactions briefly described earlier (section 1.1.2). Both dipole and induced dipole interactions are inversely related to the sixth power of the intermolecular distance and small molecules should make excellent stationary phases if only they could be made to stay stationary at the operating temperature. Indeed, several of the desirable characteristics of the stationary phases are contradictory to one another and the choice is often a compromise between the various theoretical and practical considerations. In addition to the selectivity of retention of the solute molecules, the stationary phase should have a low vapour pressure (non-volatile), remain a liquid over the normal temperature range of the GC (−60 to +400°C) and be thermally stable. For packed-column GC, it is also desirable that the stationary phase have a low viscosity at the operating temperature. While no single liquid meets all the requirements, the desirability of low volatility and low viscosity at high temperatures is often satisfied by several polymeric substances which have found use in GLC.

#### *4.2.3.1 Common liquid phases*

The chemical nature of the common stationary phases varies from hydrocarbons, fluorocarbons, poly(phenyl ethers), polyethylene glycols and polyesters to fused inorganic salt eutectics and molten organic salts. These stationary phases differ from one another in their polarity and selectivity of interaction with (and their retention of) different solutes. Since dispersion is the only type of interaction possible on the hydrocarbon phases, non-polar solutes are eluted in the order of their volatility and polar solutes according to their hydrophobic character. Though squalane is the least polar of the stationary phases and is the reference phase for most polarity scales, it is seldom used in practice as a stationary phase in view of its very low usable temperature limit of 120°C. Apiezon greases, obtained by high-temperature treatment and molecular distillation of lubricating oils, are of indefinite structure and purity, requiring extensive chemical treatment and purification before use. Apolane-87 is a specially synthesized branched-chain aliphatic hydrocarbon with 87 carbon atoms (molecular weight, 1222) and an operating temperature limit of 280–300°C. The hydrocarbon phases, in general, are susceptible to oxidation and the column life is relatively short.

Several fluorocarbons have been examined for use as stationary phases in GC, but have not been generally useful owing to poor support-wetting properties, inadequate column efficiencies and low usable-temperature limits. Fomblin YR is poly(perfluoroalkyl ether) which can be used at up to 255°C. The meta-linked poly(phenyl ethers), represented by OS-124 (5 rings) and OS-138 (6 rings), are non-volatile, low-viscosity materials of well-defined structures but are of very poor thermal stability and column efficiency. Among polyesters of dihydric alcohols and dicarboxylic acids, poly(diethylene glycol succinate) is the best known for its use in the analysis of fatty-acid esters. However, in view of their very poor chemical and hydrolytic stability, they cannot be used when the solutes contain reactive functional groups; e.g. amines, isocyanides, epoxides, halo acids and anhydrides. Poly(ethylene glycols) and poly(propylene glycols) have been used for a wide variety of polar compounds. Retention on these phases is essentially determined by the concentration of hydroxyl groups and, to a lesser extent, by the molecular weight distribution of the material. Carbowax 20M (MW=14,000) and Superox-4 (MW, approximately 4 million) can be used up to 225 and 300°C, respectively, and are the best known materials of this class.

Volatile metal halides have been analysed on fused inorganic salt eutectic mixtures; separation is essentially based on the ability of the solute to form reversible complexes with the stationary salts. Organic solutes are better separated using molten quaternary ammonium and alkylpyridinium salts, which show high selectivity for compounds containing polar and hydrogen-bonding functional groups. Ethylpyridinium bromide and tetrabutylammonium tetrafluoroborate are examples of this type of stationary phase.

The most important stationary phases in GLC, however, are the polysiloxanes [14,15]. The general structure of these materials may be represented as:

–O–(R)Si(R)–O–(R)Si(R)–O–

The polymer backbone is –Si–O–Si–, with the other two valencies of siloxane satisfied by different R groups which may be methyl, phenyl, haloalkyl or cyanoalkyl group, etc., and which impart different polarities to the liquid phase. They are prepared by acid hydrolysis of appropriately substituted dichlorosilanes or by catalytic polymerization of cyclosilanes using small quantities of hexamethyldisiloxane for chain termination. The reasons for their immense popularity and wide applicability are their high thermal stability, wide liquid range and, above all, their availability in different polarities ranging from the hydrocarbon-like non-polarity of the dimethylsiloxane to the very polar dicyanoethylsiloxane. The inorganic part of the polysiloxane molecule may be responsible for the high thermal stability of these phases while the organic pendant group provides the selectivity through varying polarity.

Dimethyl polysiloxane gums and fluids have been the most widely reported stationary phases for a long period. Their chemistry, manufacture and use have been discussed in detail by Trash [16]. Dimethyl polysiloxanes containing 2 to 2000 monomer units with viscosities ranging from 0.65 to 1,000,000 centistokes are available; fluids of viscosity 12,500 cSt are more commonly used. The viscosity of these phases changes very little with temperature, making them particularly suitable for programmed-temperature GC. This property may be due to the helical structure of the polysiloxanes. At higher temperatures, the helical structure expands, causing increased intermolecular interaction; the resultant increase in viscosity offsets the temperature effect, which generally causes a decrease in the viscosity of fluids. Other substituted polysiloxanes also generally exhibit temperature independence of viscosity, but to a lesser extent. Several of the commercially available dimethyl polysiloxanes (e.g. the silicone elastomer SE-30 from the General Electric Co., OV-1 and OV-101 of Ohio Valley Co. and the Supelco Phase SP-2100) differ from each other in their molecular weights and molecular-weight dispersion.

Substitution of 5 to 75% of the methyl groups in dimethyl polysiloxanes by phenyl groups generally increases the polarity of these phases and also imparts a certain degree of selectivity. OV-17, with 50% phenyl substitution, has been one of the most frequently used stationary phases of this type in packed-column GC. However, its poor film stability precludes its use in open-tubular GC. A stationary phase of comparable polarity (OV-1701) and with appropriate film stability for open-tubular GC is obtained by reducing the phenyl substitution to 7% and introducing 7% of cyanopropyl substitution. Indeed, cyanoalkyl substitution appears to be ideal for tailoring the polarity of the stationary phase without adversely effecting the surface tension and, consequently, the film stability. A series of very polar polysiloxane phases, containing increasing proportions of cyanopropyl and cyanoethyl substitution, were commercially introduced by Silar Laboratories. With increased polar character of these phases, their selectivity for aromatic hydrocarbons and olefins relative to alkanes increases markedly, probably due to the formation of a $\pi$-complex between the cyano groups and the $\pi$-electrons of the unsaturated compounds. These phases also exhibit significant selective retention of *cis*-olefins in a *cis*/*trans*-mixture. While many of the polar stationary phases have been conveniently used in packed columns, preparation of capillary columns could be problematic. As the polarity increases, their adsorption on the de-activated column wall becomes weak. In addition, the usable temperature range is also narrower compared to several apolar phases; e.g. Silar 10C has a usable range of only 140°C (100–240°C).

A number of linear polymers, incorporating both carborane (consisting of 75.8% boron) and siloxane units, are available under the trade name 'Dexsil'; they have unique thermal stability and some of them (e.g. Dexsil 300 GC, which consists of one carborane group for every four units of dimethyl siloxane) are stable even at 625°C. The carborane-containing phases have very modest polarity and are generally non-selective. They are also very sensitive to oxygen. The main advantage of their use lies in their exceptional stability to heat, making them particularly suitable for high-temperature GC.

#### *4.2.3.2 Liquid-phase characterization*

The very large number of stationary phases introduced required some sort of characterization so that their utility could be predicted. A polarity scale is probably the easiest to use but is unsuitable in view of the lack of clear definition and expression. It is also inadequate since, as noted earlier, there are a large number of different intermolecular interactions that determine retention. An adequate index (polarity-based, or otherwise) should express the solute–solvent interactions on a single numerical scale without exceptions or regard to the experimental conditions. The Kovats retention index, $I$, is an important contribution in this direction [17,18].

The Kovats retention index is based on the fact that for a homologous series of compounds, the logarithms of the adjusted retention times are linearly related to the carbon number. Thus, the Kovats index, $I$, is given by the expression:

$$I = 100z + 100\left[\frac{\log t'_{R(x)} - \log t'_{R(z)}}{\log t'_{R(z+1)} - \log t'_{R(z)}}\right] \tag{4.1}$$

For a given solute $x$, two adjacent normal alkanes with $z$ and $z+1$ carbon atoms are chosen so that, for a given stationary phase and at a given temperature, they elute before and after $x$; that is, $t'_{(z)} < t'_{(x)} < t'_{(z+1)}$. Thus, $I_{100}{}^{\text{squalane}}$ for dioxane is given as 651, to mean that it elutes midway between $n$-hexane ($I = 600$) and $n$-heptane ($I = 700$). Obviously, for any $n$-alkane of $z$ carbon atoms, the retention index under the given set of stationary phase and temperature, is $z \times 100$. The retention index for a polar solute on a given stationary phase can be a measure of the latter's polarity. Thus, the indices for dioxane on the three ester-type phases, di-isodecylphthalate, neopentyl glycol succinate, and diethylene glycol succinate are 779, 1080 and 1363, respectively. The polarity of a stationary phase relative to squalane is expressed by the difference ($\Delta I$) in the retention indices of the chosen compound on the two phases.

For meaningful comparison and reproducibility, a completely non-polar (squalane), a 100% polar (β,β-oxydipropionitrile) stationary phase, a polar solute, and the test conditions, need to be standardized. However, no single solute would reflect all the various intermolecular interactions. Rohrschneider [19], therefore, selected five different solutes to describe the different polar interactions. They consist of an aromatic hydrocarbon (benzene, $\Delta I = x$), an alcohol (ethanol, $y$), a ketone (ethyl methyl ketone, $z$), an electron acceptor (nitromethane or chloroform, $u$) and an electron donor (pyridine or dioxane, $s$). By determining the Kovats indices of these compounds on different stationary phases at 20% loading and at 100°C (sample size: 0.1 to 0.3 $\mu$l), 22 phases were characterized. The unique features of this sound, precise and easily reproducible approach are the use of several pure solutes as probes and the expression of the retention index difference ($\Delta I$) to reflect the intermolecular interactions. For this purpose, $\Delta I$ was redefined as in eq. (4.2):

$$\Delta I = ax + by + cz + du + es \qquad (4.2)$$

It was assumed that the various intermolecular polar forces are mutually independent and that the interaction energy is proportional to the values $a$, $b$, $c$, $d$ and $e$, characteristic of each probe, and $x$, $y$, $z$, $u$ and $s$, characteristic of the liquid phase. Several modifications have been suggested later, but essentially based on the same principles. McReynolds [20] characterized more than 200 phases using benzene ($x'$), butanol ($y'$), 2-pentanone ($z'$), nitropropane ($u'$), pyridine ($s'$) and several other solutes as probes. The choice of the probes was such that the bracketing $n$-alkanes ($z$ and $z+1$) are liquids at ambient temperatures for convenience. An important feature reflected by this approach is that a given stationary phase differed little from its neighbour in the polarity scale; it is thus possible to reduce the number of stationary phases needed for a majority of applications.

#### *4.2.3.3 Preferred liquid phases*

The need for reducing the number of usable stationary phases is well recognized. A notable effort in the direction was undertaken by the Hawkes committee [21]. According to them, almost all the GC analyses can be performed on one or other of the six common polymeric phases; namely, dimethylsilicone (SE-30 or OV-101), 50% phenylmethylsilicone (OV-17), polyethylene glycol (e.g. Carbowax 20M), poly(diethylene glycol succinate (DEGS), cyanopropylsilicone (e.g. Silar 10C), and trifluoropropylmethyl silicone (e.g. OV-210). They also suggested a further list of 18 polymers to fill the polarity ranges in more detail and a third list of 13 for use in case of special problems, e.g. when a stationary phase with special affinity to alcohols is needed. Even these listings are considered inadequate for separation of chemically unstable compounds, positional isomers or when the GC is to be carried out above 250 or below 50°C.

Leary *et al.* [22] had earlier suggested a list of 12 stationary phases, which covers a wide range of polarities and which is based on the statistical nearest neighbour technique applied to McReynolds' long list. The list includes a hydrocarbon (squalane), seven polysiloxanes (SE-30, OV-3, DC-710, OV-7, OV-22, QF-1 and XE-60), a polyethylene glycol (Carbowax 20M), two polyesters (DEGA and DEGS) and a highly polar compound (TCEP). More recently, Perry [23] has recommended substitution of some of these with more stable and better defined ones. Thus, the dimethylpolysiloxanes SE-30 or OV-1 (OV-101 or SP-2100) can be used in place of squalane, which has a very low temperature limit of 150°C. The carborane-dimethyl polysiloxane, Dexsil 300, with a temperature limit of 450°C could replace both OV-3 and OV-7, and similarly, OV-11 could substitute DC-710. OV-210 and OV-225 are specially purified versions of QF-1 and XE-60. Carbowax 20M is prepared by joining polyethylene glycol 6000 with a diepoxide. At high temperatures and in the presence of traces of oxygen in the carrier-gas, it may decompose to some extent to acetaldehyde and acetic acid; Silar-5CP was therefore recommended as a substitute. The polyesters DEGS and DEGA as well as TCEP may also be replaced by Silar-10C and OV-275 in view of their better thermal stability. The chemical nature, temperature limit and the polarity indices of the preferred stationary phases are given in Table 4.3; of the eight stationary phases suggested by Perry, OV-11 may be replaced by the slightly more polar, more stable and more commonly used OV-17 (or its equivalent SP-2250). Silar-10CP is also replaced by the more stable SP-2340 of very similar polarity index.

**Table 4.3** — Properties of recommended stationary phases for GC

| Name | Chemical nature of polysiloxane | Max. temp. | McReynolds' constants | | | | | $P'$* |
|---|---|---|---|---|---|---|---|---|
| | | | $x'$ | $y'$ | $z'$ | $u'$ | $s'$ | |
| SE-30 | Dimethyl | 350 | 15 | 53 | 44 | 64 | 41 | 104 |
| Dexsil300 | Carborane-dimethyl | 450 | 43 | 64 | 111 | 151 | 101 | 226 |
| 0V-17 | 50% Phenyl methyl | 375 | 119 | 158 | 162 | 243 | 202 | 406 |
| OV-210 | 50% Trifluoropropyl | 270 | 146 | 238 | 358 | 468 | 310 | 722 |
| OV-225 | 25% Cyanopropyl–25% phenyl | 250 | 238 | 369 | 338 | 492 | 386 | 836 |
| Silar-5CP | 50% Cyanopropyl–50% phenyl | 275 | 319 | 495 | 446 | 637 | 531 | 1111 |
| SP-2340 | 75% Cyanopropyl | 275 | 520 | 757 | 659 | 942 | 804 | 1678 |
| OV-275 | Dicyanoallyl | 250 | 629 | 872 | 763 | 1106 | 849 | 1919 |

*$P'$=Polarity distance from squalane, measured as the square root of the sum of the squares of the McReynolds constants.

It may be noted that all these materials are polysiloxanes. The popularity of the polysiloxane phases is mainly due to the fact that other phases are less universally applicable; e.g. the 87-carbon Kovats phase is particularly suitable only for retention index determination. Even among the polysiloxanes, the apolar ones offer the broadest applicability range of −30 to +370°C; i.e. a range of 400 degrees. Their other useful features are low column bleed, chemical inertness to all classes of organic compounds, except to acids stronger than fatty acids and bases stronger than primary amines, and stability to oxygen. They are also easy to immobilize (see below) over a wide range of film thickness from 0.02 to 8.0 $\mu$m, representing a ratio of 1:400. The main disadvantage of these phases appears to be the lack of selectivity exhibited by some of the polar phases, but the lack of selectivity is more than compensated by the high resolving power of the open-tubular columns for all but a few classes of compounds. As a general rule, it is always advantageous to first explore an apolar column for any unknown sample and use a polar column only when necessary.

A stationary phase of different chemical functionality from that of the solutes exhibits better selectivity and is likely to be more effective in achieving the separation. Non-selective and highly polar interactions may be avoided while selecting the stationary phase for a given separation problem. The McReynolds' constants can be also be used for the selection. For example, one of the ten McReynolds' probes closest in chemical nature to that of the solute is chosen and its retention indices on different stationary phases is studied. The stationary phase with the largest retention index for the probe has the highest selectivity for the type of compound under question. When more than one functional group is present in the solute, more than one probe may have to be considered and the stationary phase with the highest values for all (or as many as possible of) the probes may be selected. The American Society for Testing and Materials (ASTM) periodically publishes GC data compilations. The relative retention data for two compounds can be located in the compilation and from this, the number of theoretical plates required for the separation can be calculated for the stationary phase. Computer programs are also available that can search and identify the appropriate stationary phase and calculate the required number of theoretical plates.

#### *4.2.3.4 Mixed stationary phases*

Mixed stationary phases can sometimes achieve selectivity and resolution of certain difficult-to-separate mixtures that cannot be resolved on the homogeneous phases. Three different methods can be used to combine the separation power of two or more stationary phases; namely, coupling the columns, separately coating the different liquid phases on solid supports and packing the mixed beds and, thirdly, mixing the stationary phases and coating the mixture on the support. The pressure drop across coupled columns presents certain difficulties of predictability; also, performance of these columns depends on the order in which the columns are connected. The mixed-bed columns are generally stable and are not influenced by non-ideal miscibility effects. However, transport of the more volatile stationary phase along the column may be appreciable. Homogeneously mixed packings are prepared by dissolving two liquids of different polarities in a volatile liquid and using the solution to coat the support, as will be described later. The retention behaviour on these phases understandably depends on the nature and proportion of the two components of the stationary-phase mixture; the relationship is non-linear, but predictable if the partition coefficient of the solute in the two phases is known [24]. Addition of a small quantity of a non-polar stationary phase to a polar phase causes a dramatic decrease in the retention of a polar solute, but the reverse phenomenon is less pronounced. Mixed stationary phases are rarely needed or used in open-tubular GC.

### 4.2.4 Immobilized or bonded stationary phases

The main limitation of several of the usable liquid phases in GC is their labile nature at higher temperatures. The lability may be due either to their thermal instability or to their volatility, which causes the stationary phase to leave the solid support and bleed into the detector. The latter problem can often be overcome by certain techniques of cross-linking or chemically bonding the stationary phase to the solid support. Covalent bonding of the stationary phase to the activated functionality on the support particles has an additional advantage in that a uniform mono-molecular layer of the stationary phase is obtained which makes chromatography more efficient owing to the minimal resistance to mass transfer. The relatively flat van Deemter curve permits high flow-rates and fast analysis without significant loss of efficiency.

Retention by the immobilized stationary phase depends on the functionality of the organic chain, the degree of surface bonding, and the size and polarity of the solute molecule. Adsorption may be on either the bonded group or the support, if the bonded phases contain short alkyl chains. Long-chain alkyl-bonded phases exhibit less spatial ordering and behave more like the liquid phases.

Some of the common liquid stationary phases, such as polyethylene glycols, polyesters and polysiloxanes, can be directly attached to the silica surface by high-temperature treatment in an autoclave. Monomeric stationary phase film may be formed by reacting appropriate isocyanates and isothiocyanates with silanol groups of the silica to form urethane bonds; after the treatment, the unbound and excess reagents are removed by exhaustive extraction with an appropriate solvent [25].

The increased efficiency of the bonded phases makes them particularly attractive stationary phases for high-performance open-tubular GC; several bonded-phase fused-silica columns are commercially available. A majority of the commercially available cross-linked stationary phases are again based on polysiloxanes in view of their very desirable chromatographic properties such as high thermal stability, permeability to solute molecules and minimal viscosity change with temperature. Immobilization of these phases on the support further enhances all these characteristics. The immobilized stationary phases are also resistant to wash-out by organic solvents and this feature is extremely important for on-column injection using large quantities of the solvent.

There are two basic approaches for the production of fixed (non-extractable) polysiloxane stationary phases in open-tubular columns; one of them leads to the formation of Si–O–Si bonds and the other, Si–C–C–Si, usually between the methyl groups attached to the silicon atoms. The Si–O–Si type of immobilized phase is formed by first preparing the monomer by hydrolysis of the appropriately substituted dichlorosilane, coating the silica column with the partially polymerized hydrolysate, filling the column with ammonia gas and heating the sealed column to a high temperature for 24 h [26]. The product is a linear polymer (with five to ten monomer units) and may have a cyclic structure; the high-temperature treatment is expected to cause some rearrangement of the cyclic oligomers. Alternatively, the polysiloxanes may be synthesized *in situ* by treating the glass column with tetrachlorosilane followed by the monomeric or partially polymerized disubstituted silanol. The column is then sealed and heated to over 300°C for 20 to 48 h. The high degree (10–50%) of cross-linking obtained by this method can cause two problems. One is the decreased solubility of the solutes, and second is the unapproachability of the residual silanol groups for deactivation resulting in very active solid surfaces. However, the cross-linked polysiloxanes are very stable and controlled polymerization (longer polymer chains and lesser cross-linking) should lead to better stationary phases.

Immobilization of the stationary phases through Si–C–C–Si bond formation using a free-radical mechanism is more common [27]. The radical initiators commonly employed include: *t*-butyl peroxide (TBP), azo-*t*-butane (ATB), azo-*t*-octane (ATO), azo-*t*-dodecane (ATD), benzoyl peroxide (BP), dicumyl peroxide (DCP) and 2,4-dichlorobenzoyl peroxide (DCBP). In the case of the first three initiators, which are liquids at room temperatures, the columns coated with the appropriate liquid phase are saturated with the reagent vapours and then purged with nitrogen for about 2 h. The solid reagents may just be added to the liquid phase solution prior to coating the columns. The columns are then sealed at both ends and heated in a programmed manner from 40 to between 120 and 200°C at a rate of 4°C/min. The azo compounds are the most popular of the radical initiators noted above. They cause no polarity changes, decomposition or oxidation that lead to increased column activity.

Also, unlike most of the other reagents, they can be used for adequately cross-linking both polar and non-polar stationary phases. Phenyl and cyanopropylpolysiloxanes are cross-linked using a percentage of methyl, vinyl or tolyl groups. The cross-linking of polysiloxanes by γ-irradiation from a cobalt-60 source has been suggested as an alternative to radical-initiated procedures, but the technique is not of general applicability.

While a variety of stationary phases have been prepared in the immobilized form, about five of them, representing a wide range of polarities, may satisfy the requirements for a majority of applications [28]; they are dimethylpolysiloxane, methyl(phenyl)polysiloxane, polyethylene glycol (Carbowax 20M), trifluoropropylpolysiloxane and cyanopropylpolysiloxane. These are obviously based on the popularity of the corresponding phases in the traditionally coated packed-column GC. In general, the demands on reproducibility, coatability and thermal stability are much higher in open-tubular columns and some of the phases (e.g. Carbowax 20M) are available in refined versions, especially suitable for capillary GC (Superox 20M). A new generation of stationary phases, based on substituted phenylpolysiloxanes, is now under development to enhance the selectivity of the phases [29]. The polarizability of the stationary phase could be increased by substituting the phenyl group by biphenyl, 4-methoxyphenyl, 3,4-dimethoxyphenyl, 3,4,5-trimethoxyphenyl, pentafluorophenyl, 3-nitro-4-cyanophenyl, etc. Co-polymers of silarylene and methyl- or phenyl-substituted siloxanes may be of interest; the arylene moiety enhances the thermal stability of the polymer, while the siloxane substituents provide different selectivities.

### 4.2.5 Speciality stationary phases

The stationary phases discussed above are essentially of the general-purpose type in the sense that the interactions between the solute and the stationary phase, though somewhat selective, are non-specific. The operating intermolecular forces may be considered as relating to the gross structure of the molecule, either polar or non-polar. It is possible to take advantage of the more specific interactions based on the molecular geometry of the solute molecules and their ability to interact with appropriate stationary phases by complex formation, etc.

#### *4.2.5.1 Metal complexes*

Transition metals are characterized by the presence of unfilled lower energy orbitals in their atomic structures and by their ability to accept electrons from electron-rich organic compounds containing double bonds and hetero-atoms like oxygen, nitrogen and sulphur. Formation of such donor–acceptor complexes can be used in GC to impart a certain degree of selectivity to the stationary phase. The metal salts and their complexes can be used either as such or coated on a solid support, with or without a second (liquid) phase; the metal atom may also be a part of a polymer chain. Use of zeolites containing metal ions as adsorbents for GSC is well known. Inorganic ion exchangers (e.g. zirconium and titanium carbonates) have been used to separate carbon monoxide, carbon dioxide and lower hydrocarbons. These materials in the potassium ion form have also been used to separate aromatic hydrocarbons, aliphatic amines and aldehydes. Organic cation exchangers (based on polystyrene–divinylbenzene) carrying alkali or alkaline earth metal ions have also been useful. Oxides of transition metals (silver, copper, zinc, iron, etc.) can separate permanent gases and light hydrocarbons. Molten non-volatile metal halides have been used to analyse volatile metal chlorides, taking advantage of the latter's ability to form complexes with the chloride ion from the stationary phase.

Among liquid phases containing transition metal ions, a glycol solution of silver nitrate was the first to be used to separate alkenes from alkanes and subsequently for the separation of the *cis* alkenes from the *trans*. The stability of the metal complexes and hence the effectiveness of retention are influenced by steric and electronic factors. Thus, the *cis* alkenes are better retained than the *trans*; substitution in the vicinity of the double bond renders complex formation difficult. Since the ability to donate electrons is a prerequisite for complexation, attachment of electron-withdrawing groups to the unsaturated centre decreases the stability of the complexes. The Hg(II) ion forms strong complexes with alkenes and aromatic compounds, while Ni(II), Pd(II), Pt(II) and Cu(II) salts have affinity for nitrogenous compounds. Zr(IV), Th(IV) and Co(II) salts of di(2-ethylhexyl)phosphoric acid have been used as stationary phases for the separation of a variety of alcohols, ketones and aliphatic and aromatic hydrocarbons. The strength of the interaction of organic donors with the transition metals usually increases in the following order: alkanes $<$ ethers $<$ aldehydes $<$ esters $<$ cyclic ethers $<$ tertiary alcohols $<$ secondary alcohols $<$ primary alcohols [30].

The possibility of using organic polymers containing metal atoms as part of the polymer chain has also been examined [14]. Particularly attractive from the thermal stability point of view are the polysiloxanes in which one, or more, of the dialkyl silicon units is substituted by a divalent or trivalent metal atom. The molecular weight, stability and chromatographic properties depend on the conditions of synthesis of the polymer and the nature and amount of the metal ion used. Substitution by a metal more electropositive than silicon enhances the Si–O polarity and thus the thermal stability of the polymer. These polymers are generally prepared first by hydrolysing phenyltrichlorosilane to the trihydroxy compound and treating its monosodium salt with the appropriate metal salts. The immobilized metallo-organic polysiloxanes are prepared by first treating silica gel with thionyl chloride followed by the metallo polymer; the residual silanol groups are capped using trimethylchlorosilane. Polysiloxane stationary phases containing the following metals (in the order of decreasing affinity to alkenes) have been prepared this way: Cu $>$ Cr $>$ Mg $>$ La.

#### *4.2.5.2 Liquid organic salts*

Two of the major drawbacks of the molten inorganic salts have been their poor wettability properties and the limited solubility of organic solutes in these phases. Both the problems could be overcome by the use of liquid organic salts [31]. A variety of tetralkyl (mainly tetrabutyl) ammonium salts, and alkyl pyridinium salts, have been examined for use at a temperature of 50–110°C as stationary phases in GLC. Salts based on phosphonium or arsonium cations may also be useful. The anions may be halides, nitrates, picrates, alkyl- or aryl-sulphonates or certain carboxylates, but they do not appear to influence the retention. The retention mechanism may vary from total partition to total adsorption depending on the solute–salt combination, but the former predominates in most cases. Where mixed mechanisms operate, the relative importance is usually temperature dependent, higher temperatures generally favouring partition.

The usable temperature range of these compounds is determined by their melting point and thermal stability and extends from about 50 to 150°C from the melting point. The efficiency of the liquid organic salt column of 2 mm diameter (10% loading) may be in the range of 1000–3000 plates/metre, which is comparable to most other liquid phases but with greater selectivity to certain types of compounds. The retention selectivity for the McReynolds' probes appears to be similar to that exhibited by the polar phases such as dicyanoethylpolysiloxane (OV-275) but the retention times are three times longer. The retention of the probes is more than an order of magnitude greater compared to Carbowax 20M. The retention appears to increase as a function of the number of carbon atoms attached to the cation of the organic salt. Dispersive interactions (benzene or 2-octyne) are generally weaker compared to proton-acceptor interactions (alcohols) or the much stronger orientation forces (nitropropane).

Since the organic salts exhibit a wide range of acidity and nucleophilicity, chemical transformation of the solutes under the conditions of chromatography cannot be ruled out. Indeed, nucleophilic reactions do predominate and tetrabutylammonium methanesulphonate columns are known to rapidly degrade when alkyl halides are injected. Excessive and irreversible retention of phenols, alkylamines and certain alcohols is frequently reported. Proton-transfer reactions with basic solutes is also common. As can be expected, the results of chromatography are greatly influenced by the column temperature and the amount of the compounds injected. There is also a general decrease of column efficiency with increasing column temperature owing to the decline in viscosity, causing changes in the film homogeneity. However, the possibility of synthesis of stationary phases with well-defined structures and properties makes the liquid organic salts attractive polar reference phases.

#### *4.2.5.3 Liquid crystals*

Liquid crystals, as the term indicates, form the mesophase between the solid (crystalline) and the liquid (isotropic) state. They show some of the anisotropic (orientation) properties of the crystalline state as well as the mechanical properties characteristic of liquids. Thermotropic liquid crystals enter the mesophase (liquid crystalline state) when the solid is melted, and show properties intermediate between those of the two distinct phases over a certain range of temperatures. On further heating to the so-called clearing temperature, complete transition to the liquid phase occurs. The temperature range between the melting and the clearing points is called the mesophase range. Lyotropic liquid crystals, on the other hand, show the characteristic properties over a wide range of concentrations in a suitable solvent and are obtained by mixing two or more components, usually surface-active agents or polymers. Molecules that show liquid crystalline properties generally have elongated (rod-like) and rigid shapes, usually with polar terminal groups. The liquid crystalline properties are exhibited by the tendency of these molecules to orient themselves parallel to each other along the linear axes. The thermotropic-type liquid crystals are more frequently used as stationary phases in GC than the lyotropic phases. They may be classified into three distinct types, namely, smectic, nematic and cholesteric. The smectic phases possess a two-dimensional layered structure, whose thickness is determined by the length of the molecules. The nematic-type liquid crystals also align themselves parallel to each other but, owing to the absence of the layered structure, are more mobile within the limits of parallel orientation. Chiral liquid crystals are generally of the cholesteric-type, with a twisted nematic structure.

Liquid crystals that are used as stationary phases in GC should have a high thermal stability and a low vapour pressure at the temperature of operation. They should also have (particularly when temperature-programming is required) a wide mesophase range, since their use is not advantageous outside the mesophase range. The nematic-type of liquid crystals offer a wider range of operating temperatures, and are therefore preferred for use as stationary phases in GC. The mesophase range can sometimes be extended by using eutectic mixtures of two or more liquid crystalline phases; the mixtures also generally show better chromatographic properties than the single phases. Understandably, the retention properties vary considerably from one liquid-crystal phase to another, and with temperature. The best separations are usually obtained at the lowest temperature or, better, in the supercooled mesophase, where the degree of ordering is the highest. The chromatographic properties depend not only on the nature of the liquid-crystal phase but also on the support and the concentration of the stationary phase on the support. At low stationary-phase loading and with large surface area of the solid support, adsorptive effects of the solid may dominate and the retention data may be unpredictable. The reproducibility may be improved by using high stationary-phase loading on supports with low surface area. The supports and the methods of preparation of the columns are otherwise the same as those for the normal liquid phases.

Liquid crystals are ideally suited for separation of substances which differ in their molecular shapes. Their effectiveness is based on the principle 'like-dissolves-like' and liquid crystals preferentially retain molecules with an elongated linear axis. Thus, *para*- and *meta*-xylenes (boiling points, 138.5 and 139.5°C, respectively) which are very difficult to separate on normal liquid phases are easily resolved using liquid crystalline stationary phases; *p*-xylene is more strongly retained. Isomeric mixtures that are more readily separable on liquid crystalline phases include substituted phenols, phthalic acid esters, vinylbenzenes, alkyl, hydroxy, halogeno and amino naphthalenes, bile acids, chlorinated biphenyls, quinones, polycyclic aromatics, methylpyridines, *cis* and *trans* unsaturated fatty acids and alcohols and pheromones. Chromatography on liquid-crystalline stationary phases could also be used for investigation of the physicochemical properties of the liquid crystals, particularly in the transition zones. The various liquid-crystal stationary phases, their properties, retention mechanisms and their applications in chromatography have been reviewed [32].

#### *4.2.5.4 Cyclodextrins*

Cyclodextrins are cyclic oligomers of α-(1,4)-linked D(+)glucopyranose units, obtainable by enzymatic degradation of starch. α, β and γ-cyclodextrins contain 6, 7 and 8 monosaccharide units. The cyclic structure of these compounds possesses a cavity of 5–8 Å (0.5–0.8 nm) and permits formation of inclusion complexes with compounds of appropriate size and shape. For example, α-cyclodextrin can form inclusion complexes with inert gases such as krypton and xenon, but not with helium, neon or argon. It also forms very stable complexes with oxygen, carbon dioxide and chlorine at high pressure, as well as with several hydrocarbons and their halogenated derivatives. These properties of cyclodextrins are retained in their crystalline monomeric form and in aqueous solutions as well as in the form of their polymers (either homopolymers or co-polymer with bifunctional molecules). They also retain their capability to form inclusion complexes when they are immobilized by covalent bonding or otherwise on polymeric supports.

In view of the reversible interaction of cyclodextrins with compounds of appropriate size and geometry, they can be used as stationary phases for selective retention of these compounds. That geometrical inclusion is the major operative mechanism is evident from the fact that 1,2-dichloroethane is retained on α-cyclodextrin ten times more strongly than on β-cyclodextrin; similar differences exist in the retention of hydrocarbons. Linear compounds are retained more strongly than the branched-chain ones, whose geometry does not correspond to that of the cyclodextrin cavity. The fact that the guest polarizability rather than polarity determines the strength and stability of the inclusion complex indicates that van der Waals' forces are operative in the retention, though hydrophobic interactions can also occur.

α-Cyclodextrin and β-cyclodextrin acetates, and β-cyclodextrin propionate, butyrate and valerate have reasonable thermal stability (220–236°C) and have been used as polar stationary phases in GC for the separation of a variety of α-olefins, aldehydes, alcohols and esters. The retention, however, is based mainly on polarity and not inclusion. Cyclodextrins themselves and their methylated derivatives may be used for inclusion-based retention in GC. Aliphatic and aromatic hydrocarbons, halogen derivatives and alcohols have been analysed using α- and β-cyclodextrin-polyurethane resins. Chromatography is also an effective method for studying the inclusion phenomena in general. It has been demonstrated that macroporous polymers with in-built cyclodextrin molecules, as well as cyclodextrins deposited on solid supports, retain their inclusion capabilities even at high temperatures [33].

#### *4.2.5.5 Chiral stationary phases*

It is well known that optical isomers (enantiomers) have very similar chemical and physical properties except that their molecular structures are mirror images of one another; this feature is reflected in their ability to rotate the plane of polarized light in opposite directions. A prerequisite for this phenomenon is that the molecular (or crystal) structure be asymmetric (chiral). Their physical properties and chemical reactivity with achiral reagents are indistinguishable and this makes their separation extremely difficult. Traditionally, they are separated by forming diastereoisomeric derivatives by reaction with an appropriate reagent, followed by fractional crystallization or, more recently, by chromatography. More relevant to the present discussion is the approach involving reversible interaction with a chiral stationary phase. During the past 30 years, a variety of chiral stationary phases have been prepared and used. However, the resolution and selectivity is generally poor and the high-efficiency open-tubular columns are more effective.

Understandably, early attempts to develop chiral stationary phases involved the use of *N*-trifluoroacetates of amino acid esters, in view of their ready availability in optically pure forms. The N–H and C=O groups were considered to take part in hydrogen-bonding interactions with the appropriate groups in the solute molecules.

On the assumption that an additional hydrogen donor in the stationary phase would increase the probability of specific interaction, a variety of dipeptide esters have been introduced. However, detailed investigations on the effect of structural changes in the stationary phase and the solute, as well as examination of $^{1}H$ and $^{13}C$ magnetic resonance and x-ray analysis, revealed that the dynamic diastereoisomeric complexation between the stationary phase and the solute originated from dipole–dipole interactions and dispersion forces. For best separations, the peptide ester should be an *N*-trifluoroacetyl dipeptide with bulky side groups and a bulky ester group such as *t*-butyl, isopropyl or cyclohexyl ester; the column should be operated at the lowest feasible temperature. *N*-Trifluoroacetyl-L-valyl-L-valine cyclohexyl ester was the first stationary phase of this type to be introduced and remains a typical example.

Diamides of amino acids, particularly L-valine, exhibit greater efficiency, selectivity and reduced retention times. A number of these derivatives have been prepared and tested. Among these, *N-n*-docosanoyl-L-valine-*tert*-butylamide proved most satisfactory owing to its high resolution factors and low polarity which permitted faster elution at relatively low temperatures. To obtain better thermal stability, L-valine-*t*-butylamide was coupled to the carboxyl group of the co-polymer of dimethylsiloxane and 2-carboxypropylmethylsiloxane, obtainable by alkaline hydrolysis of the appropriate cyanopropyl methyl polysiloxane. The material is available under the brand name Chirasil-Val (Applied Science Labs, U.S.A.). The ratio of the dimethyl siloxane and the chiral siloxane units in the polymer is between 5 and 7 and this is found to be the optimum to permit the separation by intermolecular hydrogen-bonding alone and prevent intramolecular hydrogen-bonding that would make the chiral phase ineffective. These stationary phases are claimed to be stable up to about 260°C.

The so-called ureides or, strictly, carbonyl bis(amino acid esters), are a third group of chiral phases, which are liquid crystalline. They show so high a resolving power in their mesophase that separations could also be achieved on packed columns. For a detailed review of the chiral stationary phases for GC, the separation mechanisms and applications, see Ref. 34.

## 4.3 PACKED COLUMNS

As noted earlier, the columns that have been used in GC vary widely in their dimensions; the length could be a few centimetres to a hundred metres or more, and the diameter, from about a few micrometres for the narrow-bore open-tubular columns to several centimetres for process-scale applications. The packed columns for analytical gas chromatography are normally of an o.d. of 1/4 and 1/8 inch (6 and 3 mm i.d.). The column material may be glass, stainless steel or, occasionally, plastic. Though increasingly replaced by the much more efficient, high-resolution, open-tubular columns in recent years, the packed columns may be preferred for routine use, where the requirements are ease of handling, inertness, moderate efficiency, selectivity and disposability.

The packed columns are prepared by first depositing the stationary phase on a powdery solid support and then packing the impregnated powder into the appropriate column. The column efficiency increases with decreasing support-particle size and narrow size distribution. However, the high head pressure required to obtain reasonable flow-rate and the consequent pressure drop across the column, also increases as the particle size decreases and this would make it difficult to maintain optimal mobile-phase linear velocity along the column. The particle size is, therefore, normally held to not less than about 100 mesh (usually 80–100 mesh), corresponding to 180–150 μm diameter. The size distribution can be narrowed to 90–100 mesh using appropriate sieves. For a given particle size, size distribution and stationary phase, the efficiency of the column increases with decreasing weight per cent of loading of the liquid phase on the support; thinner coating facilitates mass transfer from and to the stationary phase. Very low loading may, however, increase the contribution from adsorptive effects of the solid support. Low stationary phase loading also requires smaller sample size and, for a given retention-time difference, longer columns. 0.5 to 1% loading is reasonable, though even 10 to 20% loading is not uncommonly found in the older literature.

### 4.3.1 Solid supports

The solid support, which holds the stationary phase in the form of a thin film and exposes it to the carrier-gas and thus to the solute molecules, can be important (and even crucial) as the separation becomes difficult. The ideal support would have a large surface area but would be totally inert so that it could have no interaction with the solute molecule; in other words, it would have no surface activity. However, it must have sufficient energy and wettability to be able to hold the liquid phase. The solid support must also be thermally and mechanically stable and be a good conductor of heat. The nature and preparation of the common support materials and their physical and chemical properties relevant to their use in chromatography have been recently described in detail by Perry [23].

Diatomaceous earth, which was used by James and Martin in their first experiments on GC, remains the most widely used support material. Also known as diatomite or kieselguhr, it is strip-mined from the large deposits of the remains of the numerous microscopic, unicellular marine or fresh-water algae, known as diatoms. The primary (skeletal) structure of the diatom is based on silica and contains holes of about 1.5 μm diameter, which may possess a secondary structure of 50 to 100 smaller holes. Two types of supports, prepared for chromatography, may be distinguished. The type I (Chromosorb P) is made by crushing the originally greyish diatomaceous earth and calcining it to above 900°C into a red brick, which is then ground. The type II (Chromosorb W), is made by mixing the earth with a flux (sodium carbonate) and then heating the mixture to above 900°C; Chromosorb G is a new kind of type II that combines the properties of Chromosorb P and W. The original skeleton of the diatoms remains almost intact in the type I support, while in the type II, it is fused together by a glass; the uniform porous structure of the former facilitates diffusion, a property of extreme importance in chromatography.

While Chromosorb P may offer excellent diffusion and chromatographic efficiency, adsorption of polar solutes on the large surface could be a major source of the undesirable tailing and, occasionally, total retention. The adsorptive effects of the silanol (Si–OH) groups in the diatomaceous earth can be largely eliminated by silylation; that is, treatment with hexamethyldisilazane (HMDS) or dimethylchlorosilane (DMCS). Mineral impurities are removed by acid-washing (AW) before silylation. Residual adsorptive effects, if any, may be suppressed by addition of a surface-active agent to the stationary phase. Another effective, but infrequently used, method of eliminating adsorption is to add a site deactivator, such as steam, formic acid, ammonia or carbon dioxide, to the carrier-gas, provided the detector is insensitive to the additive. Repeated injection of water, or even the solute that is being adsorbed, may also help. The type I supports may retain unacceptably high adsorptivity even after the above treatment; the type II-AW-DMCS supports, in view of their low adsorption, are generally more useful. Representative of type II materials are Chromosorb W and G; the latter is more efficient, particularly at low stationary-phase loading (<2%).

There are several supports which are less frequently used. Chromosorb T (Teflon), because of its non-polar nature, shows very little adsorption but it is very difficult to pack and cannot be used above 250°C. Textured glass beads are probably the closest to ideal for GC. To facilitate distribution of the stationary phase evenly over the surface, the surface of the glass beads is roughened by leaching. Silanized and roughened sodium silicate glass is the most suitable; the advantages are low adsorptivity and workability at very low loadings (about 0.05%), permitting chromatography of low-volatile compounds. However, the material is not readily available. Also, porous polymer beads can be used to support the stationary phase, but are often used bare. In fact, increased stationary-phase loading decreases the retention time, indicating that the partition is effected more efficiently by the beads themselves rather than by the added liquid phase.

### 4.3.2 Preparation of packed columns

The first step in the preparation of packed columns for GLC is the coating of a thin, uniform layer of the stationary phase on the solid support. While the several procedures available for the purpose may be simple, the operator's skill is often reflected in the performance of the packed column [23]. In the simplest method, the neat liquid is added to the support and the mixture is gently tumbled in a closed jar for a few minutes. Complete redistribution of the stationary phase may take several weeks at room temperature but equilibration may be speeded up at elevated temperature. The method may be used for liquids of low surface tension and viscosity; in any case the initial efficiency of the column may be poor.

In the slurry method, a known weight of the support is mixed (slurried) with a solution of a known weight of the stationary phase in a volatile solvent like dichloromethane. Alternatively, the stationary phase may be added to a slurry of the support in the solvent. The solvent is then slowly removed either in a rotatory evaporator or by gentle warming in an open dish under a current of nitrogen and with constant and gentle stirring. Disintegration or breaking-up of the particles into smaller ones may occur during the operation and re-sizing of the coated particles is advisable; the wider size distribution of the particles would otherwise cause channeling and thus decrease the efficiency. Among the diatomaceous earths, Chromsorb W is the most friable and fine-producing while Chromaosorb G is the least. Solvent removal by fluidization (i.e. by a gentle, upward current of gas, which lifts and tumbles the packing) prevents abrasion and formation of fines; this method is particularly efficient for low loading of the stationary phase. The need to prevent abrasion is particularly great when the support is graphitized carbon, in view of its very poor mechanical strength. In this case, the sized support is spread to a height of about 0.5 mm in a flat dish and a solution of the stationary phase in a volatile solvent is slowly added. The solvent is then allowed to evaporate at room temperature without shaking or stirring [10].

In the filtration procedure, the slurry is prepared as above but with known excess of the stationary phase and the solvent over the required amounts. The wet and nominally impregnated support is then filtered under suction. The quantity of stationary phase held on the packing is calculated from the volume of the filtrate or the weight of the stationary phase in the filtrate. More accurately, a known weight of the coated packing is extracted with a volatile solvent in a Soxhlet apparatus and the extract, after removal of the solvent, is weighed.

Before packing, the column, if metal, is cleaned with acetone or, if glass, silanized by rinsing with a solution of dimethylchlorosilane in toluene. The packing is poured into it using a matched funnel in small portions, while the column is gently vibrated or lightly sucked by a vacuum pump at the far end. When the column is filled, the ends of the column, about 1 cm on either end, are loosely plugged using silanized glass wool. Straight metal columns may be conveniently packed first and then coiled to suit the configuration of the GC instrument (oven); here again, graphite columns may crack and it is preferable to pack coiled columns.

The columns packed as described above may contain some trapped volatiles and low molecular weight analogues of the stationary phase, and these may *bleed* into the detector during chromatography. The newly packed columns should, therefore, be *conditioned* prior to use. This is simply done by connecting the column to the inlet (injector) port, leaving the other end to the vent (not connecting to the detector) and heating the column oven overnight under the normal flow of the carrier-gas. The temperature of the oven is maintained about 25°C above the normal operating temperature, taking care not to exceed the maximum recommended temperature for the stationary phase.

### 4.3.3 Micropacked columns

The terms capillary packed columns [35], or micropacked columns [36], generally refer to columns of about 1 mm diameter, loaded with packings similar to those described above for the conventional packed columns. The distinction based on the column diameter may be arbitrary, but appears justified on the grounds that several of the operating parameters are different from those of the conventional columns.

Thus, the length of the micropacked columns is generally in the range of 5 to 20 m, requiring special packing procedures, usually involving forcing the stationary phase through the coiled column under progressively increasing head pressure. There are two ways of increasing the efficiency of micropacked columns: (a) using a small particle diameter, or (b) increasing the column length; a 30–35 μm fine-grain packing could generate about 10,000 plates/metre but would require 25 atm. of head pressure to attain an optimal mobile-phase velocity in a 1.5 m column [37]. Coarse particles and longer columns are, therefore, generally preferred. The increased length also causes significantly higher pressure drop (3 to 5 atm.) across the column. However, the longer columns can generate a larger number of theoretical plates (say, up to 60,000) and, compared to open-tubular columns, can hold larger amounts of the stationary phase. The micropacked columns can thus accept larger sample sizes and can be used for trace analysis. The analysis time is significantly shorter compared to both conventional packed and open-tubular columns. Owing to their linear dimensions and small diameters, they offer a more stable temperature regime in temperature-programming and reproducibility is high. Thus, micropacked columns combine the advantages of both conventional packed columns and open-tubular columns. High-efficiency micropacked columns have been used for analysing complex hydrocarbons from petroleum, trace analysis of pollutants in water and air, and for detection and determination of impurities. They are uniquely suited for use in industrial process control where high efficiency, reliability and long column life are needed. However, the requirements of special packing procedures, high operating pressures and modifications in injection and detection systems prevented their widespread application in the laboratory. Most of the advantages of the micropacked columns are now offered by the so-called wide-bore open-tubular columns to be discussed below.

## 4.4 OPEN-TUBULAR COLUMNS

### 4.4.1 Porous-layer open-tubular (PLOT) and support-coated open-tubular (SCOT) columns

Most of the GC columns of capillary dimensions (i.e. of less than 1 mm i.d.) now in use are the wall-coated open-tubular (WCOT) type, without any additive to support the stationary phase. But, historically, two other types may be noted; they are: porous-layer open-tubular (PLOT) columns and support-coated open-tubular (SCOT) columns. In the PLOT columns [38], the inner surface is extended by substances such as fumed silica or elongated crystal deposits, whereas in the SCOT columns, there is usually an added porous substance to support the stationary phase. The distinction between the three types of columns is not always clear-cut; some of the treatments that are normally given to the inner surface of the WCOT columns make them resemble the PLOT or the SCOT columns. While the terms PLOT and SCOT are sometimes used interchangeably, not all PLOT columns are SCOT columns; the converse, however, is true. The PLOT and SCOT columns may be about 30 m long while the WCOT columns could extend to over 100 m.

The PLOT columns are generally prepared in two steps by first depositing the porous layer on the inner surface of the column wall and then coating it with the appropriate stationary phase. A larger-diameter glass tube is filled with a suitable material so that when the tube is drawn into a capillary, the solid particles are partially embedded on the inner surface of the column. Diatomaceous earth, powdered quartz or graphitized carbon black, which can be obtained in the microparticulate form and which do not melt at the softening point of the glass, can all be used for the purpose. Permeability of the columns may be improved by suspending a wire centrally along the linear axis while drawing the capillary. More efficient PLOT columns may be obtained by chemical (e.g. hydrofluoric acid) treatment of the drawn capillary column. The SCOT columns may be prepared in one step by coating the wall-treated open-tubular column in the usual way using an appropriate mixture of the stationary phase and a microparticulate support material such as Silanox 101 (trimethylsilylated silica with a particle size of about 7 nm). If the stationary phase is not sufficiently surface-active, a binder (e.g. benzyltriphenylphosphonium chloride) may be used. Silanox 101, being a hydrophobic material, is not suitable for polar stationary phases. Diatomaceous earth, unsilanized fumed silica, graphite or kaolin may be used in such cases [38].

The main advantage of the PLOT and the SCOT columns is that the extended surface area can hold more of the stationary phase, permitting larger sample sizes, albeit at the cost of some of the efficiency. They are thus akin to the micropacked columns but with an added advantage that, unlike the latter, high head pressures and special instrumental designs are not required.

### 4.4.2 Wall-coated open-tubular (WCOT) columns

The wall-coated open-tubular columns are currently the most popular in high-resolution gas chromatography. Though it is estimated that over 80% of GC is still being carried out using packed columns [39], there is an unmistakable trend toward replacement of these by the WCOT columns. This is due, in no small measure, to the commerical availability of bonded-phase, fused-silica open-tubular columns and their very high flexibility, reliability and convenience. It may be noted, however, there has also been considerable discussion on the relative merits of glass and fused silica as well as on the advisability or otherwise of making the columns in-house. These and the precedures for 'making and manipulating capillary columns for gas chromatography' are presented in great detail and in an excellent manner by Grob in a recent monograph [40]. As will be seen, there is no clear preference for either of the two column materials and it may be convenient to use the commercially available fused-silica columns when the requirement is small and the ready-made columns are adequate.

Use of glass capillary columns may have certain practical advantages such as clear visibility (of any cracks or of any non-uniformity of the coating, etc.), and the convenience of several routine operations like connecting and disconnecting pre-columns and traps. The slight inconvenience of the need to straighten the ends for connecting to the detectors, etc., can be overcome by coupling an empty fused silica section at the end of the glass column. The glass columns can be relatively easily prepared in the laboratory using a glass-drawing machine. Making of the fused-silica capillaries, on the other hand, is far more demanding owing to the much higher drawing temperature (1800–2200°C, compared to 600–750°C in case of glass). Also, owing to the very thin walls of the fused-silica capillaries (25 m thickness compared to 250 m of the glass capillaries), they are very sensitive to lateral forces unless coated with a suitable organic polymer (polyimide) immediately after drawing. Glass capillaries with the same wall thickness are also very flexible, but cannot be used owing to insufficient tensile strength. Polymer-coated glass open-tubular columns are also now available.

The polymer coating has a limited stability to oxygen at high temperatures and, therefore, these columns are not suitable for prolonged high-temperature use, unlike the glass-capillary columns. The polyimide coating has an upper-temperature limit of about 350°C while certain polysiloxane stationary phases can be used up to 450°C. Covering the columns with aluminium (Quadrex) or use of metal columns themselves may be attractive alternatives. To prevent reaction of the solutes with the hot metal surface, the inner surface of the columns may be coated with elemental silicon. Nickel, stainless steel and copper columns have been suggested for use in this manner.

Among the various types of commercially available glasses, borosilicate glass (81% silicon dioxide, 13% boron trioxide and the rest, oxides of aluminium, sodium and potassium), available under the brand names Pyrex and Duran, is the most suitable for making glass-capillary columns; of the two, Duran has the better defined and reproducible properties. The borosilicate glass is more amenable to surface deionization and deactivation by acid leaching than the cheaper alkali-glass columns. Since the glass surface could be easily modified, the glass open-tubular columns were considered preferable for coating with stationary phases of a wide range of polarities. Fused silica which is very pure silicon dioxide gives rise to a metal-free and relatively inert surface that does not require deactivation by acid leaching. Most of the apolar stationary phases can thus be directly coated onto clean fused silica; however, the more polar phases tend to form droplets under (e.g. thermal) stress. For this reason, the fused silica columns are also currently surface treated in much the same manner as the glass capillaries. As a result of these treatments, it is now possible to coat the fused-silica columns with as many liquid phases as the borosilicate glass columns.

Preparation of the capillary columns obviously requires a number of materials and tools, the most important among them being the glass tubing and the glass-drawing machine. Duran tubing of 8 mm o.d. and 0.8, 1.2, 1.7, 2.2 and 3.0 mm i.d. is generally used to give capillaries of 0.8 to 0.85 mm o.d. and 0.08 to 0.3 mm i.d. Prior to drawing, the glass tubing should be cleaned to remove organic and inorganic impurities that may be present on the inner surface. This is accomplished by rinsing the tube with dilute acids, bases, water, methanol, acetone and dichloromethane. Fused-silica tubing is normally treated with a dilute (5%) hydrofluoric acid solution, to remove any cracks and pits, and thoroughly rinsed with water.

The glass-drawing machine typically consists of a tube conveyor made up of two pairs of wheels or rollers, a furnace and another set of wheels to bend and support the coiled columns. As noted above, the furance temperature is between 600 and 750°C for glass and between 1800 and 2200°C for fused silica, depending on the bore. In view of the very high temperature required, the fused-silica capillaries are generally produced using fibre-optics technology and clean-room atmosphere. Several companies offer a range of drawn capillary tubing with different dimensions, outer coatings and glass compositions.

After drawing the capillaries and prior to coating the stationary phase, the inner surface needs to be appropriately treated to deactivate the polar silanol groups and modify the surface wettability to suit the stationary phase. However, these two requirements being contradictory, it is difficult to accomplish both the goals simultaneously. A polar stationary phase with high surface tension can be coated on an active surface, while non-polar stationary phases can be coated only on deactivated surfaces. However, the deactivated surfaces have insufficient energy and wettability for medium-polar stationary phases. Surface wettability in glass columns is generally increased by roughening the inner walls by treatment with certain corrosive chemicals like hydrochloric or hydrofluoric acid gas and with barium carbonate, sodium chloride, sodium hydroxide or silica [4,40]. The corrosion methods are not generally used with fused-silica columns since the thin walls may become brittle.

Both surface wettability enhancement and deactivation of the active sites are possible by thinly coating the column with a polyethylene glycol (Carbowax 20M, Carbowax 1000 or Carbowax 400), heating the column to 280°C and washing off the extractable part of the material with an organic solvent. Several other polar stationary phases like ethoxycarbonylpolyphenylene and alkoxypolysiloxanes have also been used. It is often advantageous to cover the silica surface with a polymeric substance of the same polarity as the stationary phase to be finally coated so that incompatibility problems are avoided. High-temperature treatment of the columns with cyclic siloxanes with the same substituents has been used for coating several polysiloxane stationary phases.

As noted above, both the borosilicate and the fused-silica columns are routinely subjected to hydrothermal treatment with dilute hydrochloric acid to remove the metal ions from the glass and siloxane bridges from the fused-silica surfaces. About 18% HCl solution is used for glass and 2% acid strength for fused silica. The partially acid-filled columns (6% to 25% of the column length) are kept at 150 to 220°C for 6 to 12 h; the higher end of the temperature range with shorter treatment time is generally used for fused silica capillaries. The hydrothermal treatment is followed by rinsing the capillaries with very dilute acid solution and dehydration (under vacuum at around 250°C for 1–2 h) ato strengthen the mechanical properties of the inner wall of the capillaries.

The hydrothermal treatment is most commonly followed by silylation, for which a variety of reagents are available [40]. Depending on the polarity of the stationary phase to be coated, the silylating agent may be hexamethyldisilazane (HMDS, for apolar phases like SE-30 and OV-1), diphenyltetramethyldisilazane (DPTMDS, for apolar to slightly polar phases like OV-73 and SE-54) or dicyanopropyltetramethyldisilazane (CPTMDS, for medium polar phases like XE-60 and OV-225). The conditions of treatment are 400°C for 12 h in case of glass and 330°C and 6 h for fused-silica columns. Static pyrolysis of trimethylchlorosilane in the column produces a thin film of elemental silicon with high surface energy. A thin layer of pyrocarbon (an allotrope of carbon) is formed when gaseous hydrocarbons are pyrolysed at around 1200°C. The pyrocarbon-coated columns are essentially inert and have low enough surface energy to be wettable by apolar and medium polar stationary phases.

**REFERENCES**


[1] K. Grob and G. Grob, *J. High Resolut. Chromatogr. Chromatogr. Commun. (JHRCCC),* **2**, 109 (1979).
[2] A. Nonaka, *Adv. Chromatogr.,* **12**, 223 (1975).
[3] G. Wilke, *Angew. Chem. Int. Ed. Engl.,* **17**, 701 (1978) and the accompanying papers.
[4] E. Klesper and F. P. Schmitz, *J. Chromatogr.,* **402**, 1 (1987).
[5] P. J. Schoenmakers and F. C. C. J. G. Verhoeven, *Trends Anal. Chem.,* **6**, 10 (1987).
[6] T. H. Gouw and R. E. Jentoft, *Adv. Chromatogr.,* **13**, 1 (1975).
[7] K. E. Markides and M. L. Lee, in *Advances in Capillary Chromatography,* J. G. Nikelley (ed.), Huethig, Heidelberg, 1986; see also, M. L. Lee and K. E. Markides, *Science,* **235**, 1342 (1987).
[8] C. L. Guillemin, *J. Chromatogr.,* **158**, 21 (1978).
[9] R. M. Barrer, *Zeolites and Clay Mineral as Sorbents and Molecular Sieves,* Academic Press, New York, 1978.
[10] A. Di Corcia and A. Liberti, *Adv. Chromatogr.,* **14**, 305 (1976).
[11] G. E. Baiulescu and A. V. Ilie, *Stationary Phases in Gas Chromatography,* Pergamon, Oxford, 1975.
[12] C. F. Poole and S. A. Schuette, *Contemporary Practice of Chromatography,* Elsevier, Amsterdam, 1984.
[13] J. A. Yancey (ed.), *Guide to Stationary Phases for Gas Chromatography,* 12th edn., Analabs, North Haven, Connecticut, 1979.
[14] J. K. Haken, *J. Chromatogr.,* **300**, 1 (1984).
[15] L. B. Itsilson, V. G. Berezkin, and J. Haken, *J. Chromatogr.,* **334**, 1 (1985).
[16] C. R. Trash, *J. Chromatogr. Sci.,* **11**, 196 (1973).
[17] E. sz. Kovats, *Adv. Chromatogr.,* **1**, 229 (1965).
[18] J. K. Haken, *Adv. Chromatogr.,* **14**, 367 (1976).
[19] L. Rohrschneider, *J. Chromatogr.,* **22**, 6 (1966).
[20] W. O. McReynolds, *J. Chromatogr. Sci.,* **8**, 685 (1970).
[21] S. Hawkes, D. Grossman, A. Hartkopf, T. Isenhour, J. Leary, J. Parcher, S. Wold, and J. Yancey, *J. Chromatogr. Sci.,* **13**, 115 (1975).
[22] J. J. Leary, J. B. Justice, S. Tsuge, S. R. Lowry, and T. L. Isenhour, *J. Chromatogr. Sci.,* **11**, 201 (1973).
[23] J. A. Perry, *Analytical Gas Chromatography,* Marcel Dekker, New York, 1981.
[24] R. J. Laub, in *Physical Methods of Modern Chemical Analysis,* T. Kuwana, (ed.), Academic Press, New York, Vol. 3, 1983.
[25] G. E. Pollock, D. R. Kojiro, and F. H. Woeller, *J. Chromatogr. Sci.,* **20**, 176 (1982).
[26] L. Blomberg, J. Buijten, K. Markides, and T. Wannman, *J. Chromatogr.,* **208**, 231 (1981).
[27] K. Grob and G. Grob, *J. Chromatogr.,* **213**, 211 (1981).
[28] J. A. Yancey, *J. Chromatogr. Sci.,* **24**, 117 (1985).
[29] L. G. Blomberg, *Trends Anal. Chem.,* **6**, 41 (1987).
[30] W. Szczepaniak, J. Nawrocki and W. Wasiak, *Chromatographia,* **12**, 484 (1979); **12**, 559 (1979).
[31] C. F. Poole, K. C. Furton, and B. R. Kersten, *J. Chromatogr. Sci.,* **24**, 400 (1986).
[32] Z. Witkiewicz, *J. Chromatogr.,* **251**, 311 (1982).
[33] E. Smolkova-Kuelmansova, *J. Chromatogr.,* **251**, 17 (1982).
[34] R. H. Liu and W. W. Ku, *J. Chromatogr.,* **271**, 309 (1983).
[35] V. G. Berezkin, L. A. Shkolina, V. N. Lipavsky, A. A. Serdan, and V. A. Barnov, *J. Chromatogr.,* **141**, 197 (1977).
[36] C. A. Cramers and J. A. Rijks, *Adv. Chromatogr.,* **17**, 101 (1979).
[37] J. F. K. Huber, H. H. Lauer, and H. Poppe, *J. Chromatogr.,* **112**, 377 (1975).
[38] L. S. Ettre and J. E. Purcell, *Adv. Chromatogr.,* **10**, 1 (1974).
[39] M. A. Kaiser and M. S. Klee, *J. Chromatogr. Sci.,* **24**, 369 (1986).
[40] K. Grob, *Making and Manipulating Capillary Columns for Gas Chromatography,* Huethig, Heidelberg, 1986.
[41] G. Schomburg and H. Husmann, *Chromatographia,* **8**, 517 (1976).

# 5

# Gas chromatography III: detection systems

## 5.1 CHARACTERISTICS OF CHROMATOGRAPHIC DETECTORS

The function of a detector in any analytical instrument is to respond appropriately to the presence of a substance so that the response can be used for qualitative and quantitative analysis. The sample being the effluent from the column, the chromatographic detector has to respond almost instantaneously and continuously, unlike certain other analytical instruments where the sample is stationary and can be measured relatively slowly. To be generally useful, the chromatographic detectors have to satisfy certain requirements relating to their sensitivity, stability, selectivity, linear dynamic range, detector-cell volume and response time.

There have been some variations in the definitions of the above terms; the commonly understood terminology is used here. The detector output is normally recorded as a two-dimensional plot (the chromatogram), showing the detector response *vs.* time. Under the normal conditions of operation of the chromatograph, the plot usually shows some noise (unsteady or wavy baseline) which is due to several factors not always easily controllable; these include electrical fluctuations, detector instability to changes in column temperature, and carrier-gas flow-rate, etc. The detectability ($D$, also called the minimum detection level, MDL) of a detector is the minimum amount (g/sec or g/ml) of sample that can cause a response equal to twice the noise level, and the sensitivity ($S$) is defined as the ratio of detector response to the quantity detected; most often the terms are used interchangeably. The signal-to-sample size ratio is also called the response factor (RF). The exact or the quantitative definitions of these terms depend on the nature of the detector. For example, the flame-ionization detector responds to the mass-flow rate, while the thermal conductivity detector responds to the concentration change of the solute in the mobile phase; the third most popular detector, i.e. the electron-capture detector, can be either, depending on the manner in which it is used. The response factor is thus expressed as the ratio of the peak area to the mass of the solute injected ($A/M$, where $A$ is the peak area and $M$ is the mass) in the case of mass-sensitive detectors and as $AF/M$ for concentration-sensitive detectors ($F$ being the flow-rate). The FID is a very sensitive detector for any organic compound that burns in the flame. The universality of detection of the TCD, on the other hand, extends to inorganic materials (e.g., the permanent gases) and is therefore most commonly used for the purpose, notwithstanding its rather low sensitivity. The sensitivity of ECD to certain types of compounds (e.g., polychlorinated pesticides) far exceeds that of the FID. The relatively high response of the ECD to certain specific types of compounds is an example of selectivity of detection as against the universality of response of the FID and TCD. Several selective detectors are in use for specific applications. The selectivity or specificity is measured as the ratio of hydrocarbon to the amount of sample that gives the same response; the value may vary from 1000:1 to 1,000,000:1.

The most important parameter that determines the utility of a particular detector for quantitative analysis is its linear dynamic range. This is defined as the range of concentration (or amount) of the sample over which the detector response is equal to that calculated from the response factor; that is, the plot of the concentration *vs.* the signal is a straight line. By definition, the linearity is said to be lost when the plot deviates from the straight line by 5%. Wide linear dynamic range, stability of the detector to changes of column operating parameters (e.g. temperature and mobile-phase flow-rate), fast response time, low detector-cell volume and long life are some of the desirable characteristics of chromatographic detectors.

Understandably, no single detector satisfies all the requirements. The choice of detector is naturally dependent on the purpose of the analysis. Thus, qualitative analysis of a sample would require a high-sensitive and preferably selective detector, particularly if the constituent is poorly resolved from the background. On the other hand, a universal detector is indicated if all the constituents need to be detected. Predictability of specific response and obtainability of diagnostic information (e.g. spectra) are the requirements of the detector if the object is to identify or confirm the identity of a compound. Similarly, quantitative analysis requires the detector to have a predictable or stable response factor, a wide dynamic range and, in case of incomplete column resolution, be specific to the compound to be analysed and insensitive to accompanying materials.

A large number of GC detectors (about 50 different types) have been introduced. Many of these remain mainly of academic interest and the field is dominated by those based on ionization of the sample by flame, beta-radiation, or some other form of energy, followed by measurement of the current. The basic principles of operation and function of some of the more common detectors and their applications are briefly discussed in this chapter. The recent general review by O'Brien [1] discusses the topic in greater detail while the monograph by Dressler [2] contains an elaborate treatment of the selective detectors. In addition to the specifically designed GC detectors, the independent spectrometric techniques, i.e. mass spectrometry and Fourier-transform infrared spectrometry, have come so much into routine use in conjunction with GC, that they may be considered regular GC detectors for both universal and specific detection [3–5].

## 5.2 THERMAL CONDUCTIVITY DETECTOR (TCD)

The TCD (also known as the katharometer or the hot-wire detector) was the first universal, non-destructive and widely used detector in GC even prior to the advent of gas–liquid chromatography. As noted earlier, the gas–solid chromatography, which preceded GLC, could be used mainly for the analysis of permanent gases for which the TCD is the detector of choice even today. The main limitation for the use of TCD in organic analysis has been its low sensitivity. Over the years, however, improvements in cell design and its components have resulted in the production of more stable and sensitive thermal conductivity detectors.

The TCD is a bulk-property detector in the sense that it is based on an overall physical property of the mobile phase rather than (and unlike several other common detectors) on the chemical property of the solute. As the name indicates, the principle of operation of this detector is based on the change in conductivity of a filament with temperature which, in turn, is brought about by the presence (in the effluent) of a solute whose heat capacity is different from that of the carrier-gas. The filament forms one (or two) of the four arms ($r1$, $r2$, $r3$ and $r4$) of a Wheatstone bridge (Fig. 5.1), connected to a power supply and a meter. The meter shows no deflection when the resistance of the filaments is such that $r1/r2 = r3/r4$. Since this condition is difficult to achieve in practice, two balance potentiometers (one coarse and one fine control) are used. The input current is controlled by a rheostat and measured by an ammeter. In the TCD, the four filaments are heated by supplying adequate current (100 to 1000 milliamps) and the sensitivity of the TCD is proportional to the bridge current; an increase in the bridge current from about 130 milliamps in the older instruments to 1 amp at present increased the sensitivity 500-fold. The resistance of the filament increases with temperature and any change in the temperature of one of the arms is thus indicated by a voltage imbalance across the bridge. Metal-oxide thermistors have long been used as temperature sensors and are particularly convenient for use with capillary columns in view of their fast response, and low cell volumes; they are also better suited for operation at low temperatures. Lately, however, they have been largely replaced by the hot-wire (platinum, tungsten, nickel or their alloys) detectors in most routine applications in view of their better sensitivity and ease of optimization.

When the TCD is used as a detector in GC, the carrier-gas is divided into two streams, one of which is passed over the reference filaments $r1$ and $r4$, while the other goes through the column and over the sample filaments, $r2$ and $r3$; use of all the four filaments doubles the sensitivity. When no sample is being carried by the mobile phase, the filament temperatures and hence the resistances are balanced. As the sample is eluted from the column and carried to the detector, the heat dissipation from the filament to the surroundings (and thus the filament temperatures) changes. Usually the thermal conductivity of the gas containing a solute is poorer than the pure carrier-gas in the reference circuit and this causes the filament temperature (and thus the resistance) to rise. The response of the TCD is proportional to the temperature difference between the filament and the detector cell wall, requiring that the filaments be maintained at a high temperature. But, since high temperature shortens the life of the filaments, the detector temperature is maintained just about 20°C above the column temperature to prevent condensation of the solutes on the filaments. The TCD is naturally very sensitive to changes in the ambient and the column temperatures unless properly insulated. Another common problem with the design of the TCD is to match the detector cells. In the more recent modulated TCD design, the reference and the carrier gas alternately flow through a single detector

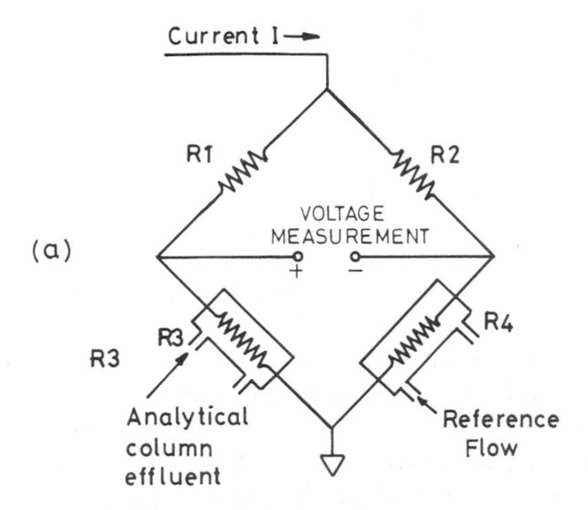


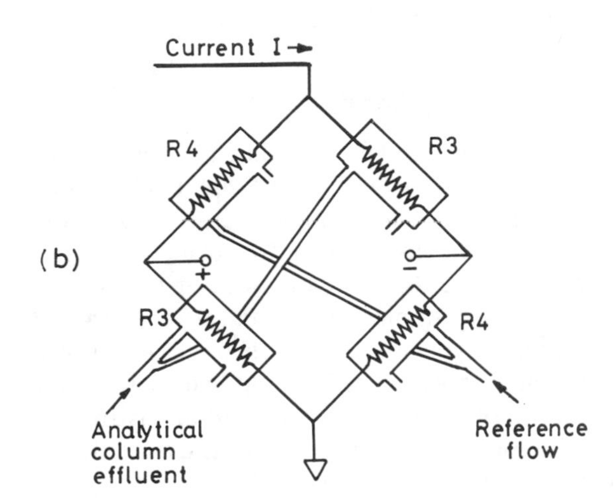


Fig. 5.1 — Schematic diagram of a thermal conductivity detector: (a) bridge circuit of a two-cell TCD and (b) a four-cell detector. Reproduced with permission from Ref. 1; copyright (1985) John Wiley & Sons, Inc.

cell at a rate of 10 cycles/sec, which is faster than any changes due to thermal fluctuations. These detectors are thus easier to stabilize and there is much less drift.

As already noted, the thermal conductivity of the pure gas is generally higher than that containing the solute and the signal is proportional to the conductivity difference. It is thus imperative that the carrier-gas thermal conductivity be large relative to that of the solutes. The thermal conductivity of some common gases ($\lambda \times 100{,}000$) are: hydrogen 45.8, helium 36.9, methane 8.6, oxygen 6.6, nitrogen 6.4, ethane 5.5, and propane 4.5. Hydrogen, or helium, are then the carrier-gases of choice when TCD is used. Hydrogen, being reactive with the metal-oxide thermistors, cannot be used when these are the sensors. Quantitative analysis using TCD and helium is very convenient since the peak areas are directly (within limits) proportional to the amount of the sample. The response factors for most compounds are close to that of iso-octane except in the case of organometallics and halogenated compounds where the response factors are very low. Compounds of very low molecular weight (less than 35), on the other hand, tend to have very high response factors. The response factor is independent of the type of detector (filament or thermistor), cell and sensor temperature, the concentration of the sample, the flow-rate, and the detector current. It must be noted, however, that the TCD signal is concentration-dependent and is sensitive to fluctuations in the flow-rate; indeed, the response is inversely proportional to the flow-rate. The flow-rate, therefore, has to be maintained constant and as low as possible, preferably close to the optimum value determined by the van Deemter plot. The flow of the gas in the reference cell need not equal that through the column and may be adjusted to minimize the fluctuations and drift in temperature-programmed GC. The sensitivity (10 ng/sec) and the linear dynamic range (five orders of magnitude) of TCD is much lower compared to the other detectors but, nevertheless adequate for most purposes.

## 5.3 FLAME-IONIZATION DETECTOR (FID)

The FID is the most commonly used detector in GC. The desirable characteristics of the detector which contributed to its status include: (a) its high and nearly uniform sensitivity (1 pg/sec) to most organic compounds, (b) insensitivity to the common impurities such as carbon dioxide and moisture in the carrier-gas, (c) minimal fluctuations due to changes in the flow, pressure or temperature, (d) wide linear dynamic range of eight orders of magnitude, and (e) the virtual lack of response in the absence of the sample, giving rise to a stable baseline. Its lack of response to the permanent and inorganic gases (e.g. oxides and hydrides of sulphur and nitrogen), and to most carbon compounds devoid of the carbon–hydrogen bond (e.g. carbon monoxide, carbon disulphide, etc.) may be considered both an advantage and a disadvantage depending on the application.

The basic features of an FID are shown schematically in Fig. 5.2. The carrier-gas from the column emerges from the tip of a jet, admixed with hydrogen, which is burnt in air. The atmospheric air being insufficient, it is also introduced into the flame chamber independently of the carrier-gas and hydrogen. The air-to-hydrogen ratio is maintained around 10:1 to sustain the flame and the flow-rate of hydrogen is optimally 30 ml/min to obtain maximum sensitivity. If the flow of hydrogen is too low, all the available gas is burnt up and the flame gets extinguished and, if it is too fast, the flame is lifted off the jet and turbulence would set in. Unless the balance between the gas flows is maintained, the ionization becomes less efficient and the sensitivity is very low. The best sensitivity to the solute molecules is obtained when the volume of the carrier-gas approaches that of hydrogen. While this volume of carrier-gas flow is common in packed-column GC, the flow-rates in open-tubular GC is one to two orders of magnitude less and this requires that a make-up gas (usually nitrogen) be also introduced at a point between the column outlet and the flame jet so that the total volume of the carrier gas and nitrogen is equal to that of the hydrogen

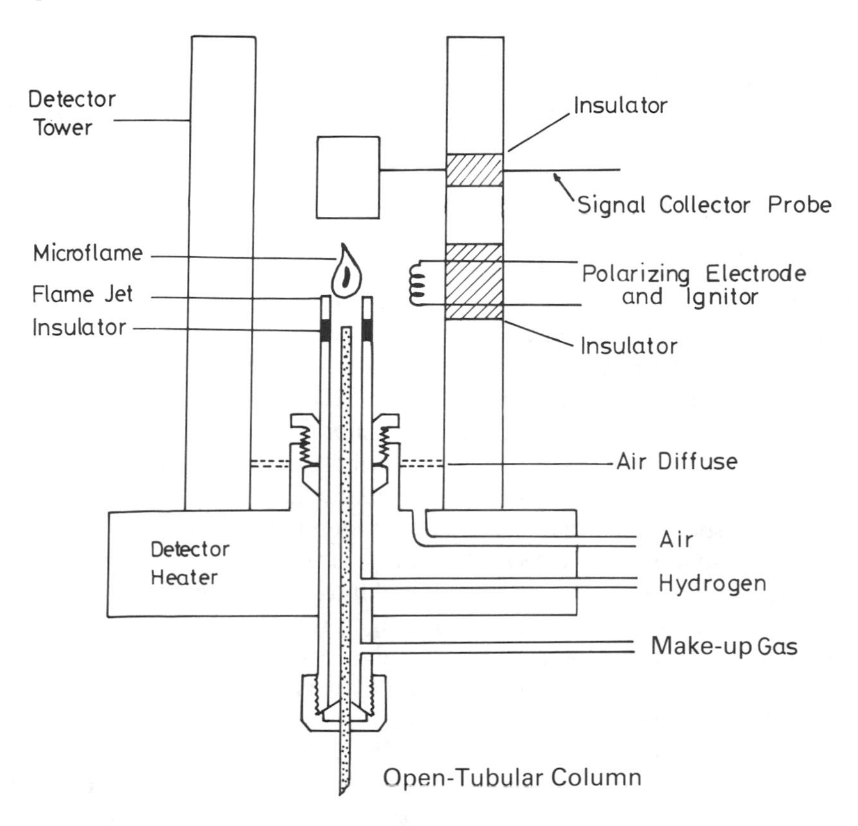


Fig. 5.2 — Schematic diagram of a flame-ionization detector. Reproduced with permission from Ref. 5; copyright (1984) John Wiley & Sons, Inc.

used for the flame. The dilution does not affect the sensitivity of the detector since the FID responds to absolute mass of the solute and not the concentration.

The reaction of hydrogen with oxygen to form water generates a lot of heat but is non-ionic, with a plasma consisting of H, O, OH and OOH radicals. When an organic compound containing C-H bonds enters the flame, positively charged ionic species and electrons are formed. The formation of the ions is detected by the passage of current across the two electrodes, the negative electrode usually being the jet itself. The shape of the positive electrode, and the distance between the two electrodes, is important and determines the sensitivity, linear dynamic range and the selectivity of the detector. It is often a ring, a wire mesh or a cylinder, held about a centimetre above the tip of the jet.

It may be noted that normally only one in 100,000 carbon atoms introduced into the flame gives up a detectable electron. Though this should make the technique of flame ionization detection very inefficient, it is still the most sensitive for the compounds containing the C—H bonds. The ratio of the number of electrons generated to the number of the hydrogen-bearing carbon atoms is so stable that the response factor can often be calculated from the carbon number of the molecule. The presence of heteroatoms and other functional groups in the molecule often modifies the *effective carbon number*, which is otherwise equal to the number of carbon atoms present in the molecule. The effective carbon number for heptane and benzene is thus 7.00 and 5.95, respectively. The effective carbon number contribution of a carbonyl group is zero while that of a nitrile group is 0.3 and that of an ether linkage is −1.0. The response factors for complex molecules are best determined experimentally for quantitative accuracy.

Several modifications are possible to render the FID sensitive to species to which it normally does not respond. Thus, at high hydrogen flow-rates and with an electrode geometry that permits the electrodes to be heated by the flame, carbon disulphide gives a response equivalent to 1/1000 to 1/100 of that of hydrocarbons. The increased hydrogen flow also causes the FID to respond to carbon dioxide, carbon monoxide, nitrogen oxides, hydrogen sulphide and even helium and oxygen, albeit at lower sensitivity; the response to nitric oxide under these conditions, however, was found to be equal to that of methane [2].

Selective response to organometallic compounds is obtained in the *hydrogen-atmosphere flame-ionization detector* (HAFID) in which the flow-streams of hydrogen and air are reversed; that is, air, or preferably a 1:1 mixture of nitrogen and oxygen, is introduced into the jet and hydrogen is allowed to flow around it. The relative volumes of flow of the two gases are also reversed, so that the hydrogen flow (which is typically around 1600 ml/min but may vary from 1–3 l/min) is an order of magnitude higher than that of air. Interestingly, it has been found that introduction of silicon compounds (e.g. from the septum or the stationary phase, carried to the detector by the carrier gas) is essential for the desired response and the HAFID is designed to deliberately introduce a few $\mu$l/min (10–35 ppm) of silane along with hydrogen [6]. The optimum concentration of silane depends on the detector design, particularly the distance between the collector electrode and the jet; the larger the distance, the lower is the optimum concentration of silane. The response of the detector to organometallics is directly related to the distance of the electrode, which is typically between 50 and 70 mm. A negatively charged electrode is more selective to organometallics than the positively charged one. Depending on the nature of the metal, the, response of the HAFID to organometallic compounds is several orders of magnitude higher than that to hydrocarbons. The detector is most sensitive to manganese and aluminium compounds and the minimum mass flow-rate for these compounds (and for those of iron, tin and chromium) is much lower than that of FID for hydrocarbons; e.g. manganese can be detected at a level of 17 fg/sec. The molar response of the compounds is proportional to the number of heteroatoms, while retaining the effective carbon atom response as in the FID. For a detailed account of the selectivity of HAFID to different metallic compounds and the effect of various parameters such as the detector design and flow-rates on its performance, see Refs. [2 and 6].

Just as introduction of silicon into the flame renders the HAFID selective to certain organometallics, introduction of the latter, commonly ferrocene, makes the detector selective to silicon and can thus be called HAFID-Si. Again, as in the case of the normal HAFID, the response and the peak shape depend, to a large extent, on the detector geometry (i.e. the diameter of the jet and the depth of introduction of hydrogen *vis-a-vis* the jet level), the electrode potential and the flow-rates of hydrogen and oxygen, and the concentration of the organometallic compound. The positive response to silicon reaches a maximum when the concentration of silicon reaches around 5 ppm; around 12 ppm, it passes into a negative response, again reaching the maximum at a concentration of around 35 ppm. Though the negative response is about three times greater, the detection limit is actually smaller by a factor of 20 owing to the concomitant increase in the noise level. The selectivity of response of Si to C is about 10,000:1. The HAFID-Si shows similar selectivity to phosphorus, iron and chlorine-containing compounds. The response of HAFID to these compounds is high, even in the absence of the doping agents which, in addition to ferrocene, include aluminium, nickel, chromium, molybdenum, brass, platinum, copper and magnesium. The detection limit of tetraethylsilane to HAFID is 4 ng compared to 50 pg of HAFID-Si [2].

A flame-ionization detector with a hydrocarbon background (i.e., a FID into which a few hundreds of microlitres/min of methane or acetylene are introduced), gives a negative response to several inorganic gases to which the common FID is generally insensitive. At appropriate flow-rates of hydrogen and hydrocarbon (which may be mixed with either the carrier gas or hydrogen), hydrogen sulphide, sulphur dioxide, carbon disulphide and nitrous oxide could be detected at levels of 10 nanomole. The response was somewhat lower for oxygen and the oxides of carbon. Even nitrogen and helium could be detected at micromole levels using perfluoromethane as the carrier gas.

The sensitivity of FID to halogen compounds may be increased by using high hydrogen flow-rates and positioning the electrode close to the flame. The sensitivity is approximately two orders of magnitude higher when hydrogen is the carrier-gas and oxygen is used instead of air. The response is directly proportional to the number of halogen atoms in the molecule.

## 5.4 ALKALI FLAME-IONIZATION DETECTOR (AFID)/THERMIONIC DETECTOR (TID)

The alkali flame-ionization detector (AFID), also known as the thermionic detector (TID) or the nitrogen–phosphorus detector (NPD), is essentially the flame-ionization detector modified by introduction of an alkali-metal vapour into the system so that its response to nitrogen- and phosphorus-containing compounds is enhanced by several orders of magnitude. In its simplest form the modification is achieved by fusing an alkali-metal salt to the negatively charged electrode ring. Such a detector is 600 times more sensitive to phosphorus and 300 times more sensitive to nitrogen-containing compounds relative to the unmodified FID. The selectivity may be further enhanced by placing a second flame detector above the first, but separated from it by a salt-coated platinum wire mesh. In this design, the first flame burns the organic compounds and vaporizes the alkali-metal salt; the vapours are then transferred to the second flame where the ionization occurs. Further improvements in specificity and detector design resulted in commercial versions with selectivity enhancement of five to six orders of magnitude for phosphorus and four orders of magnitude for nitrogen. In most of these designs, the alkali metal is located around the burner tip in the form of the salt beads or ceramics impregnated with the salt [2]. In this form, the effectiveness of the detector could be prolonged, equilibrium established rapidly and the sensitivity enhanced. In the more recent versions, the flame is often substituted by electrical heating of the volatile alkali-metal activator and hence the name thermionic detector.

In addition to the detector design, the response of the AFID depends on all the operating conditions, including the flow-rate of hydrogen (which determines the temperature of the flame) and the nature of the alkali-metal salt. The response generally increases with the increasing atomic weight of the metal. Though the AFID is well known for its very high sensitivity for compounds containing phosphorus and nitrogen, its response to several other heteroatoms is also significantly higher than the normal FID, while its sensitivity to hydrocarbons is at best the same as FID. The signal for sulphur is negative and is 20 times that given by FID, while that for arsenic is between five and 30 times. For halogen compounds the sensitivity is an order of magnitude higher than that of FID and it is a maximum for bromine when sodium is the alkali metal used. The AFID is marginally more sensitive (relative to FID) to boron, tin, lead and silicon.

Several explanations have been offered for the increased sensitivity of the AFID to phosphorus and nitrogen but none very satisfactorily. It appears that the ionization is facilitated by collision of PO, OPO, CN and halogen radicals with the excited alkali-metal ions or radicals formed by the application of thermal energy to the metal salt.

## 5.5 FLAME PHOTOMETRIC DETECTOR (FPD)

The FPD is not an ionization detector in the sense that the detection is not based on the principles discussed above but on the photometric detection of the species formed in the flame. However, ionization does occur in the flame and the detector can also be used as an FID. Selectivity is based on the fact that characteristic emission spectra are exhibited by several heteroatoms such as phosphorus, sulphur and different metals. The appropriate mode is selected by choosing the right filter which is placed between the flame and the photo-multipliers. A hydrogen-rich flame generates characteristic chemiluminescent species such as $HPO^*$ for phosphorus and $S_2^*$ for sulphur. Detectors are available which can detect both phosphorus and sulphur simultaneously.

The basic features of the FPD are shown schematically in Fig. 5.3 [7]. The effluent is mixed with air or oxygen and fed to the detector while hydrogen and air are fed directly to the detector. The radiation generated by the decomposition of the solutes is reflected by a concave mirror to the photo-multiplier tube through an appropriate interference filter.

As in the case of most other detectors, the response of the FPD is a function of the carrier gas flow-rate, the structure and concentration of the solute, and the detector temperature. The optimum flow-rates of the different gases most often have to be determined for the particular detector. The oxygen/hydrogen flow ratio is generally around 0.25 but may vary from 0.1 to 0.5. At optimum flow-rates, the minimum detectable level for P is given as 0.1–2 pg/sec and for S as 2–50 pg/sec. While the response may be expected to be proportional to the number of P and/or S atoms present in the molecule, it may vary widely for a given number of heteroatoms, based on the efficiency of production of the active species, and therefore depends on the

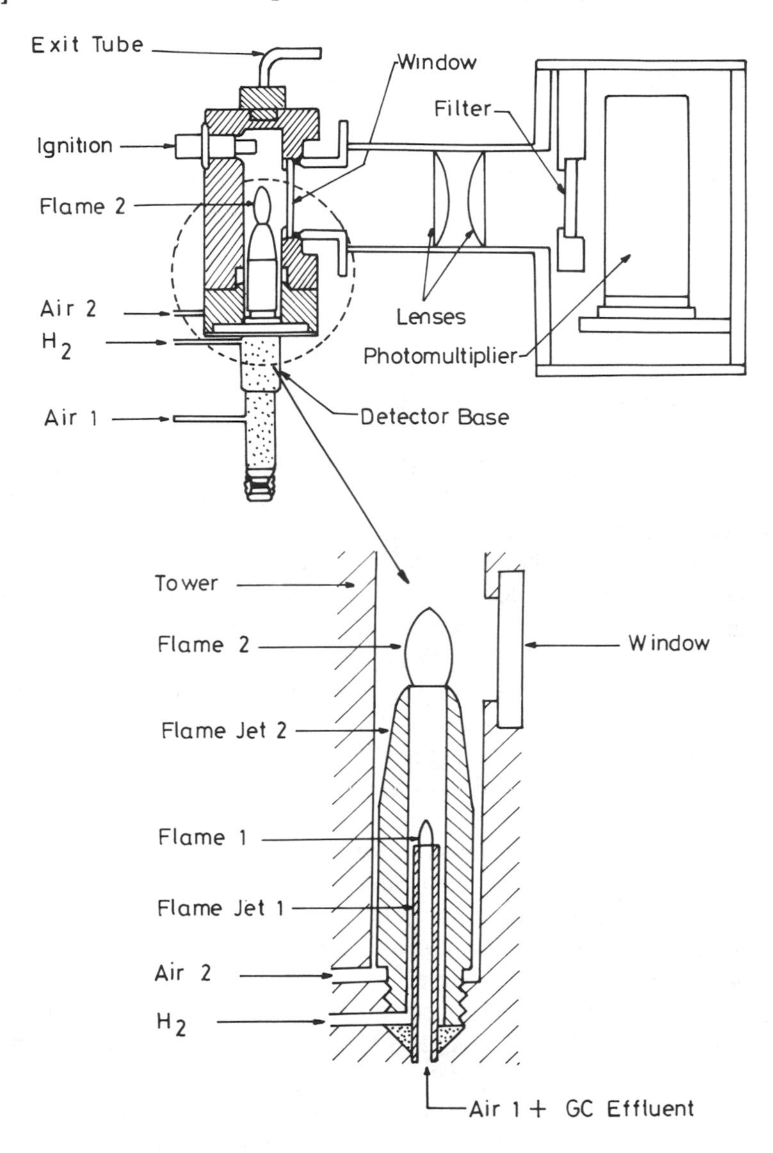


Fig. 5.3 — Schematic diagram of a flame photometric detector. Reprinted with permission from Ref. 7; copyright (1978) American Chemical Society.

functional group. The FPD response to phosphorus increases with temperature but so also does the detector noise; on the other hand, its response to sulphur decreases with rise in temperature [2]. Hence, the lower the temperature, the better.

Phosphorus compounds generally give a linear response (526 nm) over four orders of magnitude, but the sulphur signal (394 nm) is often non-linear probably because the formation of the active radical species is *via* a second-order reaction. The linearity of the sulphur signal is improved by dosing the carrier-gas with a constant concentration of a volatile sulphur compound. The selectivity ratio of the detector for phosphorus is 100,000 : 1 and that for sulphur, 10,000 : 1. By using appropriate filters, specific detection of boron (546 nm), chromium (425.4 nm) and nitrogen (690 nm) is possible. Halogens may be detected by the green (526 nm) colour produced when a copper wire mesh is incorporated into the burner; use of indium halides in the same way gives characteristic emission spectra for the different halogens.

## 5.6 ELECTRON-CAPTURE DETECTOR (ECD)

The ECD is the best known of the detectors that use a beta-ray (electron) emitting radioactive source. It is also the most sensitive of the GC detectors for certain types of compounds that are capable of capturing the electrons. The principle and applications of the ECD have been reviewed by Lovelock [8], who is credited with its invention and many of the later developments. The complex mechanisms that operate and the practice of ECD in chromatography have been discussed in detail in a recent monograph [9].

As the name indicates, the principle of detection in an ECD is the capture of electrons (generated by suitable means) by certain groups or atoms present in the solute molecules. The most common electron source is a radioactive material that emits low energy beta-particles; for example, tritium ($^3H$) or $^{63}Ni$. The radiation ionizes the carrier-gas into ion pairs of positive ions and electrons. Either nitrogen or argon may be used as the carrier-gas, but the latter is normally preferred in view of the high diffusivity for electrons in argon; the carrier-gas is admixed with 5–10% of methane to reduce the concentration of the metastable argon and promote thermal equilibrium of the electrons. Possessing only the thermal energy, the 'thermal' electrons do not combine with the positive ions but stay free to be 'captured' by those solute molecules with adequate electron affinity. For higher efficiency of separation in the open-tubular GC, hydrogen, or helium, may also be used as the carrier-gas, provided nitrogen or a mixture of argon and methane (10%) is used as the make-up gas, at a flow of not less than 20 ml/min.

The ECD also contains two electrodes across which a potential is applied periodically and very briefly to collect and measure the electrons. During the period of time when they are not being collected, the electrons can be captured by any molecule or group that has affinity for electrons. These molecules then become negatively charged and combine with the positive ions, causing a decrease in the measurable current, which is reflected in the ECD signal. The pulse-operation is necessary to prevent a number of undesirable effects of the simpler DC mode, i.e. formation of space charges, unwanted ionization effects, etc. In the DC mode, the electron energy varies with the applied potential and the affinities of different molecules for the electrons vary with the electron energy. A constant electron energy, therefore, needs to be maintained and this can be done simply by allowing the reaction between the electron and the molecules to occur in a zero-field condition. Since each pulse reduces the concentration of the electrons to zero, the longer the interval between the pulses, the more electrons are present and the greater is the sensitivity.

Early ECDs used a tritium source which had a low temperature limit of about 200°C and a parallel-plate electrode design which was prone to contamination. These defects, which were particularly serious when the solute was relatively less volatile and had high electron affinity, could be corrected using a $^{63}$Ni source that would withstand a temperature of 400°C. Several detector designs have been suggested; two of the common designs are shown in Fig. 5.4 [10].

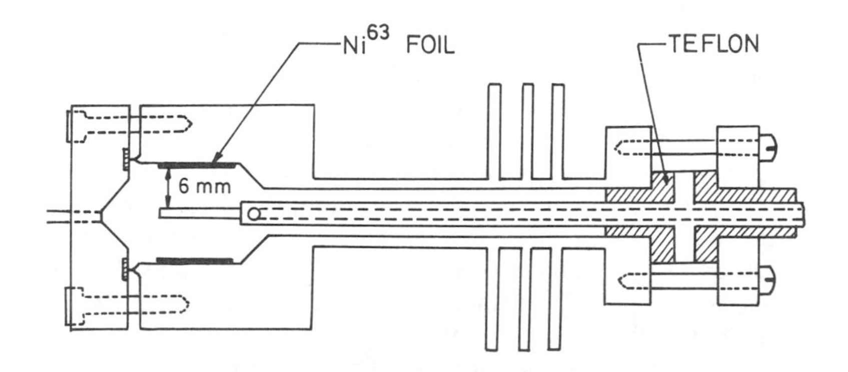


Fig. 5.4a — Schematic diagram of a co-axial high-temperature electron-capture detector. Reprinted with permission from Ref. 10; copyright (1971) American Chemical Society.

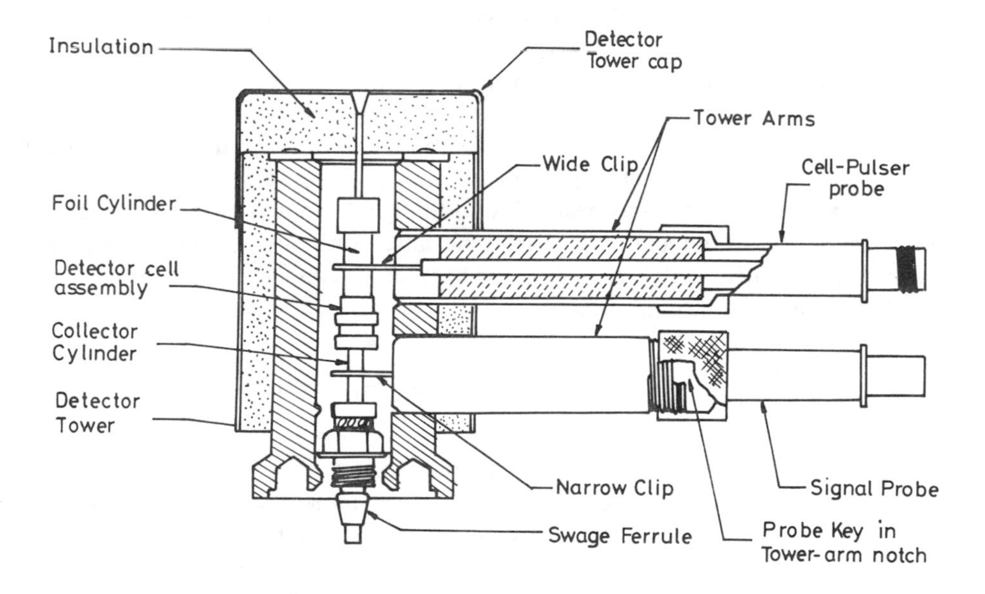


Fig. 5.4b — Schematic diagram of a displaced co-axial cylinder electron-capture detector; by courtesy of Varian Associates, U.S.A.

Use of radioactive materials as sources of electrons has certain disadvantages, such as contamination of the source by the GC effluents or column bleed, possible reactivity of the source with the electron-capturing species, and the well-known hazards of the radioactive materials in general. ECD detectors using non-radioactive $^{53}$Fe, thermionic barium zirconate cathode and the principle of photo-ionization have been described [2]. The thermionic-type detector offers fg level sensitivity and a wide linear dynamic range (six orders of magnitude). These features, coupled with the ruggedness of the detector, should enhance its utility.

The ECD mechanisms are highly temperature dependent and therefore the detector temperature substantially affects the sensitivity of detection. Thus, the detection limit of ethylene dichloride is 1 ppb at 350°C while it is only 1 ppm at 80°C. It follows that the detector temperature should not only be high but must be steady within the limit of 0.1 to 0.3°C for a quantitative precision of 1% [2].

The selectivity of the ECD for certain types of compounds depends on the fact that the electron-capture coefficient, $K'$ (the ratio of the decrease in the observed current to the standing current), can vary widely from 0.01 for aliphatic hydrocarbons, ethers and esters to about 10,000 for certain organomercurials. Between the two extremes lie compounds containing nitro and halogen groups. The ECD response increases with the size of the halogen atom ($I > Br > Cl > F$) and is proportional to the number of halogen atoms. The sensitivity is higher if the electron-capturing atoms are on the same carbon atom; thus, carbon tetrachloride can be detected at 1 fg/sec.

The structure–activity relationship makes it possible to tailor derivatives for selective detection by the ECD. Apart from introducing the ionophore, the derivatization should also ensure sufficient volatility and thermal stability of the solute. Halo, haloacyl, nitro, pentafluorophenyl and boronic acid derivatives are the preferred derivatives for ECD. Though fluoro compounds are less effective electron-capturing groups among the halides, they are more volatile, and multiple substitution enhances sensitivity. An enhancement of four orders of magnitude in the sensitivity is possible by reaction of chloro compounds with sodium iodide ahead of the ECD. The usefulness of the ECD lies in its extreme sensitivity to several environmental hazards such as pesticides and polychlorinated biphenyls. ECD has also been used for detection of organometallic compounds; under favourable conditions, the sensitivity (e.g., 0.4 pg for beryllium) can surpass that obtained by neutron activation or atomic absorption.

## 5.7 HALL ELECTROLYTIC CONDUCTIVITY DETECTOR (HECD)

Detection of gas chromatographic effluents by electrolytic conductivity poses two major problems, i.e. non-availability of gas-phase electrochemical sensors and non-responsiveness of most organic solutes to electrochemical or electrolytic sensors, even if the first problem is overcome by dissolving the column effluent in a suitable solvent. Most conductivity detectors solve the latter problem by either oxidative or reductive (or, sometimes, pyrolytic) treatment of the solute to convert it into an electrolytically active species. The decomposition products are absorbed in water and the conductivity measured. The most popular among the various detectors of this type is the Hall electrolytic conductivity detector [11].

The sample decomposition for the conductivity detection is usually carried out in a low-volume tube furnace, made of quartz or nickel at temperatures ranging from 500 to 1000°C. Air, oxygen or hydrogen, depending on the choice of the reaction mode (i.e. oxidation or reduction), is mixed with the column effluent. The carbon, hydrogen, halogens, nitrogen, sulphur and phosphorus in the solute molecules are oxidized respectively to carbon dioxide, water, halogen acids, nitrogen oxides, sulphur dioxide and phosphorus pentoxide in the oxidation mode, or reduced to methane, water (from oxygen), hydrogen halides, ammonia, hydrogen sulphide and phosphine. Selectivity of detection may be improved by use of chemical scrubbers (e.g. potassium carbonate to remove hydrogen halides and oxides of sulphur, silver wire to trap hydrogen sulphide, and halides or alumina for phosphine) so that undesired products are eliminated. The reaction products are dissolved in a solvent, usually water or an alcohol, and the conductivity measured in an appropriately designed cell. Fig. 5.5 [12] shows a bipolar, pulsed, differential detector which first

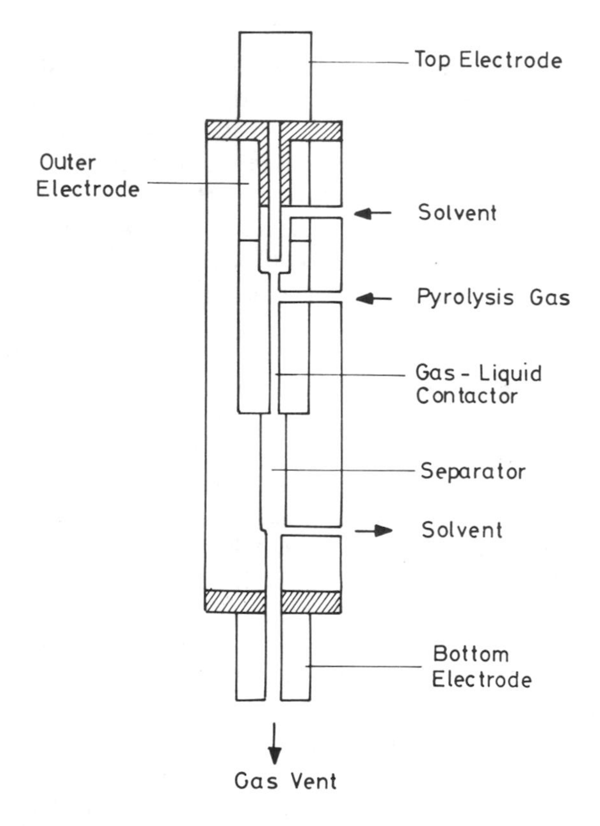


Fig. 5.5 — Cross-section of a Hall electrolytic conductivity detector. Reproduced from Ref. 12 by courtesy of Elsevier Scientific Publishing Co.

measures the conductivity of the solvent in the first portion of the cell and that of the solution of the reaction products in the second part; the difference in the conductivities is recorded. Methanol or isopropanol, either alone or admixed with water, are the preferred solvents (over pure water) in view of their greater sensitivity, selectivity and linear dynamic range. In most detector designs, the solution, after conductivity measurement, is recirculated to the cell after passing through ion-exchange resins. Modification of the solvent pH sometimes improves sensitivity of detection by increasing the ionization of the solute.

The HECD is a concentration-dependent detector, and thus its response is very sensitive to changes in the flow-rate of both the carrier gas and the solvent, and the sensitivity is higher at lower flow-rates. The detection limits for chlorine, nitrogen and sulphur are 0.1, 1.0, 1.0 pg/sec, respectively, with a linear dynamic range of three to five orders of magnitude. The selectivity ratio relative to carbon may vary from 10,000 to a billion.

## 5.8 PHOTO-IONIZATION DETECTOR (PID) AND MISCELLANEOUS DETECTORS

The six detectors described above are the most frequently used detectors for routine analysis. There are, however, several other detectors, based on widely different principles of operation that are described in the literature, for both general and specific applications. Some of these are merely of historical interest (e.g. the gas-density balance) and some others are of essentially academic interest. Several interesting detectors are under different stages of development and many are yet to be commercialized. Given below are some of the more important ones and the sources of information on them: photo-ionization detector [13], helium/argon ionization detector [1], cross-section detector [1], microwave plasma detector [14], flame emission detector [1], microcoulometric detector [15], piezolectric sorption detector [16], ultrasonic detector [17], radioactivity detector [18], Brunel mass detector [19], chemiluminescence detector [2] and ion-mobility detector [20].

Among the above, special mention must be made of the photoionization detector in which significant developments have occurred in recent years, making it a viable detection mode for capillary GC too. The principle of operation is based on excitation of a suitable discharge gas (argon, helium or hydrogen at 0.1–1 mm Hg) by UV radiation to a metastable state. The excited species emit photons of sufficient energy to ionize most compounds, permitting their detection at very low levels (minimum detection level for benzene, 2 pg). The PID may be operated in a universal or selective mode by changing the photon energy of the ionization source. The basic design of the PID is similar to that of the FID, with the photon source located between the column exit and the ion collector. An optically transparent window, made of lithium, magnesium or sodium fluorides separates the discharge compartment from the detection cell. A combination of a LiF window (transparency limit 11.9 eV) and He gas (photon energy 11.3–20.3 eV) yields high photo-ionization energy to permit universal detection. The photon energy is lowered for selective detection by changing the discharge gas to, say, hydrogen (energy, 6.9 eV) or by using a barium fluoride window (transparency, 9.2 eV). The high sensitivity, selectivity, and wide linear dynamic range of seven orders of magnitude permits a wide range of applications, including trace analysis.

In addition to the above, essentially gas chromatographic, detectors, combination of GC with the independent spectroscopic techniques such as atomic absorption spectroscopy (AAS), infrared (IR) and mass spectrometry (MS) have made the so-called hyphenated techniques extremely versatile sources of qualitative and quantitative information on a variety of samples. The GC–AAS is particularly attractive for detection of organometallic compounds at nanogram levels [21]. The GC–MS and GC–FTIR are more universal and are discussed in greater detail below.

## 5.9 GAS CHROMATOGRAPHY–MASS SPECTROMETRY (GC–MS)

Among the various possible combinations of chromatography with different analytical techniques, that of gas chromatography with mass spectrometry (GC–MS) has been the most successful, useful, and routinely applicable. The reasons for this are not hard to see. Both techniques use very small quantities (ng or pg range) of the sample, both require the sample to be volatile, and the mass spectrometer can respond as fast as the GC can elute. The only apparent incompatibility is that while the GC operates at the atmospheric pressure, the MS requires a high vacuum (about $10^{-6}$ Torr); the problem of evacuation without significant loss of the sample has been solved to a great extent. The MS scores over the other detection techniques in its ability not only to universally and quantitatively detect the sample, but also to furnish information on the molecular weight, the formula, and (from the fragmentation pattern) the structure of the compound. Under favourable circumstances, which is fortunately quite often, positive identification of the sample is possible. The interpretation of the mass-spectral data also lends itself to automation and the compatibility of GC–MS to on-line computers has greatly helped the data collection, reduction and processing.

The principle of mass spectrometry is very simple. Normally the compound under examination is volatilized under vacuum and the molecules in the vapour state are bombarded with high-energy (70 eV) electrons. Since 8–16 eV energy is sufficient to ionize most organic molecules, the bombardment with high-energy electrons, and the consequent transfer of energy causes the molecules not only to ionize but also fragment into smaller units. (The high-vacuum system is required to minimize the collisions between the sample ions and the background gas molecules). Theoretically, a molecule containing only three atoms, A–B–C, can yield a positive ion ($ABC^+$) and an electron by ionization, and the positive ion can split into AB + C and/or A + BC; both AB and BC can then split into A + B and B + C, giving a possible total of 6 ions. The possible number of ions rises to 10 for a linear molecule containing 4 atoms. Strictly, the loss of an electron from the molecule should yield a radical ion which, on further fragmentation should yield a positive ion and a radical. However, since the split can occur either way, often both the ions are formed. Also, as the number of atoms increase, not only the simple bond cleavages noted above but also elimination, rearrangement and secondary reactions can and do occur, thus increasing the number of fragments. Fortunately, however, not all fragments are formed to the same extent; in other words, the concentration of the different ions is not the same but is determined by the relative bond energies and the relative stability of the individual ions. Therefore, depending on the structure of the compound, different characteristic ions are formed to different extents and it is this feature that makes mass spectrometry extremely useful in structure determination of organic compounds. The fragments formed in the ionization chamber are led to a mass analyser where they are separated on the basis of their mass-to-charge ratio. The mass spectrum is conventionally recorded with the masses of the ions on the abscissa and their relative abundances on the ordinate. The mass spectrum is usually presented as a bar diagram in the normalized form; i.e. with the intensity of the most abundant ion taken as 100 (Fig. 5.6).

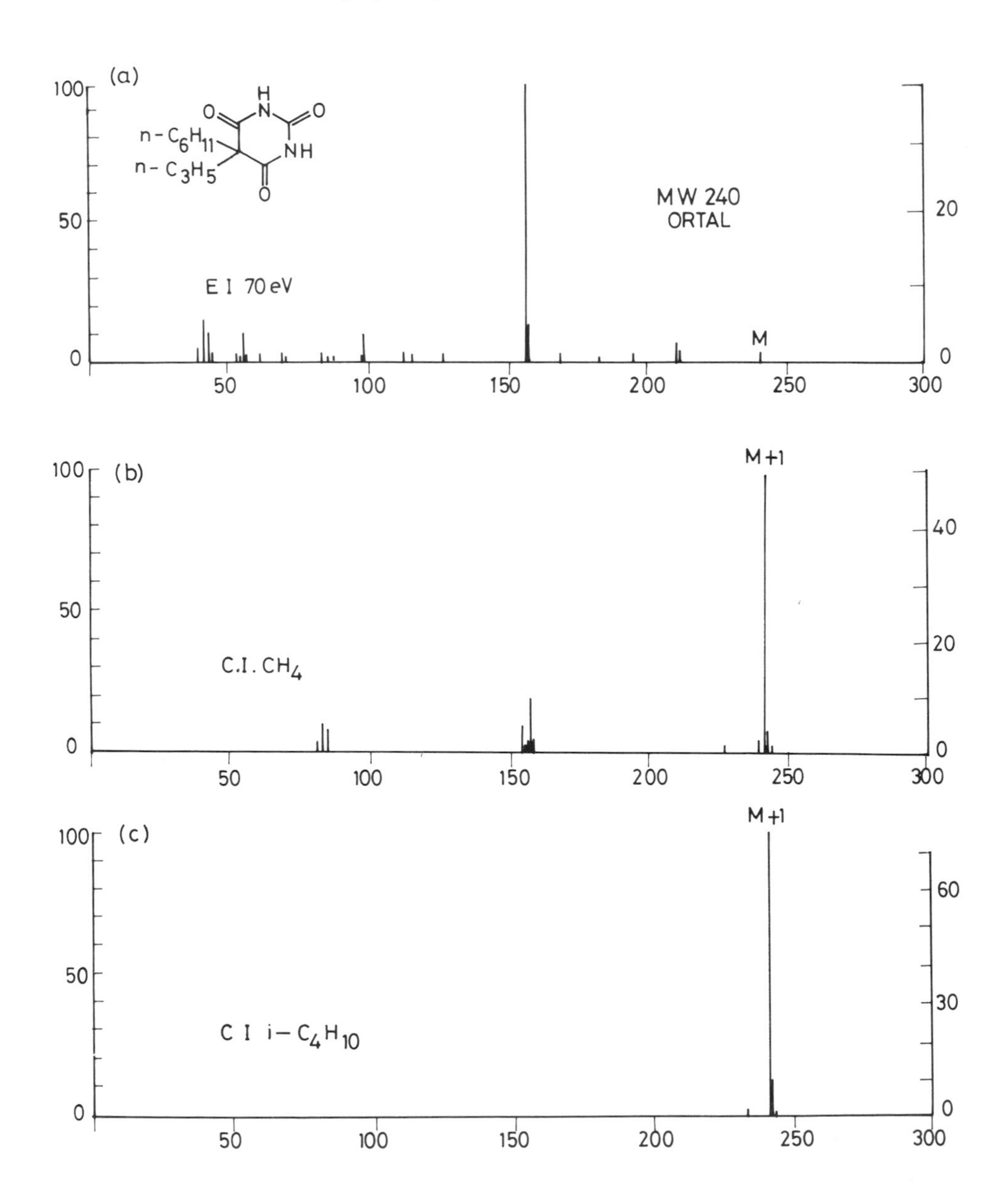


Fig. 5.6 — Comparison of EI and CI mass spectra. Reproduced from the *Journal of Chromatographic Science* [24] by permission of Preston Publications, Inc.

### 5.9.1 Ionization methods

#### *5.9.1.1 Electron ionization (EI)*

The electron ionization technique described above is the most widely used for obtaining the spectra of volatile organic compounds. Its popularity is due to many factors such as ease of operation, stability, precise control of the energy, efficiency of ionization, etc. In addition, as noted above, this method of ionization yields characteristic and diagnostic fragmentation patterns. Fig. 5.6a shows a typical mass spectrum obtained by the EI method. It shows the highest mass peak (usually the molecular ion or the ion of mass corresponding to the molecular weight of the compound) at 240 and several other peaks representing smaller (fragment) ions. The mechanism of fragmentation and the fragmentation pattern is of considerable interest to the mass spectrometrist and the organic chemist as these often lead to valuable information on the structure of the molecule [21]. However, since computer-assisted mass spectral identification is fairly well established and since computerized mass spectral libraries are readily available, it is not very necessary in routine analysis to understand these mechanisms in detail.

In Fig. 5.6a, the molecular ion ($M^+$) is relatively small. This is the case with a large number of EI mass spectra and, in some cases, the molecular-ion peak may not appear at all. Since the molecular ion directly gives the molecular weight of the compound, its appearance in the mass spectrum is very useful. It is more advantageous if the $M^+$ ion is intense since the relative intensities of the satellite peaks (i.e. $M + 1$, $M + 2$, etc.), which arise from the natural abundance of different isotopes of the elements present, are useful for calculating the molecular formula [22]. For example, the relative abundance of $^{13}C$ being 1.1% of $^{12}C$, the relative percentage of the $M + 1$ peak divided by 1.1 gives the number of carbon atoms in the molecule. The fact that the atomic weights of elements (relative to carbon as 12.0000 or oxygen as 16.0000) are not whole numbers is used in high-resolution mass spectrometry to obtain the molecular formulae directly since the accurate mass (with an accuracy of $\pm$ 0.0001) would fit only one of the several formulae possible for a given mass. For example, the nominal mass of carbon monoxide, molecular nitrogen and ethylene is 28, but the accurate mass, which differs with the elemental composition, is 27.9949, 28.0062 and 28.0313, respectively. The three compounds can thus be readily distinguished using an instrument with a resolution of 3000; higher resolution is, of course, needed for compounds of higher molecular weight with closer accurate masses. The high-resolution mass spectrometers, however, are too expensive for routine use.

#### *5.9.1.2 Field ionization (FI)*

Since the extent of fragmentation is a function of the energy given to the molecule, the fragmentation can be reduced, and the abundance of the molecular ion correspondingly enhanced, by decreasing the electron energy. As noted earlier, the molecular ionization potential of organic molecules is in the region of 8–16 eV and this technique of molecular ion enhancement has been used for a long time. Also, since the ionization potentials of certain classes of molecules are lower than others, specificity of detection is also possible. Thus, at an ionization potential of around 10 eV, aromatic hydrocarbons are selectively ionized in the presence of aliphatic hydrocarbons. The $M^+$ ion is produced more conveniently by the field ionization technique in which a high positive electric field (10–100 million V/cm) is produced by a sharp blade or wire held at a high positive voltage of 8 to 14 kV. The high electric field strength causes a potential-energy barrier in the molecule which induces electron tunnelling; the resultant molecular ion is drawn into the mass analyser. Since the available energy to the molecule is of the order 12 to 13 V, which is just sufficient for the organic molecules to ionize, there is very little further fragmentation.

#### *5.9.1.3 Chemical ionization (CI)*

An alternative to FI is chemical ionization [23]. Here, a reactant gas, such as methane or isobutane, is introduced into the ion chamber at 0.2 to 2 Torr pressure (1 to 10 ml/min); water, deuterium oxide, ammonia, hydrogen, nitric oxide and oxygen are also used for the purpose, but much less frequently. Ionization and collision of the reactant gas molecules produce metastable ionic species which act as proton donors or hydride-ion abstractors on the sample molecules to yield $(M + 1)^+$ or $(M - 1)^+$ ions. All reagent gases give a lot of peaks in the low end of the spectrum corresponding to the active species formed. Owing to its low proton affinity, methane is also likely to give rise to some fragmentation of the solute molecules. Isobutane is frequently preferred as the reagent gas. Fig. 5.6b and 5.6c show the CI spectra of ortal using methane and isobutane, respectively [24]. There is a close correspondence between the peaks detected by FID and by the total-ion current chemical-ionization mass spectrometry, using methane as the reagent gas. With isobutane, on the other hand, the ionization of aromatic and aliphatic hydrocarbons is suppressed while that of compounds containing oxygen, nitrogen and halogens remains strong.

The reagent gas may be used as carrier-gas for the GC but is often introduced directly into the ionization chamber. Because of the high flow-rates of the gases used, the instrument would need a higher-capacity or a second diffusion or turbo-molecular pump. Use of a 'carrier-cum-reagent' gas like helium also enhances fragmentation, resembling the EI spectra. The fact that different reagent gases produce different spectra from the same compound has been used on occasion as an analytical strategy, but the technique is too cumbersome for routine use. It must be mentioned that FI and CI do not replace EI but are complementary to it; EI gives valuable information on the structure of the molecules and is currently more useful for identification purposes.

Under appropriate conditions negative ions are formed by a variety of processes, the most interesting among which is electron abstraction. The electron affinity of halogens is well known and the high rate constant of this process has been used for selective detection of halogenated compounds and other environmental pollutants [25]. Use of oxygen as the reagent gas has certain advantages of providing information about the number of chlorine atoms in the molecule; also, the ion source remains cleaner, permitting higher throughput of samples. Modern instruments offer the facility of recording alternative positive- and negative-ion spectra on the same sample.

#### *5.9.1.4 Atmospheric pressure ionization (API)*

The principle of atmospheric pressure ionization is similar to the electron-capture detector in the sense that a radioactive source ($^{63}Ni$) is used to cause the ionization, but in a weak inert-gas plasma. The ion–molecule reactions occur at the atmospheric pressure, outside the mass spectrometer vacuum system and hence the name. The sensitivity under appropriately chosen conditions may be in the range 0.1–1.0 fg. The sample may be directly introduced using a solvent and is thus more attractive in LC–MS (see later); no restriction is obviously needed between the GC column and the ion source. Both positive and negative ions are produced by a series of reactions including hydride abstraction, electron-capture, charge-transfer and complex formation.

### 5.9.2 Mass analysers

The ions generated in the ionization chamber are accelerated and directed to a mass analyser where they are separated on the basis of their mass-to-charge ratio ($m/z$); since the ionic charge ($z$) is usually 1, the the separation is on the basis of the mass. There are several types of mass analyser among which three are more suitable for GC–MS; they are known as the magnetic sector, time-of-flight and quadrupole mass analysers.

When an accelerated ion enters a magnetic field, it takes a curved path, the square of whose radius is proportional to the mass-to-charge ratio. If the radius and the accelerating voltage are fixed, the mass of the ions collected at an appropriately situated slit is proportional to the square of the applied magnetic field. The resolution obtainable on these instruments varies inversely with the slit width and is adequate for most GC–MS applications. The resolution, defined as $m/\Delta m$ (where $m$ is the mass and $\Delta m$ is the distance between two adjacent peaks at mass $m$ and resolved at 10% valley) may be in the range of 300 to 400. High-resolution instruments (resolution $>$ 10,000, suitable for determination of accurate mass and molecular formulae) use a double-focusing technique, with both electrical and magnetic field.

The time-of-flight (TOF) instruments measure, as the name indicates, the time (1–50 sec) taken by the ions to pass through a 1-metre drift tube; the velocity of the ions of a given mass-to-charge ratio is proportional to the square-root of $2V$, where $V$ is the accelerating voltage. In other words, the time taken by the ions (of unit charge) to travel the 1 metre distance under a constant accelerating voltage is proportional to the mass of the ions. It can be calculated that if a hydrogen ion ($m = 1$) takes 1.7 sec to travel 1 metre under an accelerating voltage of 2000 V, an ion of mass 2500 would take 50 sec. It may thus be possible to record up to 20,000 spectra per second. The speed and resolution (500–600) contributed to its early popularity in GC–MS systems. Though over-shadowed by the quadrupole systems for some time, the TOF technique appears to be staging a comeback. Since the TOF instruments operate at much lower temperatures (100–150°C) compared to the ion-source temperatures of 200–250°C used in the magnetic-sector instruments, the molecular-ion abundance could be higher.

The quadrupole mass filter consists of four circular (or hyperbolic) parallel rods, with the opposite rods electrically connected. The applied voltages consist of a DC component and a radio frequency oscillator. The ions entering the region oscillate between the adjacent electrodes of opposite polarity. At a given radio frequency, only ions of certain mass attain stable oscillation between the electrodes and reach the detector, while all others collapse into the electrodes. The peaks in a quadrupole mass analyser have the same peak width and the resolution (mass/peak width) thus varies with the mass. The resolution of the quadrupole mass analysers is, therefore, normally stated as unit mass resolution up to to a specified mass; unit mass resolution up to a mass of 1500 to 2000 daltons is typical. The quadrupole analysers show considerable discrimination against ions of $m/z$ larger than 250, their intensity being considerably lower than in the other mass analysers. The linear mass scale, fast peak switching and digital control may be advantageous in automated instruments which are now common. Magnetic-sector instruments are preferred in case of higher mass ranges (MW > 1000) and in high-resolution analysis. Both the time-of-flight and quadrupole analysers are particularly suited for selected-ion monitoring (see below).

### 5.9.3 The GC–MS interface

As noted above, while the two techniques, GC and MS, are both very powerful in their respective spheres, i.e. separation and structure determination, and are mutually compatible in most respects, the main complication in combining the two techniques arises from the fact that the former operates under atmospheric pressure and the latter, under high vacuum. Several strategies have been employed for preferential removal of the carrier gas and introduction of the sample into the mass spectrometer [26]. The effectiveness of the GC–MS interface is determined by two factors; the enrichment factor and the efficiency. The enrichment factor is given by the ratio of the concentration of the solute after passing through the interface to the original concentration, and the efficiency is given by the percentage of the compound in the GC effluent entering the mass spectrometer. While achieving the maximum enrichment and efficiency, the ideal interface is expected not to interfere with the performance of either the gas chromatographic detector or the mass spectrometer, be devoid of any memory effects and to be non-discriminative with respect to the chemical nature of the solutes. There is expected to be no chemical change in the solute between the time it leaves the column and the time it enters the ion chamber. If all these characteristics are satisfied, the chromatogram obtained by plotting the total ion current in the mass spectrometer against time should resemble that obtained using a universal detector such as the FID.

The simplest method of combining a gas chromatograph and a mass spectrometer is by direct coupling (Fig. 5.7a), which is also the most efficient; this is feasible in conjunction with open-tubular columns where the flow-rates are low (usually less than 4 ml/min) so that the diffusion pump can maintain the vacuum fluctuations to within tolerable limits. Because one end of the column is under vacuum, the pressure gradient across the column might cause a decrease in the column efficiency but this is generally not very significant. A major problem in introducing the open-tubular column directly into the ion source of the mass spectrometer is that the column change becomes very laborious and time-consuming. With the advent of fused-silica columns, this is not a major problem. Commonly, one end of a short piece of the empty fused-silica column is left in the ion source and the other end is connected to the separation column outlet *via* a butt connector (consisting of a double-tapered Vespel ferrule housed in a stainless-steel body, and a nut to provide a gas-tight seal) or an open-split interface [27]. In the latter system, the outlet end of the GC column and the inlet tubing of the mass spectrometer are separated by a narrow slit and helium is allowed to flow through the splitter device

| TYPE OF INTERFACE | FLOW-RATE RANGE ml/min * | EFFICIENCY ENRICHMENT ** | OPERATING TEMP., °C | DECOMPOSITION ABSORPTION EFFECT *** |
|---|---|---|---|---|
| a 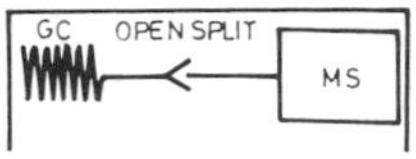 GC OPEN SPLIT MS | 1 - 100 | 1-90 % – | < 400° | – |
| b 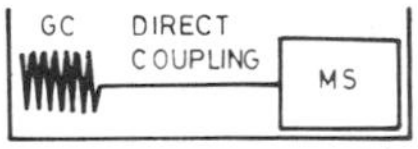 GC DIRECT COUPLING MS | ⩽10 | 100 % – | < 400° | – |
| c 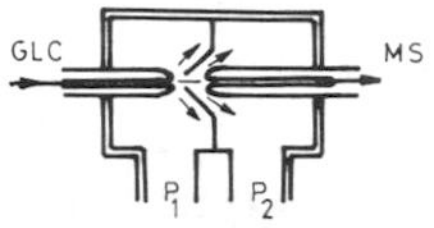 GLC MS $P_1$ $P_2$ | 10-80 | ≈ 50 % ≈ 50 | < 400° | + |
| d 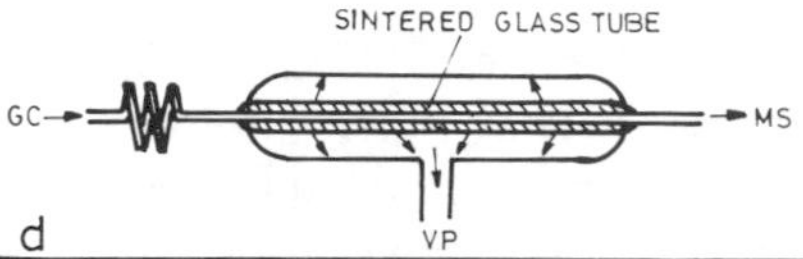 SINTERED GLASS TUBE GC MS VP | 15-100 | ≈ 40 % 20-400 | < 400° | + + |
| e 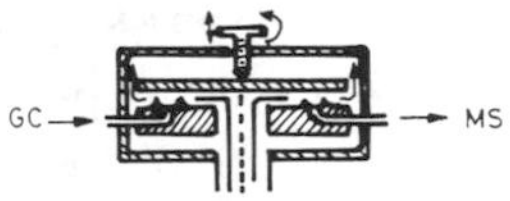 GC MS | 0-80 | 10-100 % 10-60 | < 400° | + |
| f 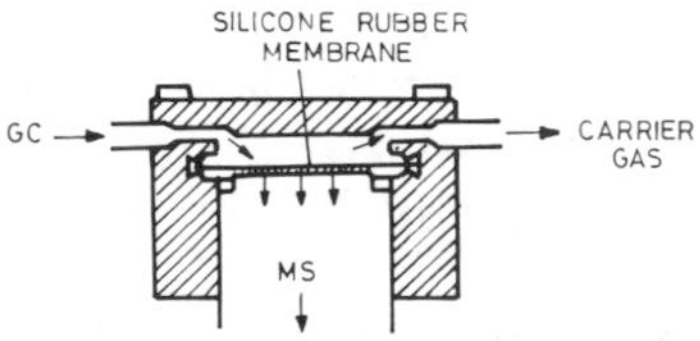 SILICONE RUBBER MEMBRANE GC CARRIER GAS MS | 1-50 | ≈ 90 % $10^4$ | < 180° | + |
| g 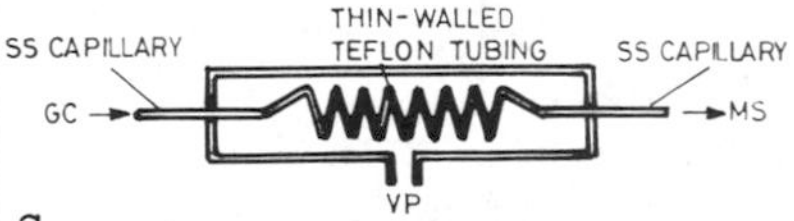 SS CAPILLARY THIN-WALLED TEFLON TUBING SS CAPILLARY GC MS VP | ⩽ 20 | ≈ 80 % 8 | 270-330° | +++ |
| h 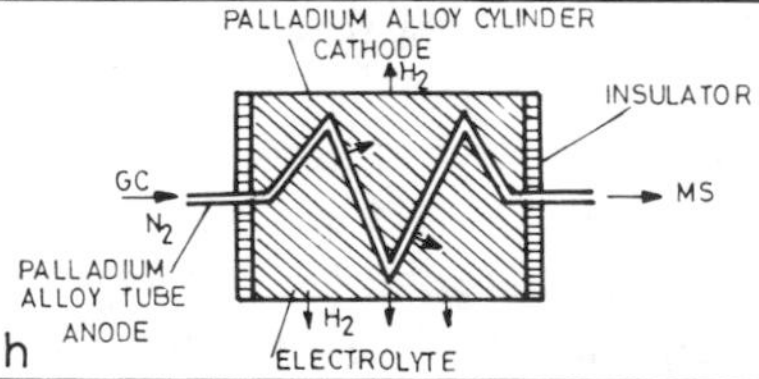 PALLADIUM ALLOY CYLINDER CATHODE $H_2$ INSULATOR GC $N_2$ MS PALLADIUM ALLOY TUBE ANODE $H_2$ ELECTROLYTE | ⩽5 | ≈100 % $10^6$ | ≈300° | +++ |

i
* Depends on type of column and on 1 or 2 stage version of separator
** Depends on flow-rate range and molecular weight
*** — no effects + neglectable
++ small effects +++ pronounced effects

Fig. 5.7 — Carrier-gas separators for GC–MS. Reproduced from Ref. 26 by courtesy of Elsevier Scientific Publishing Co.

encasing the slit. The flow into the mass spectrometer is fixed by the diameter of the inlet tube and when the carrier-gas flow-rate is larger, the column effluent is suitably split and the excess sample is vented.

Direct coupling of the gas chromatograph and mass spectrometer is also possible with packed columns by splitting the effluent, but since the split ratio has to be large and there is no enrichment, the efficiency is poor (1–10%). Fig. 5.7 shows the most common carrier-gas separators used in GC–MS and their operating parameters. The Watson–Biemann separator (Fig. 5.7d) is made up of a cylindrical fritted glass tube with a pore size of about 1 $\mu$m diameter, through which the carrier-gas can selectively diffuse out. The efficiency is about 40%, with an enrichment factor between 20 and 400. A similar arrangement but with a silver-palladium membrane (Fig. 5.7h), maintained around 350°C has the unique property of being permeable only to hydrogen. When hydrogen is the carrier-gas, this separator can offer 100% efficiency and an enrichment factor of over a million. In both cases, adsorption and decomposition of polar compounds has been observed.

The most commonly used interface is the all-glass jet separator (Fig. 5.7c), which has a 50% efficiency and can achieve 50-fold enrichment. The carrier-gas flow rate of up to 15 ml/min can be tolerated by a single-stage separator; a two-stage separator can tolerate up to 80 ml/min of flow. The device (also known as the Ryhage separator) may have up to four orifices of 0.1 to 0.3 mm diameter. The GC effluent enters and diffuses into the chamber, which is connected to a vacuum pump *via* $P_1$ and $P_2$. While the lighter carrier-gas diffuses away from the core of the flow, the heavier solute molecules are expected to travel straight through, to be further enriched on passage through additional similar devices. In the Llewellyn or the membrane separator (Fig. 5.7f), the semipermeable silicone membranes allow organic molecules to pass but not the carrier gas. The membrane offers very high efficiency (90%) and enrichment of four orders of magnitude but is prone to harden and lose permeability above 180–200°C.

Since a majority of the separators are based on differential diffusion, the lighter gases, hydrogen and helium are carrier-gases of choice. Hydrogen must be used in the case of the Ag–Pd separator, but in all other cases, helium is preferred for reasons of safety and for obtaining the total-ion chromatogram (see below). Reactant gases such as methane or isobutane may be conveniently used as the carrier-gas in connection with open-tubular columns and direct inlet systems when the chemical ionization technique is employed.

Using an open-tubular GC column, a quadrupole mass analyser and helium as the carrier-gas, very inexpensive, dedicated GC–MS systems have been designed and are commercially made available by Hewlett-Packard (mass-selective detector) and Finnigan-MAT (ion-trap detector). They are suitable for most of the routine GC–MS applications.

### 5.9.4 The SFC–MS interface

As noted earlier (section 4.1.3), supercritical fluid chromatography (SFC) is uniquely suited for the analysis of non-volatile, thermally labile, polar compounds that are not amenable to traditional GC and cannot be easily detected by the common LC detectors. It is normally operated at pressures ranging from 50–500 atm., and the flow-rates in packed-column SFC, as well as the fluid densities, are also more akin to LC than GC. The instrumentation and the strategies for interfacing the packed-column SFC with MS are also similar to those in LC–MS (see section 7.5.6), which is certainly more problematic than GC–MS. The more recent open-tubular column SFC, on the other hand, is more akin in instrumentation and operation to GC.

While coupling of SFC with MS has all the problems associated with that of LC and MS, the problems are less severe owing to the much higher volatility of the mobile phase. The main problem in SFC–MS arises from the requirement of keeping the mobile phase under high pressure until the end of the column and quickly decreasing the pressure by several (ten !) orders of magnitude. Such rapid expansion of the fluid would invariably cause cluster formation. This problem is partly overcome in the CI mode by the colliding shock waves which are produced by the interaction of the supersonic jet of the expanding fluid and the ambient reagent gas [28]. Coupling of the SFC to EI is very difficult owing to the relatively low ion-source pressures unless very low mobile-phase flow-rates are used and the ionization chamber is modified by placing a heated expansion chamber prior to the ion source in order to minimize the cluster formation.

With the introduction of narrow-bore (25–100 μm i.d.) fused-silica columns with bonded stationary phases, SFC with very low mobile-phase flow-rates (5–50 μl/min, required for direct fluid introduction into the mass spectrometer) has become feasible. In addition, several mobile phases with convenient supercritical properties make excellent reagent gases for CI–MS; e.g. methane, ammonia, *n*-pentane and isobutane. Use of the mobile phase as the reagent gas in CI–MS helps to improve detection sensitivity and selectivity and to maintain constant CI conditions. The basic instrumentation for SFC includes a high-pressure programmable syringe pump and an injection port that uses a typical HPLC valve with volumes from 0.2 to 0.6 μl; there is also a flow splitter with split ratios in the range of 1:4 for 50 μm or larger-bore columns to 1:80 for fast separations on a 25 μm column. The interface between the column and the ion source is a drawn and tapered fused-silica restrictor of 4–8 μm i.d. which is heated to compensate the cooling effect of the expanding fluid. The mass spectrometer is the conventional quadrupole instrument with a two-stage pumping system. Typical detection limits range from 0.1 to 10 pg depending on the compound and the experimental conditions [29].

### 5.9.5 Special techniques

Some of the techniques of data manipulation are unique to GC–MS. At the outset, four different combinations of GC and MS may be distinguished. They are based on the low resolution (LR) and high resolution (HR) capabilities of the two instruments giving rise to LRGC–LRMS, LRGC–HRMS, HRGC–LRMS and HRGC–HRMS. As noted earlier, LRGC or the common packed-column GC may be adequate for most routine analysis but is being fast replaced by the more efficient HRGC or the open-tubular GC. HRMS, on the other hand, is expensive and relatively slow and LRMS remains the most frequently used mode in GC–MS. In all the cases, the vast amount of data that needs to be collected (i.e. the large number of GC peaks, often closely placed and some of which may be overlapping, each peak giving rise to its own mass spectrum and each mass spectrum consisting of a large number of peaks that need to be normalized for convenient reading) requires the facility of a computer for data collection, reduction, storage, retrieval and for carrying out several of the manipulations to be discussed below. The modern GC–MS systems are almost invariably supplied with an on-line computer, which controls both the operation of the equipment and processing of the data. The computerized operation of the mass spectrometer offers an inherent advantage. The MS data for the same compound may vary widely depending on a number of instrumental parameters such as the energy of the ionizing electrons, the repeller and the analyser voltages, the arrangement of the ion-focusing systems and the scanning speed. Exact manual reproduction of the parameters being difficult and time-consuming, microprocessor calibration of the parameters against a standard (e.g. perfluorotributylamine, PFTBA) is necessary for obtaining reproducible data.

With the aid of the computer and the quadrupole mass analyser, it is now possible to routinely acquire the mass spectra in the 1 to 1000 daltons range from nanogram quantities of mixtures containing hundreds of compounds. An additional feature of the combined GC–MS–DS (DS = data system) is the facility for recording the mass spectra continuously (say, in intervals of 5 sec) throughout the chromatographic analysis. The mass spectra are thus recorded even when no peaks are visible (baseline) but such a record is useful for background subtraction from the mass spectrum of the peak of interest. While it also ensures that no peak is inadvertently missed and the data acquisition can be carried out unattended, the main advantage of the process lies in the possibility for post-run manipulations of several kinds [30].

#### *5.9.5.1 Computerized mass-spectral identification*

The mass spectrum lends itself to easy computerization since each peak is completely describable by a set of two numbers, i.e. the mass and the abundance. Computerization of the mass spectral data library for retrieval, comparison and identification of the unknown compounds is both desirable and possible. Still, the large amount of the MS data for each compound needs reduction and is done by selecting only some of the most abundant and important peaks. Facilities for large libraries and quick identification are available in several countries, where the computers can be reached by telephone. Depending on the storage capacities of the on-line or the in-house computers, libraries for a smaller number of different types of compounds can be purchased and used in the laboratory; at present libraries with over 46,000 mass spectra are available. The libraries can also be prepared by individual laboratories as required. A number of algorithms (i.e. the ways of handling a certain problem in a finite number of specific steps) have been devised for the comparison of the spectra of the unknown compounds and these have been described in detail in the literature [31].

Two types of library searches are possible. The forward search traditionally attempts to match the unknown spectrum with one of those in the library, assuming that the sample is homogeneous. The reverse search, which has overtaken the older method, compares the known, reduced mass spectrum with the unknown. The advantage is that it can locate the relevant peaks even if the unknown is a mixture, so that hidden or unresolved peaks can also be identified [32]. The earlier algorithms required matching of six to 12 of the most abundant or the most unique, peaks of the unknown with the reduced spectra in the library. In the KB system of Klaus Biemann [33], the two largest peaks in each 14 unit interval (actually 6–19, 20–33, 34–47, etc., rather than 1–14, 15–28, 29–42, etc.) are selected and compared in order not to split common-ion clusters. The algorithm of Heller [34], called the PEAK, is a conversational user-interactive option of the mass spectal search system (MASS); it is interesting in that it involves the user in asking the computer to progressively reduce the number of possible compounds by selecting specific peaks in the specified abundance range for comparison. In recent years, the completely automated probability-based matching (PBM) system has become more popular [31]; it involves the probability *weighting* of the two principal types of data; namely, the masses and the abundances on the basis of their *uniqueness*.

#### *5.9.5.2 Total ion-current chromatogram*

The total ion current is the summation of the ion currents of all the ions present in the ion chamber. When helium (ionization potential = 24.6 eV) is used as the carrier-gas and the electron ionization energy is set around 20 eV, most of the other substances in the chamber are ionized. The plot of the ion current against the retention time is the total ion-current chromatogram, which is similar to the chromatogram obtained using a universal detector such as the FID. A good total ion-current chromatogram, as well as the standard EI mass spectrum, can be obtained by means of a switching circuit to operate the electron gun alternately at 20 eV and 70 eV. As already noted, the total ion-current chromatogram may be recorded in the chemical ionization mode also but, except in the case of methane, the picture may vary significantly from that obtained using the FID. The data can be used for both qualitative and quantitative analysis just as any universal detector, with the added advantage that selectivity can be achieved by appropriately choosing the conditions, or even after the GC–MS run is completed. Indeed, the post-run manipulations with the aid of the computer, permits among other things the so-called *extracted ion-current profile* of selected masses for very specific detection. Plots of the ion current of specific ions, or sets of selected ions, which can be obtained in this fashion are variously known as mass fragmentograms, mass chromatograms or reconstructed ion-current plots.

#### *5.9.5.3 Selected-ion monitoring (SIM)/mass fragmentography*

Originally, when the scanning of the complete mass spectrum was not fast enough, the strategy was to select a specific ion, presumably characteristic of the compound of interest, and to run the chromatogram by monitoring the particular ion. While the total ion-current chromatogram described above would give peaks due to all the compounds present in the sample, the selected-ion monitoring (SIM) would show only those peaks whose mass spectra contain the specific (selected) ion. The process may be repeated using different mass numbers. Though this method is no longer needed to obtain mass-spectral information about the compounds, the technique of SIM remains popular in view of its usefulness in enhancing the sensitivity of detection of certain compounds in the presence of several other compounds. The signal enhancement arises from the possibility of using very slow scan speeds and the consequent increase in the time spent on each selected ion. The data-processing capabilities of the modern GC–MS system permits simultaneous monitoring of several (up to 20) selected ions for the analysis of compounds of interest in a pre-set mode or by the reconstruction technique. The SIM technique is routinely used for monitoring suspected drugs, metabolites in biological samples, and pesticide residues in agricultural products.

## 5.10 GAS CHROMATOGRAPHY-INFRARED SPECTROSCOPY (GC–IR)

Trapping of GC effluents and examination of the compounds by various spectroscopic techniques like ultraviolet (UV), infrared (IR), nuclear magnetic resonance (NMR) and mass spectrometry (MS), etc., has been practiced for a long time to obtain qualitative data for identification of the compounds. Barring the MS, the IR spectroscopy gave the most positive information on the identity of an unknown compound using the smallest quantity of the sample (about 10 $\mu$g). The IR spectrum of a compound is considered its fingerprint in the sense that, if two compounds have identical IR spectra, the compounds can be said to be identical. The energy of the infrared radiation (wavelength 2–15 $\mu$m) being very small, does not ionize the molecule as in the case of MS, nor does it excite the electrons as does the UV radiation, but it reflects the vibrational frequencies of the different interatomic bonds in the molecule. The IR spectrum of a compound reflects the functional groups present in the molecule and, often, their inter-relationships. The fact that IR spectra can be recorded in almost any physical state (i.e., solid, liquid, gas or gum) is also a great advantage though the spectra recorded in the gas or vapour phase could be significantly different from those recorded in the other phases. The most popular form of sampling for IR spectral recording has been to mix the sample with or condense it on potassium bromide powder and to make it into a thin pellet; IR spectra can also be recorded using the sample in the form of a solution in a solvent such as carbon disulphide or carbon tetrachloride.

An on-line GC–IR combination was not possible for a long time owing to the relatively slow nature of the recording of the IR spectrum compared to the normal rate of emergence of the GC peaks. Recent developments in the use of interferometers and computer-aided Fourier-transform (FT) and time-averaging techniques made it possible to record the IR spectra of very minute samples (10–50 ng) in less than a second [35]. Using an appropriately designed light pipe as the interface, it is now possible to record the spectra of the GC effluents from even capillary columns 'on the fly'. In a typical GC–FTIR instrument [36], a collimated light beam from a Nicholson interferometer is focused on to a gold-coated light pipe (2 mm x 16 cm, allowing multiple reflections of the IR beam and high transmittance) and the output is sensed by a liquid nitrogen-cooled cadmium telluride detector. The volume of the light pipe and the transfer lines should be small enough (0.1 to 1 ml) to prevent excessive band broadening of the GC peaks; they are kept at a high temperature (up to about 250°C) to prevent condensation of the vapours. The data collection by the computer is triggered electronically whenever the GC peaks exceed a pre-selected threshold ('on the fly' mode) or is done continuously in sets of pre-selected numbers of scans (every mode). A restricted form of every mode is to collect the data on parts of the chromatogram that exceed certain thresholds (every threshold mode). The presentation of the data may be in the form of the complete spectrum recorded 'on the fly' or in the form of an infragram (akin to the mass fragmentogram discussed above), obtainable by averaging the spectra collected over a small range, or a single frequency or wavenumber, and reconstructing the gas chromatogram. The chosen wavenumber, can be either for general absorption (3200–2800/cm) or for specific groups like 2000–1600/cm for the carbonyl, or 3600–3400/cm for the hydroxyl, etc. Once the time scale of the GC elution is determined, summation of the corresponding interferograms yields the IR spectra. Where the peaks are suspected to be overlapping, subtraction techniques can be used. Unresolved peaks can be detected by recording and superimposing infragrams of specific frequencies, characteristic of the suspected compounds.

The operating parameters that greatly affect the GC–FTIR experiment include the light-pipe temperature, sampling technique, make-up gas (total flow-rate around 5 ml/min), data acquisition rate, etc., in addition to the capillary column characteristics. As with GC–MS, several algorithm routines permit identification of the compounds by direct comparison of the spectra with those stored in the computer library [37]. Vapour-phase IR spectral libraries containing over 8000 spectra are commercially available. Depending on the nature of the solute and the experimental conditions, there is a wide variation in the detectability thresholds (40–400 ng) of different compounds. The technique of GC–FTIR has been used in several areas of analytical chemistry including environmental analysis and pyrolysis of polymers. Polar compounds have been analysed both in the free and in the derivatized forms. Free radicals and other reactive species could be studied using matrix-isolation techniques in which the sample is frozen with a large excess of the matrix gas (argon or nitrogen) on to a cold surface; very sharp IR spectral features are obtained because intermolecular interactions and molecular rotations are prohibited [38].

## 5.11 GAS CHROMATOGRAPHY–FOURIER TRANSFORM INFRARED SPECTROSCOPY–MASS SPECTROMETRY (GC–FTIR–MS)

It is interesting to note that while the mass spectra reveal a lot of information about the structure of the compound, it is often not possible to distinguish between closely related geometrical isomers. The IR spectrum, on the other hand, is inadequate as a stand-alone source of structural information but, in conjunction with the mass spectrum, can be a powerful tool in the structure determination. The GC–IR and the GC–MS data on the same sample may be obtained on the two instruments separately, but it is more convenient to couple the two techniques in a combined GC–FTIR–MS instrument which is now commonly available, though somewhat expensive at the moment [39,40].

The possibility of multiple detection in GC using different combinations of the various available detectors, and the strategies for achieving it, have been recently reviewed [41]. Two or more detectors may be connected to the GC column effluent either in series or parallel. Connection of two detectors in series requires that the first detector be non-destructive and the detector volume, and that of the connecting tubing, etc., be so small that there is no significant band broadening. Parallel connection using a column effluent splitter permits direct attachment of two or more detectors to the GC. The splitters may be of either the fixed ratio or the variable split type, made of stainless steel, glass-lined, all glass, ceramic or fused silica. Fixed ratio type splitters are preferred for programmed-temperature GC since the other type are prone to change the split ratio with temperature. Variable splitters, on the other hand, permit optimization of each of the detectors depending on the nature of the compound and the detector sensitivity.

The possibility of comparing the complementary structural information from the two sets of spectral data after a single injection is a major advantage of combining GC, IR and MS; but the fact that the IR is about two orders of magnitude less sensitive than the MS makes it necessary to use high capacity columns with high stationary-phase loading, sometimes at the cost of resolution. Wide-bore (0.32 mm or larger i.d.) columns with a stationary-phase film thickness of 1–5 $\mu$m are convenient. A FTIR instrument may be conveniently coupled to an existing GC–MS system *via* an interface, which may be in parallel or in series. Parallel interface requires an effluent splitter, with the major portion diverted to the IR spectrometer; the remainder may be further split between a FID and the mass spectrometer. In the serial interface, the column effluent naturally enters the IR spectrometer first and then the mass spectrometer. In either case, the chromatographic resolution and the requirement of high vacuum in the mass spectrometer, and ambient pressure in the IR instruments, need to be maintained by suitably optimizing the carrier and the make-up gas flow rates.

**REFERENCES**

[1] M. J. O'Brien, in *Modern Practice of Gas Chromatography*, R. L. Grob (ed.), Wiley, New York, 2nd edn., 1985.
[2] M. Dressler, *Selective Gas Chromatographic Detectors*, Elsevier, Amsterdam, 1986.
[3] G. M. Message, *Practical Aspects of Gas Chromatography/Mass Spectrometry*, Wiley, New York, 1984.
[4] S. Smith, in *Advances in Capillary Chromatography*, J. G. Nikelly (ed.), Huethig, Heidelberg, 1986.
[5] M. L. Lee, F. J. Yang, and K. D. Bartle, *Open Tubular Column Gas Chromatography*, Wiley, New York, 1984.
[6] H. H. Hill, Jr. and W. A. Aue, *J. Chromatogr.*, **122**, 515 (1976).
[7] P. L. Patterson, R. L. Howe, and A. Abu-Shumays, *Anal. Chem.*, **50**, 339 (1978).
[8] J. E. Lovelock, *J. Chromatogr.*, **99**, 3 (1974).
[9] A. Zlatkis and C. F. Poole (eds.), *Election Capture — Theory and Practice in Chromatography*, Elsevier, Amsterdam, 1981.
[10] D. C. Fenimore, P. R. Loy, and A. Zlatkis, *Anal. Chem.*, **43**, 1972 (1971).
[11] R. C. Hill, *CRC Crit. Revs. Anal. Chem.*, **8**, 323 (1978).
[12] C. F. Poole and S. A. Schuette, *Contemporary Practice of Chromatography*, Elsevier, Amsterdam, 1984.
[13] P. Verner, *J. Chromatogr.*, **300**, 249 (1985).
[14] W. R. McLean, D. L. Stanton, and G. E. Penketh, *Analyst*, **98**, 432 (1973).
[15] J. Sevcik, *Detectors in Gas Chromatography*, Elsevier, Amsterdam, 1976.
[16] H. Wohltjen and R. Dessy, *Anal. Chem.*, **51**, 1458 (1979).
[17] T. Todd and D. DeBord, *Am. Lab.*, **2**(12), 56 (1970).
[18] L. Schutte and E. B. Koenders, *J. Chromatogr.*, **76**, 13 (1973).
[19] S. C. Bevan, T. A. Gough, and S. Thorburn, *J. Chromatogr.*, **43**, 192 (1969).
[20] T. W. Carr (ed.), *Plasma Chromatography*, Plenum, New York, 1984.
[21] L. Ebdon, S. Hill, and R. W. Ward, *Analyst*, **111**, 1113 (1986).
[22] F. W. McLafferty, *Interpretation of Mass Spectra*, 2nd edn., Benjamin/Cummings, Reading, Massachusetts, 1977.
[23] G. Harrison, *Chemical Ionisation Mass Spectrometry*, CRC Press, Boca Raton, Florida, 1983.
[24] R. M. Milberg and J. C. Cook, Jr., *J. Chromatogr. Sci.*, **17**, 17 (1979).
[25] R. C. Dougherty, *Anal. Chem.*, **53**, 625A (1981).
[26] M. C. ten Noever de Brauw, *J. Chromatogr.*, **165**, 207 (1979).
[27] J. F. K. Huber, E. Matisova, and E. Kenndler, *Anal. Chem.*, **454**, 1297 (1982).
[28] P. J. Arpino, J. Cousin, and J. Huggins, *Trends Anal. Chem.*, **6**, 69 (1987).
[29] R. D. Smith, B. W. Wright, and H. R. Udseth, in *Advances in Capillary Chromatography*, J. G. Nikelly (ed.), Huethig, Heidelberg, 1986.

[30] M. A. Grayson, *J. Chromatogr. Sci.*, **24**, 529 (1986).
[31] F. W. McLafferty and R. Venkataraghavan, *J. Chromatogr. Sci.*, **17**, 24 (1979).
[32] F. P. Abramson, *Anal. Chem.*, **47**, 45 (1975).
[33] H. S. Hertz, R. A. Hites, and K. Biemann, *Anal. Chem.*, **43**, 681 (1971).
[34] S. R. Heller, *Anal. Chem.*, **44**, 1950 (1972).
[35] J. R. Ferraro and L. J. Basile (eds.), *Fourier Transform Infrared Spectroscopy: Applications to Chemical Systems*, Academic Press, New York, Vols. 1 & 2, 1978 and 1979.
[36] K. Krishnan, R. Curbelo, P. Chiha, and R. C. Noonan, *J. Chromatogr. Sci.*, **17**, 413 (1979), and accompanying papers.
[37] P. R. Griffiths, S. L. Pentoney, Jr., A. Giorgetti, and K. H. Shafer, *Anal. Chem.*, **58**, 1349A (1986).
[38] S. Smith, in *Advances in Capillary Chromatography*, J. G. Nikelly (ed.), Huethig, Heidelberg, 1986.
[39] J. C. Demirgian, *Trends Anal. Chem.*, **6**, 58 (1987).
[40] C. L. Wilkins, *Anal. Chem.*, **59**, 571A (1987).
[41] I. S. Krull, M. Swartz, and J. N. Driscoll, *Adv. Chromatogr.*, **24**, 247 (1984).

# *Part III*
# *Liquid chromatography*

# 6

# Liquid chromatography I: principles and methods

## 6.1 INTRODUCTION

Liquid chromatography (LC) has had a much longer history and a far wider application compared to GC. Though overshadowed to some extent by the rapid development of GC techniques and instrumentation in the 1950s and 1960s, LC has regained its dominant position in the 1970s and is now even replacing GC in some of the latter's traditional areas of application, following comparable developments in instrumentation and detection systems. It has always been recognized that LC is much more versatile than GC. Unlike in GC, where the mechanism of retention is predominantly partition, one could exploit several other solute, sorbent and solvent interactions in LC, as noted briefly in Chapter 1 and as will be discussed in detail in the following sections. Apart from the wider variety of stationary phases and mechanisms of their interaction with the solutes, selectivity in LC can also be achieved by varying the mobile-phase composition. In addition, LC obviates one of the most serious limitations of GC, namely, the requirement of sample volatility and, therefore, the need for derivatization of a majority of analytes. A large number of natural and synthetic polymers, biochemicals and inorganic compounds can thus be more conveniently analysed using LC. As the vast majority of LC is carried out at ambient temperature, formation of artefacts is avoided; the low-temperature operation is also less hazardous in the laboratory or industry. Another important advantage of LC is the convenience of effluent collection without having to cold-trap or prevent aerosol formation, and such other problems associated with GC; since most of the commonly used LC detectors are non-destructive, the system is amenable to quantitative and automatic fraction collection. For these reasons, LC is certainly preferable over GC for preparativc and proccss-scalc scparations. Thc convenience of TLC, a major area of LC, is also well known.

The theoretical or the physicochemical principles governing the retention of solutes in LC, and the inter-relationships among the various parameters, have been discussed in Chapters 1 and 2 in general terms. This part of the book is divided into three chapters. The present chapter deals mainly with the various principles and more specific mechanisms operative in different LC modes and their application to different types of compounds. Instrumentation in LC and planar chromatography form the subject matter of the two subsequent chapters.

The discussion in this chapter is rather traditionally biased with respect to practical aspects. However, no distinction is made in the application of different LC principles since such a distinction would be artificial. Nevertheless, some important differences between traditional and instrumental techniques of column liquid chromatography (CLC) may be pointed out. Speed and reproducibility of the latter are immediately apparent. The older technique involves manual packing of the column, sample application, fraction collection and analysis of the effluents [1]; the success and reproducibility of the operation depend largely on the individual skill. The packed-column beds are often used only once. The HPLC methods, on the other hand, commonly employ pre-packed and re-usable columns, sample injection through a syringe or an automatic injector, on-line detection and data processing. The speed of such operations enabled easy optimization of the chromatographic conditions and led to some very useful theories of solvent selectivity, particularly in liquid–solid chromatography. These features notwithstanding, traditional CLC is still routinely used in most laboratories for preparative separations of less complex mixtures. The operational aspects that are common to all traditional LC modes are discussed first. Other sections are arranged in order of increasing complexity of the retention mechanisms and techniques.

## 6.2 TECHNIQUES

### 6.2.1 Bed Preparation

The success and reproducibility of LC depend to a large extent on the quality of the chromatographic bed. The stationary-phase bed is held in a long cylindrical column, made of Teflon or glass. Where glass columns are used, the 'wall effects' may be suppressed by exposing the inner surface to a silylating agent. This is conveniently effected by pouring a hot (about 60°C) solution of dichlorodimethylsilane in toluene into the column, allowing it to stand for a few minutes and then decanting the solution off; this process may be repeated, if needed. The column is finally allowed to dry in a hot-air oven before use. The size and shape of the columns may vary considerably, from a disposable pipette or a laboratory burette to specially made columns with Teflon stopcocks to regulate the flow-rate, and sintered glass discs or frits at the bottom to support the stationary phase. Wide and short columns may be convenient for crude separations whereas long and narrow columns may be necessary for compounds that are difficult to separate. The ratio of the diameter to the length of the column may vary from 1:1 to 1:10 in the former case, and from 1:10 to 1:100 in the latter. The recommended diameter-to-length ratio also varies considerably (from 1:5 for ion exchange to 1:50 for partition techniques), depending on the method of separation. The ratio of the sample size to the amount of the stationary phase is also dependent on the method and may extend from 1:20 in ion exchange (stoichiometric) to up to 1:2000 in partition chromatography. It is also subject to variation by a factor of 10, depending on the separation factor ($\alpha$), the lower values of $\alpha$ requiring larger ratios.

Narrow capillaries or flexible Teflon tubing with a screw clip may be used at the lower end of the column to regulate the flow of the mobile phase through the bed. Glass stopcocks, which may require grease, may be avoided. Frits and stopcocks are also not generally favoured as they increase the dead-volume and result in some mixing of the effluent after it leaves the bed and before collection. A simple and effective way of minimizing the column dead-volume between the bed and effluent collection point is shown in Fig. 6.1a [2]. One end of the column is drawn into a

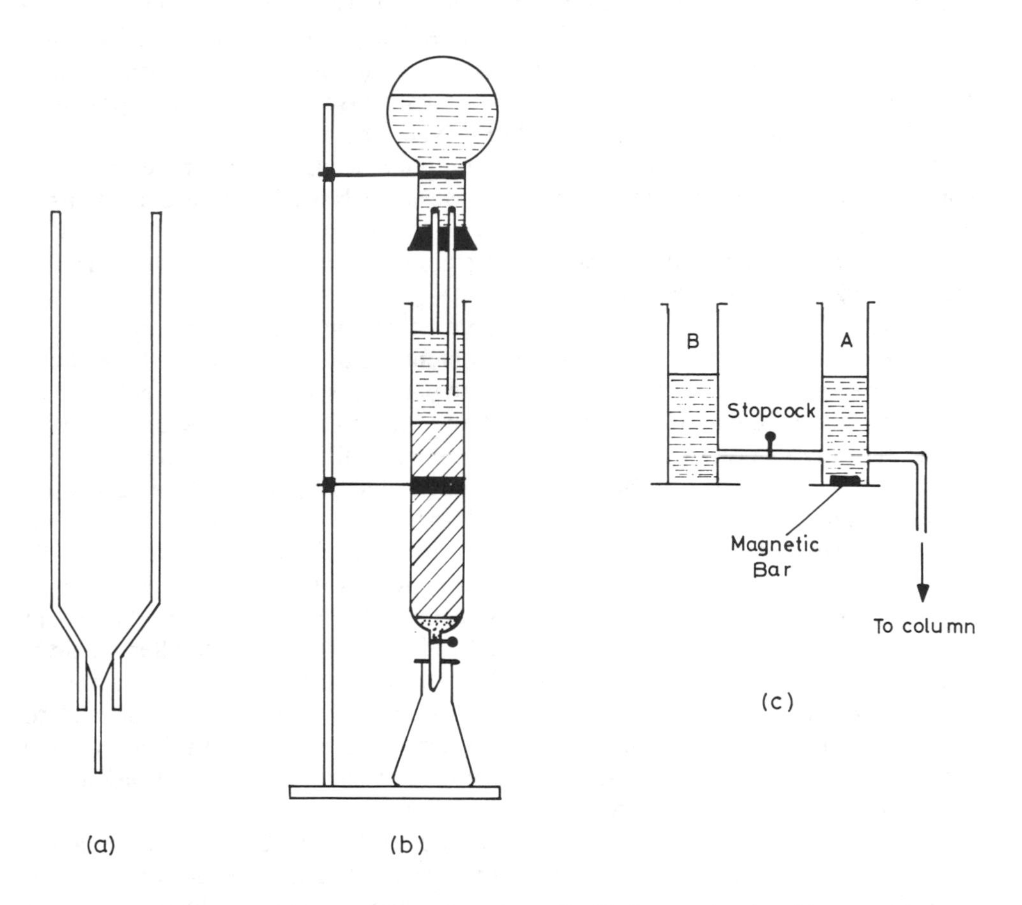


Fig. 6.1 — Arrangements for traditional column liquid chromatography: (a) empty glass column with a zero dead volume Teflon restrictor. Adapted with permission from Ref. 2; copyright (1978) American Chemical Society; (b) solvent delivery system at atm. pressure; (c) a simple gradient former.

narrow cone and a short (2 mm i.d., 8 mm o.d.) capillary tube is fused to it. A 2 mm o.d. Teflon tube (about 5 cm in length) is widened at one end using a hot pin and dropped into the column so that the other end passes through the capillary tube. It is then pulled down a bit so that the widened portion of the Teflon tube sits tightly in the capillary tube.

Depending on the principle of separation, the stationary phase to be packed in the column may vary from a polar (in the normal-phase chromatography) or non-polar liquid (reversed-phase chromatography) bonded or coated on an inert support, to a variety of adsorbents like silica gel, alumina, polyamide, charcoal, inorganic salts, polystyrene beads, ion-exchange resins, carbohydrates, and specially prepared packings for affinity chromatography. The mobile phases used for packing the column (for the wet-packing procedures described below) and elution are also equally varied. The packing procedure may vary somewhat depending on the nature of the stationary phase and individual preference. The procedures described here are fairly general but are elaborated with reference to the common normal-phase chromatography using silica gel as the stationary phase. They may be applied to other LC modes with occasional modification where necessary.

The column may be either dry-packed or prepared using a slurry of the stationary phase in a suitable solvent. The stationary phases, which swell to different extents in different solvents (e.g., some size-exclusion and ion-exchange packings), are invariably slurry packed. The slurry packing is also generally preferred as it yields a more uniform bed and reproducible results.

In the following description, it is assumed that the right stationary phase is available or prepared and the appropriate solvent for the preparation of the slurry has been selected. The solvent for slurrying and packing is normally the weakest in its ability to elute the compounds to be loaded on to the column (e.g. hexane for silica gel) so that the solutes are retained in the first few millimetres (or plates) of the column. To be able to record and reproduce the results, the particle size and porosity (or surface area) of the packing material needs to be specified. Commercial silica gels are available in a variety of modifications with surface areas ranging from 300 to 800 $m^2/g$. To a large extent, the flow-rate depends on the mesh size of the particles, which may vary from 100–200 (75–150 $\mu$m) for gravity flow to 200–400 (37–75 $\mu$m) if a slight pressure (conveniently from a gas cylinder) can be applied; still smaller particles would require solvent-pumping systems (Chapter 7). For best results, the narrowest possible size distribution is chosen; it pays to sieve the commercial adsorbents to obtain a narrow size distribution and to avoid very small or fine dusty particles which may clog the column. The flow-rate for a given particle size and head pressure is inversely proportional to the length of the column.

The column is clamped vertically and its lower end closed by a suitable arrangement. A slurry of the packing material is prepared in the chosen solvent, the ratio depending upon the swelling and solvent-gaining characteristics of the packing. The object is to obtain a well-equilibrated, pourable slurry. If the slurry is too thick, there may be some trapped air bubbles in the bed and, if too thin, size separation of the particles may occur and these could adversely affect the results. In some cases, as, for example, with several size-exclusion packings, it is advisable to allow the slurry to stand overnight to complete the swelling and equilibration. The column is filled to about a third of the height with the pure solvent and, if there is no base frit to support the bed, a small plug of cotton or glass wool is pressed to the bottom of the column using a glass rod. A few millimetres of purified sand may then be added, but is not always necessary. The slurry is then poured into the column and, as the bed settles down, the stopcock or the screw clip is slightly opened to drain the solvent and help the bed settle. As the solvent drains, more of the slurry is added, taking care to see that the bed in the column is not completely settled before more of the slurry is added; otherwise, density layers may form along the length of the bed. It is preferable to pour the whole of the slurry in one operation (uninterruptedly) but this is not always practicable. Care should be taken to see that no air bubbles are formed during or after the bed preparation; degassing the solvent may sometimes be necessary. It may help to gently tap the column wall around the area where the adsorbent is sedimenting; a rubber stopper attached to a wooden stick, or a glass rod, or a vibrator may be used for the purpose. The solvent should not be allowed to drain below the bed level. The column is repeatedly filled and drained to allow the bed to settle and when the bed is completely settled, it is ready for use. If a denser packing is desired, air pressure from a cylinder may be applied at the top. Several variations of this typical packing procedure are possible. A common variation is to pour the sieved adsorbent in a continuous stream into the column (which has been filled to about two-thirds length with the solvent) while maintaining a constant flow of the solvent through the column. Automated systems permit uniform and simultaneous packing of several columns. Pre-packed columns of different sizes and with different stationary phases are also commercially available.

Dry packing of the column is carried out by a similar process, except that the dry sorbent instead of the slurry is transferred to the column without any solvent. Uniform packing can be obtained by constantly tapping the sides of the column during the filling. Easy transferability of TLC results to preparative column liquid chromatography and faster separations are the main advantages of *dry column chromatography*. In *flash chromatography*, the dry packed column is flushed with the eluting solvent under pressure (to remove any trapped air) prior to sample application. By forcing the eluant through the column at the rate of 2 ml/min, 0.1 to 10 g of the sample, containing compounds with an $R_F$ value difference of about 0.15 could be separated in 15 min; using TLC-mesh silica gel, compounds with an $R_F$ difference of even 0.05 could be separated [3].

### 6.2.2 Sample application

A narrow and uniform band of the sample is applied at the top of the column. Improper application often results in streaking and mixing up of the separated bands. After preparation of the bed as described above, the solvent is drained slowly until the solvent level is just above the bed. For adsorption chromatography, the sample must be dried to avoid deactivation of the column. A solution of the sample, preferably in the same solvent that is used for the bed preparation and subsequent elution, is carefully transferred to the column without disturbing the bed surface. This is conveniently done using a long dropper whose tip is held just above the bed, touching the wall of the column. The solution is allowed to sink into the bed and the top of the column is rinsed down with the solvent. The above operations may be carried out while maintaining a slow and steady flow of the solvent through the column, not allowing the column to dry up. To avoid disturbing the bed surface, a layer of purified sand, glass wool or cotton wool may be placed on the top of the bed. If the sample is insoluble or only partly soluble in the eluting solvent, it may be dissolved in a minimum amount of a stronger solvent and diluted with the eluting solvent. Partial precipitation of the sample may occasionally occur in which case, the supernatant solution is transferred to the chromatographic bed and the procedure repeated.

In another procedure of sample application, a thick filter paper (e.g. Whatman 3MM, cut into a circular disc of diameter slightly smaller than that of the column) is dipped into a concentrated solution of the sample in a volatile solvent, withdrawn and the solvent evaporated. The operation may be repeated, if necessary. The disc, with the sample deposited on it, is placed on the bed after draining off the solvent above the bed. The sample is transferred to the bed as the column is eluted with the mobile phase. This procedure is convenient if the sample is very viscous or contains components that differ widely in their polarities and solubility. If the sample is insoluble in the eluting solvent, it may also be deposited on a small quantity of the adsorbent which is added to the solution of the sample in a volatile solvent. The solvent is then evaporated (preferably at room temperature under vacuum, using a rotary evaporator) and the dry adsorbent is transferred to the column.

A third procedure, useful in the chromatography of polymeric materials, involves placing the sample solution underneath the eluant, just on the top of the bed, with the help of a syringe. The eluant above the bed is not drained off prior to sample application, but a steady flow-rate is maintained during the operation. For this procedure, the viscosity of the sample solution should be more than that of the eluant.

### 6.2.3 Elution methods

#### *6.2.3.1 Isocratic elution*

Separation of the compounds loaded on the column is usually achieved by eluting with an appropriate solvent. The choice of a solvent for elution is a relatively complex task and requires an understanding of the mechanism of retention, separation and elution if an empirical approach is to be avoided and if a similar system has not been recorded in the literature for the same or similar compounds. These mechanisms and the techniques of choosing the mobile phase are discussed in detail in connection with the individual methods below.

The technique of elution development is distinguished from the other two methods of development of the chromatogram; namely, frontal and the displacement methods. The success of the elution process depends on the ability of the solvent to selectively desorb the components retained by the stationary phase. The plot of the concentration or the amount of substances thus eluted *vs* the volume of the eluant collected leads to the familiar elution curve or the chromatogram (Fig. 1.3). The rate of flow of the mobile phase is maintained rather empirically at as low as practicable, depending on the type and scale of chromatography. The flow-rate in ml/hour is usually adjusted to be equal to the amount in grams of the adsorbent used.

The elution may be carried out either in the isocratic mode (i.e., without changing the composition of the mobile phase) or in the gradient mode, which involves changing the composition of the eluant to a more powerful one after the initial sample application and elution of the less-retained components. The isocratic elution has several advantages such as easy reproducibility, reusability of the recovered solvents, compatibility with a wide variety of on-line detectors and the ease of solvent delivery. Continuous solvent delivery in isocratic elution at atmospheric pressure can be arranged using very simple devices. Fig. 6.1b shows an inverted round-bottom flask with two tubes of unequal length. The solvent in the column is maintained automatically at the level of the end of the shorter tube. As the solvent in the column drains, more solvent is delivered *via* the longer tube until it reaches the initial level. A stoppered separatory funnel with a wide stem also serves the purpose.

#### *6.2.3.2 Stepwise and gradient elution*

Convenient though it may be, isocratic elution suffers from one major drawback. A solvent system which gives a good resolution of certain components of a sample may prove too weak to elute all the components in a reasonable length of time. Strongly retained components may not be eluted at all unless the solvent strength is increased. This situation is similar to that observed in isothermal GC discussed earlier and is illustrated in Fig. 6.2. In Fig. 6.2a, the mobile phase is optimum for the first two

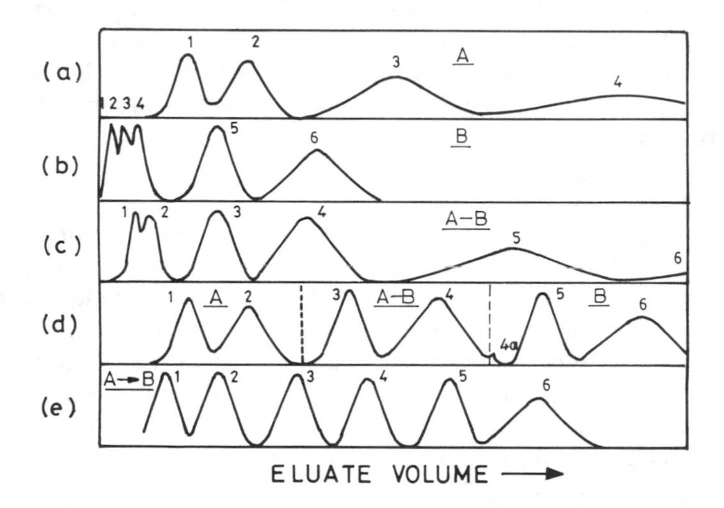


Fig. 6.2 — Illustration of the general elution problem. Reproduced from Ref. 4 by courtesy of Elsevier Scientific Publishing Co.

weakly retained compounds; these compounds are well separated and quickly eluted in narrow bands. However, with this solvent, compounds 3 and 4 are eluted much later and require much larger elution volumes as indicated by the wider bands. In Fig. 6.2b, the solvent system B is optimized for the separation and elution of the more strongly retained compounds 5 and 6. Here, compounds 1 to 4 are eluted very fast and are not well separated. A certain mixture of the solvents A and B may be optimum for the separation and elution of compounds 3 and 4 (Fig. 6.2c); but, again, compounds 1 and 2 are not well separated while 5 and 6 have rather long retention times and appear as broad bands. A strategy that can be conveniently employed in such cases is to start the elution with solvent A, allow components 1 and 2 to be eluted one after the other, switch to an appropriate mixture of A and B to separate 3 and 4 and finally elute 5 and 6 using solvent B (stepwise elution, Fig. 6.2d). For a more complex mixture, increase in the eluant strength may have to be in a larger number of steps with successively higher proportion of the stronger solvent B. This technique, which does not require any special equipment, is routinely used in traditional column liquid chromatography.

*Gradient elution* may be considered a special case of stepwise elution wherein the solvent composition is changed continuously (in the above case from 100% A to 100% B, Fig. 6.2e). Fig. 6.1c shows how this can be effected without any special equipment. Two measuring jars, containing the two solvents A and B, are interconnected through a stopcock; a magnetic bar is used in the jar containing solvent A to mix the solvents. As the level of solvent A comes down, more of solvent B gets mixed with A, forming a gradient. As the levels in the two jars have to be the same, the rate of the gradient depends on the relative diameter of the two jars; a wider jar for the solvent B yields a faster gradient. Alternatively, a solid core of suitable size may be placed in jar A to increase the rate of the gradient. For other simple devices, see Ref. 4. Gradient elution, which is treated in further detail later, is achieved more conveniently and reproducibly using the modern electronic gradient-forming pumps (Chapter 7). It should be noted that the stepwise or the gradient elution does not directly affect the resolution. The advantages of these techniques lie mainly in the increased speed of analysis through faster elution and increased sensitivity resulting from the sharper bands.

### 6.2.4 Fraction collection

To obtain the individual fractions separately, the collection flask, or tube, can be changed manually if the separated bands on the column are coloured. For colourless compounds, a number of fractions, at intervals of time or volume, are collected either manually or automatically and subsequently analysed. The volume of the individual fractions is determined by convenience and complexity of the analysis; as a rule of thumb, the volume (in ml) of the individual fractions should not exceed the weight in grams of the adsorbent (e.g. silica gel) used. Several types of automatic fraction collectors are commercially available. The older versions consisted of a turntable which could hold 100 to 200 tubes; the turntable is advanced periodically by an electrical signal, derived from a timer, a photoelectric drop counter or a siphon. Fractions can thus be collected at predetermined time intervals, or after a certain number of drops, or by the volume-equivalent of the siphon reservoir. Alternatively, the tubes are kept stationary and the outlet from the column is connected through a Teflon tube to a device which moves from tube to tube. The arrangement of the tubes in more recent versions that can hold 80 to 400 tubes is either circular or rectangular and more compact. Modern fraction collectors are equipped with an arm connected to an on-line detection system and the detector signal and the fraction change are synchronized. More recent microprocessor-controlled versions are programmable in a number of ways and combine the functions of fraction collector, recorder, integrator and data systems.

### 6.2.5 Monitoring

The effluents from the column can be monitored in several ways. On-line monitoring is needed and is routine in the modern high-speed (high-performance) LC and the principle can be conveniently used in the traditional low-pressure LC also. The column outlet is connected to a flow-through cell for monitoring the change in either bulk property of the effluent (e.g., refractive index) or the specific property of the solute (e.g. light absorption using a spectrophotometer, or optical rotation using a polarimeter). Before the on-line detectors became widely available the usual practice was to collect the fractions as described above and examine individual fractions by chemical, spectroscopic or biological means, depending on the properties of the solute and the solvent. Sometimes the fractions are appropriately treated with a chromogenic agent (e.g. ninhydrin for amino acids), and the intensity of the resulting colour is measured spectrophotometrically.

If an on-line monitor is used, the response of the detector is plotted (usually by a potentiometric recorder) against time or elution volume, and the fractions are separately collected to contain the individual bands. Alternatively, the fractions are analysed by some faster chromatographic technique like TLC or GC. For rapid analysis of a large number of fractions, a microslide (7.5 × 2.5 cm) TLC plate may be effectively used by spotting along the longer side and developing along the shorter [2]. The fractions containing identical composition are pooled and the solvent evaporated under vacuum (preferably using a rotary evaporator). The elution with a particular solvent is continued until the effluent is free of any solute and then the solvent is changed to a stronger one in a stepwise manner. Where possible, a continuous gradient may be preferred if isocratic elution is not adequate. If the eluted fractions are not homogeneous, they may be rechromatographed.

To be able to reproduce the data when necessary, a record of the following information may be maintained: weight of the sample chromatographed, internal diameter of the column and its length, characteristics of the adsorbent (type, particle size, surface area, porosity, origin, activity, etc.), amount of the adsorbent used, the solvent composition and its change (if any), the number of fractions collected and their volumes, the weights of the solute in each fractions (if present and measurable), and any other data that may be of relevance.

## 6.3 SIZE-EXCLUSION CHROMATOGRAPHY (SEC) AND RELATED METHODS

### 6.3.1 Introduction and theory

Since its introduction in 1959 by Porath, the technique of size separation of molecules is variously described as gel filtration, gel permeation chromatography, molecular-sieve filtration, molecular-sieve chromatography, exclusion chromatography, size-exclusion chromatography and steric-exclusion chromatography and so on [5]. The terms gel filtration and gel-permeation chromatography generally represented the use of hydrophilic and lipophilic gels for the separation of water-soluble and water-insoluble compounds, respectively. More recently, the term size-exclusion chromatography (SEC) has gained widespread acceptance to describe all types of chromatographic separations based on the exclusion of the solute molecules from the pores of the column packing as a function of their molecular size [6]. Some of the recent developments in the size-separation techniques based on the flow or the kinetic properties of the solute and the solvent (hydrodynamic chromatography) and on the effects of thermal, gravitational, electrical, magnetic and other fields (field-flow fractionation, FFF) are also discussed later in this section.

The technique of SEC differs from the other chromatographic methods in certain aspects of the theory and practice. Its principle is probably the simplest to understand and its results, in the absence of interfering mechanisms, are highly predictable and reproducible. As already noted and as the name indicates, the principle involves separation of molecules on the basis of their size and shape. The stationary phase consists of a gel or a micro-particulate, porous organic or inorganic solid with a defined pore-size distribution. Molecules larger than the pore size are completely excluded while those which are smaller are able to diffuse into the gel structure and are retained by the stationary phase. The retained molecules are displaced or excluded by those of the mobile phase and are eluted in the order of decreasing molecular size. Mobile-phase molecules, being the smallest, are most strongly retained. The principle is illustrated in Fig. 6.3. This is certainly an over-simplification of the events occurring in SEC, as will be apparent later.

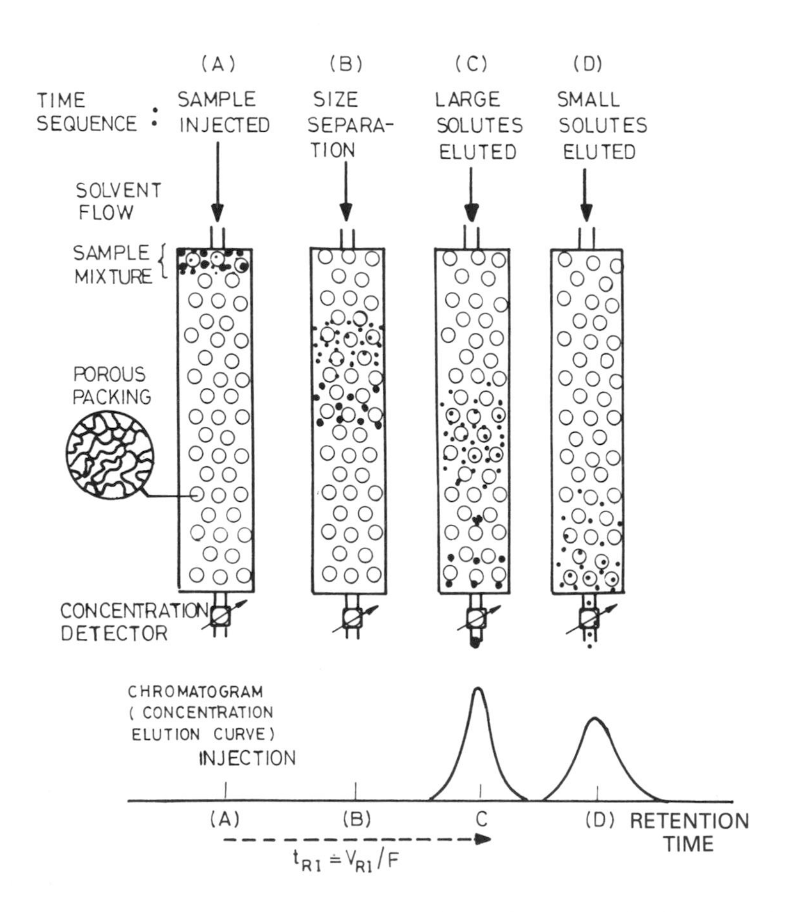


Fig. 6.3 — Illustration of the principle of size-exclusion chromatography. Reproduced with permission from Ref. 6; copyright (1979) John Wiley & Sons, Inc.

In contrast to the other LC methods, the role of the mobile phase in SEC is largely limited to swelling the gel and keeping the analyte molecules in solution. In the latter respect, it is more akin to the role of the carrier-gas in GC. The solvent inside the gel, which is relatively immobile and which can thus be considered the stationary phase, is not equally accessible to the solute molecules of different sizes and only those that can enter the pore structure can be retained. Thus, the distinguishing feature of the SEC is that the composition of the stationary phase and the mobile phase is the same and that the separation is caused by the geometry of the pore structure of an otherwise inert solid support. Retention is understood to occur when the solute molecules could diffuse from the moving interstitial liquid to the stationary liquid inside the pore structure of the packing, so that the band migration along the column is arrested until these molecules return to the mobile interstitial fluid. The retention is determined either by the hydrodynamic properties of the solute and the solvent or by the partitioning of the solute between the interstitial and stagnant pore fluids [6].

The following example illustrates the events occurring in a typical SEC experiment. Consider a column packed with a swollen gel of a definite pore size and a mixture of three compounds, A, B and C. The molecular size of A is assumed to be larger than the pore size of the gel; and that of C, very small; B is of intermediate size, capable of permeating the gel. The total volume of the bed, including that of the mobile phase outside the gel ($V_o$) and inside the gel ($V_i$) as well as that of the matrix ($V_m$) is $V_t$ so that,

$$V_t = V_o + V_i + V_m \tag{6.1}$$

Compound A, being size excluded, is eluted in the void volume, $V_o$. Unlike in the other forms of chromatography, the mobile-phase hold-up volume ($V_M$) and the void volume ($V_o$) are not the same in SEC; the former is given by:

$$V_M = V_i + V_o \tag{6.2}$$

Compound C, whose molecules are small enough to permeate into all regions of the pore structure, may have a retention volume ($V_R$) equal to the sum of $V_o$ and $V_i$, i.e. $V_M$. Compound B is neither completely excluded nor is it capable of free diffusion, and is eluted with a $V_R$ that is intermediate between $V_o$ and $V_M$.

The value of $V_i$ can be calculated by determining the solvent gain by the dry gel, but this value is not always easy to obtain accurately. $V_i$, thus calculated, is always larger than the experimentally determined value since a portion of the solvent is used up for solvation of the polymer chains and is not available for diffusion of the solute molecules. The available pore volume ($V_{av}$) for a given molecule is a function of its size and is given by:

$$V_{av} = V_R - V_o \tag{6.3}$$

The amount of the solute in the stationary phase is given by its concentration multiplied by the available pore volume, and that in the interstitial volume by a similar relation. Since the composition of the two liquids is the same, the equilibrium concentration of the solute in them is also the same, so that the capacity factor ($k'$) is given by:

$$k' = \frac{V_{av}}{V_o} = \frac{V_R - V_o}{V_o} \tag{6.4}$$

The available volume ($V_{av}$) is related to the intraparticulate volume ($V_i$) by eq. (6.5); here again, $K_o$ may be distinguished from $K$, the distribution constant which, as already noted, should be unity but for certain features discussed later.

$$V_{av} = K_o V_i \quad \text{or} \quad K_o = \frac{V_{av}}{V_i} = \frac{V_R - V_o}{V_i} \tag{6.5}$$

Since $V_i$ (and hence, $K_o$) is difficult to obtain accurately, the parameter $K_{av}$ is defined as:

$$K_{av} = \frac{V_R - V_o}{V_t - V_o} \tag{6.6}$$

$K_{av}$, which can be easily measured, is independent of the geometry and packing density of the column. If $K_{av}$ values of two compounds are known, it is possible to decide whether the two compounds can be separated by SEC. The separation volume, $V_s$, is given by the equation:

$$V_s = (K_{av2} - K_{av1})(V_t - V_o) = V_{R2} - V_{R1} \tag{6.7}$$

The practical significance of this relation is that, to increase $V_s$ one may have to use a very large bed volume or a very fine particle size, both of which are technically difficult using the traditional packings but feasible in the modern high-performance size-exclusion chromatography (HPSEC). All retention volumes lying between $V_o$ and $V_M$ is a characteristic feature of SEC, which also severely limits the number of compounds that can be separated by the method; i.e., the peak capacity of the column.

The $K_{av}$ values and the retention volumes on a given gel for a series of compounds of similar molecular shape are linearly and sigmoidally related to the molecular weight of the solutes as illustrated in Fig. 6.4 for globular proteins [7]. This property has been extensively used for the determination of molecular weights of macromolecules. Several models have been proposed describing the pore and solute geometry as cones or as a network of rigid rods, etc., and relating the inner volume of the gels to the molecular radius of the compounds. While these may not be accurate in the strict sense of the molecular geometry, they have been successfully used to predict the elution behaviour; e.g. see Fig. 6.5 [6].

It is important to realize that the solute molecules cannot occupy all the available space inside the porous particle. Assuming (reasonably) the inner space is rigid and

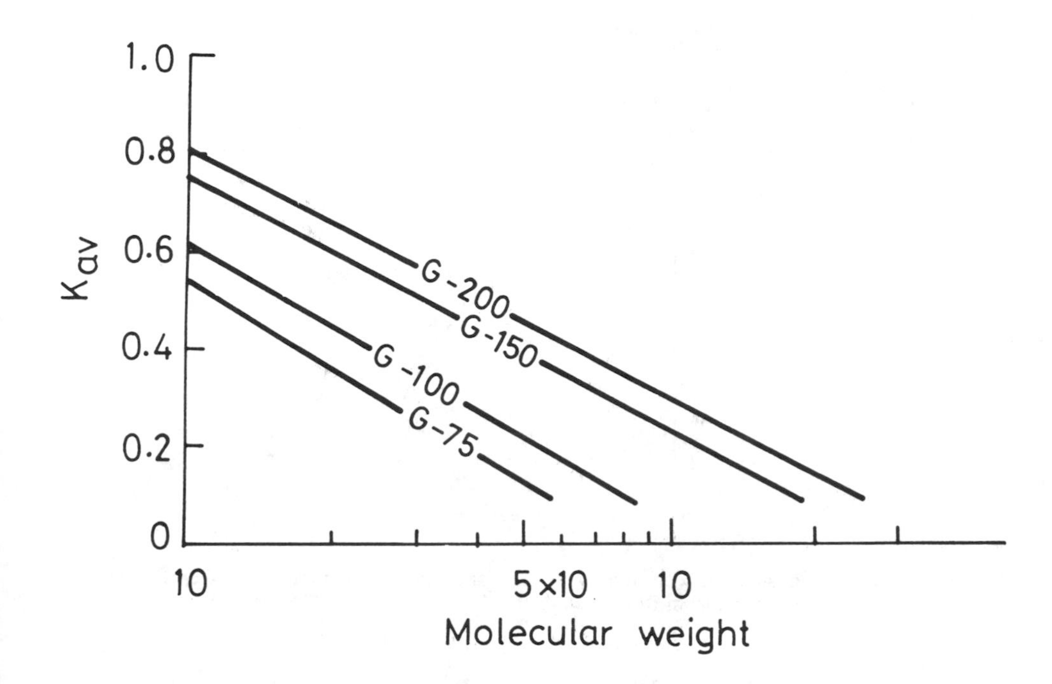


Fig. 6.4 — Relationship between elution volume and solute molecular weights in size-exclusion chromatography. Reproduced from Ref. 7 by courtesy of Pharmacia AB.

that the solute molecule has a finite shape, the relation between size and shape of the solutes and their retention has been developed using the mean external length, $L'$, to characterize the molecule [8]. $L'$ is a function of the aspect ratio of the molecule (that is, the ratio of the diameter to the length of an ellipsoid of revolution. The product $sL'$ ($s$ being a measure of the gel surface area to free volume) is approximately linearly and sigmoidally related to $K_o$ over a range of aspect ratios. Thus, the retention in SEC is a function of both the pore geometry and the solute molecular geometry at small pore diameters. The value of $K_o$ tends to unity as the ratio of the pore size to molecular size increases. $L'$, being a function of the volume of free revolution of the molecule, is directly related to the gyration radius, $R_G$ and this relationship forms the basis of the universal calibration plots which relate the retention parameters on SEC to molecular weight. For flexible polymers, the gyration radius is related to the molecular weight ($M$) by eq. (6.8), where $N$ is the Avogadro's number, $\eta$ is the intrinsic viscosity and $\xi$ is the proportionality constant

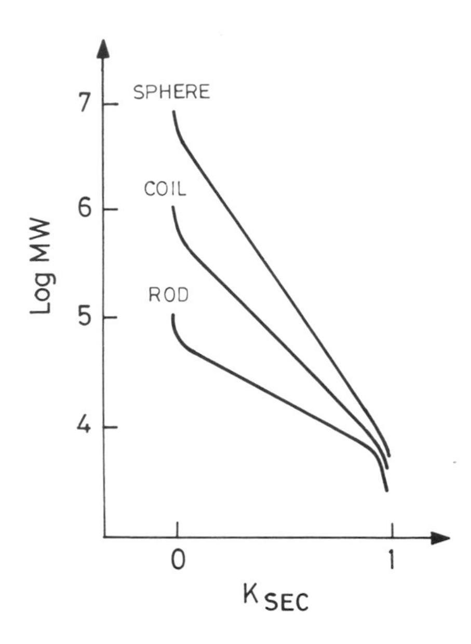


Fig. 6.5 — Relationship between solute geometry and calibration curve slope. Reproduced with permission from Ref. 6; copyright (1979) John Wiley & Sons, Inc.

relating the gyration radius to the equivalent spherical radius, $R_e$ so that $R_e = \xi . R_G$. The value of $\xi$ is approximately 0.835 in ideal solvents.

$$[\eta]M = 10\pi N\xi^3 R_G^3/3 \tag{6.8}$$

In the universal calibration plots [9], log $[\eta]M$ (i.e. the logarithm of the hydrodynamic volume, which is expressed as the product of the intrinsic viscosity, $[\eta]$, and the molecular weight) is plotted against the retention volume. The intrinsic viscosity of a linear polymer in a specific solvent is related to its molecular weight according to Mark–Houwink relation:

$$[\eta] = KM^a \tag{6.9}$$

where $K$ and $a$ are constants for a given polymer type, solvent and temperature. The calibration curves for different synthetic polymers fall into a single linear plot (Fig. 6.6). This plot is generally valid for a variety of polymers including proteins once a detergent or a denaturant like sodium dodecyl sulphate is used. The retention volume is also independent of the temperature and the flow-rate. The above relationships and assumptions have been validated experimentally in several classes of compounds and this feature indicates that the radius of gyration, $R_G$, is a basic parameter that determines the retention in SEC. Deviations from the above

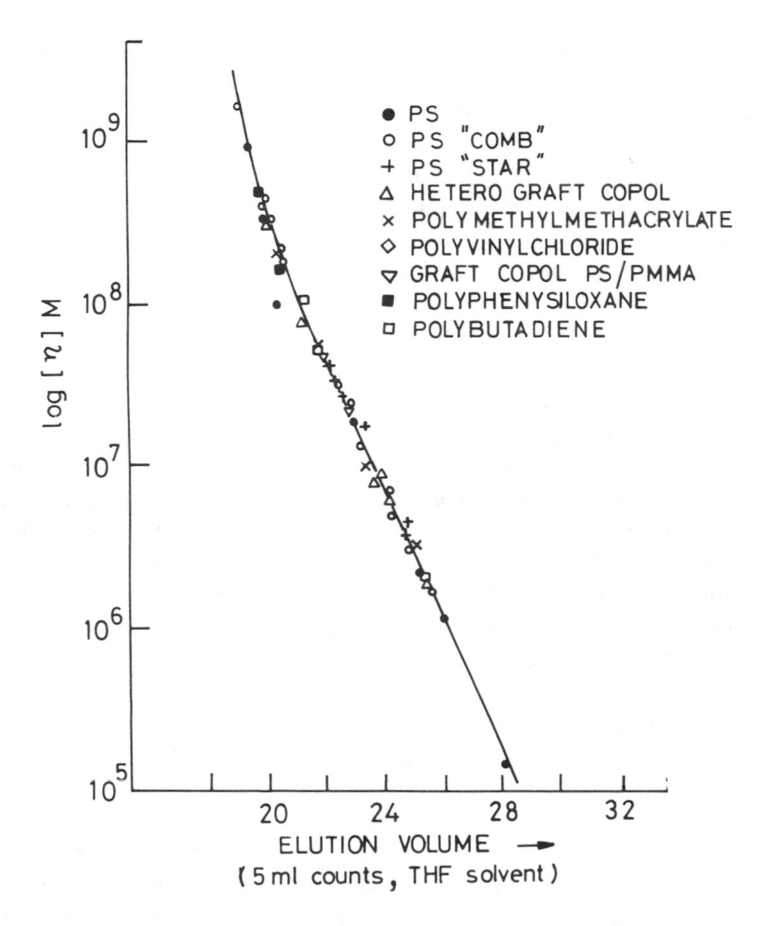


Fig. 6.6 — Universal calibration plot. Solvent: THF. Reproduced with permission from Ref. 9; copyright (1967) John Wiley & Sons, Inc.

relationships may be observed when extra-size-exclusion mechanisms like adsorption operate, but these may be modulated by changing the solvent composition.

There are several instances, particularly in the chromatography of low molecular weight compounds on gels of relatively high density and low pore volume, where the elution order is the reverse of the kind noted above. In most of these cases, $V_R$ is greater than $V_M$ and mechanisms other than size exclusion need to be invoked. The carbohydrate-derived gels (e.g., Sephadex G-25, see below), which are commonly used in SEC, may contain a small proportion of carboxylic acid groups which may retain and exchange cationic groups and exclude anionic groups (see Donnan exclusion, section 6.5). The order of retention of alkali-metal cations on Sephadex gels is: $Li < Na < Cs < Rb < K$; this order is more akin to the order found in cation-selective electrodes than that of hydrated ion radii. Also, the ion distribution between the stationary and mobile phases is dependent on the eluting anion, indicating that size exclusion does not play an important role. Partition and adsorption, which are also encountered in SEC, are more difficult to distinguish and deal with. Interestingly (and rather inexplicably), the hydrophilic dextran gels exhibit great affinity towards aromatic hydrocarbons and their derivatives; elution behaviour in such cases is influenced by the structure of the gel and solvent. The affinity of the gels to the hydroxy aromatic compounds increases if the hydroxyl substitution is on the aromatic ring and decreases if it is on the alkyl side chain. Amino groups have a similar effect as the hydroxyl groups.

The extra-steric exclusion effects operating in silica-based gels have been elaborated by Barth [10]. The ionic effects include ion exchange, ion exclusion, ion inclusion, electrostatic intramolecular effects and coulombic adsorption. All these interfering mechanisms can be suppressed by adequately increasing the ionic strength of the mobile phase. The ionization of the groups in the matrix (e.g. carboxyl groups in carbohydrates and silanol groups in silica) and thus ion-exchange interactions with the solutes can be reduced by operating at low pH. In addition to ionic interactions, hydrogen-bonding, hydrophobic and viscosity effects also complicate the results in SEC. Addition of urea or guanidine-HCl, use of organic modifiers such as ethylene glycol or of surface-coated solid supports, and lowering of sample concentration are some of the strategies suggested to overcome these effects.

Certain other interfering mechanisms of retention in SEC may also be noted. As is well known, several natural and synthetic zeolites (aluminosilicates) have very well-defined and rigid pore structures and are very effective molecular sieves. The selectivities of retention of non-electrolytes on these materials is primarily determined by steric hindrance. The crosslinked gels are generally made up of a random network of strands that can be solvated and swollen by appropriate solvents; the selectivity of retention or exclusion on these gels is therefore not as sharp as in zeolites. Also, since the system used for molecular sieving is a solvent-swollen gel, the role of the solvent and the swelling pressure are also important. The swelling of the gels in a compatible solvent continues until the equilibrium is reached; i.e. the elastic forces of contraction of the gel structure are balanced against the swelling tendency. Imaginably, the osmotic pressure is considerable in highly cross-linked gels. In the larger-pore-sized gels (e.g. Sephadex G-50 to G-200), where the solvent pressure cannot appreciably account for the solute exclusion, steric exclusion is considered to play the major role in separation.

In a multi-component solute, it is possible that two or more of them could interact chemically or associate physically to give molecular species which could show unexpected elution behaviour. Exclusion of certain solutes from the mobile phase may sometimes account for enhanced retention; repulsion of the solute molecules by those of certain components of the mobile phase could cause such a phenomenon. Indeed, the biomolecules which are subjected to chromatography contain a variety of units that may be charged groups, multipolar moieties, groups capable of forming hydrogen-bonds and non-polar (hydrophobic) groups. With such diversified structural features, it is unlikely that most of the above noted extra-size-exclusion effects can be completely prevented on any of the commonly used matrices. In fact, it is advantageous to control and coordinate the effects so that the multiple mechanisms of interactions help separation.

Of particular interest in connection with the use of SEC packings for the separation of biomolecules, are the hydrophobic interactions between the matrix and the solute. It is possible to imagine that two molecular species with large non-polar areas may both be repelled by the hydrogen-bonding polar water molecules so that the two species aggregate. It is also possible that the associating non-polar molecules could help ordering of the water molecules and thus overcome the thermodynamically unfavourable situation of dispersion of non-polar entities in a polar environment. While the exact mathematical treatment of the hydrophobic interaction is subject to debate, it is common to invoke the concept of such interactions as the most important driving force in biological processes. The theoretical basis of the various mechanisms of retention in SEC have been discussed in detail by Bywater and Marsden [11].

### 6.3.2 Column-packing materials for size-exclusion chromatography

Under ideal conditions of SEC, the role of the packing material is limited to providing a solid matrix, into the pore structure of which the solvent and the solute molecules can permeate. The major classes of materials in general use are: (1) carbohydrate-derived gels, (2) polyacrylamide gels, (3) composite gels, (4) polystyrene gels and (5) inorganic gels.

#### *6.3.2.1 Carbohydrate-derived gels*

The most popular of these materials are the Sephadex G-type gels, manufactured by Pharmacia Fine Chemicals [7]. They are based on dextran, which is a 1,6-α-linked polymer of glucose obtained by the action of the bacterium *Leuconostoc mesenteroides* on sucrose. Sephadex gels are formed by cross-linking dextran with epichlorohydrin. The chemical structure of Sephadex is given in Fig. 6.7. The pore size is inversely related to the degree of cross-linking and is determined by the concentration of the cross-linking agent and dextran as well as the molecular weight of dextran used. Sephadex G-10, G-15, G-25, G-75, G-100, G-150 and G-200 are commercially available, with the pore size increasing in that order. The fractionation range varies from a molecular weight of 700 to 600,000 for globular proteins and from 700 to 200,000 for dextrans. Because of the large number of hydroxyl groups in the structure, these gels are hydrophilic and have good swelling characteristics and chromatographic properties (the number appearing after the G indicates 10 times the water regain of the gel). In certain cases, for example, aromatic and heteroaromatic compounds, nucleosides and nucleotides, and even cyclohexane, nonsteric factors may predominate, more so in tightly cross-linked gels like G-10 and G-15. As noted above, both dipolar and hydrophobic interactions are invoked to explain the strong retention of these compounds by the gels.

Alkylation and acylation of the hydroxyl groups in Sephadex results in materials which have good swelling properties in organic solvents in addition to water. Hydroxypropyl ethers of Sephadex of G-25 and G-50, designated as LH-20 and LH-60 respectively, may thus be used with both aqueous and organic solvents. These products find use in the separation of amino acid derivatives and several other natural products, including aromatic compounds like phenols, quinones, etc. A spherical modification of Sephadex LH-20 has also been introduced (Sephasorb HP Ultrafine) which may also be used for high-performance size exclusion chromatography (HPSEC). Resolution on these materials depends largely on adsorption and partition effects.

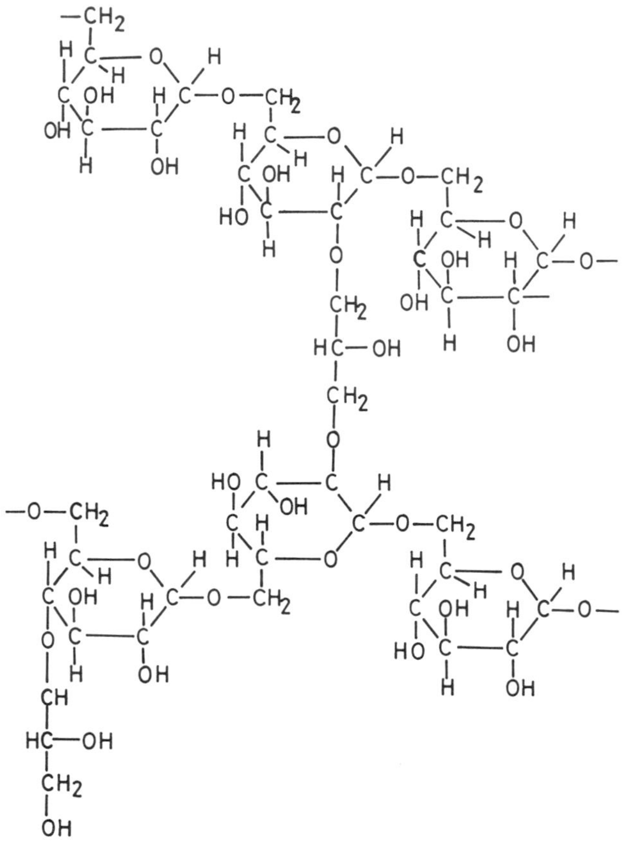

Fig. 6.7 — Partial structure of Sephadex. Reproduced from Ref. 7 by courtesy of Pharmacia AB.

Agarose (a carbohydrate polymer consisting of alternating D-galactose and 3,6-anhydro-L-galactose) occurs as a part of the complex mixture of polysaccharides known as agar, obtainable from certain seaweeds. On cooling a hot solution of agarose, the individual polysaccharide chains form double helices which subsequently aggregate yielding a stable gel. The commercial product from Pharmacia is Sepharose. The cross-linked Sepharose-CL (Fig. 6.8), with improved thermal and chemical stability but having substantially the same porosity as the parent gel, is formed by reaction of agarose with 1,3-dibromo-2-propanol under strongly alkaline conditions. Because of the stiff helical structure of the gel, its pore size changes very little in different solvents and the gels can be used with both aqueous and organic solvents. The carbohydrate-derived gels were extensively used during the 1960s for

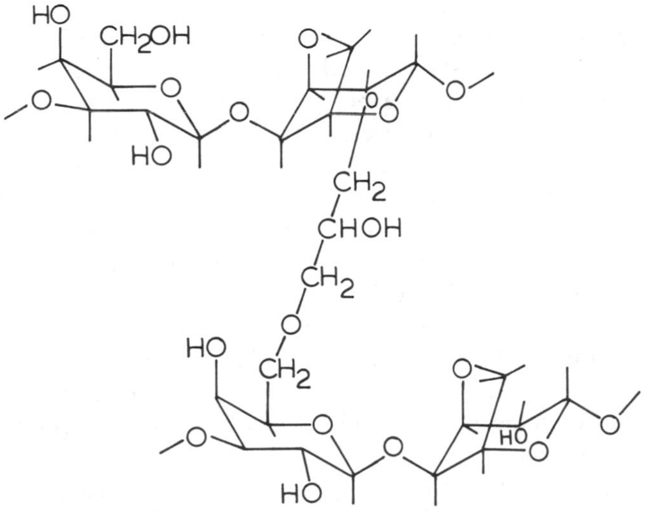

Fig. 6.8 — Partial structure of Sepharose CL. Reproduced from Ref. 7 by courtesy of Pharmacia AB.

the separation and molecular weight determination of a variety of biopolymers. They are still being widely used but, being soft, they cannot withstand high pressures and their popularity has decreased somewhat with the advent of high-performance instrumental techniques.

### *6.3.2.2 Polyacrylamide gels*

Cross-linked polyacrylamide gels are available from Bio-Rad Laboratories under the trade name Bio-Gel. Bio-Gels P-2, P-4, P-6, P-10, P-30, P-150, P-200 and P-300 have fractionation ranges varying progressively from 200 to 400,000 for globular proteins. The exclusion limit of these gels are roughly given by the number following the letter P multiplied by 1000. They are stable in the pH range 2 to 9 and their swelling characteristics are essentially unaffected by change in the ionic strength. The gels are extremely hydrophilic and compatible with dilute organic acids, 8 M urea, 6 M guanidine-HCl, chaotropic agents and even organic solvents to the extent of about 20% in water. Despite the basic difference in the structure of the monomers, the swelling properties and separation ranges of the polyacrylamide and cross-linked dextran gels are strikingly similar; e.g. the chromatographic properties of Sephadex G-15 are comparable to those of Bio-Gel P2, those of Sephadex G-75 to Bio-Gel P-30, etc. The exact swelling characteristics and chromatographic properties of these products are given in detail in the literature available free from the manufacturers, namely, Pharmacia AB, Uppsala, Sweden and Bio-Rad, Richmond, CA, U.S.A.

### *6.3.2.3 Composite gels*

As noted earlier, while SEC separations are not significantly affected by the flow-rate, full advantage of this feature could not be taken because the particle size would need to be reduced for better resolution; this would decrease the flow-rate considerably unless high head pressures were used. Several useful combinations of carbohydrates and vinylic monomers have been introduced by different companies to improve the chromatographic properties. In contrast to the soft gels, these composite gels can withstand high head pressures so that faster flow (and thus shorter analysis times) and smaller particle size (and hence better resolution) can be achieved. These may be considered intermediate between the soft gels discussed above and the rigid gels for HPSEC to be discussed below. They are, therefore, sometimes referred to as semi-rigid gels.

Sephacryl from Pharmacia is obtained by cross-linking allyldextran with *N,N'*-methylenebisacrylamide (Fig. 6.9). The gel is relatively rigid, with the polyacrylamide part forming a granular structure strung out along the matrix of dextran and with large pores between the granular regions. The rigidity of the gel reflects itself in its stability to high flow-rates.

A series of composite gels with varying concentrations of agarose and polyacrylamide are available from LKB Produkter, Bromma, Sweden, under the trade name Ultrogel. While their detailed micro-architecture remains unknown, their rigidity and wide range of selectivity curves for protein fractionation makes them useful alternatives to the soft gels.

Enzacryl, marketed by Koch-Light Ltd., Colnbrook, U.K., is a co-polymer of acryloyl-morpholine and methylenebis(acrylamide). It is equally compatible with aqueous and organic solvents and shows no hydrogen exchange.

Spheron gels, formed by co-polymerization of hydroxymethylmethacrylate and ethylenedimethacrylate and marketed by Lachema, Prague (Czechoslovakia), and can withstand pressures up to 3000 psi. They are available in a wide range of porosities and selectivities and find extensive use in protein fractionation. Their large matrix volume relative to the carbohydrate gels does not adversely affect the selectivity but reduces the sample capacity considerably.

### *6.3.2.4 Polystyrene gels*

Styrene (vinylbenzene), unless stabilized by addition of a radical scavenger, has a tendency to form a linear polymer. When the polymerization is allowed to take place in the presence of divinylbenzene (DVB), cross-linked polystyrene is obtained. The degree of cross-linking depends on the proportion of DVB used and the pore-size is inversely related to the degree of cross-linking. Styrene-DVB polymers are modified by chemical functionalization for use as ion exchangers (see later, Fig. 6.18). The highly cross-linked and chemically unmodified gels barely swell in solvents; nevertheless, they are widely used in view of their well-defined pore structure. Waters Associates (a subsidiary of Millipore, Milford, U.S.A.) market Styragels with porosity varying from 60 to 1,000,000 Å. The pore-size is not indicative of the absolute size of the pore but, by historical definition, refers to the length of the linear polystyrene molecule which is just large enough to be excluded from the pore structure of the packing. Three different particle sizes are also available for

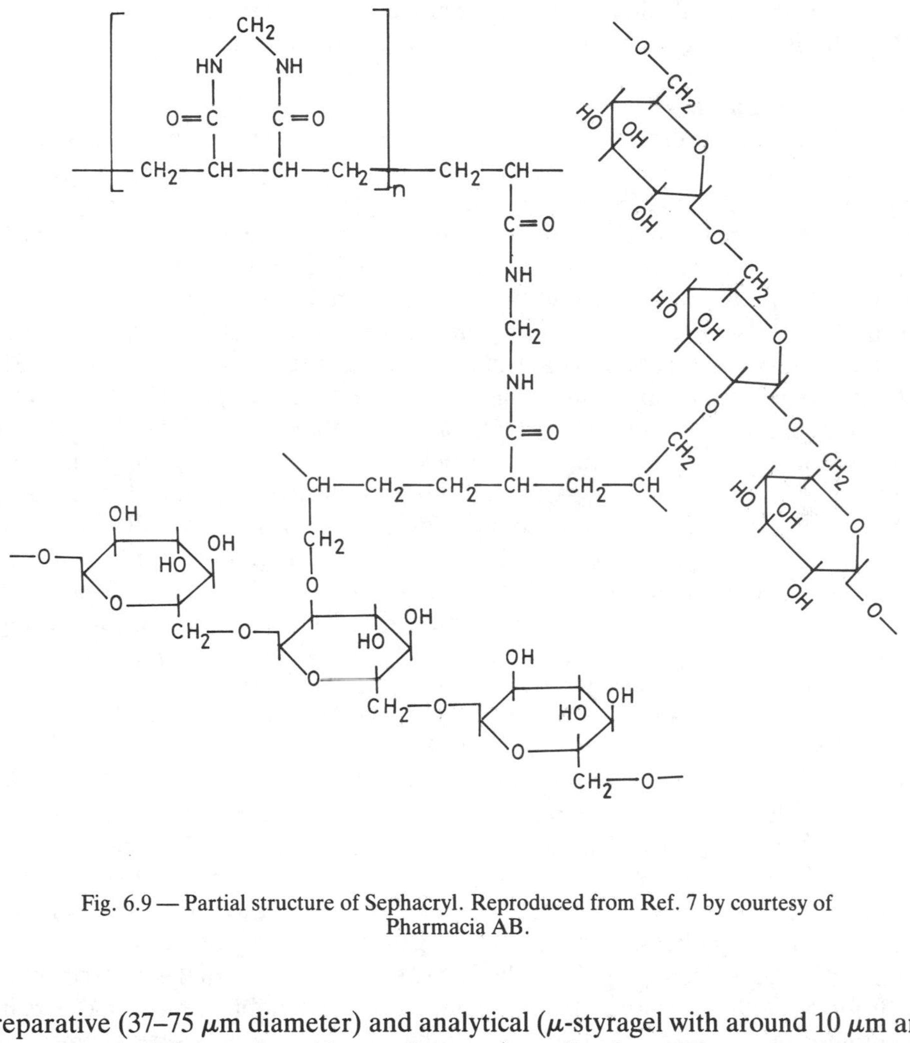

Fig. 6.9 — Partial structure of Sephacryl. Reproduced from Ref. 7 by courtesy of Pharmacia AB.

preparative (37–75 $\mu$m diameter) and analytical ($\mu$-styragel with around 10 $\mu$m and ultrastyragel with less than 10 $\mu$m diameter) applications. The molecular-weight fractionation range of these gels is very wide, ranging from 50 to 1500 for the 60 Å gel, and from 200,000 to 10 million for the 1-million Å gel. Aromatic and chlorinated solvents, dimethylformamide and tetrahydrofuran are recommended as solvents; acetone, water, alcohols and acids cause shrinkage of the gels and may be avoided. Naturally, these gels find use in the separation and analysis of organic soluble polymers. They are relatively rigid and can be used in HPLC. The columns can be used for high-temperature SEC up to about 150°C; however, the 100 Å $\mu$-styragel columns, can be used only up to 60°C.

The OR-PVA gels (Fractogels from E. Merck and EM laboratories) are copolymers of polyvinylacetate and are used mainly for the separation of synthetic polymers. Spherical particles of controlled pore size can be used in HPSEC up to a molecular-weight exclusion limit of about 1 million. They are compatible with organic solvents like alcohols and acetone and elevated temperatures of up to about 100°C. Being semi-rigid, they cannot sustain high pressures beyond 300 to 600 psi.

#### *6.3.2.5 Inorganic gels*

As is well known, the efficiency of separation in chromatography is inversely related to the square of the particle diameter; in other words, high-efficiency columns are prepared from particles of small diameter. These packings would then require high pressures to ensure reasonable flow-rates. The polystyrene gels mentioned above are rigid enough to withstand moderate pressures and are used extensively for separation of synthetic polymers that are soluble in non-polar solvents. The carbohydrate-derived gels are too soft for use under pressures of 200 psi and over. Porous microparticulate silica has now come into widespread use in connection with the SEC of biopolymers in aqueous media. Controlled-pore glasses of pore size 40 to 1500 Å units and particle size of 5 to 10 $\mu$m are available from Pierce Eurochemie, Amsterdam (or Pierce Chemical Co., Rockford, U.S.A.). They are prepared by heating borosilicate glass to separate borates and silicates and then etching out the borates, leaving the silica pore structure. The silica gel is mechanically, chemically and thermally stable and is compatible with a wide range of solvents. However, owing to the polar nature of the silanol (Si–OH) groups, adsorption and electrostatic interactions interfere with SEC applications. For this reason, the silica-gel surface is modified by chemically bonding silylpropylglycerol groups. These gels (e.g. the Glycophases from Pierce) retain the hydrophilicity but, at the same time, exhibit insignificant adsorption effects. Bondagels (from Waters Associates) consist of a monomolecular layer of a polyether, chemically bonded to silica. They can be used with both organic and aqueous mobile phases. $\mu$-Bondagel E-125, E-500, E-1000 and E-High A have exclusion limits of 50,000, 500,000, 2 million and 7 million, respectively (based on polyethylene glycol standards). They are compatible with all solvents including aqueous buffers in the pH range 2 to 8. They have been used for characterization and molecular-weight determination of a variety of polar and non-polar, synthetic and natural polymers including dextrans, polyesters, lignins, polyvinyl alcohols, polyacrylic acids, etc. To ensure analysis based on size exclusion without adsorptive effects in aqueous media, mobile-phase modifiers like sodium dodecyl sulphate (1%), inorganic salts, or organic solvents (tetrahydrofuran or methanol) may be used. Detailed descriptions and test procedures for characterization and evaluation of the HPSEC packings are available in the literature [12].

The TSK-SW and TSK-PW series of gels (Toyo Soda Co., Tokyo, Japan) have been introduced specifically for HPLC of proteins. Their exact structure is unknown but it is understood that they contain a silica or styrene-divinylbenzene co-polymer core with the surface modified by a layer of hydrophobic bonded phase or hydroxylated polyether. The G 1000 SW and G 2000 SW are suitable up to a molecular weight of 25,000 and 100,000, respectively for globular proteins while the TSK G 3000 SW and G 4000 SW have fractionation ranges of 10,000 to 600,000. TSK G 1000 PW to G 6000 PW series have progressively increasing exclusion limits extending to higher than 2,000,000. These (or similar) gels are also available from Bio-Rad and other suppliers.

### 6.3.3 Experimental techniques in size-exclusion chromatography

#### *6.3.3.1 Choice of the column*

SEC may be carried out either in the traditional open-column mode or by the high-pressure instrumental techniques. In either case, the separation system consists of a solvent reservoir, a packed column, a solvent-delivery system (either using the hydrostatic pressure, or a peristaltic pump for near ambient pressures, or a high-pressure metering pump for the HPSEC systems) and a detector to monitor some bulk property of the effluent (e.g. ultraviolet absorption for proteins or refractive index for carbohydrates and several synthetic polymers.

The soft, carbohydrate-derived gels are generally used for SEC at atmospheric or slightly elevated pressures (up to 200 psi). These gels are mechanically weak and should be packed with care. They are invariably packed wet; as their swelling properties depend on the solvent, it is recommended that the slurry be prepared in the same solvent that is to be used subsequently for chromatography. Instructions for preparation of the slurry are usually given by the manufacturer of the packing material. Swelling and equilibration occur quite rapidly and can be further accelerated by gentle heating; degassing also occurs when the slurry is brought to boiling. The slurry is allowed to attain ambient temperature before packing the column. The packing procedures have been described in the previous section. Pre-packed columns are also available commercially. For HPSEC under high pressures, the silica- or polystyrene-based gels are used in aqueous or organic solvents, respectively. Sulphonated polystyrene gels may be used for SEC in aqueous solvents.

The most important parameter to be considered in the selection of the column-packing material is the appropriate pore-size for the particular application. The suppliers provide the fractionation range and the approximate calibration plots for different packing materials. To improve the accuracy, the exact calibration showing the relation between the elution volume and the logarithm of the molecular weight will have to be determined for each column and each type of the polymer. This is done using the macromolecular standards obtainable from the suppliers of the packing materials. Compounds are best separated if the retention volumes fall in the middle of the fractionation range of the column. Single-pore-size column materials are used for the separation of samples with a narrow molecular-weight distribution. Columns of small pore-size (40–60 Å) may be used for separation of small molecules (molecular weight $< 5000$). Packings of intermediate pore-size (200–500 Å) are suitable for isolation of individual components in the molecular weight range of 5000 to 500,000. Material of a given composition can often be better resolved on a particular gel by increasing the column size. This is most conveniently achieved in HPSEC by simply coupling two or more columns in series.

For a mixture of compounds with widely differing molecular weights it may be advantageous to couple two or more columns of non-overlapping but adjacent fractionation ranges. The total fractionation range is then additive; e.g. if two columns with fractionation ranges 100 to 10,000 and 10,000 to 1 million are coupled, the range of the set is 100 to 1 million. However, resolution of compounds with close molecular weights may be less than desirable on such coupled columns. Two columns of each type may be combined to improve resolution. Fig. 6.10 illustrates the result of an appropriately coupled bimodal pore-size column arrangement using porous-silica microsphere packings [13]. A wide, linear molecular-weight calibration, needed to

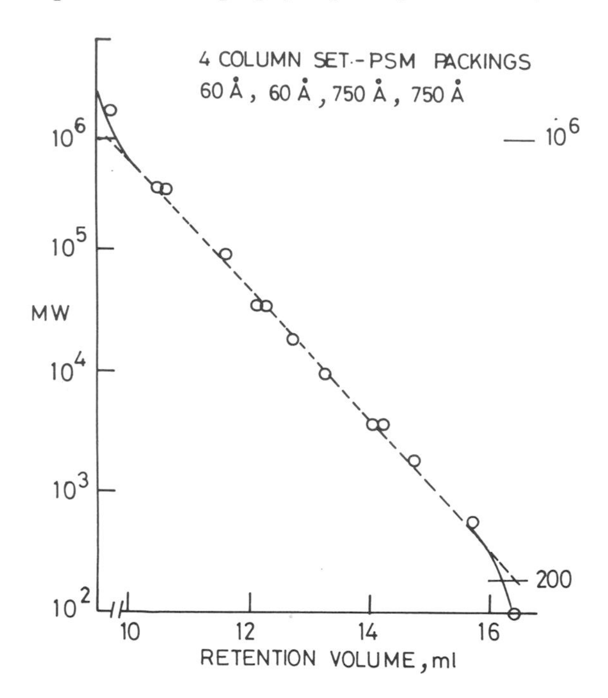


Fig. 6.10 — Bimodal coupling of SEC columns. Reproduced from Ref. 13 by courtesy of Elsevier Scientific Publishing Co.

determine the molecular-weight distribution of polymers, is achieved by combining columns that exhibit bimodal pore-size distribution; i.e. columns containing only two discrete pore-sizes with about 1 decade difference (1 decade is the molecular-weight range from $10^x$ to $10^{x+1}$) and approximately equal pore volumes so that the linear portions of the individual calibration plots are essentially parallel.

For a given column material and pore size, the efficiency of the chromatography is largely a function of the particle size. Analytical SEC using the soft gels or preparative separations on polystyrene-DVB resins and porous silica employ larger than 40 μm particles. A particle size of 5 to 10 μm is more common in analytical HPSEC. The effect of plate height on particle size is evident from the fact that, while 5 μm Styragel columns (30 × 0.85 cm) can generate 40,000 plates per metre, 10 μm particles can generate about 30,000 plates, and 20 μm particles only 18,000 plates/metre at optimal flow-rates (2.5 ml/min) for 5 μm and about 1.0 ml/min for > 10 μm particles) [14]. It was noted earlier that the flow rate has little effect on the efficiency of SEC. While this is true in connection with SEC using larger than 40 μm particles, the flow-rate can significantly alter the plate height in HPSEC. Thus, at 8 ml/min, the above columns furnished 8,000, 4,000 and 2,500 plates, respectively.

### *6.3.3.2 Sample preparation and sample size*

There may be several problems in the preparation of high-molecular weight samples for SEC. The first problem may be faced in attempting to dissolve a polymer. Yau *et al.* [6] have discussed in detail the theoretical aspects of the polymer-dissolution process, which is rather different from that for the low molecular weight substances. Many of the biopolymers can be rendered soluble by altering the salt concentration and/or pH, subject, of course, to the stability of these molecules. There is also the possibility of denaturation of biomolecules during dissolution and chromatography and this may alter or destroy the biological properties of the polymer. Change in the biological activity in itself may not be serious (unless the activity is used for monitoring the effluents) if the goal of the experiment is to determine the molecular weight or the homogeneity of the sample. Another complication results from the possibility of shear degradation of polymers in solution. Avoidance of vigorous mechanical agitation is generally recommended to minimize such shearing, but care should also be taken to prevent aggregation of molecules. Ideally, the sample is dissolved in the same solvent that is used for elution.

The sample size is generally determined in practice by the trial and error method using increasing sample volumes until the resolution deteriorates beyond acceptable limits. Sample volume should not normally exceed 1–2% of the bed volume (or 10% of the flow rate, measured in ml/min) for analytical chromatography and 25-30% of the total volume of mobile phase in the bed for preparative separations. In any case, it is advisable to keep the sample size less than the separation volume, $V_s$, as defined by eq. (6.7). It is clear that to increase the sample volume, the choice of the gel should be aimed at obtaining the largest possible difference in the $K_{av}$ values, or the pore volume of the gel is chosen to be large so that $(V_t - V_o)$ is large. For a given sample size, improved resolution can be expected for larger diameter columns. The viscosity of the sample solution is significant for polymeric samples and highly viscous samples cause significant band broadening owing to streaming or 'viscous fingering' of the solute band. The viscosity of the sample solution should be less than twice that of the mobile phase; 0.1–0.25% solutions are common.

Unlike in the other LC methods, retention volumes in SEC of polymers generally increase with increased sample concentration. The contribution of the injection volume to the retention volume could be to the extent of 50% of its former value. Since sample volume can affect the apparent retention times (or volumes), it is advisable to use a constant sample volume during calibration and analysis.

Retention volumes also increase with increased sample concentration. The extent of concentration dependence of the retention volume is larger for the higher molecular weight solutes, obviously due to the enhanced gyration radius of the macromolecular coils in the concentrated solution. To minimize the concentration dependence of the retention volume, it is recommended that the minimum sample concentration, consistent with the detector sensitivity, be employed.

#### *6.3.3.3 The mobile phase*

As noted earlier, the role of the mobile phase in SEC is limited to keeping the gel in a swollen state and the sample in solution. The solubility of the sample, in fact, dictates the choice of the mobile phase and this, in turn, determines the nature of the gel material used. Compatibility of the mobile phase with the gel is particularly important as the latter's swelling and performance characteristics depend on this. Tetrahydrofuran and most non-polar organic solvents are suitable for SEC of water-insoluble compounds on polystyrene-DVB gels. Water-soluble biopolymers may be chromatographed either on the carbohydrate-derived soft gels or on suitably modified porous silica. Silica gel is particularly versatile as it is compatible with polar solvents and water. However, aqueous systems beyond the pH range of 2 to 8 may degrade these packings. Strong bases like sodium hydroxide should, therefore, be avoided but organic bases like triethylamine are well tolerated.

If possible, a mobile phase of low viscosity at the temperature of operation may be preferred since, in that case, the improved mass transfer would lead to better resolution. To maintain better resolution, solvents with boiling points just 25 to 50°C higher than the column temperature should be used so that the viscosity of the mobile phase is less than 1 centipoise. The choice of the mobile phase is also determined by its compatibility with the detector. The refractive index or the ultraviolet absorption (the two most common modes of detection in SEC) of the solvent should, therefore, be different from that of the solution of the polymer in the mobile phase.

A number of interfering mechanisms have been mentioned earlier. An ionic strength of 0.05 to 0.1 M would eliminate or suppress most of the undesirable solute-retention effects. Sodium, ammonium, potassium and tetramethylammonium ions are preferentially adsorbed on unmodified silica and are increasingly effective in eliminating cationic adsorption effects. Adsorption is more effectively controlled by polyvalent anions such as sulphate and phosphate compared to monovalent anions. A 0.025 to 0.05 M solution of tetramethylammonium phosphate, adjusted to pH 3, is particularly useful. The low pH suppresses the ionisation of acidic functional groups such as silanol and carboxylic acid and thus controls the cation-exchange properties of the gel. Methanol may be added to the mobile phase to reduce hydrophobic retention. Addition of ethylene glycol to the mobile phase may be required to eliminate adsorption of proteins, etc.

In certain cases, controlled association of solute molecules among themselves or with the solvent molecules, may be deliberately promoted to advantage. For example, the tendency of carboxylic acids to dimerize in non-polar solvents such as benzene can be used to decrease their retention, if desired.

#### *6.3.3.4 Other variables*

SEC may be generally and conveniently carried out at room temperature. However, elevated temperatures may be needed for samples of low solubility and for solvents of high viscosity. For example, SEC of high-molecular-weight polyolefins, which are sparingly soluble at lower temperatures, requires to be carried out at temperatures higher than 100°C. When the solute is stable to higher temperatures, water-soluble compounds can also be chromatographed at higher temperatures to improve resolution and peak shape. It must be noted, however, that while the temperature significantly affects the resolution, its effect on the slope and position of the calibration curves is minimal except for a slight decrease in the retention volumes with increase in temperature. Higher temperatures also generally decrease adsorptive effects of certain packings toward low-molecular-weight solutes which also contributes to the observed improvement in the peak shape.

It has been mentioned above in connection with the role of the column parameters on the efficiency of SEC that while the mobile-phase flow-rate has minimal effect on the efficiency when larger particles are used, it could have significant influence on the plate count in HPSEC using less than 20 $\mu$m particles. If we eliminate the terms relating to the effects of stationary-phase–mobile-phase–solute interactions from the plate-height equations (Chapter 2), we have the reduced plate-height equation for SEC as:

$$h = \left(\frac{1}{2\lambda d_p} + \frac{D_M}{F\, d_p^2}\right)^{-1} \tag{6.10}$$

where $d_p$ is the particle diameter, $D_M$ the diffusivity of the solute in the mobile phase and $F$, the mobile-phase flow rate; $\lambda$ is a constant. The equation clearly shows that the plate height is less (and, correspondingly, the SEC is more efficient) when small support particles and a mobile phase of low diffusivity are used with very low flow-rates. This is particularly true as the solute molecular weight increases and solute equilibration within, and outside, the gel structure becomes slower. Therefore, when high resolution is required and the speed of the analysis is of secondary importance, mobile-phase flow requires to be optimized for the particular column. The mobile-phase linear velocity is typically around 1.2 cm/min in such cases; this corresponds to around 4 ml/min of flow-rate in a typical 7.8 mm i.d. column. A flow-rate of 2 ml/min may be a compromise between speed and resolution. The effect of flow-rate on retention volume (which increases linearly with the former) has been recently discussed [15]. The role of the individual operating variables has been discussed in great detail in Ref. 6.

### 6.3.4 Applications of size-exclusion chromatography

Applications of SEC are numerous. During the first decade of its use (beginning 1959), the technique was extensively used for the separation and characterization of biomolecules [6,7]. Subsequent years saw much wider and more varied applications, notably in the area of synthetic polymers of industrial importance [6,16]. Only the general principles and the scope of the applications are discussed here. As noted earlier, SEC applications may be classified into two main categories; gel filtration of biomolecules and other water-soluble polymers; and gel permeation chromatography of organic-solvent-soluble synthetic polymers. In either case, the utility of the SEC is in the area of molecular fractionation according to the size of the molecules for determination of their molecular weight and molecular-weight distribution, removal from small molecules, study of molecular associations, and several others.

#### *6.3.4.1 Group separations*

Separation of two groups of materials of widely differing molecular sizes can be very easily achieved by SEC employing a gel of appropriate pore size; the larger molecules are totally excluded ($K_{av} = 0$) while the smaller ones are completely retained ($K_{av}$ approaching unity). Desalting of biomolecules is one of the first applications to be extensively studied [6,17]. Owing to the large difference in the molecular weights, it is possible to select a gel (e.g., Sephadex G-25) that excludes biomolecules (e.g., proteins) and retains the salt. The difference in the elution volume ($V_s$) is equal to the inside volume ($V_i$) of the gel packing; in practical terms, this means that up to 40% of this figure may be loaded on the column for complete desalting. Besides proteins, nucleic acids and carbohydrates are routinely desalted by this method. Though dialysis can also be used, SEC is more effective and relatively much faster (a few minutes compared to several days in dialysis using cellulosic sacs). A further advantage of the SEC method is that the retained molecules can be subsequently and quantitatively recovered; under favourable conditions, these compounds can be obtained as individual fractions for further studies. Also, the gels may be reused for over 100 times while the dialysis sacs are used only once. The hollow-fibre method of desalting comes close to the advantages of the SEC method but is more expensive, and the accompanying small molecules cannot generally be recovered and separated. Desalting of smaller molecules like peptides, nucleotides and oligosaccharides can be carried out on Sephadex G-10, G-15 or Bio-Gel P-2.

The above principle is also commonly used for separating biopolymers such as nucleic acids, proteins, enzymes, natural products, co-factors, antigens, etc., which are either added to facilitate isolation of the biopolymers (e.g. in ion-exchange and affinity chromatography; see later) or which co-occur with them in nature (e.g., co-factors, inhibitors, etc.). Sephadex LH-20 may be used for separating synthetic polymers from the monomers and oligomers and resinous by-products from synthetic organic reactions. Group separation is also routinely required in a biochemical laboratory during studies on modification of proteins, or radioactive labelling, where the large excess of the reagents used must be removed after the reaction. Advantage may also be taken of the fact that certain biologically active molecules, such as hormones, are natively bound to proteins and can therefore be separated from the co-occurring small molecules by SEC; the hormone-protein complex is then broken and the products refractionated [6].

#### *6.3.4.2 Separation of permeable compounds*

In the absence of other interactions, SEC offers the mildest procedure for separation of compounds that are small enough to permeate the pore structure of the gel. As already mentioned the compounds which enter the inner space of the gel structure are displaced by the solvent molecules in the order of decreasing molecular size. Peptides, enzymes, haemoglobin and serum proteins have been fractionated using Sephadex gels of increasing pore-size. Several other classes of materials, including viruses, have been separated using appropriate gels.

A major limitation of this technique of separation of individual components is the low peak capacity, or the number of compounds that can be separated by a given column; as noted earlier, all retention volumes should be between $V_o$ and $V_M$. To obtain adequate resolution of compounds with close molecular sizes, the inner volume of the gel, $V_i$ must be increased by increasing the bed volume. There is an upper limit to this for soft gels, beyond which flow-rates severely decrease, or the gels get compressed. Microparticulate polystyrene-DVB or silica gels, which can withstand high pressures, may be used to pack columns of high capacity and resolution capability; if necessary, two or more columns can be connected in series to obtain a plate count (or the number of theoretical plates) of 10,000 to 30,000. The peak capacity of the column is given by the following relationship:

$$C_p \simeq 1 + \sigma^2 \sqrt{N} \qquad (6.11)$$

The problems of resolution and peak capacity can also be solved to some extent by recycling the effluent back into the column. Most instruments for LC are equipped with recycling capability.

While the other LC methods (to be discussed later in this chapter) may offer better resolution of compounds with close molecular sizes (e.g. isomers), there are several advantages of using SEC as the first, exploratory method of analysis of unknown samples. No time is spent in method development, the only pre-condition for SEC being that the sample be soluble in a solvent that is compatible with the chosen gel and there is no need for solvent programming; in most cases, information on the sample complexity and their molecular weights can be obtained in about 20 min. The speed of the analysis makes SEC the method of choice for routine analyses and quality control of a variety of industrial products.

The choice between the microparticulate organic gels (e.g. $\mu$-Styragel) and the rigid silica particles depends on the particular application and the two matrices are generally complementary. The former are generally preferred for the analysis of lower molecular weight compounds (MW = 100 to 1000) and the latter for samples with a wider range of molecular weights. The organic (polystyrene-based) gels are preferred for the analysis of labile compounds sensitive to hydroxylated solvents and silica, they are obviously not suitable for water-soluble solutes and for high-speed separations where high head pressures are needed.

Several strategies have been discussed earlier for eliminating or suppressing non-steric effects in SEC. These interfering interactions may also be exploited to achieve resolution of SEC columns. Examples are separation of aromatic compounds and peptides containing aromatic amino acids. Ion exclusion of negatively charged nucleotides on Sephadex gels has also been used to separate these compounds. During desalting operations, the proteins sometimes get precipitated as they are separated from the salt and this may occur at different locations on the column depending on their solubility at different advancing salt gradients. Adsorption and partition effects play an important role in separation of organic compounds on LH-20 using organic solvents. Partition chromatography using water entrapped in hydrophilic gels (e.g. Sephadex G-25) as stationary phase and organic solvent as mobile phase has been used for separation of ribonucleic acids, oxytocin, actinomycins and certain oligopeptides [6].

#### *6.3.4.3 Determination of molecular weights and molecular-weight distribution*

SEC has been used extensively for determination of molecular weights. Though retention in SEC is not strictly dependent on the molecular weight, the retention data may be related to the molecular weights with reasonable accuracy for a given class of compounds, e.g. proteins [17]. It is assumed that the column to be used is suitably calibrated for compounds of the same type and conformation, and that interfering non-steric effects are completely eliminated. The simplest method of calibration is to plot the logarithm of the molecular weight of a series of standards *vs*. their retention volumes using a gel of appropriate pore structure. The calibration curve thus obtained may be used in the region over which it is linear. Large errors in molecular weights can result if the standards and the test samples have different conformations under the experimental conditions. The so-called universal calibration plots, valid for linear polymers over a certain range, have been mentioned earlier. SEC has been used extensively for the analysis of biopolymers [18].

As noted earlier, SEC can also be used for obtaining the molecular weights and sizes of small molecules. In addition to molecular weight, the molecular length and the volume of several classes of compounds have been correlated with their retention volumes. For a more detailed discussion on these, see Ref. 6.

More recently, SEC has been used extensively to determine the molecular-weight distribution of synthetic polymers [19]. Being polydisperse, these compounds do not have a single molecular weight; also, their industrial use depends largely on the molecular-weight distribution. Indeed, their separation into individual fractions and determination of their exact molecular weights is often not necessary. They are characterized by their average molecular weight ($M$), whose value depends on the analytical method used to determine them. The weight-average molecular weight ($M_w$) is the molecular weight of an imaginary homogeneous polymer, whose light scattering and sedimenation properties are the same as the mixture being studied. The number-average molecular weight ($M_n$) is obtained using colligative methods such as osmotic measurements and group analysis. The quotient $M_w/M_n$ is a measure of the inhomogneity (or polydispersity, $\delta$) of the polymer; the value may vary from 1 for homogeneous polymers to over 10 for block polymers. The Mark–Houwink equation (6.9) is valid only for materials with a narrow molecular-weight distribution. For polymers with a broader range of molecular-weight distribution, the following relationship (eq. (6.12)) is suggested:

$$[\eta] = K\, M_w^a\, (1 - \delta) \tag{6.12}$$

### 6.3.5 Hydrodynamic chromatography (HDC)

Apart from the SEC methods discussed above there are several other variants of the so-called one-phase chromatography, using which, samples can be separated on the basis of the sizes of the components. Here again, there is no stationary phase in the traditional sense of the term, retention and separation being controlled by the hydrodynamic or flow properties of the solutes or by applied external fields.

In hydrodynamic chromatography (HDC), which may be used for the separation of insoluble or suspended particles (e.g. colloids), the particles do not enter the pore structure but are separated during their flow through the interstitial space of the packing structure [20,21]. The packing material for HDC is usually an ion-exchange resin of 15–20 $\mu$m particle size; nonfunctionalized polymer beads or glass beads may also be used. Columns of 10mm i.d., and 0.5 to 3.0 m length are generally used, generating about 70,000 plates/m. The flow profile may not be the only factor influencing the mobility of the sample particles. Both the packing and the colloids have associated double layers and the strong electrostatic interaction is influenced by the ionic strength of the mobile phase. The mobile phase is usually a buffer, containing a surfactant; a low molecular weight compound such as sodium dichromate is used as the marker. As in the other separation methods based on the molecular/particle size, larger particles elute first; but, the flow order may be reversed at sufficiently high ionic strengths; van der Waal's forces take over when electrostatic repulsion is suppressed. HDC may be used for analysing and calculating the particle size of high-molecular-weight polymers and colloids, both organic and inorganic. It is claimed to be uniquely suited for studying how colloids shrink, aggregate or otherwise alter their effective hydrodynamic diameter, depending on the environment.

### 6.3.6 Field-flow fractionation (FFF)

Field-flow fractionation may be considered an open-tubular version, there being no packing material. The concept, as propounded by Giddings [22], is based on coupling of concentration and flow non-uniformities. Separation occurs owing to differential migration of the sample components as the fluid flows unidirectionally under the effect of an applied field, which may be thermal, electrical, magnetic, steric or gravitational. The apparatus (Fig. 6.11a) consists of a pair of rectangular plates held

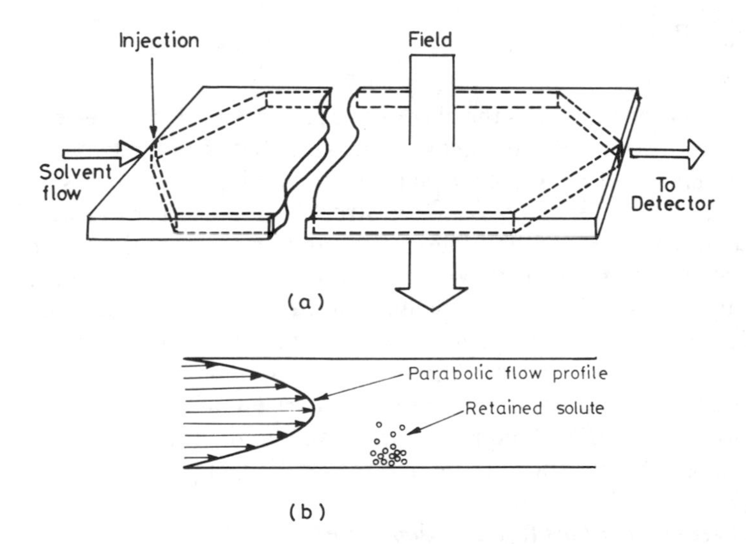


Fig. 6.11 — (a) General arrangement for field-flow fractionation and (b) parabolic flow profile. Reproduced from Ref. 22 by courtesy of Elsevier Scientific Publishing Co.

apart by a spacer of 50–500 $\mu$m thickness to support the fluid flow which has a parabolic profile (Fig. 6.11b); that is, the molecules at the centre of the tube have the highest velocity while those near the wall are almost stationary. The fluid path (channel) is usually 1 to 2.5 cm wide and about 50 cm in length. The applied field is perpendicular to the plane of the fluid channel. The sample may either be in solution or in a suspended form. The solute particles are pushed to the wall by the applied field. Under these conditions, the larger particles may settle down while the smaller particles are carried by the higher fluid velocity at the centre of the tube. Thus, the smaller particles may be eluted first. However, the situation reverses when the particle diameter exceeds about 1 μm, when the larger particles may be imagined to extend into the moving fluid and be pushed along by the moving fluid. Thus, there is a transition region (around 1 μm in which particles of different sizes may elute at the same time. This is not a serious drawback since FFF is particularly useful for particles larger than 1 μm; particles up to 100 μm size (corresponding to a molecular weight of about $10^{18}$) have been separated by FFF.

The theoretical basis and applications of different FFF techniques have been recently elaborated by Janca [23]. In general, the chromatographic constant, $R$, in FFF is given by:

$$R = 6\lambda \tag{6.13}$$

where,

$$\lambda = R^{*}T/Fw$$

$R^{*}$ being the gas constant, $T$ is the absolute temperature, $F$ is the effective force of the field on 1 mole of the solute, and $w$ is the width between the two parallel plates. The height equivalent to an effective plate, $H$, in FFF is given by $Rw$, so that the best resolution is obtained for solutes that are strongly retained (low value of $R$), when the distance between the parallel plates is the minimum and the field strength is high.

Except for the rectangular flow-through tube which replaces the column in the normal instrumental LC, the equipment for FFF is substantially the same and consists of a solvent reservoir, pump, injection and detection systems (see Chapter 7). The FFF techniques using the different fields (sometimes referred to as the FFF subtechniques) and their applications are summarized in Table 6.1. As can be seen from the Table, the FFF techniques are applicable to the separation of an exceptionally wide range of compound classes and molecular weights.

#### *6.3.6.1 Thermal field-flow fractionation (TFFF)*

TFFF, the oldest of the FFF subtechniques, is based on the principle of thermal diffusion. The TFFF channel (50–300 cm) is formed between two highly polished metal surfaces, the distance between which is maintained by a spacer (50–250 μm). The upper block is heated electrically while the lower one is cooled by water so that the temperature gradient between the walls is 20 to 100°C. The value of λ in TFFF is given by:

$$\lambda = \left[ W \frac{\alpha}{T} \cdot \frac{\mathrm{d}T}{\mathrm{d}x} \right] \tag{6.14}$$

where $\alpha$ is a dimensionless thermal diffusion factor, which is a function of the thermal diffusion coefficient, $D_T$:

**Table 6.1** — Applications of field-flow fractionation techniques†

| Subtechnique | Compound classes | Molecular weight range ($10^x$) |
|---|---|---|
| Thermal FFF | Synthetic polymers | $x=2$–7 |
| | Crude oils and asphaltenes | $x=3$–4 |
| Sedimentation FFF | Polystyrene beads, T2 virus | $x=6$–12 |
| Electrical FFF | Proteins | $x=5$–6 |
| Flow FFF | Silica beads | $x=5$–10 |
| | Polystyrene latex beads | $x=8$–11 |
| | Viruses | $x=6$–7 |
| | Proteins | $x=5$–6 |
| | SDS-proteins complexes | $x=4$–9 |
| | Sulphonated polystyrenes | $x=4$–6 |
| | Sodium polyacrylates | $x=5$–7 |
| Steric FFF | Porous glass beads | $x=12$–14 |
| Magnetic FFF | Bovine serum albumin | $x=5$ |

† Data from Ref. 23.

$$\alpha = D_{\mathrm{T}} \frac{T}{D} \tag{6.15}$$

The above relationships indicate that TFFF can be used to determine the thermal diffusion factors of polymers in different solvents. Theoretical treatment of TFFF, however, is complicated by the fact that, because of the temperature gradient, the fluid is not isoviscous and hence the flow profile is not parabolic. Nevertheless, from the practical point of view, TFFF of industrial polymers is claimed to have a number of advantages over the traditional SEC. Pressurization of the system extends the operating range of the TFFF channel and renders fractionation of the lower molecular weight samples (with molecular weights of a few hundred daltons) possible. Thus, polystyrenes of molecular weights ranging from 4000 to 7,000,000 could be fractionated in a single experiment using temperature-gradient programming.

#### *6.3.6.2 Sedimentation field-flow fractionation (SFFF)*

SFFF, which like TFFF also has had a long history, uses either a natural gravitational or a centrifugal field to effect the separation. The value of $\alpha$ in SFFF of spherical particles is given by the expression:

$$\lambda = \frac{6kT}{\pi d_{\mathrm{p}}^{3}\, Gw\, \Delta\rho} \tag{6.16}$$

where $k$ is the Boltzmann constant, $G$ is the gravitational or centrifugal acceleration and $\Delta\rho$ is the difference in the densities of the particles and the solvent. The operating variables of SFFF include programming of the centrifugal field (speed of revolution), density gradient of the solvent, its flow-rate and the width of the channel. Judicious use of these variables permits extension of the range of the particle sizes that can be used in SFFF. Polystyrene-latex particles in the range of particle diameters of 0.091 to 0.982 $\mu$m could be separated in a single run. The capability of SFFF to separate and determine the molecular weights and dimensions of biomolecules has been used to calculate the physical properties of bacteriophages; see Ref. 23.

### *6.3.6.3 Electrical field-flow fractionation (EFFF)*

The electrical field required for this subtechnique is induced by oppositely charging the two parallel plates of the column. The electrodes are separated from the fluid channel by two semipermeable membranes to permit the passage of small ions across the column (Fig. 6.12). The space between the membranes and the electrode plates is

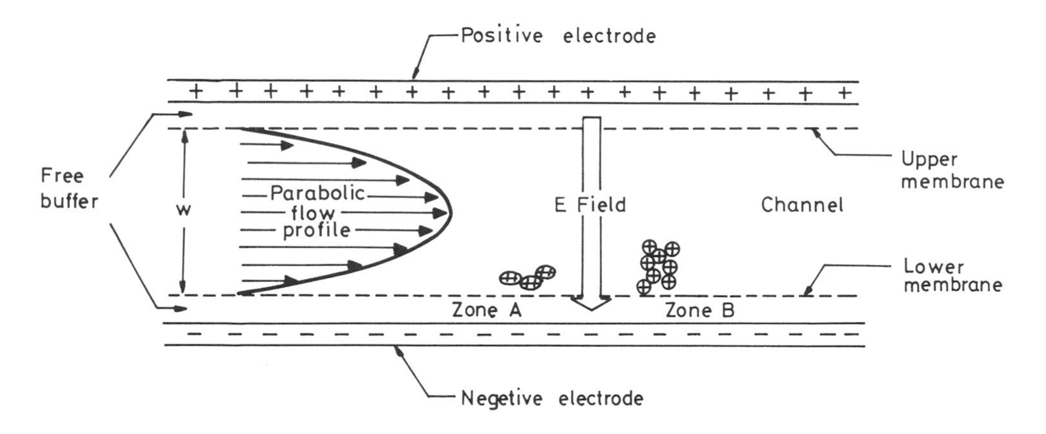


Fig. 6.12 — Arrangement for electrical FFF. Reprinted with permission from Ref. 24; copyright (1976) American Chemical Society.

filled with a suitable buffer, so that a homogeneous electrical field is induced inside the channel. The value of λ in EFFF is given by [24]:

$$\lambda = \frac{D}{\mu E w} \tag{6.17}$$

where $\mu$ is the electrophoretic mobility and $E$ is the electrical field strength. The anomalies observed in the theoretical and experimental values in EFFF are explained as arising out of electro-osmotic effects caused by the uneven expansion of the flexible membranes; stiffening of the membranes using polyethylene frits solved the problem considerably [25]. The operating variables in EFFF include the pH, ionic strength, etc., apart from the applied electrical potential. Since the mobility in

EFFF under a given field strength is determined by the ratio $D/\mu$, compounds differing in their diffusion coefficients are separable by EFFF even when their electrophoretic resolution is poor.

#### *6.3.6.4 Flow field-flow fractionation (FFFF)*

The FFFF is probably the most frequently used subtechnique. The effective field is generated by the flow of a solvent, perpendicular to the fluid channel that is moving between plates (0.038 mm thickness) or tubes (0.02 mm diameter) of semipermeable membranes. In FFFF, the field generated by the solvent of velocity, $U$, acts uniformly on all the solutes and the separation is therefore determined largely by the diffusion coefficient or friction coefficient. The parameter $\lambda$ is expressed by:

$$\lambda = \frac{R^{*}TV_{o}}{3\pi N\eta V_{c}w^{2}d} \tag{6.18}$$

where, in addition to the terms defined earlier, $V_o$ is the dead volume of the channel, $V_c$ is the flow-rate of the field-inducing solvent, $\eta$ is the viscosity of the medium and $d$ is the effective Stokes' diameter. FFFF can be used as a dialysis or ultrafiltration system for the separation of two compounds of widely different molecular sizes, by appropriately choosing the membrane, geometry and dimensions of the channel and the flow-rates. A variety of viruses, proteins, sulphonated polystyrenes and the sodium salts of polyacrylic acids have been analysed by FFFF.

#### *6.3.6.5 Steric field-flow fractionation (steric FFF)*

As already noted, the principle of steric FFF is based on compressing the solute molecules firmly to the wall, so that the larger molecules are physically pushed by the higher parabolic flow profile of the fluid. Thus, larger particles are moved faster than the smaller molecules by the higher velocity of the fluid away from the wall. While any of the above discussed types of field can be used, the gravitational field appears more practical for particles of 1 to 100 $\mu$m. The value of $R$ in this case is given by:

$$R = \lambda_a (1 - \lambda_a) \tag{6.19}$$

where $\lambda_a = a/w$, $a$ being the radius of the particles, and $w$ being the width of the channel. Since steric FFF is applicable to relatively large particles of diameter extending to hundreds of microns, its application to fractionation of cells and micro-organisms appears feasible.

## 6.4 PARTITION CHROMATOGRAPHY

### 6.4.1 General features

The principle and applications of partition chromatography, also called liquid–liquid chromatography (LLC), is akin to liquid–liquid extraction and countercurrent distribution (CCD). However, CCD is a discontinuous process in which equilibrium is apparently attained in each step, while partition chromatography is a continuous one. Separation in partition chromatography is effected by distribution of the solute components between two immiscible liquids, one of which is held stationary by coating it on a finely divided, porous-solid support and the other (the mobile phase), is allowed to pass over it. The requirement of total immiscibility between the stationary and mobile liquids restricts their choice to a few combinations. Understandably, the relative solubility of different components of the solute determines their relative mobility and retention; a distribution constant of 5 to 10 is considered ideal for separation by partition chromatography. If the distribution constants of two compounds between two immiscible liquids is known, it is possible to predict their $k$ values in a particular system and to determine if the two compounds can be separated by this method.

In recent years, LLC has been largely replaced by bonded-phase chromatography (see section 6.5, below) because of the latter's convenience in handling and the insolubility of the bonded stationary phases in a large number of solvents, permitting a wider choice of mobile phases. Nevertheless, LLC continues to be of use as it offers increased selectivity by exploiting a variety of chemical interactions between the solute and certain speciality stationary phases, a facility that is not readily available in bonded-phase chromatography (e.g. ligand-exchange chromatography, normal-phase ion-pair chromatography, extraction chromatography, etc.; see the following section). LLC is also of immense historic value, having been extensively used for the separation of amino acids and carbohydrates by paper chromatography, following its invention by Martin and Synge. The most serious drawback of the technique is that the stationary phase may be partially carried away by the mobile phase, exposing the active sites on the solid support to interaction by solute molecules. Its limitations notwithstanding, LLC remains attractive in view of its simplicity and inexpensive solvent systems as well as the sound theoretical basis which permits accurate prediction of retention behaviour. Table 6.2 summarizes the partition chromatographic systems for some important classes of compound [26].

Two types of LLC are possible; in the normal-phase LLC, the stationary phase is more polar than the mobile phase; the opposite is valid for the reversed-phase LLC. The normal-phase LLC is generally used for the separation of polar compounds which are eluted in the order of their increasing polarity. The reversed-phase LLC finds application in the separation of relatively non-polar compounds which can be well retained by non-polar stationary phases. In principle, any difference in the chemical functionality (such as the nature and the number of functional groups, length of the hydrocarbon chain, etc.) that affects the solubility of a compound can be exploited to advantage in LLC.

### 6.4.2 Column-packing materials

#### *6.4.2.1 The solid support*

Ideally, the solid support in partition chromatography should be inert, insoluble in the two liquid phases and capable of holding a certain amount of the stationary phase. Preferably, the material should have large pores to permit free access of the solute molecules to the stationary phase, but small enough to prevent removal of the stationary phase by the shearing effect of the fast-moving mobile phase. It should also have a low surface area in order to minimize adsorption of the solutes and prevent tailing of the peaks. Particles of 50–150 μm diameter and a pore size of

**Table 6.2** — Partition chromatographic systems for separation of some important compound classes†

| Stationary phase | Mobile phase | Compounds separated |
|---|---|---|
| Normal-phase systems | | |
| Dimethylsulphoxide<br>Ethylene diamine<br>Ethylene glycol<br>Nitromethane | C-5 to C-8 alkanes modified with 0–20% halomethanes/MeCN/ THF/dioxanes, etc. | Terpenoids, Steroids, etc. |
| β,β′-Oxypropionitrile | 7% Chloroform/hexane | Insecticides |
| Polyethylene glycol 600 | Hexane | Non-ionic detergents |
| Trimethylene glycol | Hexane | Pesticide metabolites |
| Tris(cyanoethoxy)propane | Iso-octane | Phenols |
| Water | *n*-Butanol | Sugars |
| Water/EtOH/Iso-octane (ternary, aqueous phase) | Organic phase | Steroids<br>Metal complexes |
| $CH_2Cl_2$/MeOH/water (ternary, aqueous phase) | Organic phase | Corticosteroids |
| Reversed-phase systems | | |
| Cyanoethylsilicone | Water | Coumarins |
| Dimethylpolysiloxane | Water + MeCN/MeOH | |
| Heptane | As above | |
| Squalane | 20% water/MeCN | Hydrocarbons |
| Hydrocarbon polymer | 3% MeCN, 0.001M EDTA, pH 4.5 | Tetracyclines |
| Polyamide | 5% Hexane/EtOH | Penicillins |

†Data from Ref. 26.

200–500 Å, which can hold up to about 50% of its weight of the stationary phase, are preferable. Smaller particles may be used for high-performance LLC. Silica gel or its variants (e.g., kieselguhr or diatomaceous earth) and cellulose are among the most suitable solid supports for LLC. Silica gel can hold up to 70% of its weight of water and still retain its free-flowing form. Both totally porous and superficially porous (pellicular) silica particles (of surface ranging from 10–250 $m^2/g$) are commercially available for use in both traditional and high-performance LLC. The pellicular particles can hold a maximum of 3% of the liquid phase (other than water) while some of the fully porous varieties can hold ten times as much.

#### *6.4.2.2 Coating of the support*

There are several ways of coating the support with the stationary liquid phase. The simplest and the most common (particularly for particles with diameter larger than 20 $\mu$m) is the solvent evaporation method, analogous to the preparation of the GC columns. A known weight of the stationary phase is dissolved in a volatile solvent such as dichloromethane and an appropriate quantity of the solid support is added. The solvent is then evaporated while stirring the mixture constantly or by using a rotatory evaporator. The column may be packed as described in Section 6.2.1.

Prepacked columns, particularly those with small-diameter supports for use in HPLC, are conveniently coated by one of the following procedures. In order to avoid air bubbles and column disturbance, these columns must be slurry packed. If the stationary phase is a low-viscosity liquid, like water or ethanol, it may simply be pumped through the column; subsequently, the column is flushed with the mobile phase which has been pre-equilibrated with the stationary phase. Alternatively, the mobile phase is first pumped through the column and small quantities of the stationary phase are successively injected into the column. The loading is complete when the stationary phase begins to elute from the column. High-viscosity stationary phases are first dissolved in a good solvent in high concentration and equilibrated with the solid support by continuously pumping the solution through the pre-packed column. A solvent that is miscible with the first solvent, but in which the stationary phase is insoluble, is then pumped through the column to precipitate the latter into the pore structure of the support. The columns coated by such *in situ* methods need to be equilibrated with the mobile phase for several hours to eliminate the excess liquid. Pumping the mobile phase, pre-saturated with the stationary phase, through the column transfers the stationary phase on to the support until equilibrium is reached. It must be noted, however, that if the stationary-phase loading is less than equilibrium concentration, tailing peaks would result owing to adsorption on the uncoated active sites. To support non-polar liquid phases on silica gel or other polar supports, the surface of the latter is rendered lipophilic di- or tri-methylsilylation. Exposure of the solid material to vapours of the appropriate chlorosilane in a closed container is usually adequate.

The sample capacity of the column is generally a function of the amount of the stationary phase on the column which may vary from a mono-molecular layer in pellicular packings to heavy loading in completely filled, porous non-pellicular packings. It is therefore preferable to increase the loading to the maximum possible level, provided the stationary phase is of low viscosity. However, heavier loading of highly viscous phases would present columns of lower efficiency owing to slow mass transfer, and this is particularly damaging with pore-filled particles.

### 6.4.3 The partitioning phases

In principle, any pair of mutually insoluble liquids can be used in partition chromatography. It is advantageous if both of them have low viscosity but a high viscosity of the stationary phase is acceptable. The choice of the partitioning phases is largely dictated by the solubility of the sample in the two phases and this property also determines whether to use normal- or reversed-phase mode; e.g. solutes soluble in polar solvents are retained better by polar stationary phases. Low sample solubility in the two phases severely restricts the 'loadability' of the column. Retention and selectivity may be achieved by appropriate choice of the mobile phase composition but the requirement of mutual solubility of the two phases limits the choice.

The polar stationary phases generally in use are: water, β,β′-oxydipropionitrile (BOP), nitromethane, dimethylsulphoxide, ethylenediamine, ethylene glycol and several polyethylene glycols, known as Carbowaxes. Of these, water, BOP and the Carbowaxes are the more frequently used. BOP is generally preferred for the separation of amines while the Carbowaxes are more selective for alcohols. Lower molecular weight stationary phases can be expected to yield more efficient columns compared to the high molecular weight phases of comparable nature (e.g. the polyethylene glycols). Aliphatic hydrocarbon solvents, occasionally modified with chloroalkanes, tetrahydrofuran or acetonitrile, form satisfactory mobile phases.

*n*-Butanol is frequently used when the stationary phase is water. Di-*n*-butyl ether and nitromethane are also popular. In the absence of information on the solubility properties and distribution coefficients, the choice of the phases is largely empirical.

Aqueous methanol or aqueous acetonitrile is generally used as the mobile phase. Hydrocarbons (C-7 and above, and squalane) and silicones (e.g. cyanoethylsilicone and dimethylpolysiloxane) may be used as stationary phases in reversed-phase LLC but it is more convenient to use the alkylated (C-2, C-8 or C-18) bonded phases, to be discussed later. However, LLC, using non-polar stationary phases is advantageous in connection with the so-called extraction chromatography of radioactive-metal ions in view of its high sample capacity, selectivity and reproducibility [27].

Use of a ternary solvent system, where the third solvent is miscible with both the liquids, is sometimes advantageous. The third solvent distributes itself between the two phases and either of the liquids may be used as the stationary phase with the other forming the mobile phase. It is often possible to vary the relative proportions of the three solvents over a range, giving rise to a number of two-phase systems. The *n*-butanol–acetic acid-water system in paper chromatography and the water–ethanol–2,2,4-trimethylpentane (iso-octane) system in column LLC are widely used.

Since the stationary phase is mechanically held on the support, the column has to be conditioned by irrigating with the mobile phase (presaturated with the stationary phase) for several hours. The mobile phase flow-rate should not be high enough to displace the stationary phase by the shearing effect of the fast moving fluid. The ratios of the stationary phase to the solute and that of column length to diameter are fairly high in LLC, the typical values being 2000 and 50, respectively. LLC is usually carried out at ambient temperature or, at the maximum, up to 50°C (if it is required to lower the viscosity of the stationary phase), provided the two phases are immiscible. The possibility of stationary phase dissolution in the mobile phase also precludes gradient elution in LLC.

### 6.4.4 Countercurrent chromatography (CCC)

The long-known method of separating compounds by distributing them between two immiscible solvents in a separatory funnel, its development into multi-stage countercurrent distribution by Craig and others and, finally, into partition chromatography by Martin and Synge, have all been discussed earlier. It has also been noted that the idea of supporting the stationary phase on a porous solid originated as a way to avoid the tedious and numerous extractions and the unwieldy size of the apparatus. An interesting development in recent years has been to reverse the trend and completely eliminate the solid support, which sometimes exhibits irreversible adsorption of certain solutes and decreases the recovery. While the traditional, multi-stage countercurrent distribution apparatus is now essentially only of historical value, the newer concepts use the same principle but in a continuous process. Two variants of the mechanics of the process may be mentioned; they are: (1) droplet countercurrent chromatography [28–30], and (2) rotatory or centrifugal countercurrent chromatography [31–34].

#### *6.4.4.1 Droplet countercurrent chromatography (DCCC)*

In this technique, a steady stream of droplets of the mobile phase is pumped through the stationary phase held in a series of long and narrow columns. 100 to 600 vertically held columns of 20 to 120 cm length and 1.5 to 4 mm (ideally 2 to 3 mm) diameter may be used; the tubes are connected in series using Teflon tubing. The columns are filled with the stationary phase and the sample is introduced, dissolved in either the stationary or the mobile phase, or in a mixture of both the liquids. While, in principle, any combination of immiscible liquids can be used, the success of the method depends on the generation of mobile-phase droplets of suitable size and mobility. The formation of stable and discrete droplets is determined by the relative specific gravity, viscosity and surface tension of the two liquids. The mobile phase is introduced at the head of the column by means of a capillary tube, so that a steady stream of droplets is formed. The droplets have to be smaller than the diameter of the columns and smaller (less than 2 mm) diameter columns severely restrict the choice of the solvent systems. The mobile-phase flow-rate must be carefully controlled by means of a pump so that only the mobile phase passes from one tube to another. A flow of 8–50 ml/hr requires operating pressures of 5–15 kg/cm$^2$. During the long passage of the droplets through the stationary phase, the turbulence within the droplets causes efficient partitioning of the solutes between the two phases. Under ideal conditions (i.e. if the droplets remain discrete throughout their passage through the columns), each droplet could represent a theoretical plate. Fig. 6.13 is a

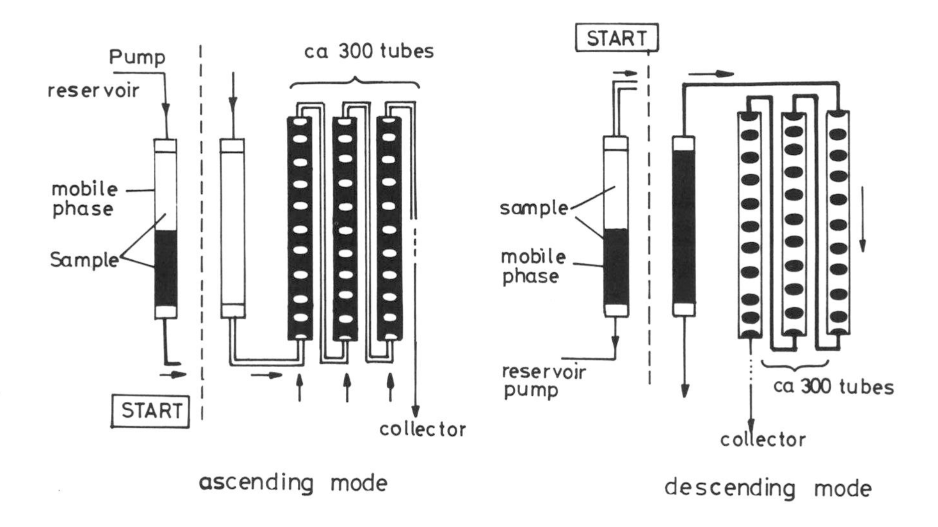


Fig. 6.13 — Flow diagram for DCCC operation. Reproduced from Ref. 28 by courtesy of Georg Thieme Verlag.

schematic representation of the apparatus for DCCC [28] that can be modified to enable the use of either a heavier or a lighter mobile phase. Either the more polar or the less polar solvent may be used as the mobile phase, but it is generally more advantageous to use the less polar solvent.

Most of the common solvent systems for partition chromatography have been examined for their suitability in DCCC but a majority of these were unsatisfactory, either due to non-formation of droplets or to the production of flow plugs that displace the stationary phase. The most popular solvent systems have been combinations of chloroform, methanol, water and/or propanol in different ratios. The technique of DCCC has been extensively used for the isolation of natural products from plant extracts. The technique is particularly useful for the separation of the more polar natural products such as the glycosides of steroids and terpenoids (saponins), flavonoids and other polyphenols and alkaloids [29]. DCCC has also been used for the separation and purification of compounds of biochemical interest; e.g. DNP-amino acids, peptides, antibiotics and lipids. Quantitative recovery of the compounds and mild experimental conditions are claimed to be the main advantages of the method. In the absence of an active solid surface, both chemical transformations and irreversible adsorption are avoided. Depending on the diameter of the columns, the scale of operation could vary from a few milligrams to several grams of the substance to be separated. The main drawback of the method appears to be that the process is extremely slow, requiring from 12 h to several days for completion of the analysis.

#### *6.4.4.2 Centrifugal countercurrent chromatography (CCCC)*

The above-described DCCC and its variants using coiled columns, namely, helical (toroidal coil) CCC and locular CCC [29], may be considered to use the principle of hydrostatic equilibrium system (HSES). These techniques share certain common features like stationary columns or coils, associated with near total retention of the stationary phase, limited flow-rates and, consequently, relatively long analysis times. The alternative, hydrodynamic equilibrium system (HDES), developed by Ito [31], uses a rotating coil to ensure faster equilibration and separation. Fig. 6.14 illustrates the difference in the mechanics of the two systems. While the events occurring in the HSES system (Fig. 6.14a) are similar to those in Fig. 6.13, rotation of the coil along its axis sets in a hydrodynamic equilibrium in which each segment (turn) of the coil is occupied by an equal amount of the two phases. Thus, when the mobile phase is introduced into the coil, filled with the stationary phase, it first displaces an equal volume of the latter, but thereafter it displaces only itself so that the remaining part of the stationary phase is retained in the coil. Obviously, the interfacial area, the degree of mixing and thus the efficiency, are far greater in HDES; the equipment and its operation are also more complicated.

The co-axially coiled column can be mounted on a holder that can support multiple layers of the coil so that the amount of the stationary phase can be increased substantially even at high flow-rates of the mobile phase. An interesting possibility presents itself when the speed of the axial rotation of the coil is increased. While the above-described (Fig. 6.14b) events (i.e. each of the two phases occupying nearly half the space in each turn of the coil) occur at low speeds, the hydrodynamic forces tend to push the heavier droplets toward the head (left-hand side) of the column, and at a certain speed complete separation of the phases within the coil occurs so that the head (or the left-hand) side is occupied by the heavier liquid and the tail side by the lighter. When the speed is increased beyond a critical point, the centrifugal forces overcome the gravitational forces and, at this stage, the heavier liquid would occupy

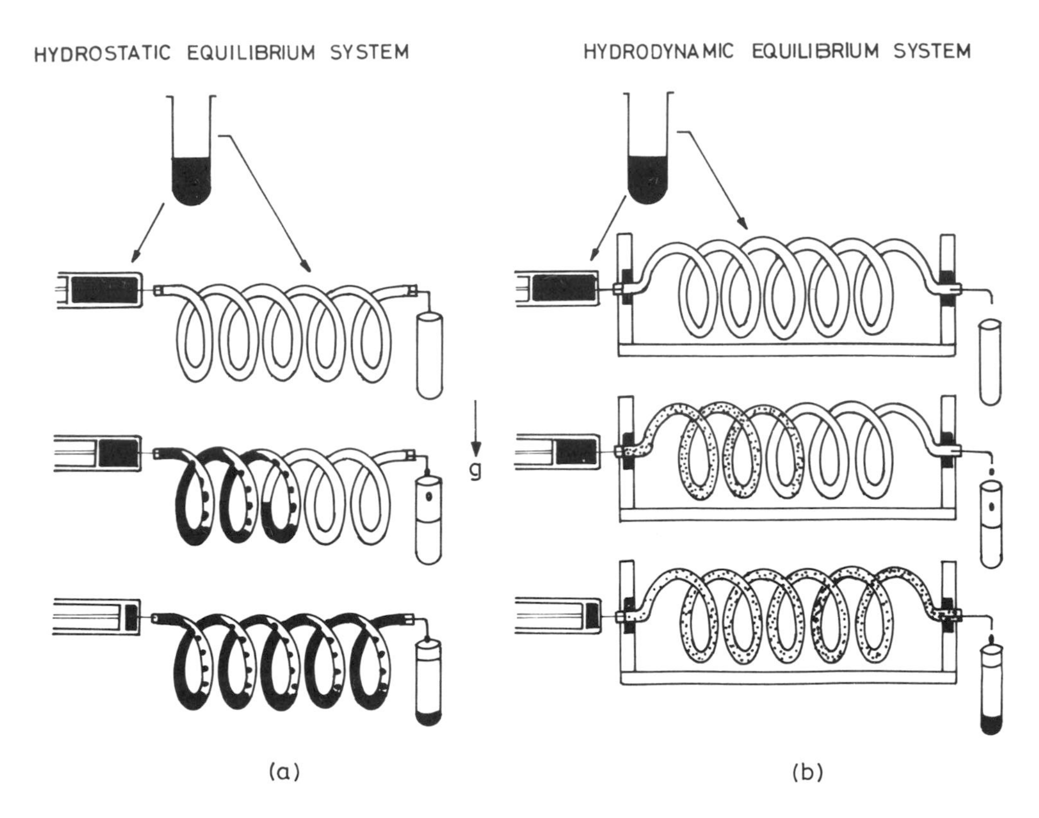


Fig. 6.14 — Schematic representation of countercurrent chromatography using (a) hydrostatic and (b) hydrodynamic equilibrium systems. Reproduced from Ref. 31 by courtesy of Elsevier Scientific Publishing Co.

the outer portion and the lighter liquid the inner portion of each coil. It is then possible to introduce both the liquids continuously and collect them at opposite ends of the coil in a dual countercurrent mode. It is also possible to introduce the sample continuously through a flow tube in the middle of the coil. There are also several possible ways of rotating the coil, distinguished as planetary (the coil rotating around its own axis as well as around a central axis) and non-planetary (rotating around its own axis as in Fig. 6.14b). The planetary motion may be synchronous (i.e. the coil rotations around its own and the central axes having the same angular velocities) or non-synchronous. The orientation of the coil holder with respect to the central axis may also vary and can be either parallel (the two rotations being in the same direction and along parallel axes), antiparallel, perpendicular or slanting. Detailed description of these arrangements and the mechanics are given by Ito [32].

The CCC systems are suitable for the extraction and clean-up of the samples in biological and natural products research, on a preparative scale as well as for analytical separations. The usefulness of the technique has been demonstrated in separations of DNP-amino acids, peptides, antibiotics, plant hormones, pesticides, etc. [32-34]. The coil dimensions are typically 1.6 mm i.d. and 130 m long (285 ml capacity). Either an aqueous or organic phase could be the mobile and the other, the stationary phase. The aqueous systems are generally 0.5 to 1 molar salt solutions containing ammonium or alkali-metal halides, acetates, sulphates or phosphates; the organic phase may be *n*-butanol, ethyl acetate or chloroform-isopropanol (3:1) and the like, or even an aqueous solution of a polymer (e.g., dextrans or polyethylene glycols) with some modification of the apparatus [34]. The high-speed versions may take a few hours for completion. Gradient elution is also possible.

## 6.5 ADSORPTION CHROMATOGRAPHY

### 6.5.1 Introduction

Adsorption chromatography is sometimes called liquid–solid chromatography (LSC), signifying the fact that it includes all forms of liquid chromatography other than liquid–liquid and gas–liquid chromatography. For convenience, one may discount the size-exclusion and ion-exchange mechanisms discussed in accompanying sections. That still leaves a number of intermolecular interactions by which a solute can be retained on a solid surface; these include electrostatic or dipole–dipole forces, hydrophobic interaction and hydrogen-bonding. Though theoretically well defined, these mechanisms (as well as partition and ion exchange) may operate simultaneously and, in actual practice, it is difficult to make a clear-cut distinction. Partition and adsorption are limiting cases of what normally occurs in LSC. As soon as a solvent or atmospheric moisture comes into contact with an active solid surface (e.g. silica), adsorption of the solvent molecules would occur; solute retention is then governed by a combination of the above-mentioned forces and partition between the solvent (or water) retained by the adsorbent and the mobile phase. Similarly, ion exchange is also the result of an extreme case of charge separation in dipolar interaction: hydrophobic interactions are at the other end of the spectrum. In other words, adsorption chromatography may be considered to include all mechanisms of retention lying from purely hydrophobic interactions, in reversed-phase partition, to ion exchange. Complications arising out of the co-existence and interplay of all these mechanisms have been mentioned at several places and are particularly evident in affinity chromatography, where, because of the size of the biomolecules, size-exclusion also plays an important role. Steric factors have been ignored in the present discussion considering the small size of the non-polymeric molecules of the common eluites in LSC relative to the pore size of the adsorbents. It may be mentioned here that, though biochemical interactions at the molecular level are strictly physicochemical, treatment of bioaffinity chromatography as a separate section is for reasons of convenience in view of the distinctly different experimental procedures and applications.

It must be apparent from the above discussion that unless care is taken to promote a particular mechanism, (e.g. hydrophobic or ion exchange) and to suppress other effects, the various solute–sorbent interactions operate to different extents depending on the structure of the solute and the nature of the sorbent. The choice of sorbent is currently narrowing down to essentially two, i.e. silica gel and octadecyl-bonded silica, predominantly the latter in high-performance column liquid chromatography and the former in open-column chromatography and TLC. Thus, the retention in adsorption chromatography is almost essentially based on the structure of the solute. Separation and elution of the individual components is, therefore, generally achieved by manipulating the mobile-phase composition and selectivity. In this respect, the situation is somewhat the reverse of that existing in GC, where the role of the mobile phase is limited and the selectivity of solute retention is controlled essentially by the stationary phase.

Being the oldest of the chromatographic techniques, adsorption chromatography has been applied to almost all classes of compounds and in different fields. Traditionally, routine applications of this method were limited to non-ionic small molecules of medium polarity and in cases where a certain degree of selectivity is needed. While non-polar compounds, such as aliphatic hydrocarbons, are too weakly retained to be effective in separation, polar (particularly hydroxylic and amino) compounds usually give rather unsatisfactory peak shapes owing to unsymmetrical adsorption isotherms. The advent of bonded-phase chromatography extended its application to non-polar as well as highly polar compounds. The recent developments in ion-pair chromatography on ionic and ionizable compounds makes adsorption chromatography almost a universal method of separation and analysis.

The present treatment does not distinguish between LSC on chemically unmodified silica and alumina (normal-phase) packings on the one hand, and that in the packings of progressively decreasing surface polarity (culminating in the essentially hydrocarbon-like octadecyl-bonded silica) on the other. However, there is some theoretical basis for drawing a distinction between the mechanism of retention on the normal-phase and reversed-packings (see later) and some authors prefer to treat the subject matter independently; e.g. see Refs. 26 and 35.

Even though, at least by some estimates, over 80% of LSC is now being carried out on bonded-phase packings, there are some applications where the common adsorbents like silica or alumina may be preferred. Thus, water-insoluble non-polar organic compounds, like polyaromatic hydrocarbons and fats, are often better resolved on the normal-phase packings, although reversed-phase, non-aqueous LSC is also claimed to be satisfactory. For example, the separation factor ($\alpha$) for picene and its isomer, 1,2,3,4-dibenzanthracene is 20 on silica gel and that for quinoline and isoquinoline is 3.45. Similarly, large separation factors are common in aromatic compounds with different functional groups, compounds differing in the number of functional groups of the same type or even *ortho*, *meta* and *para*-disubstituted benzene derivatives. The explanation for this selectivity lies in the presence of discrete adsorption sites on the surface of the sorbent which permits preferential retention of one of the isomers. In view of their low cost, the normal-phase packings also find extensive application in preparative and process-scale chromatography.

### 6.5.2 Stationary phases

#### *6.5.2.1 Individual adsorbents*

A large number of adsorbents have been introduced for use in LSC [1], several of them by Tswett himself. They may be distinguished as polar and non-polar on the basis of their surface properties. As noted earlier, polar compounds are strongly retained on polar adsorbents by electrostatic interactions and nonpolar compounds on the reversed-phase packings by hydrophobic interactions. In making these generalizations, it is assumed that the mobile phase is of the opposite (or appropriate) character. Polar adsorbents may be further subdivided into those with positive (electron-acceptor) and negative (electron-donor) surface fields. Understandably, they preferentially interact with molecules of opposite fields. These characteristics may be accentuated by suitably treating the adsorbents (see below). The polar adsorbents may also be classified as acidic (e.g. silica gel) or basic (e.g. certain forms of alumina) depending on the surface functionality. Acidic adsorbents naturally retain basic solutes more strongly, and *vice versa*.

*6.5.2.1.1 Silica gel*

Over the years, silica gel has become increasingly popular and a vast majority of the column packings now in the market are based on silica gel and its modifications [36]. The main advantages of silica gel are that it is relatively inert, has a large adsorption capacity and is easily obtainable in various pore sizes and surface areas. More than a hundred different types of silica gels are available with widely differing chromatographic properties. With so many varieties of commercial adsorbents available, it is seldom necessary to prepare them in the laboratory; nevertheless, a brief description of a few common methods may be instructive. Silica gel for chromatography is generally prepared by precipitation from a solution of water glass (sodium metasilicate) using hydrochloric acid at 50–80°C. The sodium ions from water glass may also be removed by passing through a cation-exchange resin in the hydrogen-ion form. The free silicic acid thus liberated has a molecular formula $Si(OH)_4$ in its monomeric, completely hydrated form. It quickly polymerizes to give the macromolecular polysilicic acid in the form of a gel. When allowed to stand, the gel undergoes syneresis (gel contraction with exudation of the mother liquor) which may take several days for completion. The gel is washed free of salts and dried at 120°C for several hours. During this process, further condensation of the gel may occur yielding the hard material known as the xerogel. The silica gel is ground and sized for use in chromatography; the adsorbent obtained in this manner is composed of particles of irregular shapes. Two other types of silica gel may be described. They are: (1) totally porous spherical particles, and (2) beaded or pellicular particles consisting of a hard core of about 30 $\mu$m diameter, coated with about 1 $\mu$m layer of the porous adsorbent. Spherical porous silica particles of uniform size, down to 2–3 $\mu$m are now available; they are prepared by agglomeration of submicron-range particles of amorphous silica, formed by hydrolysis of silicon tetrachloride. The silica hydrogel beads are prepared by emulsification of the silica sol in an immiscible organic liquid. The stirring speed, liquid viscosity, pH, temperature and concentration of the silica gel determine the bead size and size distribution. Microporous silica is formed by burning out the organic material from a co-polymer of the silica sol with an organic reagent. It must be mentioned that all manipulations of the preparation procedure need to be very carefully controlled as they determine not only the particle size but also the pore structure and the surface area that control the chromatographic properties of the adsorbent. The pore diameter of the commercial microporous silica may vary between 50–250 Å, with a pore volume of 0.7 to 1.2 ml/g. Usually, the smaller the pore diameter, the larger is the surface area. The surface area of the porous and the pellicular particles are in the ranges of 400–600 $m^2/g$ and 5–15 $m^2/g$, so the capacity of the pellicular particles is relatively much lower. Nevertheless, the pellicular variety does find application in certain analytical fields because of its high efficiency. It also finds favour as the packing material for pre-columns in HPLC.

*6.5.2.1.2 Alumina*

Second to silica, alumina has been the most widely used adsorbent, particularly during the early period of development of LSC. It is also easy to obtain in a variety of forms and is characterized by high adsorption capacity. It is commercially available in three chemically different forms, i.e. acidic, neutral and basic. Most of the dozen or so crystalline forms of alumina are usable for chromatography. Commercial alumina is essentially of the $\gamma$-form, with each aluminium atom surrounded by six oxygen atoms, and each oxygen by three aluminium atoms and one hydrogen. Alumina is formed by controlled dehydration (200–600°C) of aluminium hydrous oxides which are obtained by treating a solution of sodium aluminate with carbon dioxide.

The retention properties of silica and alumina are very similar, with saturated aliphatic hydrocarbons on the lower end of the scale (most weakly adsorbed), followed by olefins, aromatic hydrocarbons, halogenated hydrocarbons, sulphides, ethers, nitroalkanes and aromatics, esters, carbonyl compounds, alcohols, amines, sulphones and sulphoxides, amides and carboxylic acids [26]. Some variation in selectivity may be expected depending on the nature and activity of the particular adsorbent. The overwhelming popularity of silica over alumina is based on the commercial availability of a wider range of chromatographically useful forms, or relatively large sample capacity and, most importantly, less undesirable chemical reactivity with the solutes; there is also a much wider literature covering applications of silica gel for separating compounds of almost all classes. Nevertheless, there are certain advantages of using alumina over silica gel, at least in a few selected cases, and recently, there has been a reappraisal of the qualities of alumina [37].

As already noted, silica gel is acidic in character and shows strong affinity for basic compounds (e.g. nitrogen heterocycles), sometimes irreversibly. The surface pH of alumina, on the other hand, can be modulated to suit the application. Unlike silica whose activity is based only on the hydroxyl groups, alumina can interact either through the acidic hydroxyl groups or the basic oxygen atoms. It can also form charge-transfer complexes with electron donors (e.g. aromatic solutes). Highly polarized molecules are irreversibily retained on alumina by chemisorption, a feature that restricts its use as a general adsorbent like silica; e.g., chemisorption of water on alumina gives rise to two hydroxyl groups on the adsorbent surface. Alumina is thus more suitable for moderately polar and non-polar compounds containing double bonds and hydrogen-bonding functional groups. It has been used extensively for the separation of olefins, aromatic hydrocarbons, steroids, synthetic dyes, alkaloids and substances stable to basic media.

Alumina is amphoteric in nature and, depending on the pH, it can act either as a proton acceptor or electron-pair acceptor. This property also enables alumina to act as both a cation or anion exchanger. Washing neutral alumina with sodium hydroxide leaves chemisorbed hydrogens replaced by the sodium ions which can be exchanged for other cations. Similarly, it can be converted into an anion exchanger in the chloride form by treating with HCl. At a certain pH called the zero point of charge (ZPC), the charge on the surface is zero. At a pH below the ZPC, the net charge on the surface is positive and above the ZPC, it is negative. This property of alumina has been used for the separation of proteins. The isoelectric point (pI) of proteins is defined as the pH at which the net charge of the protein is zero. At a pH above pI the net charge of the protein is negative, and vice versa. Thus, at ZPC $<$ pH $<$ pI, the protein is positively charged and the adsorbent has a negative charge, which means the protein is retained by the adsorbent, and similarly, when ZPC $>$ pH $>$ pI, the negatively charged proteins are retained. Large variation in retention can be achieved by varying the pH around either the ZPC of alumina or the pI of the protein. Size-exclusion effects for proteins of molecular weights between 1000 and 20,000 have also been observed in the absence of electrostatic attraction.

#### *6.5.2.1.3 Magnesol*

Among the other adsorbents, which are all of relatively minor importance, mention must be made of magnesium oxide and silicate, polyamides, cross-linked polystyrenes and carbon-based materials. Synthetic magnesium silicate (or Magnesol), when properly deactivated (0–35% water) has chromatographic retention intermediate between alumina and silica gel. Magnesium oxide for chromatography is prepared by heating magnesium hydroxide at 400°C for 12 hours. It has been used extensively for the separation of carotenoids and porphyrins. Its chromatographic properties are generally similar to that of alumina but with higher affinity towards unsaturated compounds and aromatics. To facilitate the solvent flow through the column, it is normally used in combination with diatomaceous earth.

#### *6.5.2.1.4 Polyamides*

Polyamides have been popular for a long time for the separation of highly polar compounds (e.g. phenolic compounds) by elution with water, alcohols and their mixtures with some organic modifiers. In addition to alcohols, a variety of compounds containing proton-donor groups, including amino, imino, sulphonic, carboxylic and phosphoric acids, have been separated on columns (or thin layers) of polyamides. Retention of these compounds on polyamides is considered to be based on hydrogen-bond formation between the protons of the solute with the amide carbonyl of the polymer. Selectivity of elution depends on the ability of the solute to disrupt these hydrogen bonds and compete for the sites. However, the types of compounds that could be well separated include those containing highly electron-deficient groups like quinones, nitro compounds, nitriles and aldehydes, in which case the hydrogen-bond theory fails. Here, the interaction could be between the electron pair of the amide nitrogen and the electrophilic group. The fact that olefins and aromatics also have an affinity for the adsorbent indicates that, in addition to the above mechanisms, simple partition processes could also play a part in the retention. Depending on the polarity of the mobile phase relative to the polyamide surface, it could be either the normal-phase or the reversed-phase partition.

The commercial polyamides are normally based on Nylon-66 (polyhexamethylenediamine adipate) or polycaprolactam. Their chromatographic properties vary widely. Preparation of materials suitable for use as chromatographic adsorbents and evaluation procedures are described in the literature [1]. It is interesting to note that polyamides of poor adsorptivity can be dramatically improved by simply dissolving them in hydrochloric or acetic acid and precipitating with aqueous methanol. The types of compounds for which polyamide has been used include polyphenols (flavonoids, coumarins and lignins), phenolic acids, alkaloids and other nitrogen heterocycles, amino acids and peptides, glycosides, iridoids as well as steroids, bile acids, antibiotics, pesticides and dyes. Acetylated polyamides have been used for the isolation of highly electrophilic compounds like quinones and polynitro compounds.

#### *6.5.2.1.5 Cross-linked polystyrenes*

The use of cross-linked polystyrene beads was mentioned in connection with the discussion on SEC. Recently, there has also been some interest in their use as stationary phases in adsorption chromatography [38]. The main drawback is their residual swelling even at high degrees of cross-linking, which limits the head pressure that can be used. The so-called macroreticular variety of these materials (see section 6.6.2.4) are available from Rohm and Haas Co. (Philadelphia, PA, U.S.A.) under the trade name Amberlites XAD, the most widely used among them being Amberlite XAD-2. In addition to SEC, their strong affinity for non-polar compounds has been extensively used for selectively adsorbing organics from dilute aqueous solutions (e.g. flavours from liquors or soft drinks, or environmental pollutants from water [39]. The adsorbed compounds can be eluted using increasing proportions of alcohol or other organic solvents. The adsorptivity of the resins depends strongly on the number and nature of the polar groups relative to the non-polar portion of the molecules; e.g. phenol completely adsorbed from an aqueous solution, while sodium phenolate is not. Compounds with large hydrophobic portions (e.g. carotenoids and steroids) are so strongly adsorbed from methanol that they can be eluted only by mixtures of methanol and dichloromethane. Pure dichloromethane may be used for strongly retained non-polar compounds.

#### *6.5.2.1.6 Charcoal*

Charcoal has long been known to be a very efficient adsorbent for a variety of organic compounds. Adsorption on charcoal occurs by dispersion forces and is characterized by low selectivity. The chromatographic properties are greatly affected by the nature and concentration of the impurities which, in turn, are determined by the raw material used for their preparation and the method of preparation. The three main types of commercial products are: (1) the graphitized charcoal, (2) active charcoal and (3) the carbonaceous molecular sieves, formed by thermal degradation of the fibrous vinylidine chloride co-polymers with vinyl chloride or acrylonitrile (Saran-type). Only the first two types of charcoal are normally used in chromatography. The graphitized charcoal is distinctly non-polar and is formed by high-temperature carbonization (above 1000°C). On the other hand, active charcoal is obtained by low-temperature oxidation of organic matter and may contain considerable amounts of oxygen and hydrogen, as well as polar active sites. The charcoals are so brittle that, unless great care is taken in packing, the fine granules formed in the process may clog the columns. They are, therefore, normally used as 1:1 mixtures with Celite or cellulose. Their use in HPLC is limited by their inability to withstand high pressures. Porous graphitic carbon, consisting of 3–10 $\mu$m spherical particles and suitable for HPLC, has been prepared by polymerizing a phenol–hexamine mixture on porous silica gel of the desired particle size and pyrolysing it under nitrogen. The silica template is dissolved out using hot aqueous potassium hydroxide and the remaining portion is heated to above 2500°C under argon to complete the process. The structural studies on these materials and their chromatographic properties have been discussed in detail by Knox *et. al.* [40]. These are highly porous (80%) and have a surface area of around 150 $m^2/g$. They act as very strong hydrophobic adsorbents with unique selectivity. The elution pattern is somewhat similar to that observed on octadecylsiloxane-bonded phases, to be discussed below. Graphitized porous silica packings have also been introduced specifically for HPLC [41]. Being non-polar, retention on these materials is analogous to that on the polymer beads discussed above and the alkyl-bonded phases to be described below, but with some differences in the selectivity.

#### *6.5.2.2 Modified adsorbents*

The graphitized silica described above is an example of masking the adsorption properties of a solid by another functionality. The object is to use the adsorption properties of the silica (or, for that matter any of the above adsorbents) to support a reagent, which would very selectively interact with the solute to effect a separation. In a sense, the situation is similar to that occurring in partition chromatography, discussed above, except that the stationary phase in this case is generally one solid supported on another. The support in this case may not be inert as in partition chromatography but may act as a modified adsorbent. The chemically bonded phases to be discussed in the following subsection are also modified adsorbents but, again, they may be distinguished from the present subject matter by the fact that the retention of the modifier is by physical bonding in one case and chemical bonding in the other. The surface modification, in a broader sense of the term, is also practised in separations based on ion exchange and bio-affinity which are discussed later in this chapter.

The purpose of surface modification, in general, is to alter the chromatographic properties and selectivity of the adsorbent by changing its electrostatic, hydrogen-bonding, hydrophobic or electron donor–acceptor characteristics. Use of different reagents to modify the electron donor–acceptor properties of the adsorbents has been in practice for a long time and is sometimes called donor–acceptor complex chromatography (DACC) [42]. It is based on the principle that an electron-deficient molecule and an electron-rich species can interact to form a charge-transfer complex; this may involve either complete transfer of an electron to form an ionic bond or a temporary and reversible coordination. The strength of such an interaction may vary depending on the two interacting species. The affinity of the silver ion to olefins, and picric acid to aromatic hydrocarbons is well known. A large number of electron deficient polynitro and polyhalogeno aromatic compounds are offered as electron acceptors and alkyl-aryl ethers as donors for impregnation on silica or alumina. By suitable manipulation these may also be chemically bonded to the silica-gel surface for increased stability. The close similarity of these principles to ligand-exchange chromatography [43] and metal-chelate affinity chromatography [44] (to be discussed later) will be quite obvious.

Another type of complexation, more often used in TLC, is that of polyhydroxy compounds with inorganic anions (borates, tungstates, molybdates and arsenates); metal salts of these ions are used for impregnation of the adsorbent. Complexation is based on both covalent and hydrogen bonding. Derivatization of carbonyl compounds with reagents such as semicarbazide adsorbed on silica gel and subsequent decomposition of the derivative has been used for group separation of the former [45].

#### *6.5.2.3 Bonded phases*

As the name indicates, the adsorbents for bonded-phase chromatography (BPC) carry functional groups that are chemically bonded to a suitable solid support. Development of BPC is a logical consequence of the success of partition chromatography and chromatography using impregnated adsorbents. Certain drawbacks in the practice of the two relatively older techniques are well known; bleeding out of the stationary phase and severely restricted choice of compatible solvent systems are examples. Though originally designed to overcome these disadvantages, the versatility and convenience of the bonded phases has resulted in their widespread application.

##### *6.5.2.3.1 General-purpose bonded phases*

Four types of bonded phases are distinguished on the basis of their applications to different types of compounds. Modifications of surface polarity to bring about a certain degree of selectivity are discussed here. This is followed by a discussion on chiral stationary phases carrying chemically bonded peptides, crown ethers and cyclodextrins; the selectivity of interaction of asymmetric functionality is used for resolving optical isomers. Ion-exchange and affinity packings, though based on the same principle of modification of surface functionality of a solid support, merit discussion as separate sections (see sections 6.6 and 6.7, below).

Silica has been the most versatile and popular raw material or backbone for the preparation of almost all the BPC packings. Fully hydrated silica contains up to 8 $\mu$ moles/m$^2$ of silanol (Si–OH) groups of which about 55% may be chemically modified; the remainder of the silanol groups are probably sterically shielded by the reacted groups and therefore unapproachable. By suitable application of elementary chemical principles, a wide range of stationary-phase selectivity can be achieved. The key steps in the preparation of the bonded phases are given in Fig. 6.15. Some of the first bonded phases belonged to the so-called ester-type (Si–O–R), which could be prepared directly (Fig. 6.15a) by treating with the appropriate alcohol (e.g., 3-hydroxypropionitrile or Carbowax 400) or by first chlorinating them with thionyl chloride to give the chlorosilane (Si–Cl) functionality and then treating with the alcohol. Since the ester groups of the 'estersils' are not stable to hydroxylic solvents and temperature, these packings have had limited application. The chlorinated silica may be reacted with amines (Fig. 6.15b) or alkali-metal derivatives of alkanes (Fig. 6.15c) or arenes (or the appropriate Grignard reagents) to give Si–N or Si–C bonded phases which have superior hydrolytic and thermal stability. The organic chain is available for the chemical modification; e.g. the aryl-substituted material can be sulphonated to give a phenylsulphonic acid functionality that can act as a cation-exchanger. The stability of the Si–N bond, however, is limited to the pH range 4–8 and preparation of Si–C derivatives is not very easy.

The siloxane type of bonded phases, containing the Si–O–Si–R groups, offer the best compromise between stability and preparative facility. They are prepared by

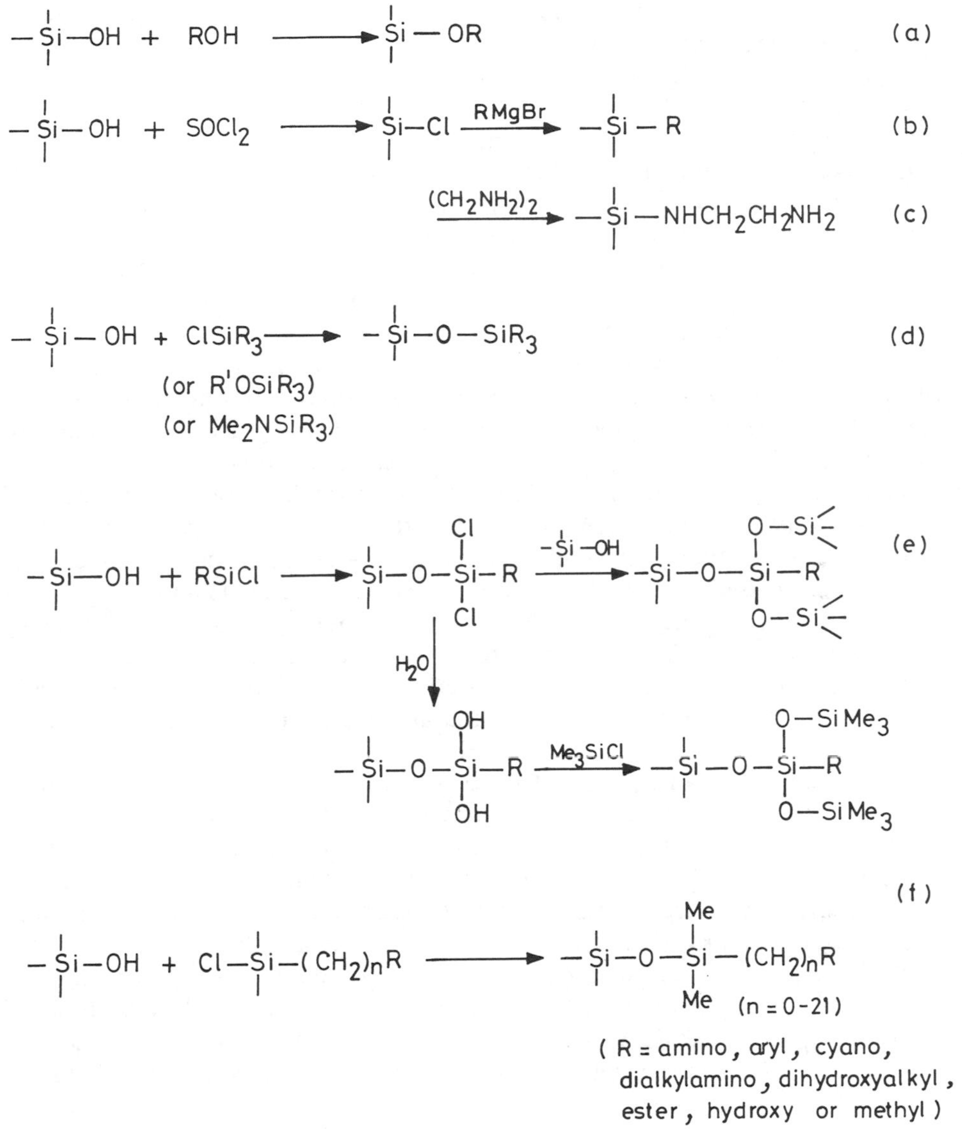

$$-\overset{|}{\underset{|}{Si}}-OH + ROH \longrightarrow \overset{|}{\underset{|}{Si}}-OR \quad (a)$$

$$-\overset{|}{\underset{|}{Si}}-OH + SOCl_2 \longrightarrow \overset{|}{\underset{|}{Si}}-Cl \xrightarrow{RMgBr} -\overset{|}{\underset{|}{Si}}-R \quad (b)$$

$$\xrightarrow{(CH_2NH_2)_2} -\overset{|}{\underset{|}{Si}}-NHCH_2CH_2NH_2 \quad (c)$$

$$-\overset{|}{\underset{|}{Si}}-OH + ClSiR_3 \longrightarrow -\overset{|}{\underset{|}{Si}}-O-SiR_3 \quad (d)$$

(or $R'OSiR_3$)
(or $Me_2NSiR_3$)

$$-\overset{|}{\underset{|}{Si}}-OH + RSiCl \longrightarrow \overset{|}{\underset{|}{Si}}-O-\overset{Cl}{\underset{Cl}{Si}}-R \xrightarrow{-\overset{|}{Si}-OH} -\overset{|}{\underset{|}{Si}}-O-\overset{O-Si\lessgtr}{\underset{O-Si\lessgtr}{Si}}-R \quad (e)$$

$$\downarrow H_2O$$

$$-\overset{|}{\underset{|}{Si}}-O-\overset{OH}{\underset{OH}{Si}}-R \xrightarrow{Me_3SiCl} -\overset{|}{\underset{|}{Si}}-O-\overset{O-SiMe_3}{\underset{O-SiMe_3}{Si}}-R$$

$$-\overset{|}{\underset{|}{Si}}-OH + Cl-\overset{|}{\underset{|}{Si}}-(CH_2)_nR \longrightarrow -\overset{|}{\underset{|}{Si}}-O-\overset{Me}{\underset{Me}{Si}}-(CH_2)_nR \quad (n = 0-21) \quad (f)$$

(R = amino, aryl, cyano, dialkylamino, dihydroxyalkyl, ester, hydroxy or methyl)

Fig. 6.15 — Preparation of bonded phases.

treating the silanol groups of porous silica with appropriately substituted mono-, di- or tri-chlorosilane, or an alkoxysilane (Fig. 6.15d). When di- or tri-chlorosilane is used, one or two of the chlorines may remain unreacted, depending on the steric factors; i.e. non-availability of silanol groups in the vicinity. These chlorosilane groups are readily hydrolysed during further processing to give silanol groups again, causing tailing of peaks and other complications unless the free silanol groups are 'end-capped' using trimethylchlorosilane (Fig. 6.15e). Therefore, trisubstituted monochlorosilanes are used more frequently. If all the three substituents on the monochlorosilane are large, steric factors would again come into play, decreasing the efficiency of coverage. Hence, monofunctional dimethylchlorosilanes are preferred. The surface of these as well as the other single-chain bonded phases can be visualized as consisting of organic chains protruding out of the silica base, as in a brush. The surface coverage in these products being a monomolecular layer, the mass-transfer properties of these packings are very good and high column efficiencies are reported.

As noted above, the most common siloxane-type bonded phases are prepared by the action of monochlorosilane or monoalkoxysilane bearing the appropriate substituents. The chloro derivatives are normally used for the preparation of the alkyl-substituted silanes and the alkoxy derivatives for those bearing polar groups. Prior to the treatment, the silica gel is hydrated by heating with 0.1*N* HCl at 90°C for about 24 h to maximise the number of silanol groups. The silica gel would then contain a large amount of water that is physically retained by the gel and this must be removed, prior to the bonding reaction by vacuum drying at 120°C for about 12 h. To obtain the maximum coverage of the surface, a large excess (at least twice that required stoichiometrically) of the reagent is used and the condensation is carried out by refluxing in a high-boiling solvent like toluene for 16–24 h. An acid scavenger (e.g. pyridine) may be used in the reaction mixture when a chlorosilane is the reagent. At the end of the reaction, the product is filtered and thoroughly washed by refluxing the product with excess methanol. To ensure complete coverage of all the accessible silanol groups, the above reaction is repeated using trimethylchlorosilane for end-capping. For details of preparative procedures and characterization, see Refs. 26 and 35].

The amount of organic phase attached to the support may be determined by infrared spectroscopy and elemental analysis.The bonded phases are characterized by a decrease in the mean pore diameter (roughly by twice the thickness of the organic layer covering the surface), the pore volume and the specific surface area. As can be expected, the decrease in the values of these parameters is more pronounced as the length of the alkyl chain increases. However, the fact that the volume of the organic phase, rather than the area of coverage, is the important parameter that determines the retention of the bonded phases is indicated by both the retention and the sample capacity increase with the increasing length of the alkyl chain. In general terms, the $k$ values and the sample loadability are doubled as the chain length is increased from C-4 to C-18.

The surface polarity of the adsorbent can be successively decreased by substituting the R group of the siloxane with aminoalkyl, cyanoalkyl, phenyl and long-chain aliphatic functionalities (Fig. 6.15f). The selectivity of the BPC packings may also be modified by attaching diol or nitro groups. These and the aminopropyl-bonded packings may be considered normal-phase adsorbents; a variety of moderately polar compounds have been separated on these adsorbents. More recently, cyclodextrin and crown ethers have been bonded to silica, yielding stationary phases with a high degree of selectivity to aromatic compounds and certain metal ions, respectively [46,47]; they are also useful for chiral separations (see below). The cyanopropyl-bonded silica is moderately polar; its normal or reversed-phase character depends on the relative polarity of the mobile phase. The available bonded phases and their applications are discussed in detail in Chapter 7.

By far the most frequently used bonded phases are the reversed-phase packings, mainly the octadecylsilane (ODS or C-18) bonded silica and, to a lesser extent, the octyl (C-8) bonded silica. Ethyl-, hexyl- and phenyl-bonded phases are also popular. So many types of compounds are retained by these materials that they may be considered nearly universal adsorbents; selectivity of elution is normally achieved by modulating the mobile phase. The mobile phase is generally water or water mixed with methanol, acetonitrile or tetrahydrofuran (in the order of their popularity) and, where required, buffers. Use of ion-pairing reagents is advantageous in separating ionic compounds (ion-pair chromatography, see later). The BPC is analogous to LLC in its ability to separate the compounds on the basis of the nature and the number of functional groups present in the solute molecules; it is very efficient in resolving low and medium molecular weight compounds (MW < 3000), particularly, homologues, benzologs and oligomers differing in the number of methylene, benzene and monomeric units.

In addition to the stability and the convenience of commercial availability of many of the BPC packings, a major advantage of this technique is the wider mobile phase option and the facility of gradient elution. Use of functionalized polystyrenes, carbohydrates and polyacrylamides in ion-exchange and affinity chromatography, known long before the development of the above BPC packings, is based on the same basic principles.

#### *6.5.2.3.2 Chiral stationary phases*

The difficulties in resolving optical isomers or enantiomers, which are chemically identical in an achiral environment but structurally differ from one another in being mirror images, have been mentioned in connection with the chiral stationary phases in GC (section 4.2.5.5). It has also been mentioned that open-tubular GC has the advantage of generating a large number of theoretical plates needed for separating the difficult-to-separate isomers. However, the applications of GC are generally limited by the requirement of solute volatility and the very low sample capacity of the open-tubular GC, which does not permit preparative (large-scale) separations. Also, the relatively high temperature of operation renders distinction between the stabilities of diastereomeric complexes (between the chiral stationary phase and the components of racemate) difficult. The feasibility of preparation of bonded stationary phases capable of chiral recognition has opened up interesting possibilities and this has been the fastest growing approach to enantiomer separation by LC during the past decade. Over 30 chiral stationary phases (CSPs) are now commercially available [48–50]. Unlike the general-purpose stationary phases and the ion-exchange resins (see below), the interaction with the solute with the CSPs is necessarily geometry oriented; i.e. stereoselective.

In the ligand-exchange chromatographic approach of Davankov [43,51,52], an amino acid such as proline (or, preferably, hydroxyproline or azetidine carboxylic acid) is covalently bonded to a chloromethylated cross-linked styrene-divinylbenzene polymer. Advantage is taken of the fact that certain transition metal ions like Cu(II) (and to a lesser extent, Ni, Co, Zn, Cd, Mn, Hg, Fe and Ag ions) are capable of forming four-coordinate or ternary complexes, with the stationary amino acid on one side and another species in the mobile phase (Fig. 6.16a). The relative stabilities of the ternary complexes with the amino acid in the solute determines the

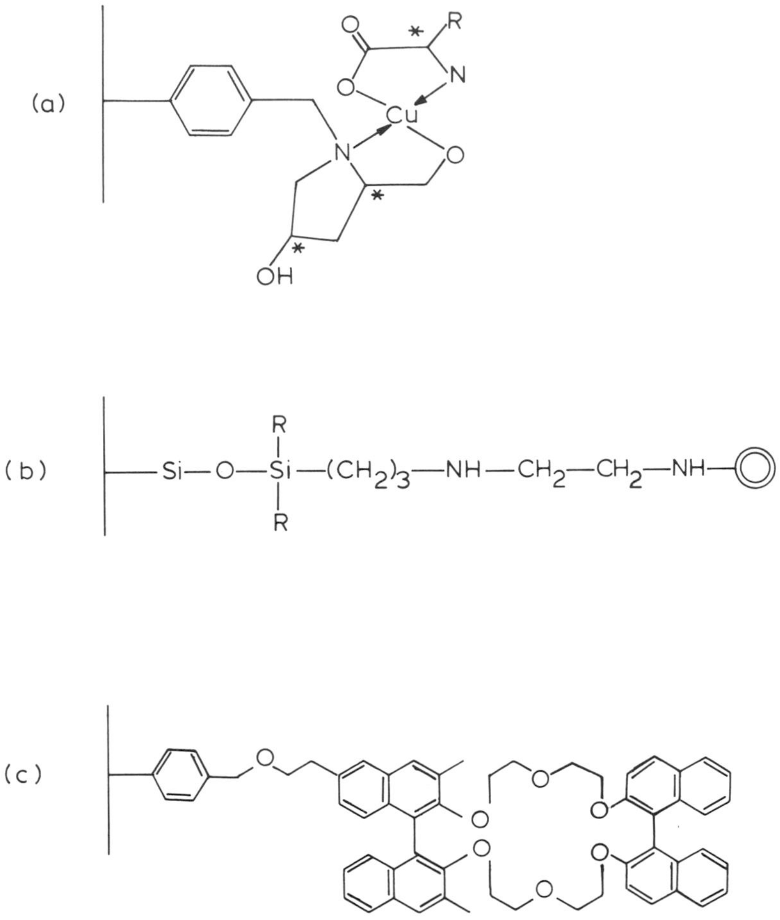

Fig. 6.16 — (a) Amino acid–copper complex in ligand-exchange chromatography; (b) attachment of cyclodextrin moiety to silica gel through a spacer; and (c) attachment of a crown ether to silica.

separation factor (α). This technique and its variations appear to be the most effective means of separation of underivatized amino-acid racemates. Over 40 chiral polystyrene-based CSPs have been prepared and investigated [52].

Two variations of ligand-exchange chromatography (LEC) are possible. In one, the metal ion is fixed on the stationary phase and the ligand is eluted by displacement using a stronger ligand in the mobile phase. When the metal ion is bound more strongly to the ligand in the mobile phase, the metal-ligand complex itself could move along the column. (The mobile phase could also be a gas.) In addition to amino acids, any solutes that can donate a pair of electrons to the metal ion to form a reversible complex may be separated by this method. They include amines, alcohols, acids, mercaptans, sulphides and compounds containing combinations of these groups such as hydroxy acids.

In practice, the sorbent-metal chelate is first prepared by equilibrating the two materials; a base (usually ammonia) is used along with the metal salt to facilitate removal of the liberated protons. Occasionally, water may compete to form coordinate complexes with the metal in which case, non-aqueous solvents may be used as the mobile phase. For this purpose the sorbent-metal chelate is thoroughly washed with water followed by an organic solvent such as acetone. In aqueous media ammonia or ethanolamine solutions may be used for elution. Variation of the pH and temperature may sometimes be advantageous, as also in the use of a metal salt to elute the ligand by forming a mobile complex. Elution of amines and sulphides can be achieved in non-aqueous hydrocarbon media, using methanol and ether, respectively.

The sorbent, binary and ternary metal chelates may have different colours, so that, when the chromatography is carried out in glass columns, it may be possible to follow the progress of separation. Amino acids have also been grafted on cross-linked polyacrylamide gels for improved permeability. For LEC under high-pressure conditions, stationary phases have been prepared by adsorbing the proline-modified linear polyacrylamide on silica. 3-Aminopropylsiloxane- or 3-glycidoxypropylsiloxane-bonded silica has also been used to covalently link proline to porous silica, to obtain stationary phase suitable for high-performance LEC. For a detailed discussion on these and the mechanistic aspects of LEC, see Ref. 52.

α-, β- and γ-cyclodextrins [cyclic α (1–4)-linked glucose derivatives formed from six, seven and eight monomer units] have been attached to silica using a 6- to 10-atom spacer (Fig. 6.16b) and used for separation of racemic dansyl and naphthyl amino acids, aromatic drugs, steroids, alkaloids, metallocenes, naphthyl crown ethers, aromatic amines, sulphoxides and polycyclic aromatic hydrocarbons. The cyclodextrin-bonded stationary phases act like hydrophobic adsorbents and the common mobile phases are aqueous buffers, modified with organic solvents like methanol or acetonitrile. Being relatively inexpensive, they are also used in preparative separations [46].

As noted earlier, certain metal ions and the ammonium ion, as well as the ammonium ion of the zwitterionic free amino acid can form complexes with crown ethers such as 18-crown-6, which is a cyclic oligomer formed from six molecules of ethylene glycol. Asymmetric crown ethers are fairly easily synthesized using a binaphthyl template and can be chemically bonded to cross-linked styrene-DVB polymer (Fig. 6.16c). These stationary phases are used in the normal-phase mode with chloroform and acetonitrile mixtures as mobile phases. High $\alpha$ values (e.g. 26 for *R* and *S* *p*-hydroxyphenylglycine) have been reported using these immobilized chiral crown ether stationary phases, making this a very attractive means of chiral separations, both on analytical and preparative scales [47,53].

It is well recognized that for chiral recognition, the interaction between the active site on the stationary phase and the solute should occur at least at three points around the asymmetric centre. The interaction could be based on dipolar attraction, $\pi,\pi$-interaction or hydrogen-bonding. A large number of amino acid derivatives have been used to prepare chemically bonded phases by reacting them with aminopropylsilanized silica. The amino-acid derivatives so bonded include: (*R*) or (*S*)

*N*-(3,5-dinitrobenzoyl)phenylglycine, *N*-(3,5-dinitrobenzoyl)leucine, *N*-(2-naphthyl)alanine, 1-(α-naphthyl)ethylamine, etc. The success of this approach has been demonstrated by the large number of compound classes that could be separated on these stationary phases [50].

Biopolymers like cellulose and proteins have built-in chirality by virtue of the asymmetric structure of the constituents and these, or their derivatives, could be used as stationary phases for the separation of a variety of *N*-acyl- and *N*-aryl-amino acids, aromatic sulphoxides, organophosphorus compounds, insecticides and drugs. The cellulose esters commonly used are the acetate, benzoate, cinnamate and phenylcarbamate; these are used as normal-phase adsorbents with hexane–isopropanol or similar solvents as the mobile phases. Cellulose tribenzyl ether and some synthetic polymers like (+)-poly(triphenylmethylmethacrylate) and (+)-poly(2-pyridyldiphenylmethylmethacrylate) are also commercially available for use with alcoholic mobile phases.

#### *6.5.2.4. Retention mechanisms*

Several physical characteristics of the adsorbents may control the chromatographic properties; namely, retention and selectivity. Under a given set of conditions, retention is proportional to the surface area which, in turn, is directly related to the porosity and inversely to the particle size. The pore size does not directly influence retention of small molecules. The active sites in a particle may not be evenly distributed on the surface of the adsorbent and their accessibility to solute molecules may sometimes be affected by the pore structure. The accessibility to active sites is also controlled in polar adsorbents by the amount of water present. Though several different types of adsorbents have been described in foregoing subsections, only two adsorbents (namely, silica gel for the normal-phase adsorption chromatography and octadecyl-bonded silica for reversed-phase LC) are taken as examples for the present discussion on the mechanism and control of retention. As already noted, a vast majority of LSC is carried out on these phases.

Unless freshly activated, silica gel contains a considerable amount of water. Since water is more strongly adsorbed on the active sites of the polar adsorbent, the activity of the adsorbent decreases as the water content increases. Activation of the adsorbent (which, in effect, means removal of water) may be carried out in a number of ways: e.g. storing over a drying agent like sulphuric acid or phosphorus pentoxide in a desiccator, washing with a water-miscible polar solvent followed by a non-polar solvent or simply by heating in a hot air oven for several hours (below 150°C). For reproducible results, the adsorbents used for chromatography must be standardised. The Brockmann scale (activity I for the most active through V for the least active, representing 0, 5, 15, 25 and 38% of adsorbed water on silica gel) makes use of the mobility of certain common synthetic dyes under standardized conditions [1].

The solute-stationary phase interactions in adsorption chromatography are still in the process of being understood. Silica gel, being the most commonly used adsorbent, has been well studied in this respect [36]. The approximate chemical structure of silica gel is shown in Fig. 6.17a. Under the normal conditions of temperature and humidity, the silica gel surface may have three layers of water (Fig. 6.17b). The outermost layer is lost on warming the gel to about 40°C and the second layer at 100–120°C. Stronger heating to about 400°C removes the bound water

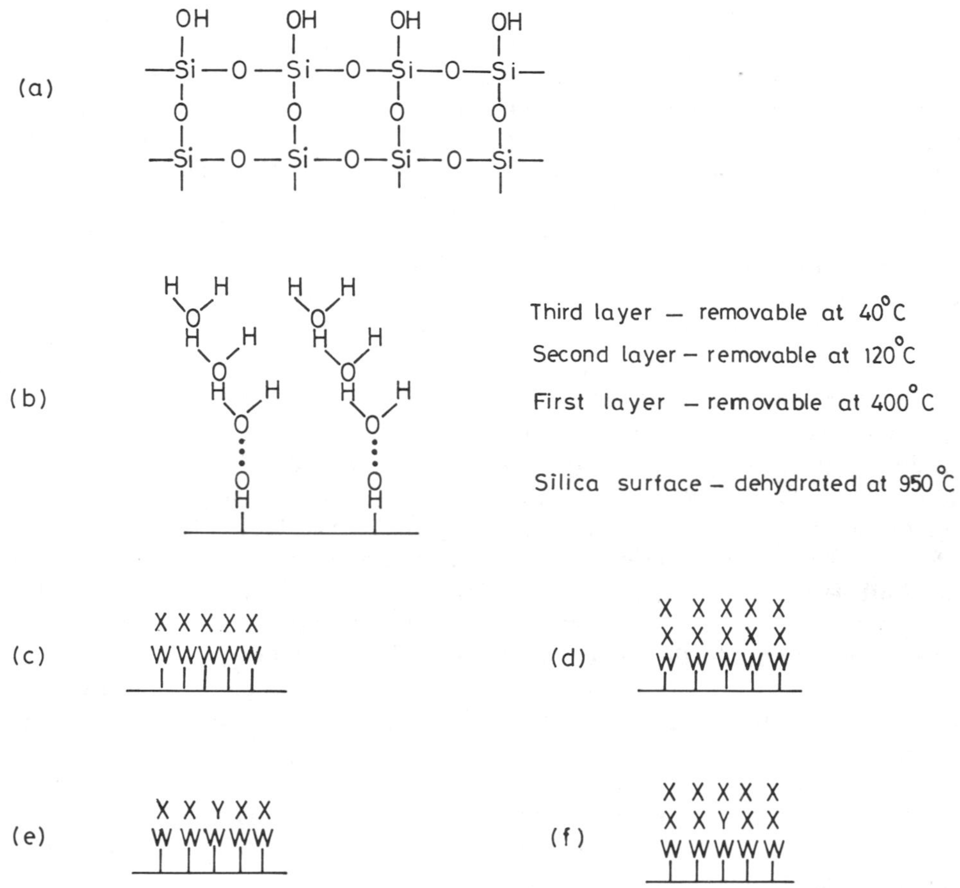

Fig. 6.17 — The structure of silica gel.

completely, but water may be regained on cooling and exposure to ambient atmosphere. On heating to about 950°C irreversible dehydration of the silanol groups is observed. By careful analysis of the amount of water lost at different temperatures, it has been calculated that there are $8.5 \times 10^{20}$ hydroxyl groups per gram of silica gel activated at 120°C, representing one molecule of strongly bound water per silanol group. The successive loss of water at 40 and 120°C accounted for two monomolecular layers of water. For all practical purposes, this activated silica gel containing a strongly bound monomolecular layer of water may be considered as the normal stationary phase in adsorption chromatography.

To understand the mechanism of retention and elution on and from the stationary phase, three types of solvent may be considered. The first are the totally non-interactive solvents such as hexane, the second are the slightly polar or polarizable solvents (e.g. butyl chloride or chloroform) and the third, polar solvents such as ethyl acetate, tetrahydrofuran and isopropanol. On contact with activated silica, the second type of solvents form a monomolecular layer, held on the surface of the silica by dispersive forces (Fig. 6.17c) and the polar solvents form bilayers (Fig. 6.17d); in the latter case, the first layer is thought to be bound by hydrogen bonding and the second, by dipolar interaction. The situation depicted in Fig. 6.17c also occurs when the mobile phase is a mixture of the first two types of solvents, X representing the polar solvent. Now, if a solute Y is added to the system and the molecule of Y can interact as strongly or more strongly with the surface as X, they can displace the molecules of X (Fig. 6.17e). Or if the molecules of X are more polar, there may be a second layer of these molecules (Fig. 6.17d) and these can be displaced by Y; this situation is similar to that occurring in partition chromatography. It is also possible that Y could be polar enough to displace X even from the first layer as shown in Fig. 6.17f, in which case, the solute is very strongly retained.

The mechanism of retention on non-polar (ODS) packings is still not very clear and is subject to considerable debate [35]. The interaction of the solute with the stationary phase is considered to be based on weak and non-selective dispersion forces of the van der Waals' type, the selectivity being determined mainly by the solvent effects. There is insufficient knowledge of the exact topography of the surface of the bonded phases and several models have been presented including that of an alkyl brush, molecular fur or hydrocarbonaceous sheath formed by aggregation of the hydrophobic chains. The non-polar part of the solute molecule may be considered as fitting into the lipophilic cavities, with the polar groups of the solute oriented towards the mobile phase. An alternative suggested for this adsorption mechanism is partition between the mobile phase and the organic modifier of the mobile phase which is adsorbed by the surface of the adsorbent; the concept of liquid–solid partition has also been invoked as a compromise between the above mechanisms. In the solvophobic theory [54], the stationary phase is considered to be merely a non-interactive support for the solutes thrown out of the aqueous mobile phase owing to their hydrophobic nature.

### 6.5.3 Mobile phases

#### *6.5.3.1 Properties of solvents*

As has been mentioned in the foregoing section, adsorption chromatography is dominated by two stationary phases; namely, silica gel (representing the normal-phase liquid chromatography, NPLC) and octadecylsilyl-bonded silica (for reversed-phase liquid chromatography, RPLC). It has also been noted that the normal-phase chromatography using essentially non-hydroxylic, less polar mobile phases is generally preferred for the separation of non-ionic, medium polar small molecules, while the reversed-phase LC is chosen for compounds belonging to either extremes of the polarity scale; the mobile phases used in RPLC are usually water-based and may contain salts and medium polar, water-miscible organic solvents. Since the retention mechanisms of the solutes are limited by the choice of either of the two most popular stationary phases, the result of the chromatographic experiment is predominantly determined by the selectivity of the mobile phase; this is particularly true of the reversed-phase adsorbents where, as noted above, the retention is controlled by the hydrophobicity of the solute in the aqueous mobile phase rather than its affinity to the adsorbent.

The choice of the mobile phase in LC is limited by the requirement that it should be a liquid of reasonably low viscosity at the ambient temperature. Even this limitation leaves an unmanageably large number of liquids (nearly a hundred) and combinations thereof. Most of the modern LC experiments use an on-line detector, particularly a light (UV) absorption detector; compatibility of the solvent with the detector is a desirable characteristic of the solvent. This and the requirements of safety and non-reactivity with the sample and instrument components would eliminate a large number of solvents from further consideration as general-purpose solvents. Table 6.3 lists some relevant properties (boiling point, polarity, viscosity,

**Table 6.3** — Properties of some selected solvents for LC†

| Solvent | UV | RI | BP | cP | $P'$ | $\varepsilon$ | $\delta$ | Group |
|---|---|---|---|---|---|---|---|---|
| (see footnotes) | 1 | 2 | 3 | 4 | 5 | 6 | 7 | 8 |
| FC-75‡ | 210 | 1.276 | 102 | 0.80 | $< -2.0$ | 0.25 | 1.9 | — |
| Iso-octane | 197 | 1.389 | 99 | 0.47 | 0.1 | 0.01 | 1.94 | — |
| *n*-Hexane | 190 | 1.372 | 69 | 0.30 | 0.1 | 0.01 | 1.88 | — |
| Cyclohexane | 200 | 1.423 | 81 | 0.90 | −0.2 | 0.04 | 2.02 | — |
| Cyclopentane | 200 | 1.404 | 49 | 0.42 | −0.2 | 0.05 | 1.97 | — |
| $CS_s$ | 380 | 1.624 | 46 | 0.34 | 0.3 | 0.15 | 2.64 | — |
| $CCl_4$ | 265 | 1.457 | 77 | 0.90 | 1.6 | 0.18 | 2.24 | — |
| Triethylamine | | 1.398 | 89 | 0.36 | 1.9 | 0.54 | 2.4 | I |
| Isopropyl ether | 220 | 1.365 | 68 | 0.38 | 2.4 | 0.28 | 3.9 | I |
| Diethyl ether | 218 | 1.350 | 35 | 0.24 | 2.8 | 0.38 | 4.3 | I |
| Benzene | 280 | 1.498 | 80 | 0.60 | 2.7 | 0.32 | 2.3 | VII |
| Dichloromethane | 233 | 1.421 | 40 | 0.41 | 3.1 | 0.42 | 8.9 | V |
| 1,2-Dichloroethane | | 1.442 | 83 | 0.78 | 3.5 | 0.44 | 10.4 | V |
| *t*-Butanol | | 1.385 | 82 | 3.60 | 4.1 | 0.70 | 12.5 | II |
| *n*-Butanol | 210 | 1.397 | 118 | 2.60 | 3.9 | 0.70 | 17.5 | II |
| *n*-Propanol | 240 | 1.385 | 97 | 1.90 | 4.0 | 0.82 | 20.3 | III |
| Tetrehydrofuran | 212 | 1.405 | 66 | 0.46 | 4.0 | 0.57 | 7.6 | III |
| Ethyl acetate | 256 | 1.370 | 77 | 0.43 | 4.4 | 0.58 | 6.0 | VIa |
| Isopropanol | 205 | 1.384 | 82 | 1.90 | 3.9 | 0.82 | 20.3 | II |
| Chloroform | 245 | 1.443 | 61 | 0.53 | 4.1 | 0.40 | 4.8 | VIII |
| Ethyl methyl ketone | | 1.376 | 80 | 0.38 | 4.7 | 0.51 | 18.5 | VIa |
| Acetone | 330 | 1.356 | 56 | 0.30 | 5.1 | 0.56 | | VIa |
| Ethanol | 210 | 1.359 | 78 | 1.08 | 4.3 | 0.85 | 24.6 | II |
| Acetic acid | | 1.370 | 118 | 1.10 | 6.0 | | 6.2 | IV |
| Acetonitrile | 190 | 1.341 | 82 | 0.34 | 5.8 | 0.65 | 37.5 | VIb |
| Methanol | 205 | 1.326 | 65 | 0.54 | 5.1 | 0.95 | 32.7 | II |
| Water | | 1.333 | 100 | 0.89 | 10.2 | | 80.0 | VIII |

† Data from Ref. 55.
‡ Fluorocarbons, available from 3M Company.
1. The solvent cannot be used for on-line detection below this wavelength (nm).
2. Refractive index at 25°C.
3. Boiling point in °C.
4. Viscosity at 25°C.
5. Solvent polarity parameter.
6. Solvent strength parameter.
7. Dielectric constant at 20°C.
8. Solvent selectivity group; see Fig. 6.19.

solvent strength and selectivity, refractive index, UV cut-off point or wavelength above which it can be used, etc.) of some of the common solvents useful for chromatography [55]. The solvents are arranged roughly in order of their increasing polarity; those above tetrahydrofuran (THF) in the list are generally used in normal-phase adsorption chromatography, while THF and those that follow are favoured in RPLC.

The intermolecular interactions that determine the solubility of a compound in a particular solvent (i.e. dispersion, dipole–dipole, hydrogen-bonding and dielectric interactions) have been discussed earlier. Two concepts; the polarity and the solvent strength, permit one to determine the strength of these interactions with different solutes as well as with other solvents. Understandably, mutual solubility of different solvents is a requirement when two or more solvents are used.

### *6.5.3.2 Choice of the mobile phase*

The purpose of method development in LC is to arrive at the proper composition of the mobile phase to elute the solute components in a reasonable period of time (say, with $k$ values between 2 and 10, and with adequate selectivity or resolution).

The polarity of solvent may be described in a number of ways but probably the most suitable in the present context is to use the parameter $P'$, which is based on the solubility data of different types of compounds [56]. Since like dissolves like, the degree of solubility of a polar substance is a direct measure of the solvent polarity. $P'$ is also a measure of the solvent eluotropic strength ($\varepsilon^0$) [56] in the sense that the greater the polarity of the solvent, the greater is its elution power in normal-phase adsorption chromatography. The $P'$ is also directly related to the Hildebrand solubility parameter ($\delta$) in most cases. Since $P'$ is a measure of solubility, which in turn determines the partition coefficient of a substance between two liquids, the $k$ values in LLC and LSC are predictable from this value. The $k$ values of a substance in two solvents with polarity parameters $P_1$ and $P_2$ are given by the relationship:

$$\frac{k_2}{k_1} = 10^{(P_1 - P_2)/2} \tag{6.20}$$

For reversed-phase systems, the above equation becomes:

$$\frac{k_2}{k_1} = 10^{(P_2 - P_1)/2} \tag{6.21}$$

The $P'$ values for common solvents vary from $-2$ for the non-polar fluorocarbons to over 10 for water (see Table 6.3). Roughly, for every 2 units of difference in $P'$, the capacity factor ($k$) changes by an order of magnitude, which means that retention of a substance on a particular column can be varied by a factor of $10^6$ simply by varying the solvent. The polarity of a binary mixture of solvents A and B is given by:

$$P' = \phi_A P_A + \phi_B P_B \tag{6.22}$$

where $\phi_A$ and $\phi_B$ are the volume fractions, and $P_A$ and $P_B$ are the polarity parameters of solvents A and B, respectively.

The variation of the eluotropic strength ($\varepsilon^0$) with increasing proportion of the more polar solvent (B) is not linear. Addition of a small quantity of B to solvent A brings about a large change in the ε° value, but further addition of B causes a progressively smaller change. The calculated ε° values for some solvent mixtures are shown graphically in Fig. 6.18 [57]. Saunders [58] derived empirical rules for the

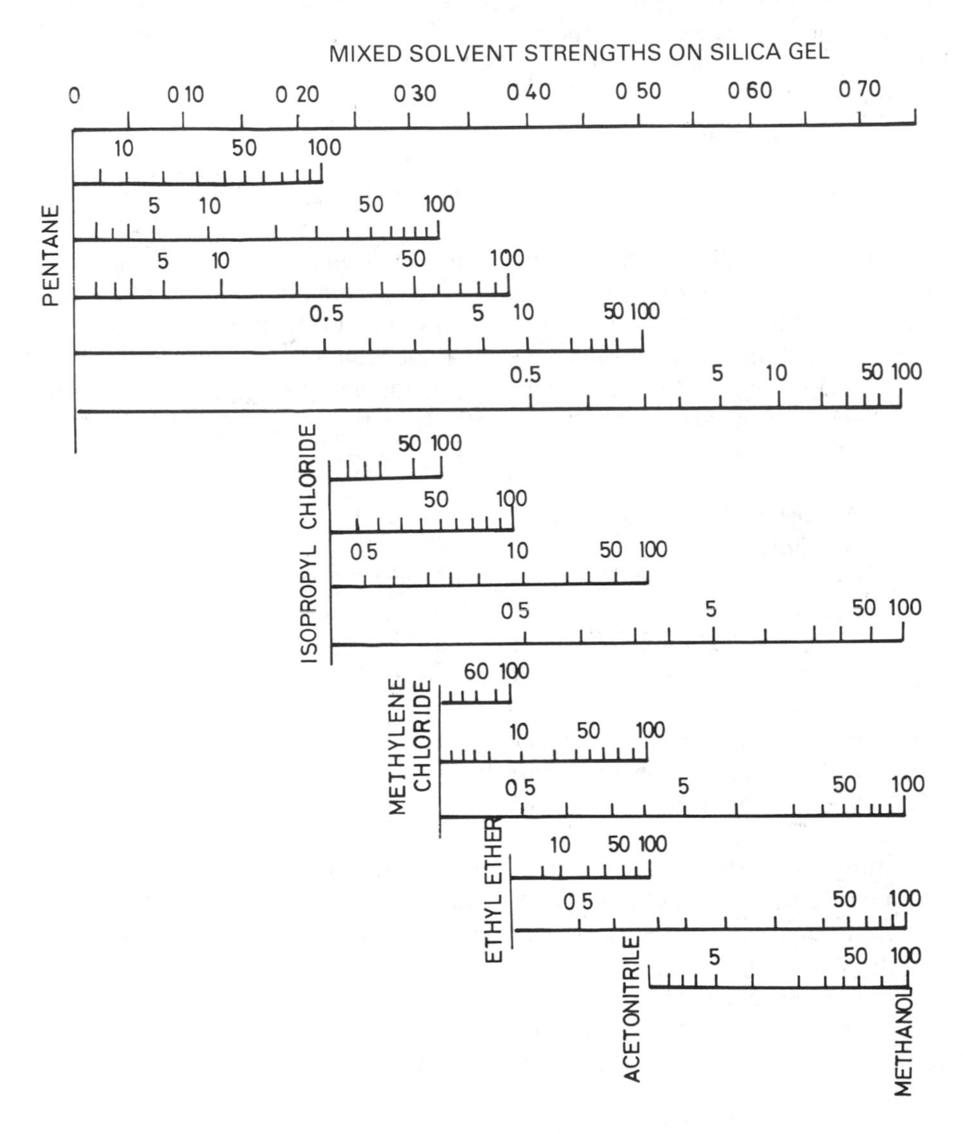


Fig. 6.18 — ε° values of mixed solvents on silica gel. The top set of five lines indicates the percentages, respectively, of isopropyl chloride, methylene chloride, ethyl ether, acetonitrile and methanol required to be mixed with pentane to give the solvent strength (ε° value) depicted on the top-most line. Similarly the solvent strengths of binary mixtures of the other solvents are obtained as shown by the lower sets of more polar solvents. Reproduced with permission from Ref. 57; copyright (1978) John Wiley & Sons, Inc.

determination of $E_3$ values (i.e. $\varepsilon^o$ value of a solvent or solvent mixture that would give a $k$ value of 3 on a given adsorbent) based on the chemical structure of simple compounds. Once the $E_3$ value is known, the appropriate solvent composition may be obtained from Fig. 6.18. For convenience, the strength of the mobile phase is adjusted to give $k$ values for the different components of sample in the region 2–5 for a simple mixture containing two or three compounds and up to 20 for more complex mixtures.

It is apparent from the above discussion that a solvent composition of desired strength can be obtained by several different binary combinations of solvents. Any of them may be useful if the compounds to be separated have large separation factors ($\alpha$). In certain cases, the $k$ values of the compounds in a given system would be very close (and hence the $\alpha$ value would be nearer to unity) so that resolution would be poor (see Chapter 2 for the discussion on the significance of these parameters). In such cases, improvement in the selectivity of the chromatographic (mobile phase) system is necessary. This would require a knowledge of the mechanisms of solute–solvent interactions, which include dipolar, dispersion, hydrogen-bonding, etc. On the basis of these and the (electron or proton) donor–acceptor properties of different solvents, the solvents are divided into eight selectivity groups, I to VIII [59]. They are:

I. Aliphatic ethers and amines.
II. Alcohols.
III. *N*-Heterocycles, amides, tetrahydrofuran.
IV. Glycols, glycol ethers, benzyl alcohol, formamide, acetic acid.
V. Methylene chloride and ethylene chloride.
VIa. Alkyl halides, ketones, esters, nitriles, sulphoxides, sulphones, aniline, dioxan.
VIb. Nitro compounds, alkyl aryl ethers, aromatic hydrocarbons.
VII. Halobenzenes, diphenyl ether.
VIII. Fluoroalkanes, chloroform, water.

By sub-dividing the polarity parameter ($P'$) on the basis of H-acceptor, H-donor and dipolar factors, and plotting them for each solvent on a triangular diagram, the inter-relationships among the various groups are obtained as shown in Figs. 6.19 and 6.20 [60]. The $X_d$, $X_e$ and $X_n$ in the figures are solvent selectivity parameters for the test solutes dioxane, ethanol and nitromethane, respectively, representing the above factors. The values of the parameters are calculated from the experimentally determined solute polarity distribution coefficients, $K''_g$, according to the following example:

$$X_d = \frac{\log (K''_g)_{\text{dioxan}}}{P} \qquad (6.23)$$

where the polarity index, $P$, is the sum of the log ($K''_g$) values of the three solutes. Now, if $\alpha$, the separation factor of two compounds is not adequate using a solvent mixture A and B (B being the minor, polar solvent) and B belongs to, say, group VIII, an alternative solvent C may be chosen from the group farthest from group VIII, either

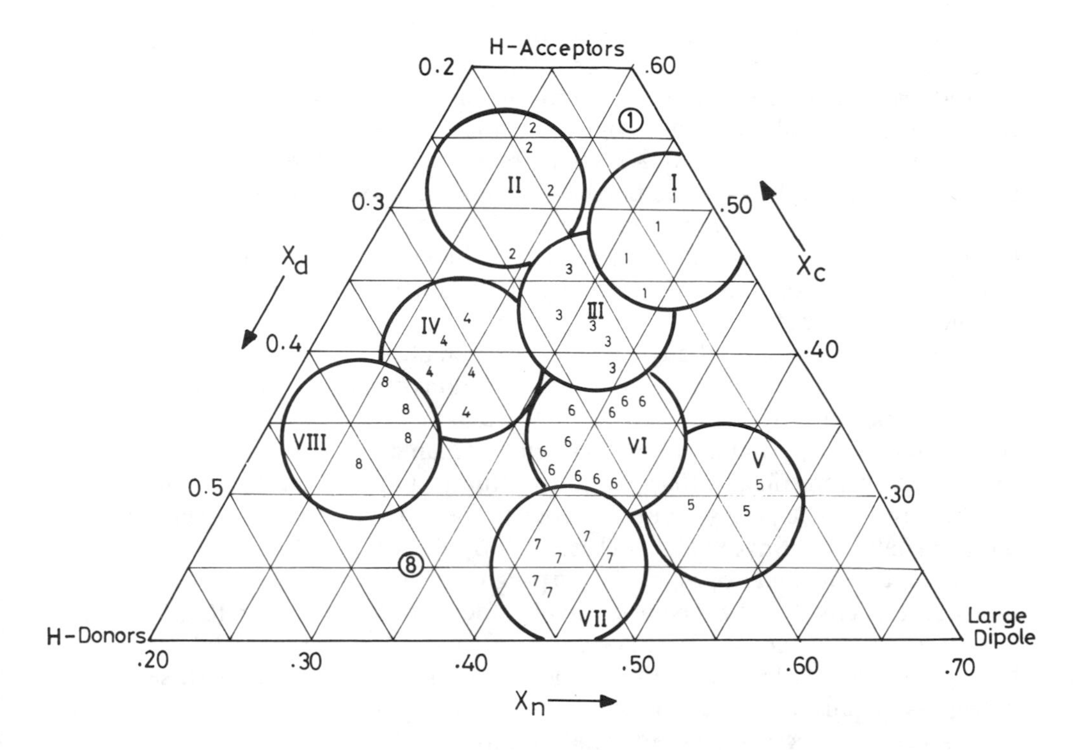


Fig. 6.19 — Solvent selectivity triangle for common solvents in LC. Reproduced from the *Journal of Chromatographic Science* [59] by permission of Preston Publications, Inc.

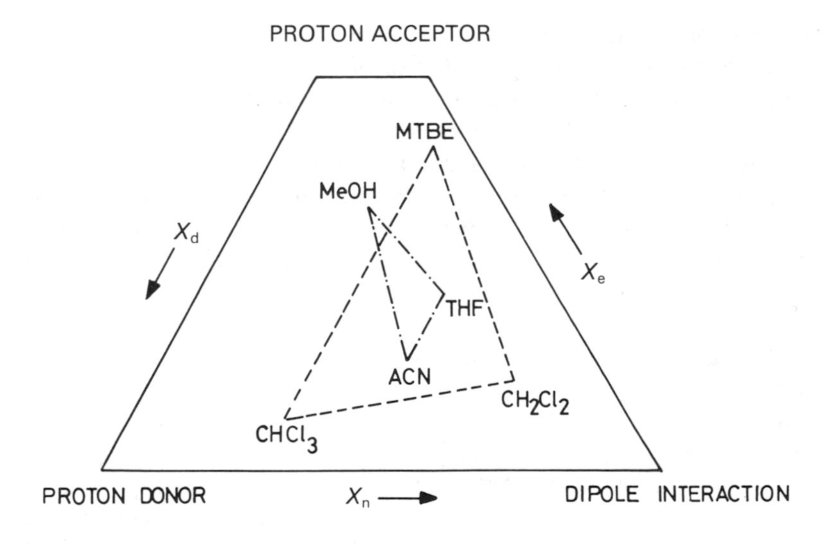


Fig. 6.20 — Solvent selectivity triangle for preferred solvents (– · – · – · –) for RPLC and (– – – –) for NPLC. Reproduced from Ref. 60 by courtesy of Elsevier Scientific Publishing Co.

group I or V; hopefully , the selectivity is better. The proportion of C required to obtain adequate polarity is calculated from eq. (6.22). If this does not yield the desired result, a ternary solvent system consisting of A, B and C may be tried. In this case, selectivity is achieved by varying the ratio B/C, while, maintaining the polarity by varying the proportion of A. Water or methanol in the normal phase and tetrahydrofuran in the reversed phase are commonly employed as the third solvent (the modifier). When water is used, it is advantageous to maintain about 50% saturation level. For preliminary method development, it is convenient to use TLC. Solvent mixtures of different selectivity, but with identical strength, are called isoeluotropic solvents.

The above discussion, though mainly based on normal-phase adsorbents, is also valid for RPLC, of course, with the polarity relations reversed. Water has the lowest eluotropic power in RPLC and the solvent strength is increased by mixing progressively higher proportions of water-miscible organic solvents such as methanol (MeOH), acetonitrile (ACN), tetrahydrofuran (THF), methyl *t*-butyl ether (MTBE) or dioxane. The solvent strength parameter ($S$) for reversed-phase systems has been defined by Snyder *et al.* [61] using the $S$ value of water as zero. The alcohols, methanol, ethanol and isopropanol have progressively increasing $S$ values of 3.0, 3.6 and 4.2, respectively. The solvent strength of the other four common RPLC solvents is as follows: acetonitrile (3.1), acetone (3.4), dioxane (3.5) and THF (4.4). The solvent strength of a binary mixture is calculated as given above and where selectivity change is required, the less-polar solvent is changed while holding the solvent strength constant according to the relationship:

$$\phi_B S_B = \phi_C S_C \tag{6.24}$$

In addition to the use of organic solvent modifiers noted above, the essentially aqueous mobile phase also permits use of salts and buffers to control retention. Addition of salts would increase retention of non-ionic solutes by the salting-out effect, while the effect of salts on the retention of ionizable solutes is more complex. The pH may be adjusted to suppress ionization of weak acids and bases. The cations most frequently used are sodium, potassium and ammonium, and the anions are acetate, citrate and phosphate. A novel approach to control of retention in RPLC (and, to some extent in normal-phase LC also) is to use the ion-pairing technique which is discussed in the following section.

#### *6.5.3.3 Ion-pair chromatography*

Ionic and ionizable groups can form tight ion-pairs with appropriate counter-ions so that the free ions of the solute are soluble in the more polar solvent while the ion pair is soluble in the less polar (organic) phase. The principle of ion-pair extraction has been one of the most useful distribution techniques in biology. It has subsequently been applied to modulate retention in chromatography (ion-pair chromatography or IPC) and has found enormous application, to the extent that it is the favoured method wherever applicable. The ion pair may be selectively retained on a chromatographic bed or eluted depending on the reversed-phase or normal-phase nature of the stationary phase. Retention in IPC, in either normal- or reversed-phase modes, understandably depends on the nature and concentration of the counter-ion. Several experimentally variable factors like the pH, the eluotropic strength of the mobile phase, its surface tension and linear velocity, the salt effects, the phase ratio and the temperature, will be discussed in the following paragraphs.

The charge on the counter-ion naturally depends on that of the solute ion. Ideally, the counter-ion is univalent, aprotic and soluble in the organic phase, and should not interfere with the detection system. It may, however, be noted that ion pairing can be used to advantage in detection by choosing a counter-ion with large absorptivity (in light-absorption detection) so that the sensitivity of detection is enhanced. If the solute is anionic, as in carboxylic acids, the counter-ion is a lipophilic cation (e.g. tetraalkylammonium or trialkylammonium or simpler alkylammonium ions). Similarly, if the solute is cationic (e.g. amine in a low pH medium), the ion-pairing reagent (also called, hetaeron) can be a halide, a carboxylate, arene- or alkane-sulphonate or several of the alkyl sulphates and phosphates. For an exhaustive list of the ion-pairing agents, see Ref. 62.

That the mechanism of separation by IPC is unclear is reflected in the fact in the fact that it is known in the literature by several names, which include: paired-ion chromatography, hetaeric chromatography (from the Greek word, *hetairon*, meaning companion and now taken to mean ion-pairing agent), soap chromatography, detergent-based ion-exchange chromatography, surfactant chromatography, micellar chromatography (when the amphiphilic hetaerons agglomerate to form micelles at sufficiently high concentration), dynamic (or solvent-separated, or solvent-generated) ion-exchange chromatography, solvophobic-ion chromatography, ion-association (or ion-interaction) chromatography, etc. The principle of ion-pair chromatography has been the subject of considerable theoretical treatment [62,63]. Several models involving reversed-phase partition of the ion pairs formed in the mobile phase, dynamic ion exchange of solute with the counter-ions adsorptively retained on the stationary phase, etc., have been proposed. A rather simplified treatment is presented here.

The equilibrium constant ($E$) of the ion-pairing reaction (eq. (6.25)) is given by eq. (6.26), so that the capacity factor, $k$, in RPLC can be derived as given in eq. (6.27). The value of $k$ in the normal phase is given by eq. (6.28). As usual, the symbols, [AB], [A] and [B] represent the concentration of the respective species in the appropriate phases. For the sake of simplicity it is assumed that the ion pair is essentially in the organic phase and the individual ions in the aqueous (or polar) phase.

$$A^+ + B^- \rightleftharpoons A^+B^- \tag{6.25}$$

$$E = \frac{[A^+B^-]}{[A^+][B^-]} \tag{6.26}$$

or, simply,

$$E = \frac{[AB]}{[A][B]}$$

$$k = \frac{V_S}{V_M} \cdot \frac{[AB]}{[A]} = \beta E[B] \tag{6.27}$$

$$k = \frac{\beta}{E[B]} \tag{6.28}$$

However, to account for effects of several side reactions and the secondary equilibria where the solute A may exist in several forms, the equilibrium constant is redefined as the conditional equilibrium constant, $E^*$, so that,

$$E^* = E \frac{\Sigma\, \alpha_{A,B}}{\Sigma\, \alpha_A \cdot \Sigma\, \alpha_B} \tag{6.29}$$

where $\Sigma\alpha$, etc., are the summation of the $\alpha$ coefficients of all the side reaction effects. Thus, the capacity factor (and hence, the distribution constant ($K$)) are a function of the concentration of the counter-ion as given in eq. (6.30).

$$K = \frac{k}{\beta} = E^* [B] \tag{6.30}$$

Assuming a non-polar, bonded stationary phase and a mobile phase, several of the following events may be understood to occur when a solute ion (A) pairs with a counter-ion (B, of opposite charge) in the aqueous phase to form an ion pair (AB). The ion pair AB may complex with the ligand, L (bound to the stationary phase), to give a species LAB or if the two ions are sufficiently lipophilic, they may be individually retained by the stationary phase as LA and LB. The rate or equilibrium constants of these interactions are given below. Again, to make matters simple, all species not associated with the ligand, L, are assumed to be in the aqueous mobile phase.

$$K_1 = \frac{[LA]}{[L][A]} \tag{6.31}$$

$$K_2 = \frac{[AB]}{[A][B]} \tag{6.32}$$

$$K_3 = \frac{[LB]}{[L][B]} \tag{6.33}$$

$$K_4 = \frac{[LAB]}{[L][AB]} \tag{6.34}$$

$$K_5 = \frac{[LAB]}{[LB][A]} \tag{6.35}$$

$$K_6 = \frac{[\mathrm{LAB}]}{[\mathrm{LA}][\mathrm{B}]} \tag{6.36}$$

In bonded-phase packings, [L] is constant and much greater than [LA] or [LAB] and chromatographic conditions can be chosen to have [B] much greater than [A]. The general dependence of the capacity factor ($k$) on the concentration of the counter-ion, [B], is obtained by combining the above equations:

$$k = \frac{k_o + x[\mathrm{B}]}{(1 + K_2[\mathrm{B}])\,(1 + K_3[\mathrm{B}])} \tag{6.37}$$

where $k_o$ is the capacity factor of A in the absence of B and

$$x = K_2K_4, \quad \text{or} \quad K_3K_5, \quad \text{or} \quad K_1K_6 \tag{6.38}$$

If $K_2$ or $K_3$ is small (i.e., $\ll 1$),

$$k = \frac{k_0 + x[\mathrm{B}]}{1 + y[\mathrm{B}]} \tag{6.39}$$

where, $y = K_2$ or $K_3$. The capacity modification factor in IPC is defined by the following equation.

$$\delta_{\mathrm{CMF}} = \frac{k}{k_o} = \frac{1 + \dfrac{x}{k_0[\mathrm{B}]}}{1 + y[\mathrm{B}]} \tag{6.40}$$

The above relationships indicate the chromatographic properties of ion pairs under different limiting situations. When the counter-ions are small and hydrophilic (e.g. phosphate ion), $K_2$ is much larger than $K_4$ and $K_3$ is small; the ion pair is then formed in the aqueous phase and $k < k_o$ and is called hydrophilic ion pairing. When the counter-ion is small and lipophilic (e.g. heptanesulphonic acid), $K_4$ is much larger than $K_2$, and $K_3$ is small again; the hydrophobic ion pairing occurs in the mobile phase but the complex binds rapidly and strongly to the stationary phase, so that $k > k_o$. $K_3$ becomes large when the counter-ion is hydrophobic and large (e.g. sodium lauryl-sulphate); in such cases, modification of the stationary phase by the counter-ion becomes important (dynamic liquid ion exchanger). The properties of the solute, counter-ion and mobile phase thus have important roles in modulating the retention in IPC. The most significant properties are the dipole moment, surface area of the solute and the counter-ion, and the interfacial surface tension and dielectric constant of the mobile phase.

The retention and the capacity factor can be changed according to eq. (6.27); however, this may not be very convenient in the normal-phase IPC as the ion is in the stationary phase and not in the mobile phase. Variation in the length of the alkyl chain of the counter-ion can cause a change in the *k* value by 2.5 for every methylene unit; thus, changing the counter-ion from tetraethylammonium to tetrapentylammonium ion can alter the *k* value by five orders of magnitude. Similarly, changing the anionic counter-ion from a chloride ion to dipicrylamine brings about a change in the *k* value by $10^{10}$. Variation of the counter-ion, understandably, does not dramatically alter the retention of fully ionized solutes. The *k* values can also be varied by varying the solvent composition, but the *k* value does not correspond to the polarity, $P'$. In addition to the polarity which controls the solubility of the solute, the solvent strength in IPC is also a function of the dielectric constant (δ) which controls the stability of the ion pair, so that the expression $(P' + \delta)$ better approximates the solvent strength. An increase in the solvent strength naturally decreases the *k* value. The pH affects the ionization of weak acids and bases and, thus, their ability to form ion pairs. In reversed-phase IPC, the *k* values of anionic solutes decrease (and those of cationic solutes increase) with decreasing pH; the order is reversed in normal-phase IPC. The solvent selectivity may be varied as discussed earlier. IPC is normally carried out at ambient temperature except that occasionally, operation at 50–70°C may be advantageous in the case of reversed-phase IPC using high-viscosity mobile phases.

While IPC may be carried out either on microparticulate silica in the normal-phase mode or the hydrocarbonaceous bonded phases in the reversed-phase mode, the latter is most often preferred in view of its very high reproducibility and sensitivity. As is well known, the capacity factor (*k*) of the solutes on the non-polar phases is in the order of increasing hydrophobicities. The hydrophobic fragmental data can be used to calculate the log *K* values; and the higher the value of log *K*, the higher is the value of *k*. If the retention is very low ($k < 1$), a hydrophobic ion-pairing agent may be used, and *vice versa*. In general, polar compounds require hydrophobic counter-ions, and non-polar solutes require hydrophilic counter-ions. The pH is adjusted to keep the solute in a predominantly ionized form. Selectivity changes are brought about by changing the nature of the counter-ion, say, from alkylsulphates to alkylphosphates, etc. The optimum concentration of the ion-pairing agents needs to be determined in each new case. A concentration of around 5 mM may be adequate on most reversed-phase separations but, in the normal-phase mode, it may need to be as high as 1 M for counter-ions with poor buffer capacity.

IPC has been extensively used in recent years for the analysis of a large variety of compounds which had otherwise unsatisfactory retention and tailing. These include a variety of drugs, nucleic-acid components, dyes and alkaloids [62]. The IPC technique is particularly suited for analysis of polar compounds like drugs and other biologically active compounds (e.g. catecholamines). A large variety of drugs useful in clinical medicine have been analysed with a very high degree of selectivity [63]. IPC offers a new approach to the separation of amino acids and peptides on reversed phases which strongly retain them by hydrophobic interactions. These interactions cannot, because of poor solubility, be reduced by increasing the proportion of the organic modifier. However, the use of hydrophilic ion-pairing agents brings about a reduction in the *k* value dramatically. On the same lines, use of hydrophobic counter-ions permits retention of polar amino acids with ionogenic groups which are poorly retained on the reversed phases. Similar approaches are useful for the separation of complex mixtures of peptides from natural sources (e.g. hormones and protein hydrolysates), or from the products of solid-phase peptide synthesis.

## 6.6 ION (-EXCHANGE) CHROMATOGRAPHY (IEC/IC)

### 6.6.1 Introduction and theory

The best known everyday use of the ion-exchange process is probably the softening and de-ionization of water for use in industry and the laboratory. Use of ion-exchangers for the replacement of potassium in molasses by calcium to aid crystallization of sugar is also well known. In addition to these long-standing applications, the ion exchangers also find several nonchromatographic uses in the chemical laboratory. The sulphonic acid-type resins are very convenient catalysts for esterification and dehydration. Ion-exchange resins carrying different functional groups find increasing application as convenient reagents in organic synthesis [64]. Other important applications are the de-ionization of non-electrolytes, and of chromatographic paper, removal of interfering ions during quantitative determination of other ions, determination of total salts, dissolution of insoluble salts and concentration of trace elements. It is also possible to convert the non-ionic compounds into ionic compounds by derivatization for separation by the ion-exchange methods. Application of IEC to amino acid analysis is well known [65]. Separation of peptides and proteins [66], nucleic acid derivatives [67], inorganic ions [68], carboxylic acids [69], nitrogenous compounds [70], carbonyl compounds, ethers and alcohols [71] and carbohydrates [72] have been extensively reviewed.

It must be pointed out that in several of the applications noted above, the mechanism of separation may be partition, size exclusion or ion exclusion and not ion exchange *per se*, as implied by the name ion-exchange chromatography. For the separation of non-ionic compounds on ion-exchange resins, advantage is taken of the latter's swelling properties and the consequent size-exclusion and partition effects. Indeed, one extreme view is that there is as much of *exchange of ions* in IEC as there is *colour writing* in chromatography. In general, the term IEC is used whenever an ion-exchange packing is used for the separation. The term, ion chromatography (IC), was originally coined to describe the IEC technique in which a low-capacity ion-exchange column was used as an ion suppressor and a conductivity detector was used for the determination of inorganic ions. The term IEC is now aptly used whenever chromatographic analysis of ions using ion-exchange resins is involved [73].

The history and the basic principle of IEC have been discussed very briefly in Chapter 1. The ion exchangers typically carry a covalently bound, fixed ionic functionality, the charge on which is neutralized by a counter-ion of opposite charge. The counter-ion can be exchanged for an ion of the same charge in the environment (i.e. the mobile phase in IEC). The process involved is a chemical reaction, as represented by eq. (6.41). When the fixed ion is negatively charged, the exchangeable ion is positive and the material is called a cation exchanger; similarly, anion exchangers carry a fixed positively charged functional group and a negatively charged exchangeable ion. Several cation and anion exchangers are available for routine use in both chromatographic and non-chromatographic applications (see below).

$$\mathrm{Mx{-}A^-B^+ + C^+D^- \rightleftarrows Mx{-}A^-C^+ + B^+D^-} \qquad (6.41)$$

The success of IEC depends on the relative affinities of the exchangeable ions B and C toward the fixed ion A and the co-ion D. When an ion-exchange resin is allowed to swell in water, there is a dynamic equilibrium, and as the water enters the pore structure of the resin, the concentration of the exchangeable groups and the counter-ions decreases. If an ionic compound is introduced into the medium, distribution of the original and the new ions occurs until thermodynamic equilibrium is reached while maintaining electrical neutrality (Donnan equilibrium). The equilibrium quotient ($Q$) of the reaction is given by:

$$Q = \frac{[\mathrm{B}][\overline{\mathrm{C}}]}{[\overline{\mathrm{B}}][\mathrm{C}]} \qquad (6.42)$$

The bars over the symbols represent the relevant ions in the resin phase. The true equilibrium constant, $K$, is given by substituting the concentrations by the activity coefficients of the ions in the respective phases, as in eq. (6.43).

$$K = \frac{\gamma\mathrm{B}\cdot\gamma\overline{\mathrm{C}}}{\gamma\overline{\mathrm{B}}\cdot\gamma\mathrm{C}} \qquad (6.43)$$

The practical utility of the above equation, however, is limited since the activity coefficients in the resin phase are unknown. Therefore, assuming that the activity coefficients in the resin phase are constant, it is the normal practice to use the selectivity or the distribution ratio, $D$.

$$D = \frac{[\overline{\mathrm{C}}]}{[\mathrm{C}]} \qquad (6.44)$$

The competing ion (B) and its concentration affect the values of $Q$, $K$ and $D$. Assuming B is singly charged, $D$ is proportional to $[\mathrm{B}]^n$, where $n$ is the charge on the solute ion, C. The retention volume of C on a resin of weight, $w$, is given by $(Dw + V_o)$, where $V_o$ is the void volume. The value of $D$ can be calculated by the expression $(QX/[\mathrm{B}])$, $X$ being the capacity of the resin bed, defined as the number of milli-equivalents (meq) of exchangeable ions per gram of the resin.

The theory of selectivity in IEC has been the subject of considerable discussion. The selectivity probably depends on the electrostatic field strength around the fixed ions and this is inversely proportional to the radius of the spherical ion. Solvation of the ions by the mobile phase molecules also exerts considerable influence on the electrostatic interactions. The sulphonate ion, which has a large radius and low field strength binds alkali metal ions in the order: Cs > Rb > K > Na > Li. Alkaline earth metal ions are held more strongly than the alkali metals and the order decreases with decreasing atomic weight: Ba(II) > Sr(II) > Ca(II) > Mg(II). In general, the affinity of the ions for the resin is inversely proportional to the radius of the hydrated ion and directly proportional to the ionic charge. The selectivity order among anions on strongly dissociated ion exchangers is: citrate > sulphate > oxalate > iodide > nitrate > chromate > bromide > thiocyanate > chloride > formate > fluoride > acetate [1].

It should be possible to understand the selectivities of the various ions toward different fixed ions on the basis of the hard and soft acids and bases (HSAB, pronounced '*hassab*') principle [74]. The molecules (both ionic and non-ionic) may be considered to be made up of two parts, the positive acid (acceptor) and the negative base (donor). Both parts are categorized as hard or soft. The hard acids (or bases) are generally characterized by a small radius, high effective charge and low polarizability; the soft ones have the opposite properties. The softness of a base is also associated with low electronegativity, easy oxidizability or empty low-lying orbitals. According to the principle, *hard acids prefer to bind with hard bases, and soft acids with soft bases*. Though the hardness of a particular ion is influenced by several ambient factors, it is apparent that ion-exchange selectivities can be a testing ground for the validity of the concept as well as for quantification of hardness–softness scales. The concept also points to the possibility of selective (affinity) elution of ions by varying the co-ions. The mechanism of ion-pair chromatography (section 6.5.3.3) may also be rationalized on the basis of this principle.

Several competing and interfering mechanisms contribute to the failure of the theory when applied to certain relatively large ions; e.g. the selectivity of Cs over Rb ions on inorganic ion exchangers is also imperfectly understood. In the case of multivalent ions, chelation or complexation may change the sorption process. The selectivity of organic ions is more difficult to predict since dispersion forces in un-ionized residues (e.g. adsorption by hydrophobic interaction) also operate. Hydrogen-bonding between the hydration shell of the ions and the exchanging group could reverse the selectivity orders noted above. The ionization (and thus retention) of weak acids and bases is understandably dependent on the pH.

### 6.6.2 Ion-exchange materials

#### *6.6.2.1 Cross-linked polystyrene resins*

The polystyrene-divinylbenzene cross-linked resins are the most extensively used ion-exchange materials. They can be prepared as spheres of fairly uniform size and their chemical functionality can be easily modified to suit different applications [75]. They have high exchange capacities and rates. They exhibit superior physical strength and reasonable stability to pH variation. As noted earlier, styrene readily undergoes radical-initiated polymerization to yield a linear polymer (Fig. 6.21a) and if the polymerization is carried out in presence of divinylbenzene (DVB), cross-linking occurs (Fig. 6.21b). The amount of DVB used determines the degree of cross-linking which, in turn, determines the properties of the resin such as pore-size, swelling capacity and rigidity. Polystyrenes with 2–24% cross-linking are commercially available, but those with 4–8% cross-linking are commonly used. The higher percentage of cross-linking yields tighter (harder) polymer structures that can support a larger number of fixed ions per unit volume; however, the permeability of these resins to ions is also poorer. Low cross-linking permits fast diffusion of the ions, but the resins swell or shrink appreciably with change in the solvent composition;

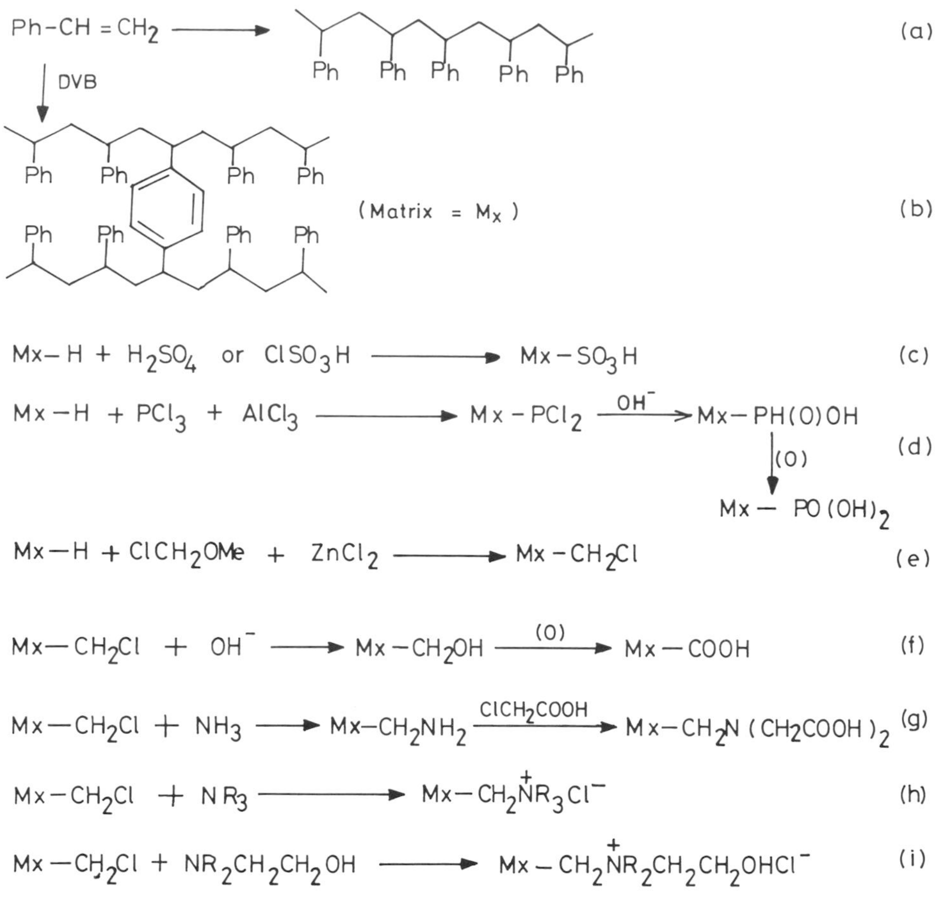

$$Ph\text{-}CH{=}CH_2 \longrightarrow \text{polystyrene chain (Ph substituents)} \quad \text{(a)}$$

$$\downarrow DVB$$

$$\text{DVB cross-linked polystyrene chains (Ph substituents)} \quad (\text{Matrix} = M_x) \quad \text{(b)}$$

$$Mx\text{-}H + H_2SO_4 \text{ or } ClSO_3H \longrightarrow Mx\text{-}SO_3H \quad \text{(c)}$$

$$Mx\text{-}H + PCl_3 + AlCl_3 \longrightarrow Mx\text{-}PCl_2 \xrightarrow{OH^-} Mx\text{-}PH(O)OH \xrightarrow{(O)} Mx\text{-}PO(OH)_2 \quad \text{(d)}$$

$$Mx\text{-}H + ClCH_2OMe + ZnCl_2 \longrightarrow Mx\text{-}CH_2Cl \quad \text{(e)}$$

$$Mx\text{-}CH_2Cl + OH^- \longrightarrow Mx\text{-}CH_2OH \xrightarrow{(O)} Mx\text{-}COOH \quad \text{(f)}$$

$$Mx\text{-}CH_2Cl + NH_3 \longrightarrow Mx\text{-}CH_2NH_2 \xrightarrow{ClCH_2COOH} Mx\text{-}CH_2N(CH_2COOH)_2 \quad \text{(g)}$$

$$Mx\text{-}CH_2Cl + NR_3 \longrightarrow Mx\text{-}CH_2\overset{+}{N}R_3Cl^- \quad \text{(h)}$$

$$Mx\text{-}CH_2Cl + NR_2CH_2CH_2OH \longrightarrow Mx\text{-}CH_2\overset{+}{N}R_2CH_2CH_2OHCl^- \quad \text{(i)}$$

Fig. 6.21 — Structure and synthesis of polystyrene-based resins for ion-exchange chromatography.

they are also soft and collapse under higher pressures. The 8% cross-linked resin is a good compromise and is the most frequently used. Small spherical particles of the resin, suitable for most chromatographic purposes, are formed by emulsification at 90°C of a mixture of styrene, DVB and a catalyst in an aqueous solution of the emulsifying agent. The mesh size of the particles depends on the concentration and nature of the emulsifying agent and the efficiency of stirring; 100–200 mesh (75–150 $\mu$m) and 200–400 mesh (37–75 $\mu$m) particles are commonly used as the matrix for preparation of IEC packings.

The resins obtained by the above process are internally homogeneous and are called microporous. The macroporous variety is prepared by using a much higher proportion (about 50%) of the cross-linking agent in a solvent in which the monomer dissolves but in which the polymer has low solubility. As the polymerization proceeds, small particles (a few hundred nanometres in diameter) precipitate out and aggregate into loosely bound spheres of diameter equal to that of the drops in the stirred liquid. Ion exchangers made from resins of about 1 mm diameter are used in industry as they combine fast ion-exchange with good flow properties. Their use in chromatography, however, is minimal owing to the large particle size.

For the preparation of strong cation-exchange resins, the phenyl groups of a polystyrene chain are sulphonated using sulphuric acid or chlorosulphonic acid (Fig. 6.21c). Almost all the phenyl groups in the chain can be sulphonated, giving a capacity of about 5.2–5.4 meq/g of the resin. Resins of lower strength are obtained by treating the polystyrene-DVB resins with phosphorus trichloride, hydrolysis and oxidation to phenylphosphonic acid groups (Fig. 6.21d). Weakly cationic resins can also be prepared by chloromethylation of the resin (Fig. 6.21e), hydrolysis and oxidation to carboxylic acid (COOH) groups (Fig. 6.21f), but carboxyl-type resins are usually prepared by polymerization of methyl methacrylate in the presence of DVB. Chelating resins carrying iminodiacetic acid groups (Fig. 6.21g), selective to certain metal ions, are obtainable from the chloromethylated resins by treatment with ammonia and chloroacetic acid; the selectivity is highly pH dependent (particularly between pH 2 and 4).

The chloromethylated polystyrene-DVB is also the starting material for preparation of both strong and weak anion exchangers. Treatment of the chloromethylated resin with ammonia, methylamine, dimethylamine and trimethylamine yields the corresponding primary, secondary, tertiary and quaternary ammonium salts (Fig. 6.21h) with increasing anion-exchanging ability. Use of ethanolamine derivatives yield the corresponding hydroxyethylamino derivatives (Fig. 6.21i). The anion-exchange resins in the hydroxide form and the cation exchangers with sulphonic acid groups are rather unstable to heat; the resins, therefore, are normally supplied in the chloride and sodium ion forms, respectively, and are converted to the desired ionic form prior to use.

The dry resins swell considerably on equilibration with water, the extent of swelling depending on a number of factors such as the nature and number of functional groups, the percentage of cross-linking, etc. The ion exchangers are also characterized by such functional properties as exchange capacity, pH-titration curves and size of the particles. These parameters can be experimentally determined [1,26], but such information is often supplied by the manufacturer.

#### *6.6.2.2 Carbohydrate-based ion exchangers*

Polymeric carbohydrates such as cellulose can be modified to serve as ion exchangers. The primary hydroxyl group of cellulose can be directly oxidized to COOH groups, but the reaction is difficult to control. Carboxymethyl cellulose (CM-cellulose) is prepared by treating cellulose with sodium hydroxide and sodium chloroacetate; sulphoethylcellulose is similarly produced. These and cellulose phosphate have ion-exchange properties analogous to the corresponding polystyrene-based products. Among the anion-exchange celluloses, the most common is diethylaminoethylcellulose (DEAE-cellulose), which is obtained by treating cellulose with diethylamino-2-chloroethane and sodium hydroxide. Reaction of epichlorohydrin and triethanolamine yields ECTEOLA-cellulose of undefined structure; it is known to contain weakly basic amino groups. Other substituted aminoethylcelluloses have also been prepared for use in IEC. Pharmacia (Uppsala, Sweden) markets functionally similar products based on agarose (Sepharose) and dextran (Sephadex) besides cellulose (Sephacel) [76].

The main advantage of the carbohydrate-based materials over the polystyrene-based ion exchangers is that their very porous and hydrophilic nature permits diffusion of larger species such as proteins and nucleic acids (MW = 1 million). In addition, they also exhibit adsorption and partition effects which sometimes help separation. HPLC packings based on carboxymethyl and diethylaminoethyl functionalities are also available.

#### *6.6.2.3 Inorganic ion exchangers*

Though zeolites (aluminosilicates), which were the earliest ion exchangers used, had been largely replaced by organic ion exchangers for most routine analyses, interest in atomic energy in the 1950s revived the use of inorganic ion exchangers mainly because of the instability of polystyrene-based materials to heat and irradiation, both of which are unavoidable in the nuclear plants. Also, the inorganic exchangers are generally selective to Cs ions, a property useful in dealing with products from nuclear reactors.

Hydrous oxides of Cr(III), Zr(IV), Sn(IV), Th(IV) and Sb(V) are amphoteric in nature; they show anion-exchange properties in acid media, and cation-exchange properties in basic media. Each of these salts has its own isoelectric point. Silica and alumina have also been found to have selectivity for certain ions (e.g. Zr, Pu and U).

Among acid salts, the most widely used is zirconium phosphate. It is precipitated by adding a solution of zirconium chloride in aqueous hydrochloric acid to phosphoric acid. The composition and ion-exchange properties of the product depend on the method of preparation and drying. The rapidity of exchange increases with the amount of water present, but a higher temperature of drying increases selectivity. The capacity increases with the P/Zr ratio, but there appears to be an upper limit of around 2. Quantitative separations of alkali and alkaline earth metals and of uranium and plutonium have been achieved using this material.

Analogous salts such as antimonates, arsenates, arsenophosphates, molybdates, tungstates, and vanadoarsenates of zirconium, titanium, tin, antimony, bismuth, thorium and nickel have also been used with varying degrees of selectivities for different ions. Salts of heteropolyacids, among which ammonium phosphomolybdate is the best known, also have useful ion-exchange properties. The cation-exchange properties of ammonium phosphomolybdate depend on the ability of other ions to replace the ammonium ions; the material is especially selective for the Cs ion. The phosphorus in the molecule may be replaced by arsenic or silicon, and the molybdenum, wholly or partly, by tungsten [77].

#### *6.6.2.4 Liquid ion exchangers*

Organic bases, such as amines, readily form addition salts (ammonium salts) on reaction with acids. Depending on the size of the addition complex, it may be soluble in either water or an organic solvent. Thus, a trialkylamine containing a total of 18–27 carbon atoms can extract even strong acids such as nitric acid into an organic phase. These anions of the acid can exchange with other anions in the aqueous phase and can, therefore, be used for preferential extraction of the ions (extraction chromatography [27]). The amines that can be used are primary, secondary or tertiary; quaternary ammonium salts are generally less desirable in view of their tendency to form emulsions when the two layers are mixed.

Acids that are soluble in organic solvents but not in water can be used as cation exchangers on the same principle. The monobasic phosphoric and phosphonic acid esters are more useful; e.g. $(C_8H_{17})PO(OC_8H_{17})OH$. The liquid ion-exchangers mentioned above are useful for effecting partition of ions between two immiscible phases. Their use (either as such, or dissolved in a non-polar, non-volatile liquid) in partition chromatography has been referred to earlier.

To be useful as liquid ion-exchangers, the compounds must have high solubility in non-polar organic solvents and very low solubility in aqueous media. Low molecular weight amines, carboxylic acids and sulphonic acids are, therefore, not suitable. Since most liquid ion exchangers are monofunctional, the high molecular weight of these compounds would mean low specific exchange capacity. The advantage of the liquid ion-exchangers, on the other hand, is their high selectivity, fast exchange rates, permselectivity (i.e. selective permeability to a certain type of ion) and the facility of a true countercurrent process. These features are extensively used in the preparation of ion-selective electrodes and in the recovery of uranium from low-grade ores.

#### *6.6.2.5 Silica-based ion exchangers*

Two types of silica-based ion exchangers are in common use, mostly in high-performance IEC. Pellicular ion exchangers consist of small glass particles, about 40 μm in diameter, coated with an ion exchanger having a cross-linked polystyrene backbone. The other variety is similar to bonded-phase microparticulate silica described earlier; the particle diameter may be around 10 μm. Alkyl- and aralkyl-substituted chlorosilanes are reacted with silica gel and the side-chain is subsequently modified to carry either a sulphonic acid or quaternary amino group. Aminopropyl-bonded silica acts as a weak anion exchanger.

Compared to the cross-linked polystyrene gels, the silica-based materials have much less exchange capacity. The exchange capacities of pellicular silica, bonded silica and polystyrene gels are, respectively, of the order: 0.01–0.1, 0.5–2 and 3–5 meq/g. In addition to the low exchange capacity, the restricted operating pH range (2–8) of silica-based ion exchange is a major drawback. However, their stability to high pressures and very fast exchange rates are considered advantageous in HPLC.

### 6.6.3 The practice of ion-exchange chromatography

#### *6.6.3.1 The mobile phase*

The role of the solvent in IEC is limited to providing a suitable vehicle for the salts and buffers which control the retention and elution of the substances being chromatographed. As such, water is invariably the major constituent of the mobile phase. In most cases, the ionic strength of the mobile phase determines the retention on ion-exchange columns. The $k$ value for monovalent ions is inversely proportional to the concentration of the salt or, strictly, to that of the ion of the same charge. This is understandable as the ions of the same charge in the mobile phase compete with the solute ions for retention on the column. The ionic strength of the mobile phase can be increased by either increasing the buffer concentration or by adding a salt such as sodium nitrate. It should be noted that the nature of the co-ion plays a role in competitive retention of the solute ions. At a given concentration of the various ions, the strength of the mobile phase decreases in the order: citrate > sulphate > oxalate > iodide > nitrate > arsenate > phosphate > molybdate > chromate > bromide > thiocyanate > chloride > formate > acetate > hydroxide > fluoride. In cation exchange, the solvent strength depends on the nature of the cation and decreases in the order: Ba(II) > Pb(II) > Sr(II) > Ca(II) > Ni(II) > Cu(II) > Co(II) > Zn(II) > Mg(II) > Ag(I) > Tl(I) > Cs(I) > Rb(I) > K(I) > $NH_4^+$ > Na(I) > H(I) > Li(I). As noted earlier, retention of the ions under the given set of conditions is in the reverse order; tervalent ions are more strongly retained than the bivalent ions and so on. Occasionally, the nature of the co-ion also contributes to selectivity of separation.

The pH of the mobile phase plays a very important part in sample retention. It is well known that change of pH over the isoelectric point would change the ionic form of zwitterionic species, such as amino acids. Ionization of the weak acids is increased by increasing the pH and that of weak bases by decreasing the pH; the retention on the appropriate stationary phase also changes accordingly; elution of the ions is facilitated by changing the pH in the opposite direction.

Mineral acids, such as halogen acids and sulphuric acid form complexes with metal ions with a high degree of selectivity. These complexes are often in the neutral or anionic form and are thus rapidly eluted from cation-exchange resins. Thus, Hg(II), Bi(II), Cd(II), Zn(II) and Pb(II) form bromide complexes and are eluted in that order when 0.1–0.6 M HBr is used as the mobile phase; other metal ions that do not form such complexes are retained on the column. Al(III), Mo(VI), Nb(V), Sn(IV), Ta(V), U(VI), W(VI) and Zr(IV) form anionic fluoride complexes and are rapidly eluted from a cation exchange column in the hydrogen form using 0.1–0.2 M hydrofluoric acid. Selective elution of Nb(V), Ta(V), Mo(VI), W(VI) and V(V) is also possible using a dilute solution of hydrogen peroxide which forms complexes with these metal ions. Tartrate complexes of Sb(V), Mo(VI), Ta(V), Sn(IV) and W(VI) are formed in a 0.1 M tartaric acid and 0.01 M nitric acid mixture, while Pb(II) and several other metal ions do not react and are thus retained. Most of the above complexation reactions are used for group separations rather than for resolution of the individual metals, so that only short columns are needed. Use of 2-hydroxyisobutyric acid solution as the mobile phase for the separation of the individual rare earth ions is an example of the use of organic complexing agents to achieve selectivity.

The scope of ion-exchange separations can be increased tremendously by modifying the mobile phase by addition of an organic solvent. For example, the blue anionic cobalt(II) chloride is formed in acetone with very dilute aqueous HCl while, in the absence of acetone, 4–5 M HCl may be needed. Similarly, Zn(II), Fe(III), Co(II), Cu(II) and Mn(II) can be eluted one after another at the same concentration of HCl by increasing the proportion of acetone from 40–95%. Use of organic modifiers in IEC is now a common practice. Up to 10% of a water-miscible organic solvent is sometimes added to increase selectivity; solvents generally used for this purpose are methanol, ethanol, acetonitrile and dioxan.

#### *6.6.3.2 Experimental techniques*

The choice between cation and anion exchanger is clear-cut when the solute is of a particular charge type. Zwitterionic compounds such as amino acids and peptides, on the other hand, change their ionic characteristics with pH. At a particular pH, called the isoelectric point (pI), they have no net charge and, as such, are not retained by any type of ion exchanger. The pI values of a large number of peptides and proteins are on record [78]. Below the pI, the protein has a net positive charge and can be adsorbed on a cation-exchange resin. Similarly, an anion-exchange resin can retain the protein above its pI. Thus, in priniciple, one could use either a cation or an anion exchanger by selecting a suitable pH. But, in practice, many biological materials are denatured outside a certain pH range. For example, if a protein is stable in the pH range 5–8, and its isoelectric point is 5, an anion exchanger has to be used.

Strong ion-exchangers retain their charge over a wide pH range and are therefore preferred in the case of weakly ionic solutes which require extreme pH conditions for ionization and retention. On the other hand, weak ion-exchangers are to be chosen while separating highly ionic and very labile compounds. Sephadex ion exchangers, which have good capacities at high ionic strengths, may be used when handling molecules which are unstable at low ionic strengths.

Ion exchange can be carried out either batch-wise or by the column chromatographic method; the latter is more common and convenient for resolution of complex mixtures. The range of stationary phases available and their specificities have been discussed above. Polystyrene resins are widely used for a variety of applications, but the carbohydrate-derived packings, in view of their high porosity, are preferable for biopolymers such as proteins and nucleic acids. The silica-based IEC packings are preferred in HPLC, obviously for their stability to high pressures and faster exchange rates.

IEC is generally performed at temperatures ranging from ambient to about 85°C. A rise in temperature decreases both viscosity of the mobile phase and the diffusion coefficient of the solute in the mobile phase; the result is general reduction in the retention and increase in efficiency. Operation at 50–60°C is often advantageous. A change in temperature also frequently results in a dramatic change in the separation selectivity in IEC.

The recommended sample size is 1–5% of the total exchange capacity of the column. It is recommended that the sample volume be less than a third of the volume of the first eluted peak to prevent excessive band broadening; when this is difficult to achieve, the solvent strength can be decreased, or the pH varied, to increase retention.

The length of the column is not very critical in IEC and is generally shorter than it is in the other methods. Column packing is also relatively easier. Before packing the column, the gel is equilibrated with the starting buffer on the requisite ionic strength. The pH and the ionic strength are chosen to facilitate retention at the time of sample application and to obtain adequate resolution at the time of elution. The starting pH is maintained roughly one pH unit above the pI for anion-exchange chromatography, and one unit below for cation exchange. The adsorbed solute may be eluted from the column by bringing the pH of the mobile phase close to the pI. At a given pH, low ionic strength of the mobile phase favours retention and high ionic strength causes elution. Use of ionic strength and pH gradients may often be advantageous. However, many separations can be achieved isocratically (i.e. without changing the mobile-phase composition) by optimizing the buffer concentration and the pH.

#### *6.6.3.3 Suppressor ion chromatography (SIC)*

In the chromatography of ions, it is reasonable to assume that conductometry could be used to monitor the effluents. However, advantage could not be taken in IEC of this apparently universal property of ions, until recently, because of the high ionic strength of the mobile phase and the consequent high background conductivity of the eluent itself. Stevens *et al.* [79] developed the technique of ion suppression, permitting sensitive detection of solute ions by conductometry. In this technique, a low-capacity (typically in the range of 0.02–0.05 meq/g) and high-efficiency pellicular ion exchanger is used for the separation of the solute ions. The effluent is then passed through a second, high-capacity ion-exchange column to alter the mobile-phase ions and thus suppress its conductivity.

When the separator is an anion exchanger, the suppressor must be a cation exchanger. To avoid frequent regeneration of the column, the suppressor column must be of high capacity and with a large bed volume. However, the ratio of the bed volumes of the suppressor and the separator columns should be as small as possible so that the separation efficiency of the analyser column is not decreased by zone spreading.

The chemistry of ion suppression is simple to understand. For example, a mixture of anions, say as potassium salts, loaded on to an anion-exchange column may be eluted with aqueous sodium hydroxide and then passed through a suppressor column packed with a sulphonic acid resin. Both potassium and sodium ions in the effluent are exchanged for hydrogen ions so that the eluent contains only the conjugate acids of the separated anions. The sodium hydroxide in the mobile phase is completely converted into water which is devoid of any conductivity, and the conductivity of the separated acids is monitored. However, in actual practice, the hydroxide ion has very poor eluting power and normally a mixture of sodium carbonate and bicarbonate is used. The advantages are its non-toxicity, low conductivity of the suppressed product (carbonic acid) and the selectivity which can be easily achieved by varying the bicarbonate/carbonate ratio. Sodium phenoxide is another popular reagent. When sodium halides are used for elution, the halide ions may be removed from the effluent by precipitation as the silver salts.

Mineral acids are usually used as mobile phases in the analysis of cations so that after their anions are exchanged for the OH ions in the suppressor column, the background conductivity is zero. High mineral acid concentration may be required to elute certain strongly bound cations; but this may be difficult in view of their corrosive nature and short suppressor life. A mineral acid salt of an organic base (e.g. *m*-phenylenediamine dihydrochloride) may be used in such cases; the free amine liberated by the suppressor column is absorbed by the porous polymer, rendering the effluent nearly neutral.

The need for periodic regeneration of the suppressor column is obviated by use of a hollow-fibre suppressor made of a sulphonated polyethylene tube [79]. The fibre wall is permeable to only cations, anions being excluded by the Donnan exclusion principle (i.e. impermeability of the resin of a certain charge on the fixed ion, to ions of the same charge). The apparatus consists of a tube, packed with a bundle of eight low-density fibre tubes. A dilute solution (0.02 N) of sulphuric acid is passed through the tube so that its hydrogen ions can exchange with the cations of the mobile phase. When the mobile phase is sodium carbonate solution, this process yields carbonic acid of low conductivity. The solute anions similarly yield the corresponding acids. The sulphuric acid is converted into sodium sulphate which is replaced. The hollow-fibre (or the membrane) technique can be used for indefinite periods without loss of effectiveness since the suppressor is never exhausted. Several advantages, such as improved baseline stability, are claimed but a major disadvantage of the early membrane technology is the large dead volume and the consequent poverty of resolution. Because of the poor efficiency of the transport across the fibre wall or the membrane, incomplete conversion of the eluent to the protonated form is another drawback.

Mass transport across the membrane is improved by using narrower inner tubes, or by the turbulence generated when the tubes are packed with inert (plastic) beads of diameter about 60% of the inner diameter of the tube ($d < 2$ mm) [80]. Smaller beads may reduce the void volume further but would decrease the radial transport and also increase the head pressure causing bulging or even rupture of the tubes. Packing both inside and outside the tubes with the beads would render the system more stable. Alternatively, a nylon filament may be inserted along the length of the tube to achieve the same effects [81]. Turbulence is increased by coiling the fibres along with the filament inside. The efficiency of the suppressor system is improved further in the ultra-high-capacity AMMS unit of the Dionex Corporation (Sunnyvale, U.S.A.). The AMMS unit is constructed using two membrane sheets with the regenerant flowing on both sides. The very efficient ion suppression that is possible on this system allows even gradient elution which is otherwise uncommon in ion chromatography.

A wide variety of inorganic and organic ions have been analysed by the above-described technique of suppressed ion chromatography [73]. Depending on the complexity of the sample, different columns, eluents and even additional mechanisms like ion exclusion and ion exchange may be combined. Relatively few interferences from accompanying ions are observed; however, when a large excess of metal ions are present, they are advantageously removed prior to analysis to avoid overloading of the separator column, or their precipitation as hydroxides or carbonates.

#### *6.6.3.4 Single-column ion chromatography (SCIC)*

While the suppressed-ion chromatographic methods described above have been extensively used, the requirement of a second suppressor column severely limits the types of eluents that can be used and also adds to the complexity of instrumentation. The common ion-exchange resins are generally of very high capacity and require high salt concentration so that, unless a suppressor system is used, the background conductivity will be too high for sample ion monitoring. However, use of an ion-exchange column of very low capacity (say, 0.007–0.04 meq/g) and an eluent of low conductivity should permit direct detection of the effluent ion by conductometry

[82]. Very dilute solutions (0.0001–0.0005 M) of sodium, potassium or ammonium salts of benzoic, phthalic or *o*-sulphobenzoic acid were used as the eluent to achieve very sharp separation of several common ions.

The carboxylate anions mentioned above have very high affinity toward the resin. This feature permits the use of very dilute solutions of the carboxylate salts (or the acids) and, therefore, the background signal is small. The retention times of ions and the selectivity may be changed by using mixtures of mono- and divalent ions. Sodium borate may be used to adjust the pH of the eluent containing a carboxylic acid, or sodium borate itself may be used for elution. The gluconate–borate anion complex has a low conductivity and is an effective eluent. The citrate ion, with a charge of '3 − ', may be expected to be an effective eluent but the sensitivity of detection of sample anions in its presence is relatively poor.

The sensitivity of IC increases as the eluent dissociation decreases; however, the eluting power of the mobile phase also decreases correspondingly. In such cases, it may be advantageous to increase the eluent concentration to increase the eluting power without substantially increasing the conductivity. The background conductance of sodium salts of weak acids is substantially lower than the corresponding potassium salts and hence may be preferred. Extension of the argument should make lithium salts even better but no data is available. The background conductance may be further reduced by using an additional cation exchange suppressor column, this is described in the previous section. As may be expected, the sensitivity of detection of anions of weak acids is much lower than those of strong acids which are completely dissociated and whose conductance is much higher. The sensitivity also depends on the eluting ion. Thus, the detectability of acetate ions (relative to the chloride ion as 1.00) in the background of succinic, benzoic and nicotinic acids is 0.15, 0.09 and 0.03, respectively. In absolute terms, and for comparison, the chloride ion can be detected at a level of 5 ppb in 1mM nicotinic acid solution and at 26 ppb in succinic acid. For a detailed description of the technique of SCIC, see Ref. 73.

### 6.6.4 Ion-exclusion chromatography

As noted earlier, ion-exchange resins have also been used for the separation of non-ionic or weakly ionic compounds, taking advantage of the extra-ion-exchange properties of the resins. Here, the use of ion-exchange resins is essentially based on their swelling properties. An important mechanism that operates is the so-called Donnan exclusion. While ions of opposite charge to that of the fixed ions are attracted (or exchanged for) by electrostatic forces, those of the same charge type are repelled and thus excluded. On the other hand, non-ionic or weakly ionic compounds can penetrate the pore structure of the resin and are subject to a combination of steric exclusion, adsorption and partition effects. The technique is variously described as *ion-exclusion chromatography*, *Donnan-exclusion chromatography*, *ion-moderated partition chromatography* or *salting-out elution chromatography*; the last-mentioned term is used especially when high concentrations of salt solutions are employed for elution.

Ion-exclusion chromatography has been used for separation of strong acids as a class from weak acids, as well as for obtaining complete separation of a variety of acids [83]. The anions of strong acids are repelled by the fixed ions on a cation exchanger and are quickly eluted. Weaker acids, on the other hand, are partitioned between the water occluded in the resin and that in the interstitial mobile phase. When a cation-exchange resin in the H-form, and water as mobile phase, were used, elution was in the order of the pK values of the solutes. When ionization of the weak acids is suppressed by using a dilute mineral acid as mobile phase, the order was independent of the pK values indicating that mechanisms other than ion exclusion are also operating. In a homologous series, the elution order follows the acid strength and aqueous solubility; i.e. formic acid, followed by acetic acid, propionic acid, etc. Branched-chain acids elute before the normal acids while unsaturated and aromatic acids are retained more strongly. Dicarboxylic acids are less retained relative to monobasic acids. As may be expected, amines are separable on a strong anion-exchange resin [84].

The partition effects, and thus the resolution, can be greatly enhanced by addition of an organic modifier or salt to the aqueous phase. Ethers, carbonyl compounds, esters and amines have been separated by this method.

A 7%-cross-linked polystyrene-DVB sulphonic acid resin in the Ca(II) form was used for the resolution of polyols and carbohydrates using pure water at 85°C ; selectivity was shown to depend on the nature of the metal ion, suggesting the involvement of a ligand-exchange mechanism. Oligosaccharides are eluted in decreasing order of molecular size. Thus, a powerful combination of ligand-exchange, partition and size-exclusion effects operate in these cases [78].

## 6.7 AFFINITY CHROMATOGRAPHY

### 6.7.1 Introduction

Strictly speaking, all chromatographic separations occur because of the differences in relative affinities of different solutes towards the sorbent and the solvent. The affinity could be physical, chemical or biospecific. Strictly, again, biospecific interactions at the molecular level may be categorized as physicochemical. However, by convention, separation or purification of biomolecules on the basis of their biological function is referred to as affinity chromatography; terms such as bioaffinity chromatography and biospecific adsorption chromatography, though probably more accurate, have not become very popular.

It is well known that individual members of several classes of biological macromolecules such as enzymes, immunoglobulins, nucleic acids, carbohydrates, etc., have closely related chemical structures and are with difficulty, if ever, separable by the traditional chromatographic methods discussed in the foregoing sections. In affinity chromatography, advantage is taken of the fact that these individual entities have highly specific natural affinities for certain, often small, molecules or substrates. In principle, of the two (or more) biologically interacting molecules, such as enzyme–substrate, antibody–antigen or receptor–hormone, one is bound (often covalently) to a solid matrix and this material is used for specific and reversible adsorption of the complementary molecule. Affinity chromatography can thus be used for specifically separating certain compounds from complex mixtures of biological matcrials, for isolating small quantities of biomolecules from large amounts of inactive substances, or even for separating the native from the denatured form of the same substance. The requirement, however, is that a biospecific *ligand* (or *affinant*) is available and that it can be covalently bound to a solid *matrix* without impairing its biospecific binding ability. A convenient, non-destructive method of eluting the adsorbed biomolecule should also be available.

The essential steps and sequence of affinity chromatography are summarized in Fig. 6.22. After the appropriate matrix is selected (see below) its functional groups

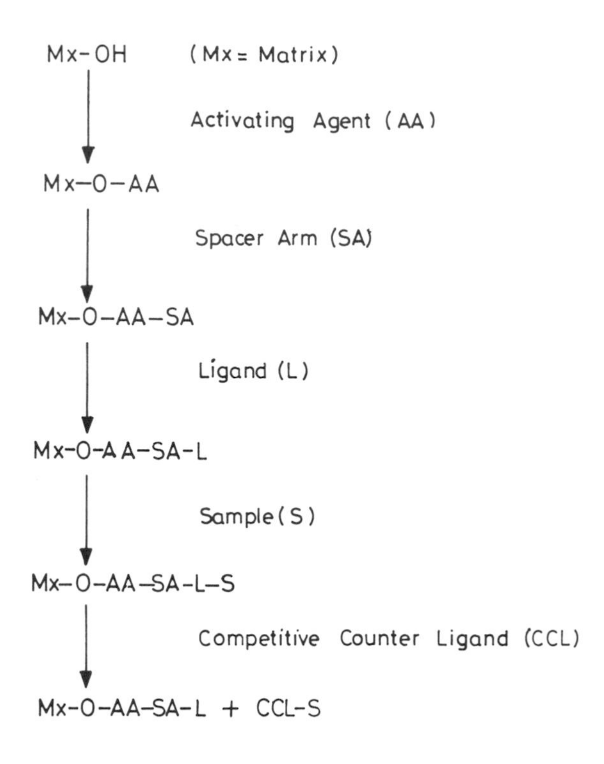


Fig. 6.22 — Basic features of affinity chromatography.

are activated for further coupling. The activated matrix can be directly reacted with the appropriate ligand but, usually, a short hydrophilic or hydrophobic chain (*spacer arm*) is introduced. This serves to keep the orientation of the ligand flexible enough to allow unhindered approach of the biomolecule to be chromatographed. In the absence of a spacer arm, the ligand is too close to the solid surface of the matrix and steric crowding around the ligand may hinder the approach of the biomolecule, particularly if the latter's active site is deep-seated. The spacer may have an in-built reactive group for further coupling to the ligand which must also have an anchor for covalently binding to the spacer.

After coupling the ligand to the matrix, the material is ready for use in chromatography. It is then brought into contact with the material to be chromatographed and the unbound material is washed off. The affinity-bound substance is then released or eluted by one of the several techniques available for the purpose. Naturally, there are several possible variations in each of these steps and these are discussed in greater detail below. The subject is complicated by several interfering mechanisms that play a significant role, and each of the variations in either the matrix, spacer arm, ligand or solute requires a unique manipulation. The discussion that follows emphasizes the chemical principles; the experimental procedures are well described in a booklet from Pharmacia AB, Sweden [85]. The applications, which are too numerous and varied to be elaborated here, are briefly mentioned at the end of this section. For further details, readers are referred to some of the excellent monographs and reviews on the subject [86–96].

### 6.7.2 The matrix

The matrix in affinity chromatography has a very important role, while being inert. Whenever a sample solution comes into contact with a solid bed, several non-specific interactions such as adsorption, partition, ion exchange and size exclusion could operate. Affinity chromatography, by its very nature, requires that these interactions be minimal. A majority of biomolecules being large, size exclusion is one of the main limitations posed by most materials of potential use in affinity chromatography. Certain structural (chemical) features of the solid matrix can also interfere with affinity chromatography, as in the case of some polyglucanases for which the glucose units of the dextran matrix could be a substrate.

The success of affinity chromatography depends on the ability of the chromatographic system to simulate the conditions of biological interaction, a majority of which occur in solution. The biomolecule should have free access to the polymer-bound ligand to enable it to assume the proper orientation for interaction and this requires that the matrix should have a loose and porous structure. On the other hand, the ideal matrix should also be rigid and stable to the conditions of pH, temperature and ionic strength of the medium during the adsorption of the solute and its elution. Contradictory to the requirement of inertness is the requirement that the matrix should carry sufficient concentration of active functional groups which can be used to covalently bind the ligand, either directly or through a spacer. Unfortunately, none of the matrices currently in use satisfies all the desirable characteristics. A variety of matrices, based on carbohydrates, polyacrylamides and porous silica, have been examined for suitability in affinity chromatography but only a few of them are in general use.

Cellulose and its derivatives, particularly carboxymethylcellulose, were very popular matrices in the early stages of development of affinity chromatography. Their hydrophilicity, rigidity and amenability to derivatization were considered advantageous. However, its fibrous structure restricts penetration of large molecules and its non-uniform character causes microheterogeneity of ligand density, leading to undesirable variations in the capacity of the adsorbent. The hydrophobic nature and the predominance of adsorption and partition effects render the polystyrene gels unsuitable for use in affinity chromatography. Other hydrophilic gels, such as cross-linked dextran gels, have been used with some success but a serious limitation in these cases is their low porosity and the consequent steric exclusion effect. Polyacrylamide gels, which are available in the form of spherical beads of uniform porosity, have many of the desirable features noted above, but suffer from shrinkage during chemical modification, large bed-volume changes with solvent gradients and low permeability to macromolecules.

Agarose (a polymer of D-galactose and 3,6-anhydro-L-galactose, see Fig. 6.8) is the most widely used matrix in affinity chromatography. As noted earlier, bead-formed agarose gels containing 2, 4 and 6% agarose are marketed by Pharmacia AB of Sweden under the tradename Sepharose 2B, 4B and 6B. The agarose gels have many of the desirable properties of high permeability, amenability to derivatization, hydrophilicity, absence of charged groups and low non-specific adsorption; however, they break down easily at high temperatures, or on contact with a high concentration of the eluent containing urea and guanidine-HCl which are capable of cleaving hydrogen bonds. They also have a narrow usable pH range (4–9) and cannot withstand pressures over 1 psi. The 2,3-dibromopropanol-cross-linked version, Sepharose CL-6B can tolerate a pH range of 2–12 and pressures up to 6 psi.

Magnogel from LKB contains polyacrylamide, agarose and iron oxide. The iron oxide permits the use of a magnetic field to collect, retain and wash the resin without having to filter or centrifuge and is thus very convenient to handle. Several newer products, typically around 40 μm in diameter, can withstand moderate pressures and may be considered medium performance supports, intermediate between the agarose gels mentioned above and the high-performance, silica-based matrices to be discussed below. Pierce Chemical Co., U.S.A., market Fractogel TSK (a hydrophilic vinyl polymer, stable in the pH range 1–14 and to pressures up to 100 psi) and Trisacryl GF (a hydrophilic polyacrylamide which can tolerate pH 1–13 and pressures up to 40 psi) with several ligands attached.

As noted earlier, porous silica can be chemically modified in a number of ways and is a natural candidate for use in affinity chromatography. This material is rigid, insoluble and stable to solvents, high pressures (over 1000 psi), flow-rates, pH (2–8), ionic strength, temperature and microbial attack. It can, therefore, be reused several times without significant loss of efficiency. Its physical stability and other desirable properties make it the preferred matrix for the modern, instrumental (high-performance) affinity chromatography [93]. Uniform and precisely controlled pore size results in sharp exclusion limits, resolution and reproducibility. Commercial porous silica may have pore sizes ranging from 60–4000 Å. The normal HPLC packings have pore sizes in the region of 60–100 Å. The solutes in affinity chromatography being, relatively, much larger, the preferred pore size is from 300–1000 Å. Bonded-phase controlled-porosity glass is available from Pierce Chemical Co., Serva and J. T. Baker companies, U.S.A. Beckmann's Ultraffinity EP is 10 μm bonded-phase silica of 300 Å pore size and carries epoxide groups (3-glycidoxypropyl trimethoxysilane) for attachment of the ligands. The epoxide groups can be used directly for coupling the ligands or hydrolysed by acid to give the diol. The diol on periodate oxidation yields the aldehyde which can react with amines to form Schiff's bases, which are then reduced by sodium borohydride to give rise to the stable C−N bond. Several simple reactions are available for the coupling of the affinity ligands to silica (see Fig. 6.13).

Some of the newer supports include very small non-porous particles of diameters 0.7 and 1.5 μm with moderate surface areas and greatly enhanced mass-transfer properties, cross-linked agarose gels of particle size 3–10 μm that can withstand pressures of several hundred psi; Superose from Pharmacia is a size-exclusion support (pressure limit 200 psi) which can also be used as an affinity matrix. Indeed, most of the silica-based aqueous size-exclusion supports can also be used as matrices for coupling ligands for high-performance affinity chromatography [96].

### 6.7.3 Coupling techniques

The methods used for coupling the ligand to the matrix naturally depend on the nature of the functional groups on the ligand and the matrix. The functionalities most commonly found on the ligand are amino, carboxyl, hydroxyl and thiol groups. The groups on the matrix that normally take part in the coupling include the hydroxyl groups on cellulose, dextran and agarose, carboxyl in CM-cellulose, silanol groups in silica and the amide groups in polyacrylamide. These groups are not active enough for direct coupling of the ligand or the spacer and, therefore, need to be modified. In other words, the coupling of the ligand to the matrix is a two-step process, the first being activation of the functional group on the matrix using a reagent either in an aqueous or organic medium. After washing off the excess reagent, the reaction between the activated matrix and the ligand is normally performed in an aqueous medium. Activation of the matrix functional groups, attachment of the spacer arm where necessary, and covalent bonding of the ligand to it, involve deceptively simple chemistry. Each of the manipulations requires skill and careful understanding of the principles in order to avoid complications that could occur. Most complications occur because of the incompatibility of the relevant functional groups on the matrix, the spacer and the ligand, the length of the spacer, the pH of the medium that may affect the stability of the molecules or the molecular interactions, and the concentration or the density of the active groups on the surface.

The ideal activation reaction should give the coupled product in a high yield at neutral pH, and should be mild enough not to cause denaturation of the biomolecule. The bonding between the matrix and the spacer (if any), and that between the spacer and the ligand must be strong enough to survive the conditions and affinity retention and elution. In most cases involving the binding of macromolecular ligands, it is possible to involve multiple bonding of the ligand so as to enhance the stability of the binding. However, the resulting distortion of the molecular geometry of the ligand and steric hindrance would greatly reduce its ability to biospecifically interact with the solute molecules. As in the case of the matrix, the ideal coupling reaction or reagent is still to be developed. Indeed, as has been noted earlier, the affinity chromatographic principle was known for more than half a century earlier but could not be used for want of a satisfactory coupling reaction until Porath developed the cyanogen bromide method.

The most commonly used method of activating the hydroxyl groups continues to be the reaction with cyanogen bromide (CNBr) under basic conditions. Cyanogen bromide reacts with hydroxylic compounds to give the very reactive cyanate ester (–OCN), which in the absence of other reactive substrates, may react with a neighbouring hydroxyl group in a polyhydroxy compound to give a cyclic imidocarbonate (−O−C(=NH)−O−), or with moisture to form inactive carbamates. CNBr-activated agarose is commercially available in the form of a freeze-dried material (CNBr-activated Sepharose 4B) for ready use [85]. These activated gels react with amines to give iso-urea compounds (−O−C(=NH)−NHR), which may get partially hydrolysed to substituted carbamates, −O−CO−NHR.

Originally, the CNBr-activation of the carbohydrate matrices required treatment of the matrix with cyanogen bromide and a base like sodium hydroxide, sodium carbonate or, preferably, triethylamine. Handling of the toxic CNBr gas may be avoided by the use of activation agents based on CNBr, such as *p*-nitrophenyl cyanate, *N*-cyanotriethylammonium tetrafluoroborate and 1-cyano-4-dimethylaminopyridinium tetrafluoroborate.

Determination of the capacity of the activated matrix is crucial for the success of the subsequent affinity chromatography. Among the two active species formed by the CNBr-activation, namely, the cyanate ester and the imidocarbonate, the former is more reactive and stable at $pH < 4$. The imidocarbonates are more stable under basic conditions. In the freshly prepared activated agarose only about 65–80% of the groups may be available for the subsequent coupling reaction, with the two active groups in the ratio of 4:1. The imidocarbonate group is determined by acid hydrolysis and estimation of the liberated ammonia by the ninhydrin method. The cyanate ester is determined by the purple colour generated when the activated ester is treated with barbiturates in pyridine [95].

A major drawback in the direct coupling of single units of ligands or spacer-bound ligands to the CNBr-activated polymers is the leakage problem; i.e. part of the ligand keeps dissociating from the polymer during subsequent manipulations. Part of this leakage may be due to the release of adsorbed material rather than the dissociation of covalently bound ligand. Extensive washing of the gel with buffers of different pH may remove such material. Covalently bound ligands may also dissociate because of the instability of the iso-urea bonding, particularly in the presence of nucleophiles. This problem is overcome to a large extent by multiple attachment of the CNBr-activated polymer to the amino or hydrazo groups of polylysine or polyacrylic acid hydrazide, respectively. The remaining amino and hydrazo groups of these chains are useful as anchors for further coupling reactions as discussed below. As a result of the multiple attachment and the large number of free amino and hydrazo groups, both the stability and the capacity of the matrix is increased.

A large number of newer methods have been developed to overcome the drawbacks inherent in the CNBr-activation method. These methods include activation by sulphonyl chlorides (e.g. *p*-toluenesulphonyl or tosyl chloride and 2,2,2-trifluoroethanesulphonyl or tresyl chloride), 1,1′-carbonyldi-imidazole and chloroformates such as *p*-nitrophenyl-chloroformate, *N*-hydroxysuccinimidochloroformate and tri- and pentachlorophenylchloroformate.

A different approach to the solution of the leakage problem and the use of the hazardous CNBr involves modification of the agarose (or other polyalcohol matrices like cellulose) by periodate oxidation to aldehydes. The aldehydes react with amines to form Schiff's bases which can be reduced to the stable amino or aminoacyl linkages using sodium borohydride or, better, sodium cyanoborohydride.

Several bifunctional reagents such as dichloro-*s*-triazines, divinylsulphones and bis-oxiranes, and several others, have been used for coupling the matrix to the ligand. Advantage is taken of the presence of two functional groups in these reagents, one of which reacts with the hydroxyl group on the matrix while the other is available for reaction with an active (usually, hydroxy or amino) group of the ligand. However, a disadvantage of such functional groups is the occurrence of considerable cross-linking of the gel, which decreases the permeability. The most commonly used activation and coupling methods for hydroxyl-containing matrices are shown in Fig. 6.23.

Apart from agarose, cellulose and dextran, which contain hydroxyl functionality,

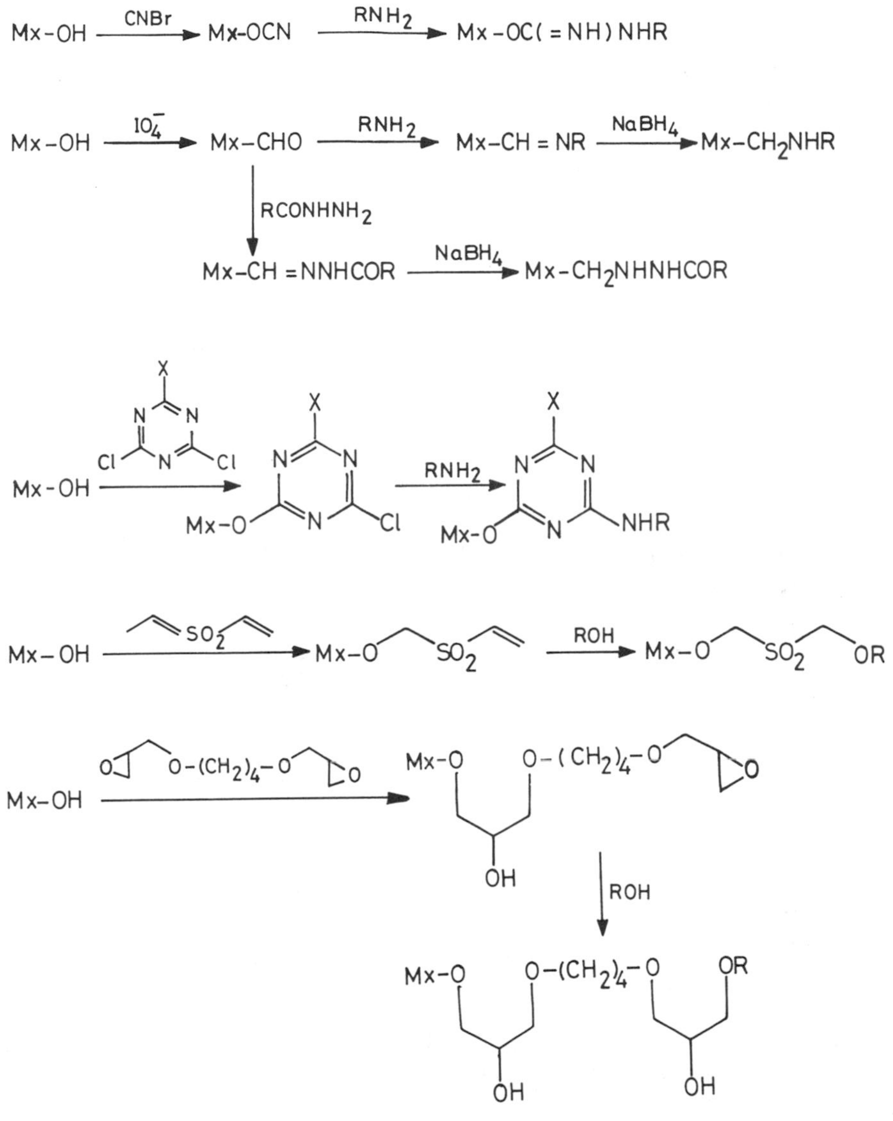

$$\text{Mx-OH} \xrightarrow{\text{CNBr}} \text{Mx-OCN} \xrightarrow{\text{RNH}_2} \text{Mx-OC(=NH)NHR}$$

$$\text{Mx-OH} \xrightarrow{\text{IO}_4^-} \text{Mx-CHO} \xrightarrow{\text{RNH}_2} \text{Mx-CH=NR} \xrightarrow{\text{NaBH}_4} \text{Mx-CH}_2\text{NHR}$$

$$\text{Mx-CHO} \xrightarrow{\text{RCONHNH}_2} \text{Mx-CH=NNHCOR} \xrightarrow{\text{NaBH}_4} \text{Mx-CH}_2\text{NHNHCOR}$$

Fig. 6.23 — Activation of hydroxyl groups for coupling of affinity ligands.

polyacrylamide and porous glass, containing amide and silanol groups, respectively are also occasionally used in affinity chromatography. The amide group of polyacrylamide can be hydrolysed to carboxylic acid, converted to hydrazide or reacted with an amine to give a substituted amide (–CONHR). These carboxyl or hydrazo groups can be coupled to ligands containing different functional groups, as shown in Fig. 6.24. Modification of the silica surface and attachment of aminopropylsilyl groups has been described earlier (Fig. 6.15); the amino functionality is very versatile and can be used in a number of ways for attachment of the ligands (Fig. 6.25).

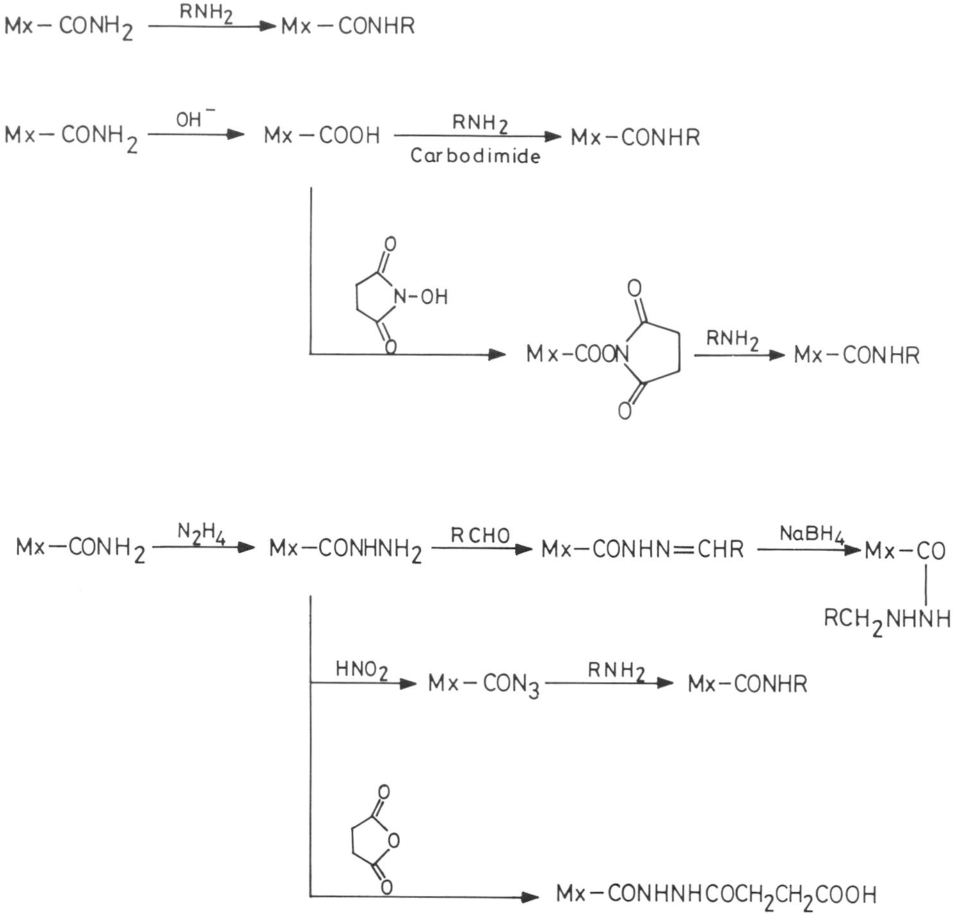

$$Mx-CONH_2 \xrightarrow{RNH_2} Mx-CONHR$$

$$Mx-CONH_2 \xrightarrow{OH^-} Mx-COOH \xrightarrow[\text{Carbodimide}]{RNH_2} Mx-CONHR$$

$$Mx-COOH \xrightarrow{\text{N-hydroxysuccinimide (N-OH)}} Mx-COON(\text{succinimide}) \xrightarrow{RNH_2} Mx-CONHR$$

$$Mx-CONH_2 \xrightarrow{N_2H_4} Mx-CONHNH_2 \xrightarrow{RCHO} Mx-CONHN{=}CHR \xrightarrow{NaBH_4} Mx-CO(RCH_2NHNH)$$

$$Mx-CONHNH_2 \xrightarrow{HNO_2} Mx-CON_3 \xrightarrow{RNH_2} Mx-CONHR$$

$$Mx-CONHNH_2 \xrightarrow{\text{succinic anhydride}} Mx-CONHNHCOCH_2CH_2COOH$$

Fig. 6.24 — Activation of amide groups for coupling of affinity ligands.

As already noted, attachment of the ligand to the matrix is often through a flexible spacer arm. The length of the spacer arm could be critical. Too short a spacer arm could be ineffective and very long carbon chains may cause non-specific hydrophobic interactions. The spacer arms usually contain two reactive groups (amino or hydroxyl, or both, separated by about six methylene groups), one of which reacts with the ligand. Though hexamethylenediamine is a common spacer, diaminopropylamine is preferable because of its less-pronounced hydrophobic properties. Other commonly employed spacers include cystamine, *p*-aminobenzoic acid and tyramine. These spacers leave an amino or carboxyl group for coupling with the ligand using any of the methods given in Figs 6.24 and 6.25. Some of the bifunctional activating groups shown in Fig. 6.23 also act as hydrophilic spacer arms. Certain spacer arms may also be first coupled to the ligand, leaving the amino end group of the spacer arm free; this is then treated with CNBr-activated polymer.

After attachment of the ligand, the remaining active groups should be blocked or

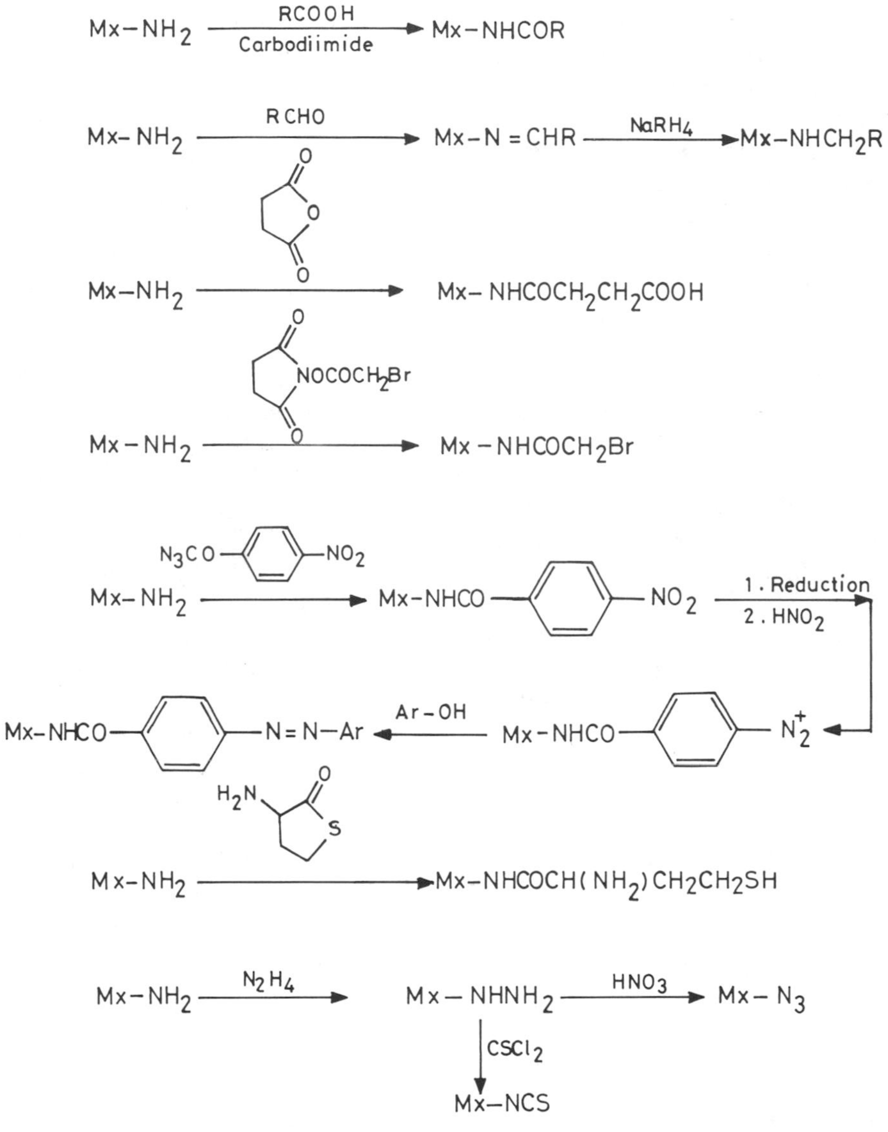

$$\text{Mx-NH}_2 \xrightarrow[\text{Carbodiimide}]{\text{RCOOH}} \text{Mx-NHCOR}$$

$$\text{Mx-NH}_2 \xrightarrow{\text{RCHO}} \text{Mx-N=CHR} \xrightarrow{\text{NaRH}_4} \text{Mx-NHCH}_2\text{R}$$

$$\text{Mx-NH}_2 \xrightarrow{\text{succinic anhydride}} \text{Mx-NHCOCH}_2\text{CH}_2\text{COOH}$$

$$\text{Mx-NH}_2 \xrightarrow{\text{succinimide-NOCOCH}_2\text{Br}} \text{Mx-NHCOCH}_2\text{Br}$$

$$\text{Mx-NH}_2 \xrightarrow{\text{N}_3\text{CO-C}_6\text{H}_4\text{-NO}_2} \text{Mx-NHCO-C}_6\text{H}_4\text{-NO}_2 \xrightarrow[\text{2. HNO}_2]{\text{1. Reduction}} \text{Mx-NHCO-C}_6\text{H}_4\text{-N}_2^+ \xrightarrow{\text{Ar-OH}} \text{Mx-NHCO-C}_6\text{H}_4\text{-N=N-Ar}$$

$$\text{Mx-NH}_2 \xrightarrow{\text{H}_2\text{N-thiolactone}} \text{Mx-NHCOCH(NH}_2\text{)CH}_2\text{CH}_2\text{SH}$$

$$\text{Mx-NH}_2 \xrightarrow{\text{N}_2\text{H}_4} \text{Mx-NHNH}_2 \xrightarrow{\text{HNO}_3} \text{Mx-N}_3$$

$$\text{Mx-NHNH}_2 \xrightarrow{\text{CSCl}_2} \text{Mx-NCS}$$

Fig. 6.25 — Activation of amino groups for coupling of affinity ligands.

deactivated to avoid undesirable chemical reactions of these groups with the biomolecules. This is often carried out by treatment with ethanolamine, tris-buffer or acetic acid [85]. In the final step, the affinity support is washed thoroughly with alternating acidic, basic and neutral buffers and equilibrated with the appropriate buffer to render the gel ready for the particular application.

### 6.7.4 The ligands

#### *6.7.4.1 Specific ligands*

Evidently, the key to success in affinity chromatography is the choice of the ligand.

To the extent possible, the ligand should be one that can specifically and reversibly bind the macromolecule of interest. For example, for an enzyme it may be the substrate, analogue, inhibitor, allosteric effector or the co-factor; for an antibody, the antigen, virus or the cell; and for nucleic acids, the complementary base sequence, histone, nucleic acid polymerase, and so on. Depending on the requirement and availability, the unique complementary molecule for specific adsorption of the particular biomolecule may be chosen. The ligand should have a functional group through which it can be coupled to the matrix without significantly affecting its affinity to the compound to be separated. In the absence of information on the structural features of the ligand–biomolecule interaction, the choice of the point of linkage of the ligand to the matrix could be empirical. High-molecular-weight ligands usually have a large number of groups through which they may be bound to the matrix without impairing the binding ability. This feature has been used to prepare several high-affinity adsorbents using immobilized proteins; the requirement, however, is that the protein should be bound to the support by the smallest number of groups so that the native conformation, and therefore the bioaffinity, remain unchanged. This can be achieved by coupling the protein to CNBr-activated agarose at low pH so that most amino groups are rendered inactive by protonation.

In most cases, attachment of a ligand to a solid surface, despite all efforts to optimize the parameters, often impairs its interaction with the macromolecules by several orders of magnitude. Owing to the limitations in attaining a high concentration of bound ligands, it has been difficult to prepare adequate adsorbents for enzyme-ligand systems whose mutual affinity in solution itself is poor. Under these circumstances, increasing the length of the column may sometimes be helpful. On the other hand, unnecessarily long columns and too high a concentration of the ligands may have a detrimental effect in high-affinity systems. In fact, in certain high-affinity systems, as in antibody–antigen interactions, it may be difficult to elute the biomolecule after affinity adsorption without extensive degradation. In such cases, it is advisable to chemically modify the ligand in order to decrease its affinity. Sometimes it may be convenient to use a substrate analogue rather than the most effective substrate, particularly when the former is more stable and readily available (see below).

#### *6.7.4.2 Group-specific adsorbents*

While highly specific ligands may be desirable and even necessary in some cases, they have several inherent disadvantages. Even when the nature of their interaction with the complementary biomolecule is known, which is in itself a rare feature, their immobilization without impairing selectivity and activity may not always be possible. Even when this is possible, the process of immobilization and optimization of chromatographic conditions on an individual basis could be very time-consuming. On the other hand, use of group-specific ligands, i.e. those which could bind a wider range of biomolecules belonging to a certain class, would be more practical and convenient. Though specificity might be lost, selectivity of adsorption may still be achieved by varying conditions such as pH, ionic strength, temperature, polarity, contact time and pore size. Separation may also be achieved by modulating the mobile phase.

Many enzymes, particularly those causing oxidations, reductions, group transfer, rearrangement and carbon–carbon bond formation, possess two binding sites, one for the substrate and the other, for the co-enzyme, which is a non-proteinous organic molecule, essential for the enzyme action. Affinity of these enzymes to two different species can be exploited by covalently binding one of these to the matrix while using the other to effect either selective adsorption or elution. The co-enzymes, which are relatively few in number, are generally immobilized for convenience and to exploit the greater affinity of many enzymes for the co-factors than for the substrates.

Several pyridine nucleotide co-factors such as nicotinamide adenine dinucleotide ($NAD^+$) have been used extensively for chromatography of complementary dehydrogenase enzymes. A number of adsorbents carrying immobilized group-specific ligands are available commercially [85]. Thus, L-lysine-Sepharose has been recommended for isolation of plasminogen, ribosomal RNA and DNA, etc. Protein A-Sepharose and immobilized lectins have been extensively used for chromatography of immunoglobulins and glycoproteins. Polyuridylic acid and polyadenylic acid-bound Sepharose find application in the separation of several nucleic-acid-dependent enzymes.

A variety of synthetic textile dyes, based on *s*-triazine and carrying anthraquinone and/or naphthalene moieties, have proved very versatile in their utility as group specific ligands. Interestingly, the protein-binding capacity of some of these dyes exceed those of the natural substrates by two to three orders of magnitude. This feature, coupled with their easy availability, low cost, stability to chemical and biological degradation, and ease of coupling to a variety of matrices, has made them very popular as affinity ligands. Additional advantages of these ligands over the biospecific ligands is their high colour value, permitting easy determination of the ligand concentration, easy identification, reusability and the ease of elution of the bound biomolecules after adsorption. Among the more popular dyes for the purpose of affinity chromatography is Cibacron Blue, a polyaromatic sulphonated dye (containing several isomeric compounds) and with selectivity to enzymes requiring nucleotide co-factors. In addition to the pyridine nucleotide-dependent oxidation-reduction enzymes, co-enzyme-A-dependent enzymes, hydrolases, phosphokinases, acetyl-, phosphoribosyl- and aminotransferases, DNA- and RNA-nucleases and polymerases, synthetases, hydroxylases, decarboxylases, sulphohydrylases, serum albumin, lipoproteins, clotting factors, complement factors, transferrin, interferon, phytochrome, etc., have been fractionated on the dye-bound matrices. Other dyes, which are also used for similar purposes, include triazine dye of the Procion series, namely, Procion Red H-E3B, Red H-8BN, Yellow MX-8G, Scarlet MX-G, Green H-4G and Brown MX-5BR. These dyes may be linked to several of the common matrices discussed above, including agarose, dextran, cellulose, metal oxides, polyacrylamide, porous silica, etc.; agarose is probably the most popular. The coupling of the ligand to the matrix may be either direct or through a spacer, but the directly coupled systems are more commonly used in view of the contradictory results obtained with the spacer-bound ligands and the simplicity of preparation of the former. Typically, the dye solution (about 1% in water), is stirred (for 30 min) with the matrix and then treated with NaCl (2%) and a base for a few hours [97].

As with the other general-purpose ligands, the effectiveness of the immobilized dyes depends on the control of operating parameters like the choice of the dye (often made after exploratory dye-screening procedures), sample size, pH of the medium (generally between 5.5 and 9), its buffer composition (5–100 mM), eluant flow-rate, ionic strength and temperature (0–50°C). Methods used to desorb the dye-retained biomolecules are similar to the other affinity chromatographic techniques; i.e. varying the pH, increasing the temperature or affinity elution.

One of the more popular group-specific ligands is heparin, which is a highly sulphated, linear polysaccharide with anticoagulant and antilipidemic properties. It has repeating units of 2-acetamido-2-deoxy-α-D-glucopyranosyl and α-D-glucopyranosyluronic acid (MW = 5,000–30,000). Being strongly acidic and highly polar, it binds a variety of proteins electrostatically. The heparin-bound protein may have either enhanced or reduced biological activity from the parent protein, but on cleavage from heparin, its original activity is restored. Heparin can be covalently linked to an agarose matrix either *via* the cyanogen bromide reaction or by reaction with cyanuric chloride (trichloro-*s*-triazine) and the matrix-bound heparin used for affinity purification of a variety of enzymes, lipoproteins, hormone receptors and blood coagulation factors (antithrombin or plasma protinase inhibitor) [94].

#### *6.7.4.3 Metal-chelate affinity chromatography*

Certain transition metals such as copper, zinc and, to a lesser extent, nickel and manganese, form stable complexes with proteins containing histidine and cysteine. This property has been used in the separation of some serum globulins by preparing modified agarose gels containing ligands which can bind these ions [44]. To prevent metal ions from getting eluted along with the proteins, the ligand–metal affinity should be stronger than the metal–protein affinity. Histidine- and cysteine-containing proteins are selectively adsorbed at pH 6–8 and desorbed without denaturization at a lower pH; at pH 7–9, the adsorption is more effective but less selective.

#### *6.7.4.4 Template chromatography*

It is well known that of the four nucleotide bases that constitute the genetic material, deoxyribonucleic acid (DNA), adenine (A) and cytosine (C) specifically interact with thymine (T) and guanine (G), respectively. The two interacting pairs (A–T and C–G) are called the complementary base pairs. This interaction occurs in nucleic acids and olignucleotides as well as single residues. Since the interaction is specific and reversible, immobilization of these materials would permit affinity resolution of complementary bases and sequences. In other words, these compounds have the same template function as DNA in biological systems and, borrowing terminology from molecular biology, this technique is called template chromatography [90,97].

Immobilization of nucleotides, nucleic acids or their fragments can be achieved in a number of ways. They may be covalently bound at multiple points or at the terminals to phosphorylated cellulose, or CNBr-activated agarose, or by UV irradiation of a solution of the material in the presence of the solid support. In the last-mentioned procedure, no covalent bonding between the solid support and the nucleotide sequence may occur; these compounds have a tendency to cross-link on irradiation, forming polymeric gels, supported on the solid matrix. Adsorption on cellulose or agarose or polymerization of acrylamide in the presence of the nucleotide may give rise to entrapped nucleic acid or nucleotide.

The technique of template chromatography has been extensively used in recent years for isolation of complementary sequences of nucleic acids, of enzymes involved in nucleic acid metabolism and for the study of the peptide-nucleotide or protein-nucleic acid interactions.

#### *6.7.4.5 Covalent chromatography*

The term covalent chromatography is probably a misnomer since no chromatography, in the strict sense of the term, is involved. In principle, the technique involves treating a compound containing a nucleophilic group (electron-pair donor) with a polymeric (insoluble) material containing a chemical moiety which is electrophilic or bears a good leaving group that can be readily replaced by the in-coming nucleophile, so that a covalent bond is formed. In other words, an insoluble derivative of the compound of interest is obtained that can be washed free of the accompanying material. The bound compound is then released from the matrix using a stronger nucleophile. Coupling of *p*-chloromercuribenzoyl group to aminoethyl-agarose gives an adsorbent with a high capacity and specificity for sulphydryl proteins (because of the chemical affinity of the sulphydryl groups to mercuric ions); solutions of cysteine and mercaptoethanol could be used as eluents. Among the other groups through which a biomolecule can be covalently retained, the most widely used is the disulphide group. The gel is modified by linking a thiol-containing alkyl group to the matrix, and is then treated with 2,2-dipyridyl disulphide (2−P−S−S−2−Py) to convert the thiol group into a disulphide (Mx−S−S−2−Py). A thiol-containing protein (ESH) can now react with the activated matrix, displacing the 2-pyridylthiol and forming the insoluble derivative, Mx−S−S−E. The enzyme (ESH) may be released by eluting with a solution containing excess of a low molecular weight thiol (RSH). Thus, no bio-affinity is involved in this technique and it can well be used for the isolation of other organic molecules except for the fact that the cost of preparation of such matrices is high and that these molecules can be conveniently separated by other simpler chromatographic methods.

### 6.7.5 Interfering mechanisms

#### *6.7.5.1 Non-specific interactions*

Apart from the biospecific retention which the technique of affinity chromatography seeks to use for the separation of bio-molecules, all the mechanisms discussed elsewhere in this chapter; namely, adsorption (electrostatic or van der Waals'), ion exchange, and size exclusion, come into operation when the solute comes into contact with the sorbent. Failure to recognize and eliminate the non-specific interactions naturally interferes with the main advantages of affinity chromatography. However, it is sometimes possible, by judicious control of the conditions, to take advantage of these interactions to supplement weak bioaffinities. But these *compound affinities* depend mainly on the gross physicochemical properties of the molecule and, therefore, methods developed for one particular enzyme are not readily applicable to homologous enzymes from another source.

Ion exchange has long been considered a major interfering mechanism because of the large number of polar groups both on the matrix and the biological macromolecules. Often, the ligands themselves are ionic and take part in ion-exchange binding. Spacer arms are also a major source of ion-exchange interactions with biomolecules, but attempts to replace them with non-polar chains resulted in what is called hydrophobic interaction (see later). Certain hydrophobic ligands attached to long and flexible hydrophobic spacer arms become ineffective because of conformational occlusion; i.e. the spacer arm attains a conformation in which the ligand curls back and gets masked by the spacer arm with which it interacts hydrophobically.

An interesting phenomenon, called ligand-dependent non-biospecific adsorption, is evidenced by the fact that while certain enzymes can be bio-eluted by competitive counter ligands immediately after adsorption, the elutability decreases with time. Apparently, the enzyme becomes progressively more enmeshed in the hydrophobic spacer arm, the longer it is allowed to be in its proximity. Occasionally, certain macromolecules exhibit weak *abortive complex affinity*; i.e. two types of weak interactions work in conjunction to cause strong adsorption. Elution in such cases is achieved by eliminating or suppressing one of them.

Many of the interfering mechanisms can be eliminated or controlled by modification of the adsorbent, the ligand, the spacer arm, or the ionic strength of the eluant. Change of pH and temperature may sometimes be effective but not generally useful as they are likely to disturb the biospecific interaction as well. For the same reason, addition of organic solvents or urea to the eluant may not be advantageous.

#### *6.7.5.2 Hydrophobic (interaction) chromatography (HIC)*

Hydrophobic interactions, which are non-covalent and non-electrostatic, play a very important role in biological systems. The hydrophobic interaction may be conceptually vague but may be understood on the basis of the principle that *like dissolves like*. Thus, the hydrophobic (non-polar) regions of one molecule may be attracted to the non-polar regions of another. Thus, lipids and proteins, which constitute the main components of the cell, are held together by hydrophobic interactions. The immunological reactions between antibodies and antigens are also considered to be, at least partly, hydrophobic.

Use of hydrophobic interaction for the separation of macromolecules has been attempted by several workers [98,99]. Under identical conditions of pH, ionic strength, buffer composition and temperature, the ability of alkylaminoagarose (agarose-NHR) to retain phosphorylase depends on the length of the carbon chain. That the amino group does not play any role in the retention was clearly indicated when a similar alkyl-chain dependence was observed with unfunctionalized hydrocarbon-bonded agarose gels. Since the retention is non-specific and since some adsorbed enzymes retain their activity, the interaction with the ligand may not be at the active site. Proteins bound on the hydrophobic supports may be eluted by decreasing the polarity of the eluant. It is possible to modulate the retention of the given solute by adjusting the chain length and avoid too strong a retention so that the protein is not denatured by the higher concentration of organic component in the eluant. Thus, a series of alkylagaroses (with the general formula $Seph{-}C_n{-}X$, where X is a functional group like R, OH, COOH, Ph, $CH{=}CH_2$, $NH_2$, etc., and $n = 1$–12) are available for achieving the separation and purification of proteins and cells. For experimental procedures of preparation and characterization as well as the criteria for the choice of these columns, see Ref. 98.

A series of exploratory columns ($0.5 \times 5$ cm, packed with homologous alkylated agarose) may be used for selecting the right column for a given application. The homologous series approach also provides an interesting basis for fractionation of proteins. Thus, sequential passage of a crude protein through a series of these columns with $n = 1$–10 may be expected to yield fractions of increasingly hydrophobic proteins which may be further purified by complementary techniques. The homologous series of columns is also used in cytology for separation of erythrocytes which can be retained and eluted from different columns with no change in the morphology. There is thus a great potential for distinguishing between closely related cells and for studying the cell surfaces.

It is also possible to elute the proteins on the basis of their relative hydrophobic character using polarity-reducing additives, mild detergents, low concentration of denaturants, or alteration of pH, temperature or ionic strength, to detach the adsorbed proteins from the column. These features are quite similar to those observed in reversed-phase chromatography. The hydrophobicity of a protein depends on its conformation and its retention on the alkylated agarose is based on the latter's lipophilicity and number of cavities of the appropriate size and shape. Therefore, the elution methods mentioned above must be able to disrupt the hydrophobic retention of the protein by changing its conformation. The term 'specific deformers' is used to describe compounds that can bring about a reversible conformational change that is rather limited and localized. The protein should revert to its original conformation on removal of the additive. Naturally different proteins are specifically deformed by different agents and these may be used for selective elution of the hydrophobically retained proteins. The effectiveness of the deformers can also be examined on the exploratory kits described above. High specificity of elution is achievable by using biospecific ligands such as co-factors, substrates (or their analogues), certain metal ions or allosteric effectors.

Hydrophobic retention of enzymes on alkylated agarose also parallels their behaviour in solution. Proteins in solution can be fractionally precipitated by increasing the salt concentration (ionic strength); the effectiveness or efficiency of precipitation (e.g. by sodium salts) decreases in the order: sulphate > chloride > bromide > thiocyanate. The same order holds for *salting-in* of the proteins on hydrophobic columns and the reverse order holds for their elution. Thus, proteins are adsorbed on the alkylated gel by increasing the ionic strength and eluted by decreasing it; when the adsorption is very strong, up to 50% of ethylene glycol may be added to the eluant. The salting-in and salting-out procedure described above makes the hydrophobic interaction chromatography a very simple practice. However, complications can be expected in view of the fact that these gels are not strictly hydrophobic but are amphiphilic, containing both hydrophobic and hydrophilic groups. Affinity elution techniques (see later) can be used to advantage in such cases of non-specific adsorption.

The high-resolution or the high-performance version of hydrophobic interaction chromatography was introduced in 1983 [99]. As usual the solid support is based on silica (5–10 $\mu$m particle size and 250–300 Å pore size), which can withstand high operating pressures. The silica gel is covered with a hydrophilic layer into which hydrophobic alkyl chains are incorporated. The hydrophobicity may be adjusted in such a way that retention of proteins could occur in the presence of 1–2 M antichaotropic ions (e.g. sulphate or phosphate). Elution is then achieved using a gradient of decreasing salt concentration.

Several proteins, *t*-RNAs, viruses and cells have been fractionated using the principle of hydrophobic retention. The very simple operating conditions; namely, adsorption at high salt concentration and elution by reverse gradient of salt concentration makes the hydrophobic interaction chromatography a very attractive technique for preparative scale protein (enzyme) purification. The technique has also been used for non-covalent immobilization of enzymes, determination of the hydrophobicity scales among macromolecules and study of the interactions in sparingly soluble substances.

### 6.7.6 Sample application

Owing to a variety of intermolecular forces (like ionic, hydrogen-bonding, hydrophobic, and van der Waal's, which operate to different extents in different systems), biological interactions are extremely sensitive to variation in temperature, pH and ionic strength. The optimal conditions for effective interaction of the biomolecule and the substrate may not be the same in solution as in the immobilized form. In general, adsorption decreases with increasing temperature and the effective temperature in affinity chromatography is generally between 0 and 10°C. The highest ionic strength compatible with the affinity complex formation is used to suppress the non-specific interactions. The optimal conditions for effective adsorption of the analyte on the adsorbent are usually determined by trial and error. The gel is equilibrated for several hours with several volumes of the suitable buffer, under the chosen conditions, before the sample is applied. Application of the sample may be made either on the column after packing or by the batch technique in which the gel is added to the solution of the sample and equilibrated. The period of equilibration (or the flow-rate if applied on the column) could be critical. The binding of the biomolecule to the surface depends on the probability of collision between the biomolecule and the ligand, particularly at the active site. This condition is difficult to achieve if the pore size of the matrix is such that the penetration and mobility of the solute inside the gel particle is restricted. Considering all these factors, the longest practically feasible contact time may be allowed.

Batchwise adsorption is preferred for the isolation of a small quantity of active substances from a crude mixture. It may be advantageous sometimes to carry out the initial adsorption batchwise and then transfer the gel along with the adsorbed material to the column. Preliminary purification by other methods like ion exchange is sometimes recommended prior to biospecific adsorption.

### 6.7.7 Elution methods

Normally, reversal of the conditions used for the sample application and retention should facilitate elution of the solute from the bed. Where affinity adsorption is weak, continuation of irrigation with the same buffer that has been used for sample application often elutes the compound. It has been mentioned earlier that low temperature favours adsorption; thus, a rise in temperature could sometimes effect elution. Similarly, a change of pH and ionic strength are also employed. Where the biospecific complex on the adsorbent is too strong to be broken by such methods, it may be decomposed by urea, guanidine salts, or chaotropic ions which denature, or conformationally modify, the biomolecule. A major drawback of this technique is the possibility of irreversible denaturation of the analyte (enzyme). However, by careful choice of concentration, temperature and exposure time, desorption of the molecules with minimal and reversible conformational change is possible. If the ligand is attached to a spacer or to the support through a bond that can be easily broken (e.g. azo, disulphide or ester), these may be cleaved to remove the complex as a whole from the adsorbent.

Occasional failure of such methods to satisfactorily elute the biomolecule after adsorption have led to the development of electrophoretic elution methods, wherein, an electric current is passed across a loaded affinity matrix at such a pH that the adsorbed material is charged; the material then gets desorbed and migrates to the appropriate electrode. The technique is mild, non-chaotropic and quantitative.

Elution with a solution of the affinity ligand has also been used for effecting desorption. Alternatively, a solution of a competitive counter-ligand or inhibitor (for an enzyme) may also be used. The counter-ligand is often structurally close to the bound ligand on the matrix or the solute, so that it could compete with the sites on the solute or the bound ligand and thus decrease the strength of retention of the solute on the adsorbent. This method is frequently used when the counter-ligand is a readily available small molecule.

In affinity elution, a non-specifically adsorbed material, such as in an ion exchanger, is selectively eluted using a specific ligand. The choice of the ligand should be such that the protein–ligand complex is unable, owing to the changed charge characteristics, to bind the ion exchanger, whereas the free protein could. While this technique has been successfully applied in several instances, often the selectivity is poor; a particular ligand could desorb more than one protein and a particular protein could be desorbed by more than one ligand. A variation of affinity elution from non-specific adsorbents is to use the salting-in and salting-out principle by changing the salt concentration. The effectiveness of this procedure on amphiphilic gels and agarose parallels the solubility properties of proteins. Since this technique depends entirely on physicochemical properties and not on biological activity, the selectivity is poor. However, addition to the eluant of a ligand that specifically binds one of the proteins could shift the solubility properties to effect selective elution.

### 6.7.8 Applications

Applications of affinity chromatography have been so numerous, varied and unique that they require special mention. Only some of the major applications can be briefly mentioned here. The vastness of applications of affinity chromatography is reflected in the fact that, in recent years, over 600 papers have been published annually on the subject. Nevertheless, agarose and cyanogen bromide remain the most popular matrix and activation reagents, respectively. Early investigations were directed mainly towards separation of biomolecules, but more recent applications include the study of biochemical interactions and mechanisms, affinity labelling and immobilized enzymes [95–100].

#### *6.7.8.1 Separation of biomolecules*

The technique of affinity chromatography has been extensively used for isolation of several biological macromolecules of different classes such as immunoglobulins (antibodies), antigens, haptens, cells and organelles, enzymes, their substrates and modified derivatives, receptors, binding and transfer proteins, polysaccharides, glycoproteins, lipoproteins and lectins, lipids, nucleic acids, viruses, interferons, etc. By immobilizing the macromolecules, the complementary small molecules have also been separated; e.g. vitamins, co-factors, hormones, inhibitors, amino acids, nucleotides, toxins, etc. These applications have been extensively reviewed by Turkova [86,89], with about 1500 examples of affinity separations. Over 52.4% of these represented application of affinity chromatographic separation of enzymes and their subunits, inhibitors and co-factors. The proportion of other classes of biomolecules (the separation of which this technique was used for) is as follows [89]: antibodies and antigens (12.2%), lectins, glycoproteins and polysaccharides (8.8%), receptors, binding and transport proteins (6.9%), nucleic acids and nucleotides (6.4%), viruses, cells and their components (3.2%) and specific peptides (2.4%). See also Table 6.4.

Among the large variety of biomolecules that have been separated by affinity chromatography, proteins form the largest single category; and, among proteins, the enzymes are the largest category [95]. Because of their discrete specificity, a large variety of potential ligands is also available. These may be the substrate analogues, inhibitors, co-enzymes, and the like. Though particular enzyme-specific ligands are employed for isolation of enzymes of narrow specificity, they may also be occasionally used to purify other enzymes of identical catalytic activity but from different organisms. General ligands (e.g. co-enzymes, which may be common substrates to an entire type of activity) may be used for the separation of a broader range of enzymes. Over 30% of the known enzymes require co-enzymes, the most common of them being NAD, NADP, ATP, and the co-enzyme-A.

Antibodies, or immunoglobulins, are another major class of protein for which (immuno)affinity chromatography is an indispensable method for isolation and purification. Specific antigens are the obvious choice as the ligands. Immobilized antibodies (both conventional and monoclonal) are also used in a one-step procedure for isolation of specific proteins from mixtures.

#### *6.7.8.2 Study of biochemical interactions and mechanisms*

As noted earlier, and as will be discussed in greater detail later (Chapter 9), a technique of separation which uses a certain principle can also be used to determine quantitatively some of the parameters in relation to the principle itself. Affinity chromatography, which is based on biospecific interaction between molecules, shows immense promise for studying such interactions quantitatively. The rate constants of association and dissociation of biospecific interaction would determine the diffusion of the solute along the chromatographic bed, and the longer the component stays on the column the more is the band broadening. Thus, dissociation constants of certain enzyme-substrate complexes have been determined both by elution and by frontal analysis methods. Dissociation constants of binary complexes of several dehydrogenases with NADH were determined, as well as the inhibition constants with various inhibitors. Several thermodynamic parameters have been determined by studying the elution profiles of biomolecules at different temperatures and concentrations. Quantitative comparison of binding of various enzymes with immobilized nucleotides (AMP) have been studied. Affinity chromatography has been used for studying the mechanism of enzyme action, for isolation of enzymes that take part in certain metabolic paths, for detection of structural differences in

**Table 6.4** — Examples of biomolecular separations by affinity chromatography†

| Substances separated | Matrix | Affinity ligands |
|---|---|---|
| Enzymes | | |
| Acetylcholinesterase | Sepharose 2B | *N*-methylacridine |
| Amine oxidase | Sepharose 4B | 1,6-diaminohexane |
| Alkaline phosphatase | Sepharose 6B | Cibacron Blue 3GA |
| Aminopeptidase | AH-Sepharose 4B | Amastatin |
| Aryl sulphatase | ConA-Sepharose | Concanavalin A |
| Aspartate amino-transferase | Aminoethyl-Sepharose | Citric acid |
| Aspergillopepsin | Sepharose 4B | Gramicidin C |
| Chymotrypsin | Spheron | *N*-Boc-Gly-Phe |
| Collagenase | CH-Sepharose | L-Argenine |
| DNA gyrase | Epoxy-activated Sepharose | Novobiocin |
| DNA polymerase | As above | Polynucleotide |
| Heparinase | AH-Sepharose | Heparin |
| IMP dehydrogenase | Sepharose 4B | Procion dyes |
| β-Mannanase | AH-Sepharose | Mannan |
| Proteolytic enzymes | Aminosilochrom | Bacitracin |
| Proteolytic enzymes | Sepharose 4B | Haemoglobin |
| Trypsin | Spheron | Ovomucoid |
| Enzyme inhibitors | | |
| α-Antitrypsin | CNBr-Sepharose 4B | Heparin |
| Catalase inhibitor | As above | Catalase |
| Trypsin inhibitor | As above | Inhibitor subunits |
| Chymotrypsin inhibitor | Spheron | Chymotrypsin |
| Antibodies | | |
| Antibodies to: | | |
| Human serum albumin | CNBr-Sepharose 4B | HSA |
| Staphylococcal eterotoxin A & E | As above | Enrterotoxin A & E |
| Antimorphine | AH-Sepharose | Morphine |
| Antispemine | CH-Sepharose 4B | Spermine |
| Antihydroglobulin | Sepharose 4B | Thyroglobulin |
| Antigens | | |
| Barley β-amylase | CNBr-Sepharose | Antibarley antibody |
| DNP-Bovine serum albumin | Sepharose 4B | Anti-DNP antibodies |
| E. coli enterotoxin | CNBr-Sepharose | Anticholera antibody |
| Polyclonal IgA | TES-propylamine CPG | Anti-IgA antibodies |
| Poliovaccine | Sepharose CL-4B | Cross-linked antibody |
| Lectins | | |
| Lectins from different sources | Spheron | D-Galactose |
| | Con A-Sepharose | Conconavalin A |
| | Sepharose 4B | Fetuin |
| | Sephadex G-100 | Dextran |
| Glycoproteins and saccharides | | |
| Plasma mėmbrane proteins | Con A-Sepharose | Conconavalin |
| Glycogen | As above | As above |
| Fibroblast interferon | As above | As above |
| Orosomucoid | As above | As above |
| Receptors, binding and transfer proteins | | |
| Glucocorticoid receptor | DNA | Cellulose |
| DNA-binding | DNA | Sephadex G-10 |
| Nucleic acids and nucleotides | | |
| Cyclic GMP | Affi-Gel 601 | Dihydroxyboronyl derivatives |
| Nucleosides | Sephadex A-25 | |
| Oligonucleotides | Ultrogel A4R | Acriflavine |
| Poly(A) | Sepharose 4B | Poly(U) |
| Other biomolecules | | |
| Influenza virus | Sepharose 4B | Neuraminic acids |
| Amnion interferon | As above | Bovine plasmalbumin |
| Fibrinogen | AH-Sepharose 4B | T-proteins of *Strep.pyogens* |
| Lipids | Glycophase CPG | Neomycin |
| Human factor VIII | CNBr-Sepharose | Factor VIII antibody |

† For more examples and references, see Ref. 89.

macromolecules such as interferons, for immunoassay, and for several other biochemical studies [89,96]. Affinity chromatography has also been used for studying the molecular structures of complex biological materials.

#### *6.7.8.3 Affinity labelling*

Determination of amino acid sequence at the active sites of enzymes is an important aspect of biochemical research. This can be done very conveniently by the technique known as affinity labelling [100], which has attracted considerable attention in recent years. An analogue of the substrate carrying a properly positioned, chemically reactive leaving group is reacted with the protein to form a complex at, or near, the active site. By virtue of the reactive group, a covalent bond is formed at, or near, the active site between the ligand and the protein. Enzyme digestion of the protein then breaks up the molecule into smaller fragments and the hydrolysate containing the labelled peptide is affinity chromatographed using the same protein immobilized on a solid support. The affinity labelled peptide is selectively adsorbed and the rest is washed off. The adsorbed peptide is then eluted suitably and the active site sequence obtained by appropriate methods.

## REFERENCES

[1] O. Mikes (ed.), *Laboratory Handbook of Chromatographic and Allied Methods*, Ellis Horwood, Chichester, 1979.
[2] H. Wyler and P. Chevreux, *J. Chem. Educ.*, **55**, 270 (1978).
[3] D. F. Taber, *J. Org. Chem.*, **47**, 1351 (1982).
[4] L. R. Snyder, *Chromatogr. Rev.*, **7**, 1 (1965).
[5] H. Determann, *Gel Chromatography*, 2nd edn., Springer-Verlag, Berlin, 1969.
[6] W. W. Yau, J. J. Kirkland, and D. D. Bly, *Modern Size-Exclusion Chromatography*, Wiley, New York, 1979.
[7] Pharmacia Promotional Literature, *Gel Filtration : Theory and Practice*, Pharmacia Laboratory Separation Division, Uppsala, Sweden, 1980.
[8] J. C. Giddings, E. Kucera, C. P. Russel, and M. N. Myers, *J. Phys. Chem.*, **72**, 4397 (1968).
[9] Z. Grubistic, R. Rempp, and H. Benoit, *J. Polym. Sci., Part B*, **5**, 753 (1967).
[10] H. G. Barth, *J. Chromatogr. Sci.*, **18**, 409 (1980).
[11] R. P. Bywater and N. V. B. Marsden, in *Chromatography*,E. Heftmann (ed.), Elsevier, Amsterdam, 1983.
[12] E. Pfannkoch, K. C. Lu, F. E. Regnier, and H. G. Barth, *J. Chromatogr. Sci.*, **18**, 430 (1980).
[13] W. W. Yau, C. R. Ginnard and J. J. Kirkland, *J. Chromatogr.*, **149**, 465 (1978).
[14] R. V. Vivilecchia. B. G. Lightbody, N. Z. Thimot, and H. M. Quinn, *J. Chromatogr. Sci.*, **15**, 424 (1977).
[15] W. Cheng and D. Hollis, *J. Chromatogr.*, **408**, 9 (1987).
[16] J. Cazes (ed.), *Liquid Chromatography of Polymers and Related Materials*, Parts I-III, Marcel Dekker, New York, 1977–81.
[17] H. Kalasz, in *Methods of Protein Analysis*, I. Kerese (ed.), Ellis Horwood, Chichester, 1984.
[18] D. Berek and K. Macinka, in *Separation Methods*, Z. Deyl (ed.), Elsevier, Amsterdam, 1984.
[19] J. F. Johnson, in *Encyclopedia of Polymer Science and Technology*, H. F. Mark, N. Bikales, C. G. Overberger G. Menges, and J. I. Kroschwitz (eds.), Wiley, New York, 1986, vol. 3.
[20] H. Small, *Adv. Chromatogr.*, **15**, 113 (1977).
[21] H. Small and M. A. Longhorst, *Anal. Chem.*, **54**, 892A (1982).
[22] J. C. Giddings, *J. Chromatogr.*, **125**, 3 (1976).
[23] J. Janca, in *Separation Methods*, Z. Deyl (ed.), Elsevier, Amsterdam, 1984.
[24] L. F. Kesner, K. D. Caldwell, M. N. Myers, and J. C. Giddings, *Anal. Chem.*, **48**, 1834 (1976).
[25] J. C. Giddings, G. Ch. Lin, and M. N. Myers, *Sep. Sci.*, **11**, 553 (1976).
[26] L. R. Snyder and J. J. Kirkland, *Introduction to Modern Liquid Chromatography*, Wiley, New York, 1979
[27] T. Braun and G. Ghersini (eds.), *Extraction Chromatography*, Elsevier, Amsterdam, 1975.
[28] K. Hostettmann, *Planta Med.*, **39**, 1 (1980).

[29] K. Hostettmann, *Adv. Chromatogr.*, **21**, 165 (1983).
[30] Tokyo Rikakikai Co., Toyama-Cho, Kanda, Chiyoda-Ku, Tokyo.
[31] Y. Ito, *J. Biochem. Biophys. Met.*, **5**, 105 (1981).
[32] Y. Ito, *Adv. Chromatogr.*, **24**, 181 (1984).
[33] Y. Ito, *CRC Crit. Rev. Anal. Chem.*, **17**, 65 (1986).
[34] I. A. Sutherland, D. Heywood-Waddington, and Y. Ito, *J. Chromatogr.*, **384**, 197 (1987).
[35] A. M. Krstulovic and P. R. Brown, *Reversed-Phase High-Performance Liquid Chromatography*, Wiley, New York, 1982.
[36] K. K. Unger, *Porous Silica: Its Properties and Uses as a Support in Column Liquid Chromatography*, Elsevier, Amsterdam, 1979.
[37] H. Billiet, C. Laurent, and L. de Galan, *Trends Anal. Chem.*, **4**, 100 (1985).
[38] M. G. McRae, R. G. Gregson, and R. K. Quinn, *J. Chromatogr. Sci.*, **20**, 475 (1982).
[39] M. Dressler, *J. Chromatogr.*, **165**, 167 (1979).
[40] J. H. Knox, B. Kaur, and G. R. Millward, *J. Chromatogr.*, **352**, 3 (1986).
[41] K. K. Unger, *Anal. Chem.*, **55**, 361A (1983).
[42] W. Holstein and H. Hametsberger, *Chromatographia*, **15**, 186 (1982); **15**, 251 (1982).
[43] V. A. Davankov and A. V. Semechkin, *J. Chromatogr.*, **141**, 313 (1977).
[44] K. K. Stewart and R. F. Doherty, *Proc. Natl. Acad. Sci. U.S.A.*, **70**, 2850 (1973).
[45] R. P. Singh, H. N. Subbarao, and S. Dev, *Tetrahedron*, **37**, 843 (1981).
[46] K. Fujimura, T. Ueda and T. Ando, *Anal. Chem.*, **55**, 446 (1983).
[47] M. Nakajima, K. Kimura, and T. Shono, *Anal. Chem.*, **55**, 463 (1983).
[48] R. Dappen, H. Arm, and V. R. Meyer, *J. Chromatogr.*, **373**, 1 (1986).
[49] W. H. Pirkle and T. C. Pochapsky, *Adv. Chromatogr.*, **27**, 73 (1987).
[50] D. W. Armstrong, *Anal. Chem.*, **59**, 84A (1987).
[51] V. A. Davankov, *Adv. Chromatogr.*, **18**, 139 (1980).
[52] V. A. Davankov, A. A. Kurganov, and A. S. Bochkov, *Adv. Chromatogr.*, **22**, 71 (1983).
[53] G. Dotsevi, Y. Sogah, and D. J. Cram, *J. Am. Chem. Soc.*, **98**, 3038 (1976).
[54] W. R. Melander and C. Horvath, in *High-Performance Liquid Chromatography: Advances and Perspectives*, C. Horvath (ed.), Academic Press, New York, Vol. 2, 1980.
[55] L. R. Snyder, in *Techniques of Chemistry*, A. Weissberger and E. S. Perry (eds.), 2nd edn., Wiley, New York, Vol. 3, Part I, 1979.
[56] L. Rohrschneider, *Anal. Chem.*, **45**, 1241 (1973).
[57] J. G. Touchstone and M. F. Dobbins, *Practice of Thin-Layer Chromatography*, Wiley, New York, 1978.
[58] D. L. Saunders, *Anal. Chem.*, **46**, 470 (1974).
[59] L. R. Snyder, *J. Chromatogr. Sci.*, **16**, 223 (1978).
[60] J. L. Glajch, J. J. Kirkland, and K. M. Squire, *J. Chromatogr.*, **199**, 57 (1981).
[61] L. R. Snyder, J. W. Dolan, and J. R. Gant, *J. Chromatogr.*, **165**, 3 (1979).
[62] M. T. W. Hearn, *Adv. Chromatogr.*, **88**, 59 (1980).
[63] M. T. W. Hearn (ed.), *Ion-Pair Chromatography: Theory and Biological and Pharmaceutical Applications*, Marcel Dekker, New York, 1985.
[64] A. Akelah and D. G. Sherrington, *Chem. Rev.*, **81**, 557 (1981).
[65] A. P. Williams, *J. Chromatogr.*, **373**, 175 (1986).
[66] T. Devenyi and J. Gergely, *Amino Acids, Peptides and Proteins*, Elsevier, Amsterdam, 1974.
[67] P. R. Brown, A. M. Krstulovic, and R. A. Hartwick, *Adv. Chromatogr.*, **18**, 101 (1980).
[68] G. Schwedt, *Chromatographia*, **12**, 613 (1979).
[69] P. Jandera and J. Churacek, *J. Chromatogr.*, **86**, 351 (1973).
[70] P. Jandera and J. Churacek, *J. Chromatogr.*, **98**, 1 (1974).
[71] P. Jandera and J. Churacek, *J. Chromatogr.*, **98**, 55 (1974).
[72] O. Samuelson, *Adv. Chromatogr.*, **16**, 113 (1978).
[73] D. T. Gjerde and J. S. Fritz, *Ion Chromatography*, Huethig, Heidelberg, 2nd edn., 1987.
[74] T.-L. Ho, *Hard and Soft Acids and Bases Principle in Organic Chemistry*, Academic Press, New York, 1977.
[75] H. F. Mark, D. F. Othmer, C. E. Overberger, and G. T. Seaborg (eds.), *Kirk-Othmer's Encylopedia of Chemical Technology*, 3rd edn., Vol. 13, Wiley, New York, 1981.
[76] Pharmacia Promotional Literature, *Ion-Exchange Chromatography: Principles and Methods*, Pharmacia Fine Chemicals AB, Uppsala, Sweden, 1980.
[77] A. Clearfield (ed.), *Inorganic Ion-Exchange Materials*, CRC Press, Boca Raton, Florida, 1982.
[78] D. Malamud and J. W. Drysdale, *Anal. Biochem.*, **86**, 620 (1978).
[79] T. S. Stevens, J. C. Davis, and H. Small, *Anal. Chem.*, **53**, 1488 (1981).
[80] T. S. Stevens, G. L. Jewett, and R. A. Bredeweg, *Anal. Chem.*, **54**, 1206 (1982).
[81] P. K. Dasgupta, *Anal. Chem.*, **56**, 96 and 103 (1984).

[82] D. T. Gjerde, J. S. Fritz and G. Schmuckler, *J. Chromatogr.*, **186**, 509 (1979); D. T. Gjerde, G. Schmuckler, and J. S. Fritz, *J. Chromatogr.*, **187**, 35 (1980).
[83] K. Tanaka, T. Ishizuka, and H. Sunahara, *J. Chromatogr.*, **174**, 153 (1979).
[84] K. Tanaka, T. Ishizuka, and H. Sunahara, *J. Chromatogr.*, **177**, 21 (1979).
[85] Pharmacia Promotional Literature, *Affinity Chromatography: Principles & Methods*, Pharmacia AB, Uppsala, Sweden, 1983.
[86] J. Turkova, *Affinity Chromatography*, Elsevier, Amsterdam, 1978.
[87] W. H. Scouten (ed.), *Solid-Phase Biochemistry*, Wiley, New York. 1983.
[88] I. M. Chaiken, M. Wilchek and I. Parikh, *Affinity Chromatography and Biological Recognition*, Academic Press, New York, 1983.
[89] J. Turkova, in *Separation Methods*, Z. Deyl (ed.), Elsevier, Amsterdam, 1984.
[90] H. Schott, *Affinity Chromatography: Template Chromatography of Nucleic Acids and Proteins*, Marcel Dekker, New York, 1984.
[91] P. Mohr and K. Pomerening, *Affinity Chromatography: Practical and Theoretical Aspects*, Marcel Dekker, New York, 1985.
[92] P. D. G. Dean, W. S. Johnson, and F. A. Middle (eds.), *Affinity Chromatography: A Practical Approach*, IRL Press, New York, 1985.
[93] P.-O. Larsson, M. Glad, L. Hansson, M.-O. Mansson, S. Ohlson, and K. Mosbach, *Adv. Chromatogr.*, **21**, 41 (1983).
[94] A. A. Farooqui and L. A. Horrocks, *Adv. Chromatogr.*, **23**, 127 (1984).
[95] M. Wilchek, T. Miron, and J. Kohn, *Meth. Enzymol.*, **104**, 3 (1984).
[96] R. R. Walters, *Anal. Chem.*, **57**, 1099A (1985).
[97] C. R. Lowe and J. C. Pearson, *Meth. Enzymol.*, **104**, 97 (1984).
[98] S. Shaltiel, *Meth. Enzymol.*, **104**, 69 (1984).
[99] Y. Kato, *Adv. Chromatogr.*, **26**, 97 (1987).
[100] W. B. Jacoby and M. Wilchek (eds.), *Meth. Enzymol.*, **46**, 3 (1977).

# 7

# Liquid chromatography II: instrumentation and techniques

## 7.1 THE HPLC SYSTEM: BASIC CONFIGURATION

The renaissance of liquid chromatography owes much to the development of stationary-phase technology and instrumentation. While the basic principles of LC, as discussed in the foregoing chapter, remained essentially the same over the years, the so-called high-performance, high-speed, instrumental or modern liquid chromatography is the outcome of the preparation of small (less than 10 $\mu$m, lately between 3 to 5 $\mu$m diameter) particles of uniform size and shape with narrow pore-size distribution, together with the development of the technology of chemical modification of the surface functionality. Unlike the traditional column liquid chromatography (CLC) using the gravity-flow or the low-pressure solvent-delivery systems that were universally used until the 1960s and are still being used in several laboratories, the use of the modern, high-efficiency packing materials of small diameter requires high-pressure solvent pumping systems that can deliver the mobile phase at uniform and reproducible flow-rates. As the system is under high pressure, sample introduction requires specially designed injection ports. The success and convenience of on-line detection in GC has led to the expenditure of considerable effort (and, fortunately, with considerable success) in developing a variety of high-sensitivity detectors. On-line detection in LC has certainly proved much more formidable as, unlike in GC, the mobile phase is neither inert nor transparent to different detection principles or systems. As the solvent removal prior to detection is not very convenient, only the differences in the bulk properties of the solvent and solution of the sample in the same solvent had to be used for detection and quantitation. Data handling from then on is very similar to that in GC. Indeed the same recorders, integrators and data processors can be used for both GC and LC. Thus, HPLC is, in principle, the same as the traditional CLC, but with improved stationary phases, coupled with sophisticated sample introduction, solvent delivery and detection systems and data analysis. Almost all the methods of LC discussed in the previous chapter have been adapted to HPLC and no distinction need be made in the treatment of the two versions. The discussion in this chapter is thus a continuation, with emphasis on instrumentation and techniques of its usage. There are several excellent texts published over the last decade which may be consulted for general information on the instrumentation in liquid chromatography [1–5].

The basic features of an HPLC system is shown schematically in Fig. 7.1. The

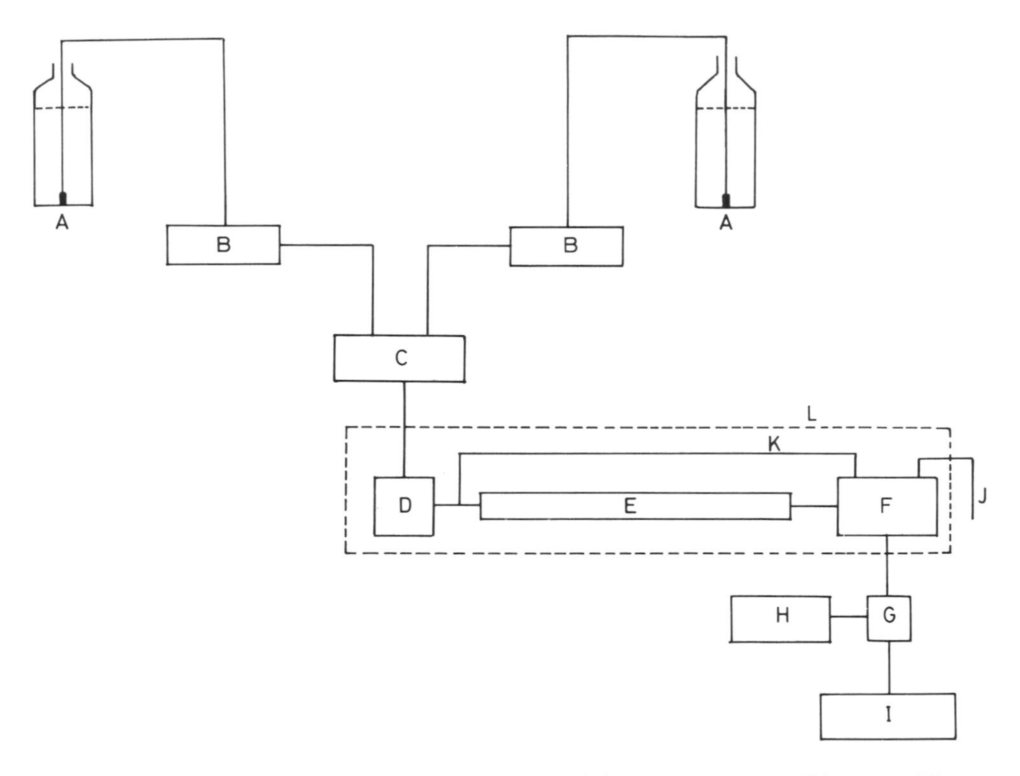


Fig. 7.1 — Basic configuration of an HPLC system. (A) Solvent reservoirs; (B) pumps; (C) solvent mixer; (D) injection port; (E) column; (F) detector; (G) amplifier; (H) recorder; (I) fraction collector.

system typically consists of a solvent reservoir (A), a pump (B) for solvent delivery to the column, an injection port (D) or valve, the column (E), a detector (F) and a recorder (H). The diagram also shows two solvent reservoirs, two pumps and a gradient former (C), suitable for gradient elution (see below). To these may be added certain convenience accessories like an auto-injector and/or autosampler, a solvent-recycling facility, more columns with switching valves, a multiplicity of detectors, multichannel recorders and integrators (I), and a fraction collector. In addition to these, the HPLC system is usually equipped with in-line filters to remove particulate matter, pressure gauges as a means to monitor any blocking of the column, pulse dampers and flow-meters. Most manufacturers supply the individual components mentioned above; these can be assembled in the laboratory without difficulty. However, the HPLC system often is obtained as a package with certain pre-selectable options and accessories. Dedicated HPLC systems for specific applications are also available. While the modular design (Fig. 7.1) is very versatile and can be altered to suit specific needs from time to time, it is difficult to automate all the operations. On the other hand, microprocessor-controlled systems (Fig. 1.5) may be convenient to handle, but are difficult to dismantle or to change the modules at will. The central microprocessor can automate and control sample introduction, column temperature, solvent selection, flow-rate, gradient programming, return to initial conditions and re-equilibration, choice of the detectors, their operating parameters and data handling. They can store sets of chromatographic parameters for repetitive and reproducible analysis. They are also useful for optimization of separation conditions. Most of the LC instrument controllers can also acquire and process the chromatographic data, much the same way as the dedicated chromatographic data systems (Chapter 9).

Precise temperature control of the system as a whole is advisable for optimization and reproducibility. As much of the HPLC is carried out at ambient temperature, air-conditioning of the laboratory is adequate. However, a thermostat or oven (30 to 100°C), accommodating the injection port, the column and the detector is advantageous in case of size-exclusion and ion-exchange chromatography, where operation at elevated temperature is helpful. Higher operating temperature also generally enhances the overall efficiency of adsorption chromatography owing to the decrease in the mobile-phase viscosity at higher temperatures. The increased dissolving power of the solvents at higher temperatures also contributes to the enhanced sample capacity and speed of analysis. Circulating air ovens or constant-temperature baths may be used for jacketing the columns. However, unlike in GC, temperature programming is rarely used in LC.

Another important accessory in HPLC is the solvent programmer or the gradient-elution accessory. Most often, HPLC can be adequately performed by isocratic elution; i.e. using a single solvent or a mixture of solvents of fixed composition. However, under the conditions of the general elution problem (Fig. 6.2), change of mobile-phase composition during elution may be necessary and this is conveniently achieved using a gradient former.

The passage of the solvent from the reservoir through the pump, injection port, column detector and the collector is achieved by the use of suitable connection tubes. The nature and diameter of the plumbing from the reservoir to the pump, and that from the detector to the waste, is not critical; a 1–2 mm, solvent-resistant Teflon tube is convenient. A stainless steel capillary tubing that can withstand high pressures is used from the pump onwards. The diameter of the connecting tube from the injection port to the column, and from the column to the detector, is very critical since radial dispersion of the solvent along the cross-section of the tube significantly contributes to extra-column band broadening [1]. The connecting tube should, therefore, be short and, more importantly, should have a very small and uniform diameter. The diameter, however, cannot be reduced beyond a certain limit because of the pressure drop across the capillary. A compromise between the pressure drop and the magnitude of band broadening appears to be an i.d. of 0.25 mm and a length of 20 to 50 mm for the connecting tubes; longer tubes are tolerated if they are wound into tight coils. Each of the above-mentioned modules is open to considerable variation in design, performance and utility and these are discussed below.

## 7.2 SOLVENT DELIVERY SYSTEMS

### 7.2.1 Solvent reservoirs

The design of the solvent reservoir may vary from a simple 1-litre Erlenmeyer flask to the very sophisticated solvent-storage system illustrated in Fig. 7.2. For the purpose

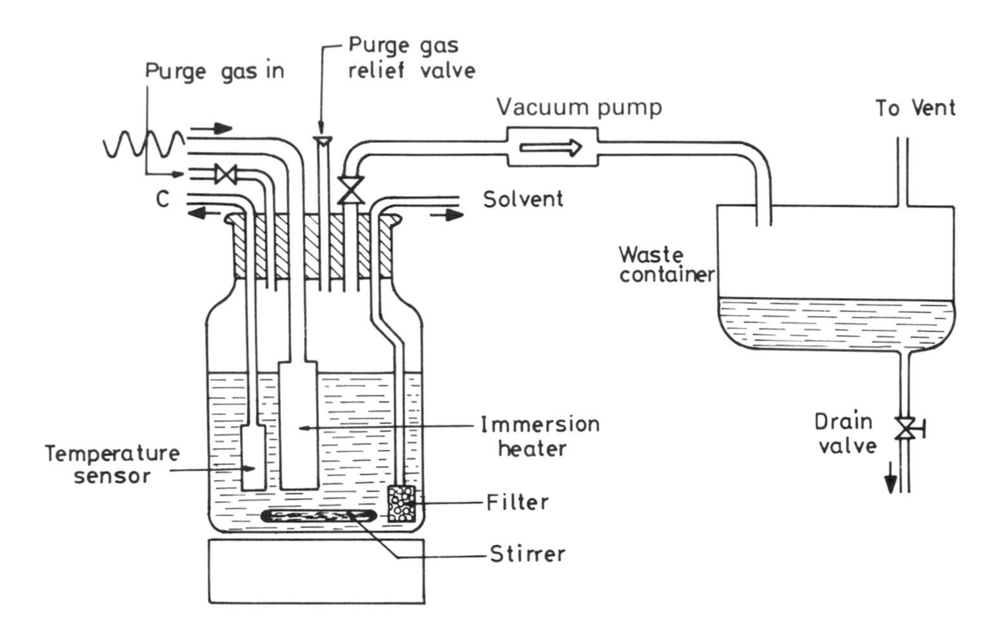


Fig. 7.2 — A solvent reservoir for HPLC. Illustration by courtesy of Hewlett-Packard, Inc.

of delivering a solvent of uniform composition, free from particulate matter and dissolved gases, the solvent reservoir may be equipped with (a) a magnetic stirrer to keep the composition from varying due to density differences in the mobile-phase components, (b) a 2 $\mu$m filter to prevent particulate matter from being drawn into the pump, (c) a temperature sensor and a heater to maintain a pre-determined temperature or to help degassing, and (d) an inert gas (helium) inlet and vacuum connection for degassing. Degassing is particularly necessary in gradient elution where the concentration of the dissolved air in the two solvents differs and when one of the solvents is water (which has a high concentration of dissolved gases relative to alcohols and other organic solvents). Removal of oxygen from the system also helps the column life by preventing deterioration of the stationary phase. While application of vacuum, heating and purging with helium may all help in degassing, change of composition of the mobile phase is possible under these conditions, as the volatile components of the mobile phase may evaporate and decrease in concentration. To ensure constant mobile-phase composition, it may be advisable to use degassed solvents and a solvent gradient mixer.

### 7.2.2 The pumps

The construction of a suitable pump for delivering the solvent to the column at sufficiently high pressure had attracted considerable attention during the early part of the development of HPLC. In fact, much of the delay in developing high-efficiency column packings may be attributed to the non-availability of pumps that could deliver precise, resettable, reproducible and pulse-free flow of solvent through the column at pressures that may range from 500 to 5000 psi. The pumps that can perform to these specifications also need to be resistant to chemical action by the mobile phases and preferably have a low hold-up volume for rapid solvent changes as particularly needed in gradient elution. Several different kinds of pumps with widely varying performance specification are available from the popular chromatographic supply houses. Because of the high-precision engineering required for their construction, these pumps are very expensive. Also, once procured, there is very little the chromatographer can do about the mechanics of the pump except taking adequate precautions to prevent its damage in use. Therefore, only the principles of operation and basic maintenance aspects are discussed here. For a detailed discussion on the description and schematics of the commonly used pumps, see Ref. [1].

The simplest form of pumping system involves the use of pressure from a gas cylinder to force the solvent (held in a coiled reservoir) through the column. It is very inexpensive and pulse-free but suffers from several drawbacks such as very limited head pressure (about 2000 psi), possible dissolution of the gas in the solvent and the consequent formation of gas bubbles in the detector, etc. Of course, use of a narrow coil to limit the area of contact between the gas and the liquid, or the use of a plunger to transfer the gas pressure to the liquid would overcome this problem. In any case, a constant head pressure and temperature (to avoid any change in the fluid viscosity) must be maintained for uniform flow. The real problem, however, is the safety hazard in the laboratory, arising out of the possibility of explosions or solvent leakage into the atmosphere.

Another simple device is the piston pump, wherein the solvent is contained in a large cylinder with a tight-fitting piston. The piston is usually screw-driven, actuated by a digital stepper motor. A very steady, pulse-free solvent delivery, irrespective of the column resistance, is an attractive feature of this system, but with the drawback that unless the quantity of the solvent held in the closed system is small, the fluid compressibility could lead to inaccuracies in the flow velocity. Yet another problem is that the time taken by the system to attain steady state of flow may also be large, depending on the selected flow-rate, compressibility of the mobile phase and the permeability of the column. While the principle is still attractive for systems like the microbore LC (see later), the popularity of the syringe pumps has considerably reduced.

In the pneumatic amplifier pump, a large-area, gas-driven piston is used to drive a small-area liquid piston. The working of the pump is shown schematically in Fig. 7.3 [5]. The forward and backward strokes of the larger piston are actuated by varying the air pressure on either side of the piston. The ratio of the inlet-to-outlet pressure

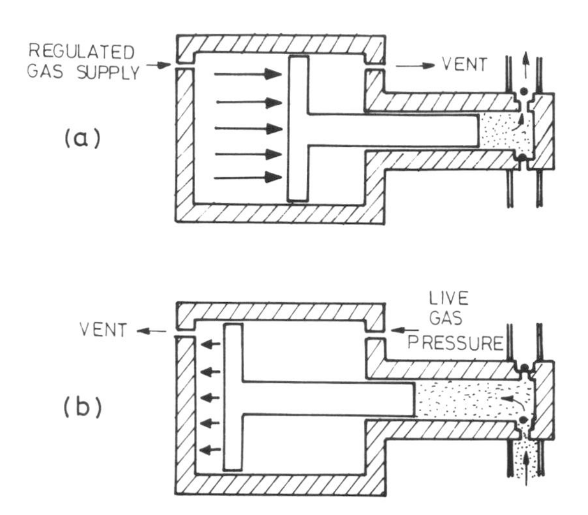


Fig. 7.3 — Schematic diagram of a pneumatic amplifier pump. Reproduced from Ref. 5 by courtesy of Elsevier Scientific Publishers Co.

on the smaller piston is given by that of the surface areas of the gas piston to the liquid piston. Thus, a relatively small inlet pressure can be used to achieve high-pressure delivery (up to 4500 psi) of the liquid with the advantage that the gas does not come into direct contact with the solvent. Precise and high flow-rates can be obtained using these systems but the disadvantage is that it is not very suitable for gradient elution. These pumps may be conveniently used in SEC and preparative liquid chromatography. Hydraulic amplification of pressure can be obtained by using a liquid instead of a gas. By using two sets of amplifiers, one for each solvent, these systems can be adapted to gradient elution also.

The more recent reciprocating pumps have several advantages over the positive displacement pumps described above. They consist of very small-volume chambers (35–400 μl capacity) into which the solvent can be drawn and then pumped into the column by backward and forward movement of a piston or a diaphragm. The flow-rate can be adjusted either by varying the stroke volume or the stroke frequency. The alternating suction and compression causes considerable pulsation in a singe-head pump. In a dual-piston reciprocating pump (Fig. 7.4), two identical single-head pumps operate 180 degrees out-of-phase so that while one piston is pumping the solvent into the column the other is drawing the solvent into its chamber. Pulsations can be further reduced by using either a pulse-damper or a triple-head reciprocating pump, phased at 120 degrees and driven by a single cam that actuates each of the pistons in turn. Though expensive, the reciprocating pumps are very popular in view of their precision, continuous solvent flow over a wide range of flow-rates and their easy adaptability to gradient elution.

Certain simple precautions can ensure trouble-free performance of the pumps. The solvent purity and freedom from particulate matter is important in HPLC, not only to preserve the pumps but also other parts of the instrument; namely, the

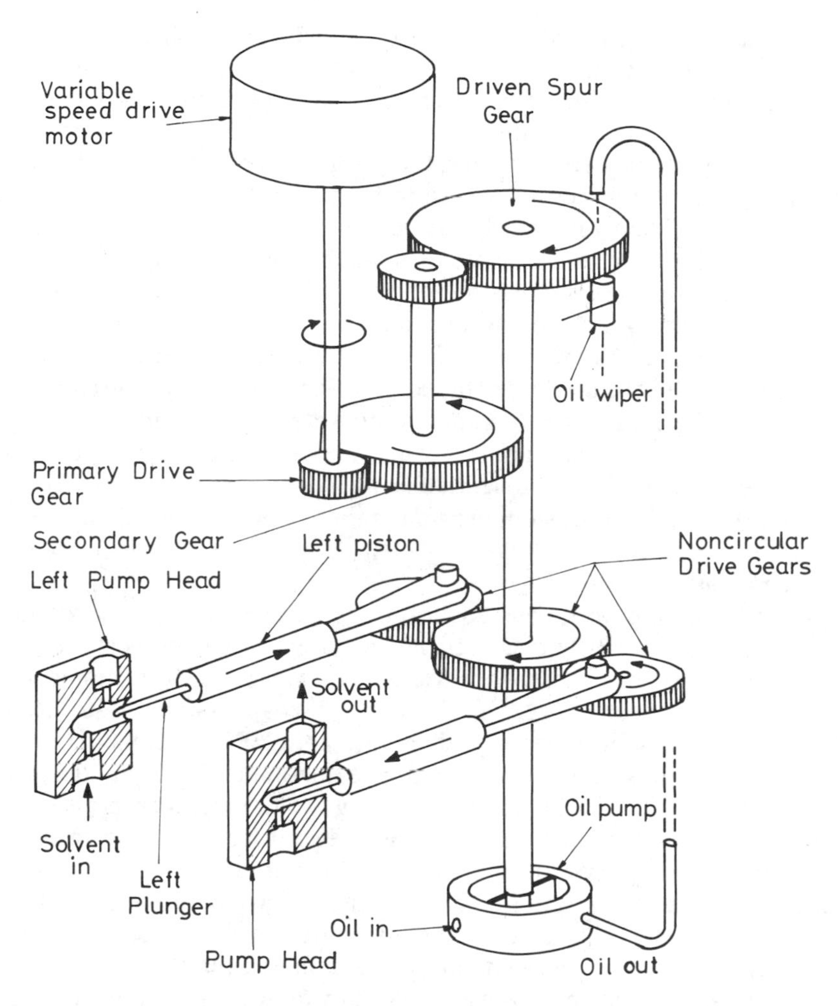


Fig. 7.4 — Schematic diagram of a dual-piston reciprocating pump. Illustration by courtesy of Waters Associates, Inc.

injection ports, the column and the detectors. A 2 μm filter is adequate for removing the particulate matter from most mobile phases. Operations of the pump without the solvent would cause friction and abrasion between the piston and the cylinder, resulting in faster wearing of the pump. The pumps should be flushed with pure water after using mobile phases containing dissolved salts or acids. Pure water on standing can cause microbial growth and this is prevented by adding some methanol or, preferably, replacing the solvent in the whole system by an organic solvent. The check valves may be periodically cleaned with 6N nitric acid. Care should be taken to see that the pumps are not subjected to higher column back pressure than that rated for the particular pump. Drawing of the liquid into the pump chamber during the refill (or backward) stroke is assisted by gravity. For proper priming, and to prevent starving of the pump, the reservoir should be generally located at a level above that of the pump and the inlet tubing should be of relatively large bore (about 2 mm i.d.). The solvent should be routinely degassed for optimum precision and to prevent formation of air bubbles.

One of the more modern and interesting approaches to solvent delivery is the use of two pump heads in series in contrast to the parallel configuration in most of the current systems. The two pump heads have different displacement volumes. The primary pump has twice the displacement volume of the secondary head. The two heads again operate 180 degrees out-of-phase. During the forward stroke of the primary head, only 50% of the solvent enters the column while the remainder goes to the secondary head. As the primary head returns to draw the fresh solvent, the secondary head delivers the fluid to the column. The secondary head is also called the accumulator. The accumulator pumping system is relatively simple both in principle and construction and is thus much less expensive without significantly compromising on the precision. Modern solvent-delivery systems are often microprocessor-controlled, for greater precision, programmability and reproducibility.

### 7.2.3 Gradient formers

The general elution problem and application of gradient elution as a solution to the problem have been briefly discussed in the previous chapter. The technique of gradient elution will be discussed again later in this chapter. The equipment available for the formation of solvent gradients is briefly discussed here; see also Refs. [2 and 6]. There are two distinct types of gradient former in HPLC; i.e. those that effect the mixing (a) at low pressures and then pump the mixture at the required high pressure into the column, and (b) at high pressures using two or more high-pressure pumps to mix the solvents under high pressure just before delivering the mixture to the column. Either of the systems may be used for obtaining stepwise or continuous change of solvent composition. The change of solvent proportion can be programmed either linearly or exponentially; more sophisticated, microprocessor-controlled programmers can generate more complex profiles. The choice of the particular type of equipment, as usual, depends on the application and economics. Low-pressure gradient mixers are generally more versatile and relatively inexpensive.

A very simple device for a linear gradient mixing at low pressure was described earlier (Fig. 6.1c); the outlet may be connected to a high-pressure pump for HPLC. More sophisticated gradient generators are, of course, needed for more complex multi-solvent programmes normally used in HPLC. The required number of reservoirs containing the individual solvents are connected to the mixer which feeds the high-pressure pump. Understandably, only the reciprocating type of solvent pumping systems can be used for the purpose. The individual solvents are drawn into the mixer by microprocessor-controlled proportioning valves. Up to 20 solvents can be mixed in different proportions using these systems. If the duty cycle of the valves and the filling period of the individual piston heads of the pump are synchronized, very precise and reproducible gradients can be obtained in the low-pressure mixers. All the solvents, however, should be thoroughly degassed before mixing; built-in continuous degassers can be conveniently used in these systems.

Alternatively, two or more high-pressure pumps can be used to draw the individual solvents in the desired proportions into a low-volume mixer prior to the delivery to the column. Any of the different types of high-pressure pumps described above; namely, the syringe, or the reciprocating or pneumatic amplifier-type may be used for the purpose. The gradient is formed by progressively increasing the pumping rate of one pump while correspondingly decreasing that of the other so that the total flow-rate is constant. This is achieved using either a dedicated solvent programmer or the microprocessor of a computer-controlled chromatograph. Since the two pumps can be independently controlled, almost any gradient can be generated. However, since the precision of the reciprocating pumps is relatively poor at very low flow-rates, reproducibility of the gradient may be less than satisfactory at the initial stages and, less critically, at the final stages of the gradient unless special devices like pulse-width modulators are used. When very low flow-rates are to be used, as in microbore LC (see later), it may be necessary to reduce the speed range of the motors driving the pumps, for example, by a factor of ten and correspondingly reduce the mixing volume by a suitable device [6].

One drawback of the high-pressure gradient system is the need for more than one of the rather expensive high-pressure pumps and this feature would make the system unattractive for ternary and more complex gradients, although the modern, microprocessor-based gradient controllers can handle several pumps simultaneously. While the use of two reciprocating pumps offers an unlimited number of profiles as well as the quantity of solvent that can be used, the differing compressibility of the two solvents at a given pressure can lead to erroneous gradient profiles and flow-rates. Another source of error in these systems is possible mismatch of the two pumps; i.e. unless the two pumps deliver exactly the same volume of the solvent when used independently under given settings, it is possible that the actual gradient may differ from the instrument settings.

## 7.3 SAMPLE INTRODUCTION DEVICES

Sample introduction into an HPLC system is more difficult, and critical, compared to GC, for several reasons. First, the former system is under high pressure at the column inlet end and the conventional injection mode is difficult. Secondly, the diffusion rate of the liquid mobile phase is five orders of magnitude lower than that of the gaseous phase, requiring the design of the injection port to reduce the dead volume and poorly flushed regions considerably. Thirdly, the mobile phase in LC, unlike in GC, is not inert and great care needs to be taken in the choice of material for construction of the injection ports. A variety of strategies are employed to overcome the special problems of sample introduction in LC, which are briefly discussed here.

### 7.3.1 Septum injectors

This method of sample introduction is very similar to that in GC and is probably the simplest of all. Indeed the design can be so simple that it can be constructed in the laboratory using readily available materials. Fig. 7.5a shows a typical on-stream septum injector for HPLC. However, since the system is under high pressure, greater care and skill are required in the injection technique. The problems with this type of injector are: (a) lack of precision, the reproducibility being rarely better than

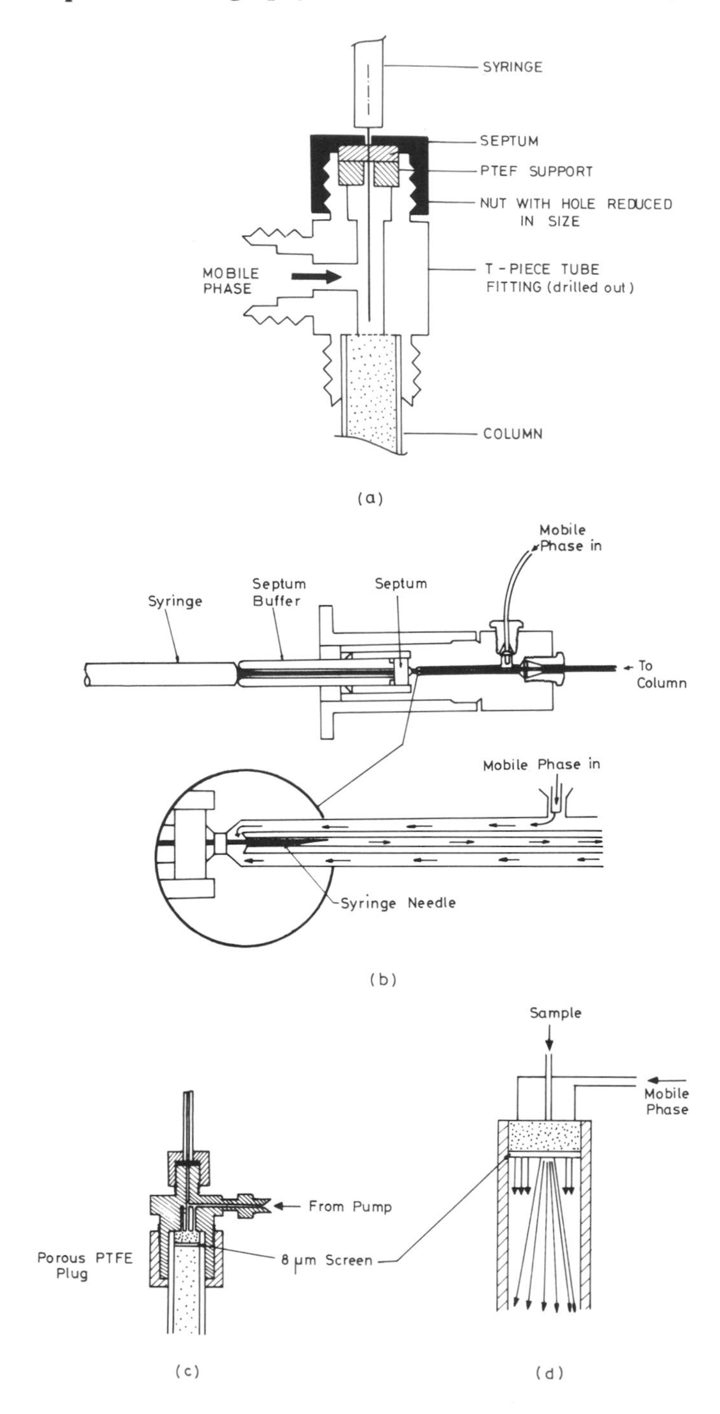


Fig. 7.5 — (a) On-stream septum injector for HPLC; (b) low-volume syringe injector (by courtesy of DuPont Instruments, Inc.); (c) co-axial sample injection; (d) illustration of co-axial sample introduction with suppression of wall effects; adapted from *Journal of Chromatographic Science*, Ref. 7, by permission of Preston Publications, Inc.

2%, (b) blocking of the column inlet with the small particles of the septum, (c) unsuitability for working under high pressures and temperatures, and (d) the possibility that the septum may be attacked by the mobile phase. Elastomeric materials such as silicone or neoprene rubber septa are used, sometimes with a solvent-resistant Teflon coating on the side exposed to the solvents. The recent popularity of reversed-phase LC with aqueous mobile phases may considerably reduce the problem of septum stability to solvents.

The precision of sample introduction is much better with on-column injection but, again, with the drawback that packing material is likely to be disturbed and transfer of the sample centrally on the bed is also not easy. This problem may be overcome by introducing a capillary tube immediately ahead of the column as in Du Pont's design (Fig. 7.5b). Alternatively, the sample may be injected into a small bed of inert glass or Teflon beads in such a way that the mobile phase is split into two co-axial streams [7]; the inner stream flushes the sample on to the column bed while the outer stream maintains the flow profile (Figs 7.5c and 7.5d). The main advantages of the syringe injection are that the sample volume may be changed if desired and the system is very inexpensive. The former feature is also true of some septumless sampling devices; described below.

### 7.3.2 Sampling valves

These sample-introduction systems are the most widely used in contemporary HPLC. The superior characteristics include: (a) usability at high pressures and temperatures without interrupting the flow of the solvent through the column, (b) rapid and reproducible introduction of sample volumes up to several millilitres with less than 0.1% error, and (c) ease of automation. The principle of operation of these sampling valves is shown in Fig. 7.6. The typical loop injector consists of a six-port valve; in the fill or load position (Fig. 7.6a), the sample is introduced into the loop while the solvent flows through the column, bypassing the loop. A simple turn of the valve changes the system into the inject position (Fig. 7.6b), bringing the sample-filled loop into the stream, between the pump and the column. Sample loops of different capacities varying from 0.5 to 5000 $\mu$l may be used in these systems. A fixed-volume (2 ml) sample loop is incorporated in the Waters Associates' U6K injector (Fig. 7.6c). In the normal load position, the flow channel bypasses the sampling loop, which is filled with the solvent and connected to the vent or drain. When the sample is introduced by means of a syringe, a volume equal to the sample volume of the solvent is displaced through the vent. Any injected volume less than the capacity of the loop (2 ml) is retained in the loop. To prevent band spreading, the valve is quickly turned to the inject position to sweep the contents of the loop on to the column.

In the loop systems, the sample volume is accurately controlled by the loop's capacity so long as the loop is completely filled with sample solution and the excess is allowed to drain. If less than the rated capacity of the loop is injected, the volume accuracy is dependent on that of the microsyringe, assuming that all the injected sample is swept on to the column. But, if the volume of the injected sample is more than half that of the loop, the possibility of a part of the sample becoming diluted and draining out exists, and the accuracy of sampling is lost. The six-port valve systems are also useful for a variety of multidimensional-column programming techniques, to be discussed later.

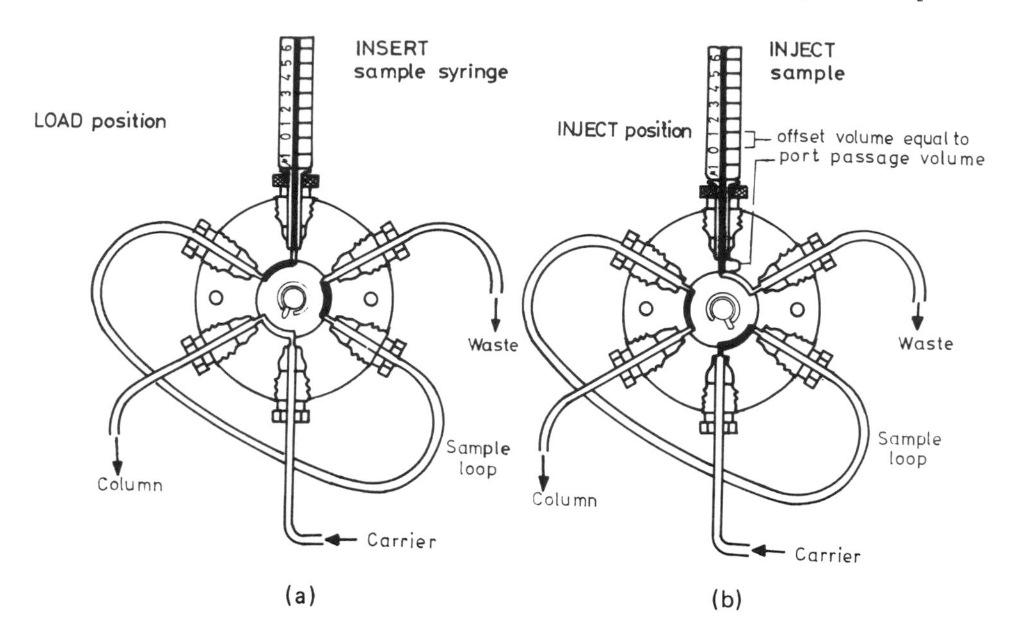


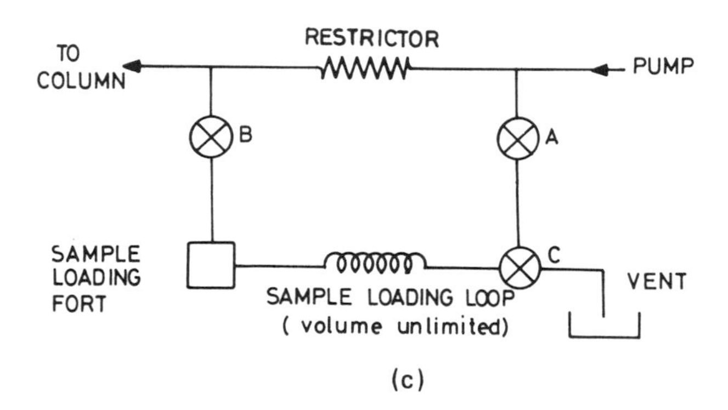


Fig. 7.6 — A 6-port high-pressure sampling valve (by courtesy of Valco Instruments, Inc.); (a) load position, (b) inject position. (c) A septumless loop injector; illustration by courtesy of Waters Associates, Inc.

### 7.3.3 Automatic injectors

The automatic injectors or autosamplers are very convenient if a large number of samples are routinely analysed, or when repetitive injection of the same sample is required (as in an unattended (computer assisted) solvent optimization programme (see later). Several versatile autosamplers are available which may be either stand-alone systems or controlled by the LC computer. Most of these are pneumatically actuated or motorized sample-introduction devices triggered by an electronic signal from the computer. The samples, up to about a hundred in number, may be loaded on to the autosampler magazine in appropriately capped vials. In some of the more sophisticated systems, pre-dissolution of the sample is also not necessary. The autosamplers may be commanded to inject variable volumes of samples at given time intervals and sequences.

## 7.4 THE COLUMN

### 7.4.1 Column materials and configuration

The column, which usually means the cylindrical tube packed with the stationary phase and complete with the end fittings, is the heart of the chromatographic system. The column material, the dimensions, the shape, the packing material, the method of packing and the geometry of the end fittings, among other things, are of decisive importance to the performance of the chromatograph. Fortunately, a large variety of ready-to-use columns of guaranteed performance are commercially available and it is seldom necessary to prepare them in the laboratory. Nevertheless, a knowledge of the factors affecting the column performance will help to derive the maximum benefit out of the column in terms of its efficiency and life.

The tube material most frequently used for the preparation of HPLC columns is stainless steel in view of its strength and inertness under the normal chromatographic conditions. Where the inertness of the stainless steel is insufficient, tantalum or titanium columns may be used. Glass-lined metal columns combine the advantages of the inertness of glass and the mechanical strength of the metal, but these have not found much favour. Polyethylene and PTFE columns are not suitable for the internally pressurized systems but find use in the radially compressed systems (see below). Fused-silica columns find use in the capillary or open-tubular LC.

#### *7.4.1.1 Standard columns*

The dimensions of the columns are generally determined by the instrument configuration, the convenience of packing and the size of packing particles. During the early period of development of HPLC, long (1–2 m) and narrow columns were favoured. Such columns were actually needed in view of the popularity and availability of pellicular packings of 30–50 $\mu$m which permitted high permeability for the mobile phase through the column. With the advent of microporous, small-diameter packings, such long columns were found not permeable enough at moderate pressures. Even then 25–30 cm columns, packed with 10 $\mu$m particles, were considered necessary. The more recent trend has been to use even shorter columns with a length in the range of 3–10 cm, depending on the diameter of the packing particles. Packings with very narrow size distribution in the range of 3–5 $\mu$m diameter are now available. However, in view of the very large back pressure that these columns tend to develop, the columns have to be shorter. The preferred length is 3–5 cm for columns with 3 $\mu$m particles, and 5–10 cm for the 5 $\mu$m packings. Larger-bore columns are mainly useful for preparative separations. Narrower columns (called the small-bore columns) are now becoming popular for routine analysis, not so much for their improved efficiency but for the impressive reduction in the solvent consumption (see later).

Indeed, the earlier notion that the smaller the diameter of the column, the more efficient it is, has been disproved in several studies. While a 2 mm diameter column was considered ideal for pellicular supports of 30 $\mu$m diameter, a column of 23.6 mm could exhibit the same efficiency [5]. In fact, in columns packed with 5–10 $\mu$m particles, the efficiency of the column increases with the diameter up to a certain point, probably because of the reduction of the relative importance of the extra-column band broadening effects. Also, the sample injected at the centre of the column may not reach the column wall and is thus not exposed to the non-uniformity in the packing or the mobile-phase streams near the column wall. Beyond about 4 mm i.d., there is little or no effect of increasing column diameter. This is probably the reason for the 4 mm i.d. columns becoming nearly universal among the commercially available pre-packed columns. The thickness of the tube wall is determined by the strength of the column material. An i.d.-to-o.d. ratio of 0.5 is sufficient for the stainless steel columns; thick-walled (12 mm o.d., 3 mm i.d.) glass tubes can withstand up to 3000 psi. Since the HPLC columns are of relatively small dimensions coiling them is not a practical necessity but U-shaped capillary tubes may be used for coupling two or more columns, e.g. in size-exclusion chromatography.

The packing material in the column is retained by means of microporous frits. The pore-size of the frits should be smaller than the smallest particle diameter; a pore size of about half the mean diameter of the packing material is usually adequate. The frit is held in place by a nut of appropriate geometry and the column is connected to the injection port and the detector by means of capillary plumbing and Swagelok ferrule-nuts. The geometry of these couplings is critical since any dead volume would adversely affect the column performance. Zero-dead-volume adaptors for all types of commercial LC systems are available.

#### *7.4.1.2 Radial compression systems*

When a solid particulate matter is packed into a rigid cylindrical column, there is inevitably some interstitial space or the void volume between the particles and the rigid column wall. A large void volume contributes significantly to column inefficiency. The radial compression system of Waters Associates is a device that minimizes the void volume with the result that significant improvement of resolution could be observed [8].

The operating principle of the radial compression system is shown schematically in Fig. 7.7. It consists of a 10 cm long, 5 or 8 mm i.d. column, made of flexible high-density polypropylene and packed with 5–10 $\mu$m spherical packing material. The column is placed in the radial compression module, which exerts hydraulic pressure uniformly around the cylindrical column. Under the pressure, the stationary-phase particles and the inner wall of the column readjust themselves, leading to a minimum possible interstitial volume. Faster analysis is possible with these columns since, unlike in the rigid columns, higher mobile-phase flow-rate does not significantly decrease the efficiency. However, owing to the subsequent introduction of high-efficiency, small-particle columns, this concept (which requires additional hardware, i.e. the radial compression module) has not become very popular. The principle of radial compression, however, finds application in the preparative-scale LC (section 7.6.6); 50 mm i.d. polypropylene columns are commercially available for this purpose.

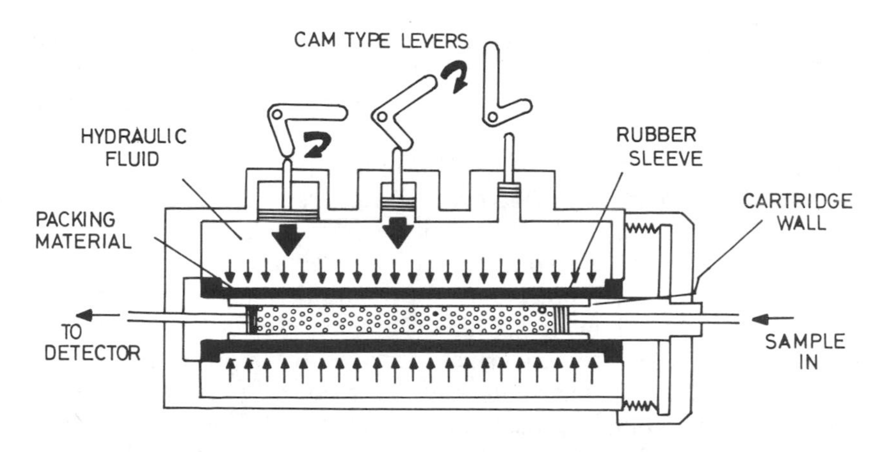


Fig. 7.7 — Schematic diagram of a radial compression system. Illustration by courtesy of Waters Associates, Inc.

### *7.4.1.3 Small-bore columns*

It is common knowledge that the efficiency of the packed column in chromatography depends on the diameter of the particles used for the packing. While 30,000 to 45,000 plates/metre is normal with the 10 $\mu$m diameter particles, 5 $\mu$m and 3 $\mu$m diameter particles could generate 60,000–80,000 plates/metre in the former case, and around 120,000 plates/metre in the latter. However, because of the high pressure drop across the columns, which increases rapidly with the column length and inversely with the particle diameter, it has not been possible to obtain ultra-high efficiencies using conventional HPLC columns. Thus, the optimum length of the columns in HPLC has generally been proportional to the particle diameter, so that the total number of plates generated have been comparable; e.g. the 10 $\mu$m particles are used to pack 10–25 cm columns, while the usable length for 5 $\mu$m and 3 $\mu$m particles is generally 5–10 cm and 3–5 cm, respectively. It may be remembered that in the discussion on the van Deemter equation for packed columns (Chapter 2), the column diameter did not appear as a factor contributing to the peak variance. Indeed, the column diameter has very little influence on the plate height and packed columns of 1 – 10 mm i.d. have shown similar efficiencies. Nevertheless, the small-diameter columns have received considerable attention in recent years and show promise for widespread application in analytical chemistry [9–14].

There has been considerable confusion on the nomenclature of the small-bore columns. Much of the developments in conventional HPLC and its instrumentation were made using columns of about 4 mm diameter. Thus, when about 1 mm diameter columns were investigated in LC, they were called microbore columns, and soon, terms like narrow-bore and packed-capillary columns appeared. Sagliano *et al.* [14] have suggested a classification of the HPLC columns on the basis of the reduced diameter, $\delta$, which is defined as the ratio of the column to particle diameters. Thus, open-tubular columns with no packing material have $\delta$ value of zero while conventional LC columns of diameter between 3 and 6 mm have $\delta$ values between 600 and

2000. All columns with $\delta$ values between 1 and 600 are proposed to be called small-bore columns; i.e. packed columns with diameters ranging from 3 $\mu$m to 2 mm. Within this group of small-bore columns, further sub-division as packed capillaries ($\delta$ = 1–100, or less than 500 $\mu$m i.d.), microbore ($\delta$ = 100–300, $d$ = 0.5–1.0 mm) and narrowbore columns ($\delta$ = 300–600, $d$ = 1–2 mm) is possible.

The small-bore columns may be made of stainless steel, plastic, glass or fused silica and can be coiled, provided the coil diameter is not less than 12 cm; a 3 cm coil gives only about 25% of the efficiency of the original column. Packed capillary columns of 100–200 $\mu$m i.d. are prepared from a thick-walled glass capillary tube (0.5–2 mm i.d.) using a drawing machine. These are packed prior to drawing with 10–100 $\mu$m alumina or spherical silica particles that can withstand the high-temperature operation; the selectivity may be subsequently modified. Open-tubular (wall-coated or bonded-phase) columns of diameters in the range 50–100 $\mu$m may be prepared as described for those in GC. A comparison of the typical characteristics and operational parameters of LC columns of different diameters is given in Table 7.1.

The advantages of reduction in the column diameter may be offset by the constraints in the status of the present instrumentation, particularly in relation to the band broadening owing to extra-column effects [13]. As noted earlier, the total variance of the chromatographic peak is an additive function of variances arising out of all the contributing factors, or broadly, the sum of the variances due to the column (c) and extracolumn (ec) effects according to the equation:

$$\sigma_T^2 = \sigma_c^2 + \sigma_{ec}^2 \tag{7.1}$$

$\sigma$ being the standard deviation and $\sigma^2$, the variance. The variance due to the dispersion inside the column is given by:

$$\sigma_c^2 = \frac{V_R^2}{N} \tag{7.2}$$

Considering a 1 mm i.d. column, if the number of effective theoretical plates ($N$) is kept constant, the numerator, $(V_R)^2$, becomes 1/400 of the value in the conventional 4.6 mm columns. Since the acceptable variance due to extracolumn effects should not exceed 10% of that from the column, the demands on the instrument design increase exponentially as the column diameter decreases. All the sources of extracolumn band broadening; namely, the injection port, the connecting tubes and the detector cell volume, as well as the detector and the recorder response time constants, need to be optimized for deriving the full efficiency of the small-bore columns.

As already noted, the efficiency of the column packed with particles of a given diameter does not increase as the column diameter is decreased. In fact, in the light of the above discussion, it must be obvious that for a given HPLC instrument, there is a minimum limit of usable column diameter below which the performance of the

**Table 7.1** — Characteristics of LC columns of different dimensions†

| Parameter | Preparative | Semiprep/ analytical | Narrow-bore | Packed capillary | Open-tubular |
|---|---|---|---|---|---|
| Column diameter (mm) | 6–60 | 3–6 | 1–2 | 0.1–0.2 | 0.05 |
| Particle size ($\mu$m) | 20–50 | 3–10 | 3–5 | 10–30 | — |
| Reduced diameter | >1000 | 600–2000 | 300–600 | 30–100 | 0 |
| Column length (cm) | 20–50 | 3–30 | 100 | 300–6000 | 500 |
| Flow-rate (ml/min) | 5–500 | 0.1–10 | 0.001–0.5 | 0.005–0.05 | 0.00001–0.01 |
| Sample volume (ml) | 1–100 | 0.001–2.0 | 0.0001–0.1 | 0.000005 | 0.00005 |
| Sample size (mg) | 100–20,000 | 0.001–100 | 0.01–0.02 | 0.00001–0.0001 | 0.00001–0.00003 |
| Elution volume ($\mu$l) | — | 100–1000 | 2–50 | 0.001–1 | 0.2–2.0 |
| Solvent consumption (ml) | 0.1–5 | 50 | 0.5 | 0.1 | 0.025 |
| Pressure drop (psi) | 300–500 | 100–3000 | 25,000 | 1500 | 100–300 |
| Detector-cell volume ($\mu$l) | 10 | 5 | 1 | 0.1 | 0.1 |
| Detector-time constant (msec) | 2000 | 2000 | 40 | 20 | 10 |

† The figures are approximate and only indicative of the relative proportions. The actual figures may vary over a much wider range.

system rapidly deteriorates. Therefore, most simple separations are routinely carried out with short and wide (conventional) HPLC columns.

For better resolution of compounds with small separation factors (α), the column length may be increased for generating high plate numbers. On the other hand, if the length is increased, the pressure drop across the column, as well as the capacity factor and analysis time, increases unless the particle size is increased. Faster analysis would require higher flow-rates and this would increase the solvent consumption. In order to reduce the solvent consumption at high linear velocities, one may use a small-bore column, packed with larger particles. Indeed, the early small-bore columns were packed with pellicular particles of larger than 20 μm diameter. These columns were of about 1 metre length and multiple columns were connected in series to generate up to 1 million theoretical plates.

Though the plate numbers achievable with the long, and less than 1 mm diameter, columns seem attractive, use of these columns requires considerable modification of the instrumental design, as discussed above. The present trend, therefore, is to use the 2 mm diameter columns (10–30 cm length) packed with 3–5 μm particles and take advantage of the intrinsic features of the small-bore columns which are impressive enough even without the high plate numbers claimed earlier. The operating parameters for these columns are obtainable simply by dividing the corresponding parameters (sample volumes and flow-rates) of conventional columns by the ratio of the cross-sectional areas. Thus, an injection volume of about 1 μl and a flow-rate in the range of 10–500 μl/min are typical. These are achievable without major modifications of the conventional HPLC systems. The standard dual reciprocating pumps, capable of delivering 0.1 to 10 ml/min are readily modified to the 0.01–1.0 ml/min range using a scaled-down function generator in the microprocessor. It was noted earlier that the main disadvantage of the pulse-free syringe pump was the limited solvent-holding capacity. Since this would not be a limitation with the small-bore columns, they may be considered a useful alternative to the expensive reciprocating pumps.

The main advantage of the small-bore LC systems appears to be the considerable reduction in the solvent consumption and the consequent saving of the high-purity solvents and expenditure in laboratories where large numbers of analyses are carried out, for example, as in quality-control and clinical laboratories. To maintain a given efficiency (linear velocity), the flow-rate (and thus solvent consumption) is reduced in proportion to the square of the column radius. Thus, about an 80% reduction in the solvent consumption is achieved in changing from the conventional 4.6 mm i.d. columns to 2 mm i.d. columns. The low solvent consumption makes it possible to use the expensive, but sometimes desirable, deuterated and chiral solvents without increasing the analysis costs significantly. The very low flow-rates that are feasible at moderate head pressures give sufficiently high linear velocities to considerably reduce the analysis time.

The very low elution volume also enhances the mass sensitivity which makes the small-bore columns ideally suited for trace analysis, particularly in biological samples where the availability of the sample may be limited. Here again, the detectability reduces in proportion to the square of the cross-sectional area. A third important feature that is not immediately obvious is the very low heat capacity of the columns owing to their geometry. The importance of this feature lies in the fact that the resistance to flow of the viscous mobile phase through the packed column generates considerable heat inside the column and there is a temperature gradient from the beginning of the column to the end, and from the centre of the column to the wall. Owing to its narrow geometry, the small-bore columns can readily dissipate heat and can be easily thermostatted. The retention data from the well-thermostatted small-bore columns are thus likely to be more accurate and precise relative to those from the conventional HPLC columns. Another feature that is often claimed as an advantage in small-bore LC is the facile interfacing with the direct liquid introduction probe of the mass spectrometer, in view of the low volumetric flow-rates. However, this advantage is real only in cases where the high mass sensitivity of these columns is needed (i.e. when the available sample size is low). Otherwise, the effluent can be easily split to achieve the same result. The techniques of interfacing the small-bore columns with the mass spectrometer, Fourier-transform infrared spectrometer and electrochemical detectors have been reviewed [15–17].

In the light of the enormous success of capillary columns in GC, it was to be expected that similar reduction of the diameter would lead to highly efficient columns. However, because of the large difference (over five orders of magnitude) in the diffusion velocities of gaseous and liquid states, open-tubular columns have not been very successful in LC. It may be recalled that the resistance-to-mass-transfer term (or the $C$ term) in the van Deemter equation is directly related to the ratio of the diffusivities of the mobile and the stationary phases. This ratio being nearly unity with a liquid mobile phase, the $C$ term is large and thus contributes significantly to the relative inefficiency. Nevertheless, theory predicts [12] that the open-tubular columns (OTCs) can be more efficient than the packed LC columns albeit under rather impractical conditions. Thus, the OTCs are four to five times faster at a given efficiency and, at a given flow-rate, they are two and half times more efficient. But then, the OTCs are really effective and competitive only when the column radius is equal to the particle diameter used for packing the columns with which it is being compared and this could lead to insurmountable permeability problems. Thus, a 3 $\mu$m i.d. and 3 m long OTC may generate over a million theoretical plates at a pressure drop of 89 atm, but elution of a compound with $k = 3$ may take 4 h; the analysis time may be reduced to 1 h if the pressure is increased to 355 atm, but with an efficiency of half the number of theoretical plates.

While the above features may reveal an interesting possibility of generating several million theoretical plates in LC, the optimal conditions of LC with smaller than 1 mm diameter may be too demanding to be practical at the present time. Thus, while the theory predicts that the optimum i.d. for an open-tubular LC column is 5–10 $\mu$m, it is difficult to prepare less than 40 $\mu$m i.d. columns. It is also recognized that the ancillary equipment needs to be significantly modified to suit the low sample and mobile phase volumes. The requirement of miniaturization envelops all parts of the chromatograph. The plumbing should be very narrow and short to minimize band broadening due to dilution and dispersion effects. Ideally, the injection and detector systems should have capacities not exceeding 1 $\mu$l and 1 nl, respectively. Because of the high linear velocities of the mobile phase, the detector response-time constants have to be in the order of 10 msec. Falling-drop spectrofluorimetry, laser-beam detection, sheath flow, micro-electrochemical detection, micro-flame-photometry, thermionic or transport detection, LC-IR and LC-luminescence are some of the possibilities.

### 7.4.2 Stationary phases and packed columns

#### *7.4.2.1 Characteristics of stationary phases*

A variety of column packings for use as stationary phases in LC have been described in the preceding chapter. For the purpose of HPLC, these materials may be categorized as rigid gels, semi-rigid gels and soft gels. Rigid gels can withstand high pressures, up to 15,000 psi, and are based on silica. The semi-rigid gels are made from cross-linked polystyrenes and can withstand moderate pressures (up to about 5000 psi, depending on the extent of cross-linking and functionality. The carbohydrate-based soft gels such as dextran and agarose, which have found extensive use in the SEC of water-soluble biopolymers, are unsuitable for use under high pressures.

The column-packing materials for use in HPLC are commercially available in different forms. These may be porous or pellicular, spherical or irregular. The effect of the size, shape and porosity of the stationary-phase particles on the performance of the chromatographic system have been discussed in Chapter 2. In general, small, fully porous particles of uniform size (3–10 $\mu$m in diameter, preferably with a size variation of the largest to the smallest particle limited to 1.5-fold) are preferred over pellicular packings. The pellicular packings, which are characterized by their hard, non-porous core, covered by a layer of the porous adsorbent, are easier to pack and have superior long-term stability but have very low capacity. They are well suited for routine analysis, preliminary investigations and for use in guard columns (see later). Both spherical and irregular packings may be used for packing columns of comparable permeability, but the spherical particles are considered less susceptible to damage in handling. While calculations show that the smallest particles are the most efficient, there is usually a trade-off between the particle size, pressure drop, column length and the sample size. This is because, for a given mobile phase flow-rate (which normally determines the efficiency of separation), the operating pressure increases sharply with the decreasing particle diameter and the increasing column length. Therefore, 3–5 $\mu$m diameter particles are used for high-resolution analysis while 20–80 $\mu$m are used for preparative-scale chromatography, where high solvent throughputs are needed.

Though silica gel has been the most popular stationary phase in traditional open-column chromatography and TLC, it has largely been overtaken by the reversed-phase chemically-bonded packings in most routine analytical applications in HPLC. The reasons for this are not hard to understand. A majority of the applications being in the area of biological sciences, the analytes are generally water-soluble, or the solute–stationary phase–mobile phase interactions could both qualitatively and quantitatively reflect their interactions in the biological systems (see, quantitative structure–activity–chromatographic mobility relationships, Chapter 9). Selectivity is more easily achieved in the reversed phases by slight variation of the carbon-chain length and may be accentuated by a variety of modifications possible in the aqueous mobile phase; namely, the variation in the pH, ionic strength and organic modifiers. The mobile phases are also relatively inexpensive, compatible with a wide variety of detectors, non-toxic, and large quantities may be handled without laboratory hazards (e.g. fire). Nevertheless, silica gel has areas of applications where it is ideally suited. For example, it is still the most favoured adsorbent for preparative or process-scale separations where large quantities of inexpensive packings are to be used; easy recoverability of the solute and the solvent from the effluent by distillation is a great cost advantage. Silica gel and other normal-phase packings are also preferred for separation of compounds by class and for separation of isomers. The reversed-phase packings, on the other hand, may be preferred for separation of homologues and for compounds which are too strongly retained, or which move too fast on silica. In general, the choice of the stationary phase should be such that it accentuates the differences that exist in the compounds to be separated. Thus, if the difference is in the nature and the number of polar functional groups, the normal-phase packings may be the first choice.

#### *7.4.2.2 Commercially available column-packing materials*

There are literally hundreds of column-packing materials that are offered for use in HPLC. A rather complicating feature, which is understandable in view of their commercial interest, is the tendency to designate the material of identical chemical nature (but probably of slightly different chromatographic performance characteristics) by different trade names. Thus, there are over 500 products (including nearly 200 size-exclusion packings of different porosities), made up of about 25 functionally different materials. Of these, there are about 60 porous silica and 40 octadecylsilyl-bonded silica phases, 20 each of cyano-, amino- and octylsilyl-bonded phases and ten each of silica- and polystyrene-based anion and cation exchangers (for a nearly complete list of the commercially available packing materials for HPLC, see Ref. 5). The chemical nature of these materials has been discussed in the previous chapter. Even among the 25 or so chemically different materials, hardly a dozen are of wide applicability and these account for over 90% of the usage. There needs to be a conscious effort to use the chemical terms alongside the trade names. Alternatively, the nomenclature of the materials should reflect the manufacturer's name, the chemical nature and the particle size, etc. A three-letter code is suggested for the commonly used packings (see Table 7.2), to be used along with the trade names; e.g. LiChrosorb-ODS and Bondapak-ODS. They may be further characterized by suffixes denoting the particle size, porosity and surface area of coverage of the bonded phases; e.g. Whatman-ODS-10 instead of Partisil-10-ODS-2. At the moment it is difficult to recognize that Chromosorb LC-6, Hitachi Gel 3030, Chromsep SI, Spherisorb S5W and VYDAC 101 TP, along with 50 others, represent silica packings. Fortunately, there is some standardisation with respect to the particle diameter and many of the packings now offered have a diameter of around 10 $\mu$m or in the range of 3–5 $\mu$m.

The packings most commonly used in HPLC are listed in Table 7.2, roughly in the order of their increasing surface polarity; i.e. from reverse-phase to normal-phase packings, followed by the ion-exchange packings. While the polystyrene gels are usable for reversed-phase LC, their main applications are in the area of non-aqueous size-exclusion chromatography. The octadecylsilyl (ODS) packings are the most widely used among the reversed-phase packings, but better peak symmetry is claimed with octylsilyl-bonded (OSB) phases, probably because of their superior wetability and solute mass-transfer characteristics. There is generally less retention on these packings, permitting the use of mobile phases containing a lesser proportion of organic modifiers. The phenyl-bonded silica (PBS) phases are similar

**Table 7.2** — Commonly used HPLC packings

| Packing material | Code† | Typical applications |
|---|---|---|
| Octadecylsilyl-bonded silica | ODS | General-purpose RPLC |
| Octylsilyl-bonded silica | OSB | Solutes strongly retained on ODS packings |
| Phenyl-bonded silica | PBS | Fatty acids, peptides, and solutes with low $k$ value on ODS packings |
| Cyanopropyl-bonded silica | CPB | General-purpose NPLC/RPLC |
| Diol-bonded silica | DBS | Organic acids and aqueous SEC |
| Aminopropyl-bonded silica | APS | Carbohydrates, ion-exchange |
| Controlled-pore glass | CPG | Aqueous SEC |
| Microparticulate silica | SIL | General-purpose, NPLC |
| Cross-linked polystyrene | CPS | Organic SEC, RPLC |
| Silica-based cation exchanger | SCX | Nucleic acids, metals, amines and nitrogenous bases |
| Polystyrene-based cation exchanger | PCX | Amino-acid analysis, carbohydrate analysis (Ca+ form) |
| Silica-based anion exchanger | SAX | Proteins, nucleic acids, phosphorylated nucleotides |
| Polystyrene-based anion exchanger | PAX | Nucleotides, organic acids, purines and carbohydrates |

† Suggested.

to the OSB packings but with a certain degree of selectivity to aromatic compounds. The cyanopropyl-bonded (CPB) and the diol-bonded silica (DBS) phases are of similar polarity, midway between the reversed- and the normal-phases and can be used in either mode by changing the relative polarity of the mobile phase; however, frequent changes from the normal-phase to the reversed-phase mode, and *vice versa*, are not recommended. They have similar selectivity in the normal phase as silica but with less retention; because of the absence of acidic silanol groups, there is less tailing. There is also more rapid response to changes in the mobile-phase composition than that observed with silica. The DBS phases also find extensive applications in the high-performance size-exclusion chromatography of water-soluble polymers. Silica gel has been the most versatile and widely used stationary phase in LC for a long time. Its suitability for a given application and the appropriate mobile phase can be conveniently determined by TLC. The aminopropylsilyl-bonded (APS) phase is a polar stationary phase that also shows weak anion-exchange properties in aqueous acidic medium but its most frequent application is in the area of carbohydrate analysis. The major drawback of these packings is their high chemical reactivity towards a variety of substances, particularly carbonyl-containing solutes and dissolved oxygen or peroxides in the mobile phase. The last four packings given in the table are strong anion and cation exchangers with a wide range of applications discussed in section 6.6.

While the stationary phases mentioned above are adequate for most purposes, there are some speciality phases for optimum resolution or performance in specific areas such as amino acid or carbohydrate analysis, and for use in clinical, environmental and quality control laboratories, which perform repeated and routine analysis of specific classes of compounds and where speed of analysis is also of prime importance.

### *7.4.2.3 Column-packing methods*

Other things being equal, the performance of a chromatographic system depends on the efficiency of the column which, in turn, depends on the way it is packed. A uniform and densely packed column is desired so that band broadening due to the mobile-phase streaming and diffusion are minimized. Packings with particle diameter over 20 μm may be dry packed but smaller diameter packings are invariably slurry packed. Whatever the procedure, a few common precautions are needed. It must be ensured that the column is scrupulously clean and is free from internal crevices, scratches or protrusions. The end fitting and the retainer frits should not be clogged and should be held in place by an appropriate arrangement (e.g. by covering with a protective cap) during the packing.

Dry-packing procedure is similar to that described for the GC columns except that vibrational techniques are not used so that the possible segregation of the particles on the basis of the size is avoided. Sizing of the particles to ensure that all the particles have diameters between 20 and 40 μm is required in any case. Gentle and frequent vertical tapping or bouncing of the column at a constant rate (about 100 times per minute) and amplitude (1 cm) while adding the packing material in small portions (about 30 mg at a time) is recommended [5]. The bouncing motion is continued for at least 5 min after completion of addition of the packing material. Using this procedure, a 1 metre column may be packed in about an hour. Rotating the column at a speed of around 180 rpm improves the packing characteristics considerably. A mechanical device that simulates the motions is available and its use makes the process less tedious.

The slurry packing of 3–10 μm particles requires skill and practice. The balanced-density method (i.e. using a solvent that is of comparable density) is used to prepare the slurry to prevent size separation of the packing material. High-density solvents like tetrabromomethane and tetrachlorethylene are suitable. As a rough indication a mixture of the above two solvents in the ratio of 3:2 is suitable for inorganic packings like silica. The main drawbacks of these solvents is that they are toxic and the packing should be followed by thorough washing to remove traces of the halogenated solvents from the column. It may be noted, however, that the balanced-density solvent is not a very important factor when a packing material of very narrow size distribution is available. Methanol, chloroform and hexane may be used with the polarity adjusted to obtain sufficient wettability of the packing material. In the down-flow method of packing, the slurry (usually 5–15% by weight) is filled into a vertical reservoir, at the lower end of which the column with appropriate end fitting is connected. The reservoir is another column, about 1.5 times the length and diameter of the column to be packed. A sudden high pressure (5,000–12,000 psi) is applied at the top of the reservoir to force the slurry into the column. Since the column needs to be filled with a high impact velocity, the usual HPLC pumps are not suitable. High-capacity, pneumatic amplifier, constant-pressure pumps are normally used. The actual filling may take only a second or two but the pressure is applied for a few minutes to allow about 100 ml of the solvent to flow. The pressure is released and applied repeatedly to consolidate the packing. The column is disconnected from the reservoir and the proper frit and end nuts are fitted. The slurry may also be pumped upwards, if desired. One metre long columns may be packed with 3 μm particles using the upward-pumping method to give up to about 120,000 theoretical plates. However, there is no special advantage of the upward-flow method when packing the normal 25 cm columns with 5–10 μm particles [2].

With some practice, high-efficiency columns of reproducible performance can be prepared using the above procedures. Equipment is available for packing several columns uniformly and simultaneously. However, the investment of time and money in establishing such a facility is not fully justified unless a large number of columns are routinely used. Almost all the packing materials listed in Ref. 5 are available as pre-packed columns of various sizes. As noted earlier, 4–5 mm diameter columns of 25–30 cm length and packed with 10 μm particles, were the most popular a couple of years ago, but are now largely replaced by 10–15 cm columns packed with 4–5 μm particles, or even 3–3.5 cm columns packed with 3 μm packings: these are adequate for most analytical purposes.

### 7.4.3 Column performance

#### *7.4.3.1 Specifications and evaluation*

The performance of the LC columns needs to be specified and periodically evaluated to ascertain their reproducibility. If the column-performance characteristics drop to undesirable levels, they may either be replaced or, in some cases, repaired. The manufacturer's specifications usually accompany the prepacked columns which may be checked soon after procurement and occasionally thereafter. There is always a difference between what is desirable and what is practical and also between what is claimed as practical and what can be achieved in practice. A 10% difference between the manufacturer's claims and the user's observation may be acceptable as it could be due to the differences in the LC systems and the method of measurements. Larger differences may have to be attributed to damage to the column during transport or handling.

The specification that is most frequently used is the plate count or the number of theoretical plates, $n$. The $n$ value for the normal 25 cm column packed with 10 μm particles may be of the order of 10,000 ±20%; similar values may be expected of a 10 cm column packed with 5 μm particles. The methods of estimation of $n$ value have been discussed in Chapter 2. As already noted, the value of $n$ for a given column depends on a number of operating parameters which also need to be specified for meaningful evaluation. These include: the mobile-phase composition, flow-rate, test-sample components, injection volume and the amount, pressure drop, peak asymmetry, resolution, selectivity (i.e. the separation factor) and the capacity factor. The $n$ values for at least three components (preferably dissimilar) components with $k$ values at, or around, 0, 3 and 10 may be given. Other specifications that may also be given are the column dead volume, mean pore-volume (particularly for the size-exclusion packings) and exchange capacity (for ion-exchange packings). Characteristics such as particle diameter, amount of packing in the column and concentration of the organic phase in bonded-phase packings may also be given but direct determination of these parameters by the user may be difficult. Often, the test samples and test chromatograms, along with the operating conditions, are supplied with the column. Otherwise, the user may select a few compounds of his interest with $k$ values in the regions mentioned above and monitor any changes during the life of the column.

#### *7.4.3.2 Column maintenance*

The performance of the column is usually the limiting factor of the performance of the LC system as a whole. In view of the fact that packing of a high-performance column could be tedious or its purchase expensive, care should be taken to prevent its damage in use. While any column has a limited life, a well-maintained column may have a working life of 6–12 months, depending on the actual usage. Observation of some elementary precautions can prolong the life of the column considerably.

The most harmful factors for the column life are the particulate matter and strongly retained impurities coming from either the sample or the mobile phase. Procedures for sample preparation and clean-up and certain special chromatographic characteristics and purification of solvents are discussed later. To the extent possible, the sample should be dissolved in the same solvent that is to be used as the mobile phase; in case of gradient elution, it should be the initial solvent composition. This would help prevention of the possible sample precipitation in the injection loop, on the column bed or in the connecting tubes and frits. The sample should be filtered through a 0.2 $\mu$m frit prior to injection. Small microfiltration kits with 0.2 to 0.5 $\mu$m pore-size filters are available from Millipore Corp., U.S.A. and several other suppliers. These are very convenient for use in conjunction with a syringe with a Luer tip. Often, it is not possible, desirable or even necessary to attempt to analyse all the components of a sample. A preliminary clean-up of the sample to remove much of the unnecessary portion not only helps to decrease the sample size, and thus aid resolution, but also, possibly to remove components which are likely to be strongly retained on the bed. Disposable minicolumns, called Sep-Pak cartridges (Waters Associates, U.S.A) packed with silica, alumina and a variety of bonded phases are convenient for such preliminary clean-up. Often, a guard column of about 3 cm length is inserted between the injection port and the analytical column. These guard columns may be fitted with the same packing material or a material of the same chemical and adsorbent properties but, preferably, of the pellicular type for easy refilling. The most common design for a guard column consists of a pre-packed, disposable cartridge, retained in a stainless-steel casing. Use of a guard column may contribute to about 5–10% of the dead volume and, therefore, to the extra-column band broadening. Even when a guard column is used it is recommended that the samples be filtered using one of the above arrangements to prevent clogging of the injection loops and the connecting tubes.

The HPLC places high demands on the purity of the solvents to be used as the mobile phases. As large quantities of the mobile phases are pumped through small quantities of the stationary phases, minute quantities of impurities can seriously affect the column performance. As a first precaution, all solvents are routinely filtered through microporous sintered-glass funnels before use. In-line filters are also used as an extra precaution. In addition to the particulate matter which can be removed by filtration, most solvents contain impurities which affect the performance of both the column and the detector. The importance of the quality of water in the reversed-phase LC has been stressed by many workers. Small quantities of organic impurities may be strongly retained on the reversed-phase columns and get eluted during subsequent analyses yielding spurious peaks or unacceptable background noise. Special water-purification systems which remove the ionic, organic and particulate matter very efficiently are available from several suppliers (e.g. Millipore Corp., U.S.A.). Organic impurities may also be removed simply by pumping the water through a large-capacity reversed-phase column such as Magnum 9 ODS-2 (Whatman, Inc., U.S.A.). The column can be subsequently regenerated by flushing with methanol. The methods of purification and precautions in handling of common organic solvents are available in most laboratory manuals and texts. The purification methods depend both on the properties of the solvents and of the likely impurities. Thus, ethereal solvents such as diethyl ether, tetrahydrofuran and dioxan may contain organic peroxides that could cause explosions. Some chlorinated solvents are highly toxic in their own right whilst others may contain phosgene which is also poisonous, methanol and acetonitrile may contain acetone and acetic acid that interfere with detection, and so on. Peroxides from ethereal solvents are removed by treatment with ferrous sulphate solution and phosgene from halogenated solvents by water. Carbonyl compounds in alcoholic solvents form non-volatile derivatives with, say, 2,4-dinitrophenylhydrazine and can be thus removed. Most polar impurities in non-polar solvents may be removed by passing the solvents through a column of alumina, followed by distillation. HPLC-grade solvents are also available commercially. It must also be remembered that most packing materials, including silica, bonded phases and ion-exchange packings are unstable beyond a certain pH range.

#### *7.4.3.3 Column regeneration and repair*

When the performance of a column has fallen below acceptable limits, it can often be brought into use by suitable cleaning and reactivation. This can be done, without disturbing the packing in the column, simply by pumping a series of solvents; usually 20 column volumes of each solvent may have to be used. The choice and sequence of the solvents for a particular column are normally given by the supplier of the column. For silica and most normal-phase packings, the following sequence may be used: heptane, chloroform, ethyl acetate (acetone), ethanol (methanol), water, methanol, chloroform, heptane. For the reversed-phase columns, the sequence may be: water, methanol, dichloromethane, methanol, water, 0.1M sulphuric acid, water.

Sometimes the above regeneration procedure may not restore the original efficiency. Peaks may be badly tailing or splitting. The reasons for such behaviour can be many. Usually, prolonged operation at high pressures and flow-rates may cause a void at the inlet side of the column. The injector, inlet tubing or the frits might have been partially clogged; or, there might have been a lot of irreversibly adsorbed material at the inlet side of the column. For repair, the inlet and fittings are removed and washed with 6N nitric acid. A part of the packing (about 1–5 mm) may be removed with a micro spatula and the void filled with either 30–40 $\mu$m glass beads or pellicular packing of the same type as the original packing. This procedure is often preferred for convenience. Alternatively, the same packing material (5–10 $\mu$m) may be used for filling the gap. In this case, a thick paste of the packing, in methanol, is transferred into the gap and the solvent allowed to drain. More slurry is added to fill the gap created and the end fittings replaced. Methanol is pumped for 30 min at 4 ml/min and the inlet side of the packing examined for any void; if a gap is formed, it is filled as just described.

The columns, new or reconditioned, are generally stored in a solvent, tightly closed by metal screw caps. The appropriate solvent is usually marked in the manufacturer's literature; normally, it is hexane for silica gel and other normal-phase packings, and methanol for the reversed-phase columns.

## 7.5 DETECTORS

### 7.5.1 Characteristics of liquid chromatographic detectors

As noted earlier, the development of instrumentation in LC was greatly inspired by the enormous popularity and convenience of the instrumentation in GC. The single most convenient feature of chromatographic instrumentation is the on-line detection, so that qualitative and quantitative analyses can be carried out without having to collect and process the column effluents. On-line detection in LC has proved to be a much more formidable problem than on-line detection in GC. The close similarity of the properties of the mobile phase and the effluent containing the solutes (unlike in GC) has rendered the development of LC detectors more difficult. One solution to the problem could be removal of the solvent after elution and this has indeed been attempted in certain cases (see later); however, this approach is fraught with practical difficulties and is not favourable. Most often, the difference in one of the physical (or, rarely, chemical) properties of the mobile phase and the solution of the analyte in the mobile phase is taken advantage of for detection.

Many of the LC detectors depend on the change in some bulk property of the mobile phase (e.g. refractive index and conductivity), while others are solute specific (e.g. light absorption and fluorescence). The former type are called bulk-property detectors and the latter, solute-property detectors. Another way of classifying detectors, which is commonly used for GC detectors, is on the basis of their concentration or mass sensitivity. However, most detectors in LC are of the former type and this classification is not very useful. Classification of LC detectors as general-purpose and specific is also not favoured in view of the fact that most of the solute-specific detectors are tunable to obtain maximum response to certain types of compounds or, by compromising on sensitivity, can be treated as general-purpose detectors. In the present treatment the bulk-property detectors are discussed first, followed by the solute-property detectors and the recent advances in the combination of LC and mass spectrometry (LC–MS).

The above differences apart, the desirable characteristics of an on-line detector are the same as in GC. These include: stability (low drift and noise level), high sensitivity to different classes of compounds (preferably with the facility of tunability to attain specificity or selectivity to a certain class of compounds, when required), wide dynamic range, fast response, low dead-volume, non-destructivity and, preferably, insensitivity to changes in the mobile-phase flow-rate and the composition. The last-mentioned property is crucial for the usefulness of a detector in gradient-elution chromatography; e.g. the refractive index detector and several other bulk-property detectors can be used only in the isocratic mode.

While the search continues for a universal detector that satisfies the above requirements, the currently available detectors use a number of different properties of the solute or the mobile phase and each of them has its advantages and disadvantages. It may be of interest to note that the four detectors; namely, the refractive index, electrical conductivity, light absorption and fluorimetric detectors, introduced during the early period of development of LC instrumentation, are still the most frequently used. Only the basic principles of the most frequently used detectors are discussed here. For more detailed information on these and several other detectors, some of the recent monographs may be consulted [18–21].

### 7.5.2 Bulk-property detectors

#### *7.5.2.1 Refractive index detectors*

The refractive index of a medium, is defined as the ratio of the speed of light in vacuum to that in the medium, and is determined by the angle of refraction as the light passes through one medium into another. The use of this principle in LC detection is based on the fact that the refractive index of the mobile phase may be different from that of the mobile phase containing a solute, and the extent of difference is proportional to the latter's concentration. The differential refractometer is a bulk-property detector and its sensitivity is dependent on the difference in the refractive indices of the mobile phase (see Table 6.3) and the solution of the analyte in the mobile phase; the larger the difference, the more sensitive is the detection.

Three types of differential refractometers may be described. The Fresnel type is based on the principle (Fresnel law) that the light reflected at an interface is dependent on the angle of incidence and the difference in refractive index across the interface. The principle of this reflection-type differential refractometer is shown schematically in Fig. 7.8a, using a Du Pont instrument. Light from the source lamp (SL) passes through a source mask (M1), an infrared-blocking filter (F), an aperture mask (M2) and a collimating lens (L1). The collimated beams from M2 impinge on the cell prism at near-critical angle to obtain maximum sensitivity and linearity. The lenses L2 focus the reflected light on the photomultiplier (D), which is connected to an amplifier and recorder. Two different prisms need to be used to cover the ranges, 1.31 to 1.45 and 1.40 to 1.55. (Some instruments, e.g. Perkin-Elmer's LC-25 may have an additional prism to cover the range 1.25 to 1.40.) Though this detector has a limited linearity range, it has the advantage of very small cell volume (3 $\mu$l), which makes it an attractive choice for use with very high-performance columns.

The deflection-type differential refractometer is shown in Fig. 7.8b, which is self-explanatory. It measures the deflection of a light beam caused by the difference in refractive index between the mobile phase and the column effluent containing the analyte. The collimated light beam from an incandescent lamp passes simultaneously through the reference and sample cells, and the reflected beams from a suitably positioned mirror pass again through the cells and the collimator lens which focuses it on to an optical zero glass deflector and a photomultiplier. The angle of deflection, arising out of the difference in the refractive index between the two parts is thus measured and recorded. Using this principle, the entire refractive index range from 1.00 to 1.75 can be measured. A thermostat is normally included in the system to overcome the changes due to any difference in temperature between the reference

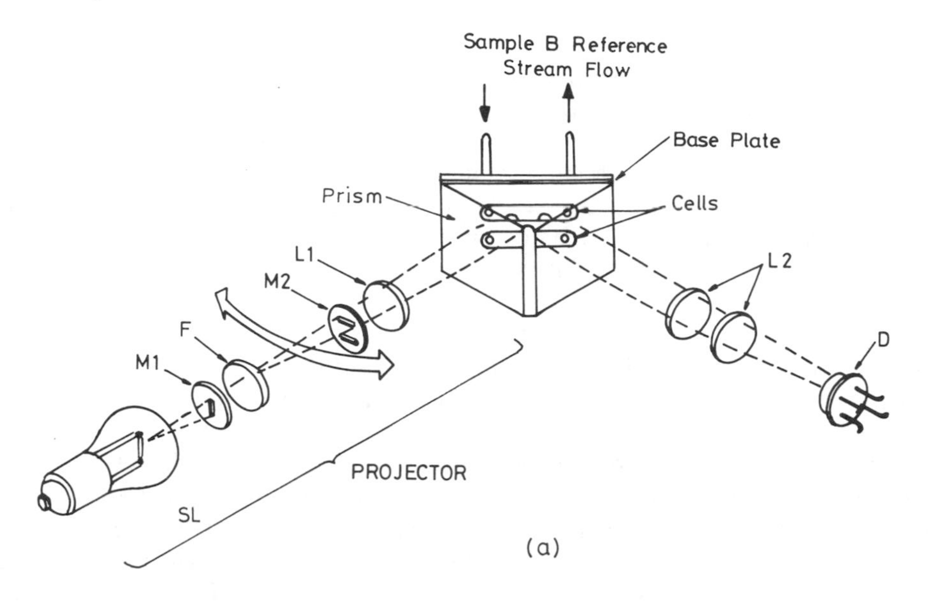


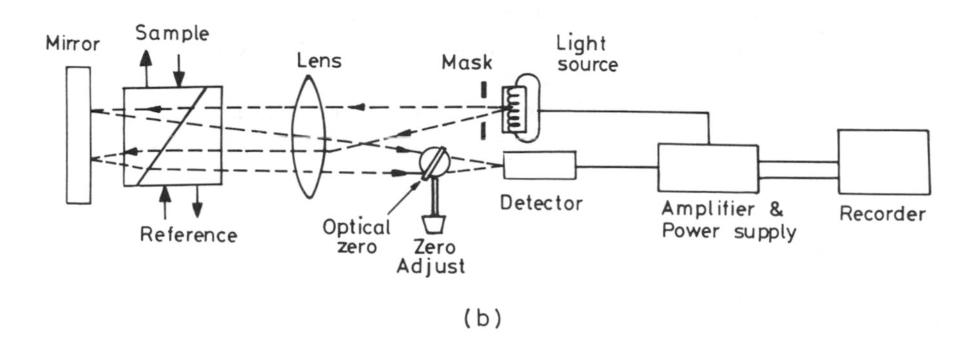


Fig. 7.8 — Refractive index detector. (a) Fresnel-type (by courtesy of DuPont Instruments, Inc.) and (b) deflection-type (by courtesy of Waters Associates, Inc.).

and the sample liquids. These detectors have a wide linear dynamic range of $10^4$ with a sensitivity of $10^{-8}$ RI units. Only one cell with a hold-up volume of about 10 $\mu$l is used.

In the shearing interferometric refractometer, the light beam is divided into two parts and focused on the sample and reference cells (5 $\mu$l). Owing to the difference in RI of the reference and the sample, the two emergent beams are out of phase. These are combined by a second lens and a beam splitter and the interference is measured. The response of this detector is claimed to be linear and the sensitivity as low as 0.1 $\mu$g of the solute, which is about ten times higher than the other RI detectors mentioned above.

The RI detectors are nearly universal in their sensitivity for detection of different classes of compounds and are therefore used frequently, particularly when the samples do not absorb UV radiation (see later). It is the preferred detector for use in the analysis of carbohydrates and in size-exclusion chromatography of polymers. In view of its universality, it is useful for preliminary screening of unknown samples. Its low sensitivity is also an advantage in large-scale preparative LC. The main disadvantages for analytical chromatography are low-to-moderate sensitivity, low stability to even minor temperature changes and unsuitability to gradient elution.

#### *7.5.2.2 Electrical conductivity detectors*

Electrical conductivity is exhibited by aqueous solutions of all ionic (organic and inorganic salts as well as strong acids and bases) and ionizable compounds (weak acids and bases). With the popularity of reversed-phase chromatography that essentially uses aqueous mobile phases, use of this property for detection of LC eluants is very attractive. Use of conductivity detection in ion chromatography has also been discussed in the previous chapter. The early attempts to use electrical conductivity were not very successful as the DC voltage that was used caused polarization to produce hydrogen and oxygen at the electrodes. Use of AC voltage, on the other hand, permitted measurement of the electrical properties of ionic solutions by suppressing the polarization effects. Ohm's law states that the current ($I$, in amperes) flowing through a medium under a given voltage potential ($V$, in volts) is inversely proportional to the resistance ($R$, in ohms) of the medium (e.q. (7.3)); in other words, the conductivity ($G$, in mhos) of a medium is equal to the reciprocal of its specific resistance, defined as the potential across the faces of a 1 cm cube when carrying 1 ampere of current.

$$I = \frac{V}{R} = GV \tag{7.3}$$

The electrical conductivity detector is relatively simple in construction, consisting essentially of a small chamber with two electrodes across which an AC voltage can be applied (Fig. 7.9). The detector cell must be of a small volume to minimize dispersion

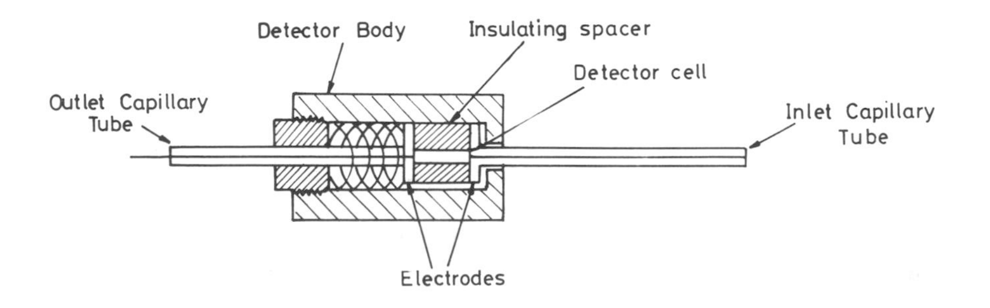


Fig. 7.9 — Schematic diagram of electrical conductivity detector. Reproduced from Ref. 20 by courtesy of Elsevier Scientific Publishing Co.

and the electrodes may be of an inert conducting material like stainless steel, gold or platinum. To prevent interferences from materials other than the sample and from temperature variations, the cell and the electrodes are well insulated and thermostatted. Interference from trace ionic impurities continues to be a problem since dissolved carbon dioxide from air can also cause significant noise. The detector, however, is reasonably stable to changes in flow-rate and pressure and can even be used in gradient elution provided the ionic concentration of the mobile phase is kept constant.

When an electric current is applied to the two electrodes, the cations in the sample move toward the cathode and the anions toward the anode. The conductivity of the solution is determined by the number, size, charge and velocities of the ions present in the medium. The velocity of the ions can be increased by increasing the field strength, and thus the signal is proportional to the applied potential. However, this potential must be carefully controlled since, above a certain level, electrolysis, oxidation (at the anode) and reduction (at the cathode) would occur and the resulting (Faradaic) impedance lowers the applied potential. A lower than optimal potential is also harmful due to the formation of an electrical double layer capacitance, which also lowers the effective potential on the sample.

The sensitivity of the conductivity detectors can be very high under ideal conditions (1–10 ng/ml of NaCl in water) but is subject to considerable interference from impurities that cannot be easily avoided. As noted earlier, it finds extensive application in ion chromatography and in the analysis of ionic components in foods, liquors and environmental samples.

When the mobile phase is a non-polar liquid that does not conduct electricity (i.e. a dielectric), the change in the capacitance owing to the presence of a solute can be monitored using a dielectric constant detector. This and several other bulk-property detectors, (e.g. thermal lens, thermal conductivity, differential viscometer, interferometer, density variation and density balance detectors are described in the literature but have not found wide application [20].

### 7.5.3 Photometric detectors

These are the most frequently used of the solute-property detectors and exploit three types of interaction of light with substances; i.e. absorption, fluorescence and scattering for detection of compounds. It is well known that when radiation passes through a medium, part of it is absorbed and the rest reflected, scattered or transmitted. The extent of absorption depends on the wavelength (or frequency) of the radiation and the property of the medium. It is also well known that the wavelength and the intensity of absorption of the radiation depends on the presence of chromophoric groups in the molecules. The common chromophores in organic compounds are unsaturated functional groups such as carbonyl, nitrile, aromatic and olefinic; these compounds absorb in the ultraviolet (UV) region (i.e. wavelength from 190 to 360 nm) and the intensity and wavelength of absorption increases if these groups are conjugated. Coloured compounds with highly conjugated chromophores, and which contain easily excitable electrons, absorb visible light (350 to 900 nm). The whole region from 190 to 900 nm is of interest to the present discussion. Radiation beyond 900 nm is the infrared (IR) region, where no electronic excitation occurs. It is absorbed by almost all compounds and results in the excitation of some interatomic dynamics such as stretching and vibrational frequencies. Certain compounds, exemplified by some polyaromatic and heterocyclic compounds, re-emit the absorbed radiation of a particular frequency in a phenomenon called fluorescence. All or any of the three modes of measurement, namely, absorption, transmittance and fluorescence, may be used to monitor the elution of the analyte from the LC column.

Light absorption is governed by the fundamental law of spectrophotometry, i.e. the Beer–Lambert law, which states that the fraction of the radiation absorbed is proportional to the concentration ($c$) of the absorbing species and the radiation pathlength through the medium ($b$), so that:

$$\log \frac{I_o}{I} = abc = A \quad (7.4)$$

where $I_o$ and $I$ are the intensities of the incident and the transmitted radiation, and $a$ is the proportionality constant; $A$ is called the absorbance. Since, in a given spectrophotometric system, $b$ is also normally constant, $A$ is proportional to the concentration and this relationship forms the basis of the use of light-absorption detectors for both qualitative and quantitative analyses. The value of $A$ at any given wavelength is characterisitic of the medium or, if the solvent is transparent to the radiation, of the solute. Each substance has a characteristic maximum absorption at a certain wavelength, depending on the molecular structure and a plot of the wavelength (or frequency) vs. the absorbance gives the *spectrum*.

#### *7.5.3.1 Fixed wavelength UV detectors*

This type of detector has been one of the first of its type (photometric detectors), inexpensive and the most frequently used. The optical system of the detector is shown schematically in Fig. 7.10a. It consists of a UV source (with or without a filter) that emits light of a certain wavelength, a collimator lens, two flow-through cells (one for the sample and the other for the reference) and photoelectric sensors. The difference in the signal output from the two sensors is amplified and fed to a recorder. The cadmium sulphide photocells used in the early instruments had a very narrow range of linearity with respect to the solute concentration, and thus required an amplifier with multiple ranges. The modern detectors use silicon diodes either singly or in arrays that offer excellent linearity of response with the intensity of the transmitted radiation.

The light source most frequently used is a low-pressure mercury lamp that predominantly emits radiation of a wavelength of 254 nm (253.7 nm to be precise) along with light of much lower intensity at 302.2 and 313.2 nm. Low-pressure zinc and cadmium lamps provide light of wavelengths 212 and 225 nm, respectively. The cadmium lamp emits high-intensity radiation of several discrete wavelengths (214.4, 226.5, 228.8, 283.6, 286.1, 298.1, 308.1, 313.2, 325.4, 326.1, 340.3 and 346.6 nm). This property may be used for independently measuring the absorption at different wavelengths using appropriate filters. The mercury lamp with its 254 nm radiation is by far the most commonly used lamp in fixed-wavelength detectors. It is essentially monochromatic with the other two bands being of much lower intensity. The

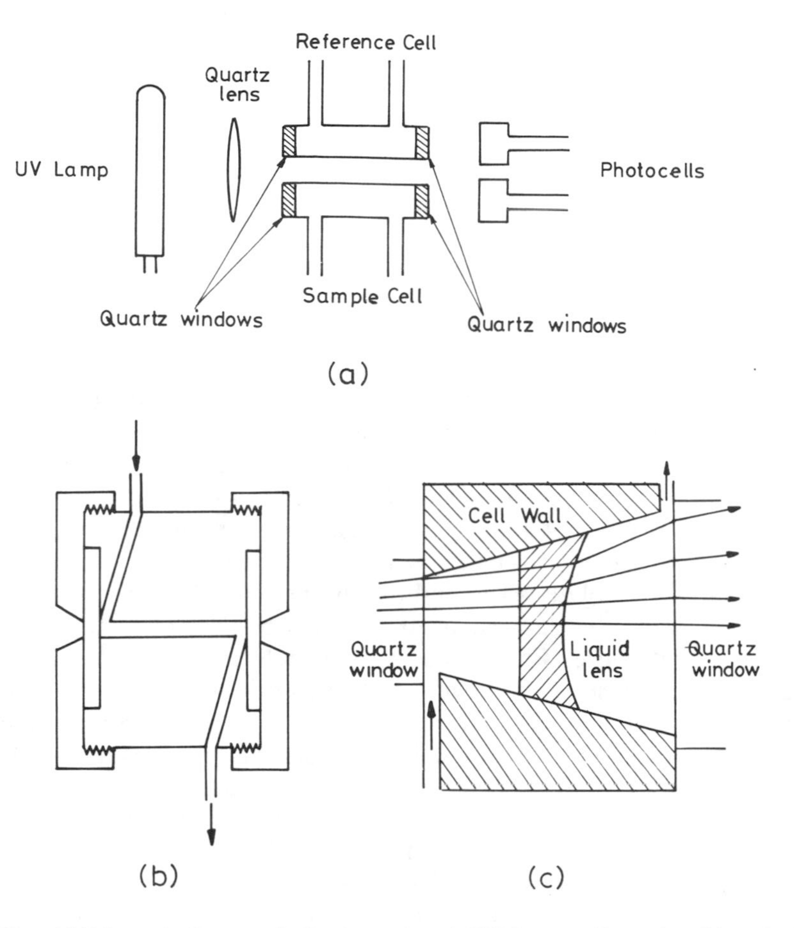


Fig. 7.10 — (a) Schematic diagram of a fixed-wavelength UV detector. Reproduced from Ref. 20 by courtesy of Elsevier Scientific Publishing Co. (b) Z-type UV cell and (c) tapered UV cell. Reproduced from Ref. 4 by courtesy of Elsevier Scientific Publishing Co.

mercury lamp also emits low-intensity radiation at lower than 200 nm but, since this radiation is often absorbed by the mobile phase, it is not transmitted through the medium.

The UV cells are very sensitive to changes in temperature and flow-rate. Excellent detector stability with low noise is obtained by suitably thermostatting the detector and bringing the temperature of the column effluent to the same temperature using appropriate heat exchangers. Thermostatting at a given temperature also has been demonstrated to render the system insensitive to flow-rate changes [20].

Considerable attention has been paid to the detector-cell geometry. Eq. (7.4) shows that the absorbance is proportional to the pathlength, which means that the geometry should provide as long a pathlength as possible. On the other hand, the requirement of high efficiency (low band spreading) calls for a low detector-cell volume. Early detectors used a narrow (1 mm diameter) tube of pathlength 1 cm, giving a cell volume of 7.5 $\mu$l. Modern UV detector cells have volumes as low as 2 $\mu$l or even 1 $\mu$l and different geometries; e.g. the Z-cell (Fig. 7.10b) and tapered cell (Fig. 7.10c). The tapered cell geometry helps to prevent radiation losses due to refraction.

#### *7.5.3.2 Variable wavelength UV detectors*

In the concept of the low-cost fixed-wavelength detector, advantage is taken of the fact that the low-pressure mercury-vapour lamp emits a nearly monochromatic radiation near 254 nm, and that most organic compounds, except the saturated hydrocarbons, absorb in that region. While strong absorption in the 254 nm region requires the presence of a conjugated chromophore, isolated carbonyl and olefinic compounds also have a finite, if weak, absorption at 254 nm and these compounds can be detected, albeit at low sensitivity. Several classes of compounds that have highly conjugated chromophores may have maximum absorption wavelengths extending to the visible region but may have very weak absorption (even a trough instead of a peak) at 254 nm. In all these cases except where the maximum absorbance of the solute is around 254 nm, the response of the detector at a given concentration of the solute is less and the full potential sensitivity of the detector cannot be realized; the 254 nm fixed-wavelength detector is thus, at best, an economical compromise. For high-sensitivity detection, it is required that the wavelength used is close to the maximum-absorption wavelength of the analyte. The multiple wavelength detection using the cadmium lamp and appropriate filters meets the situation halfway but leaves much to be desired. The desirable solution is a variable wavelength detector extending over the region 190–900 nm so that the wavelength of the incident radiation can be chosen to provide the maximum response for the given analyte, or class of compounds. Several of this type of variable wavelength detectors are now available at moderate cost.

The detector design is essentially that of a common double-beam spectrophotometer (Fig. 7.11) The light source is a hydrogen or deuterium discharge lamp for the UV (190–360 nm) and a tungsten lamp for the visible (350–900 nm) regions. The switch from the UV to visible region may be manual or automatic, depending on the selected wavelength. The light passes through a monochromator and a slit, and is then split into two beams, one of which passes through the sample and the other through the reference cell. The detector is amplified and fed to the recorder in the usual manner. The wavelength selection may be single or multiple for simultaneous detection at several wavelengths, depending on the capability of the microprocessor. It can also be manual, automatic or programmable. Depending on a known sequence of elution of the analytes and their maximum-absorption wavelengths, the detector may be programmed to change the wavelength of the incident radiation at the appropriate time. Alternatively, the instrument may be designed to scan the complete spectrum of a peak (using the stop-flow technique) and choose the appropriate wavelength for maximum sensitivity; the stop-flow technique can also be used to record the complete UV spectrum of the eluted sample.

#### *7.5.3.3 Diode-array detectors*

The continuously variable spectrophotometric detector described above is called a dispersion-type detector, signifying the fact that the incident light is dispersed prior

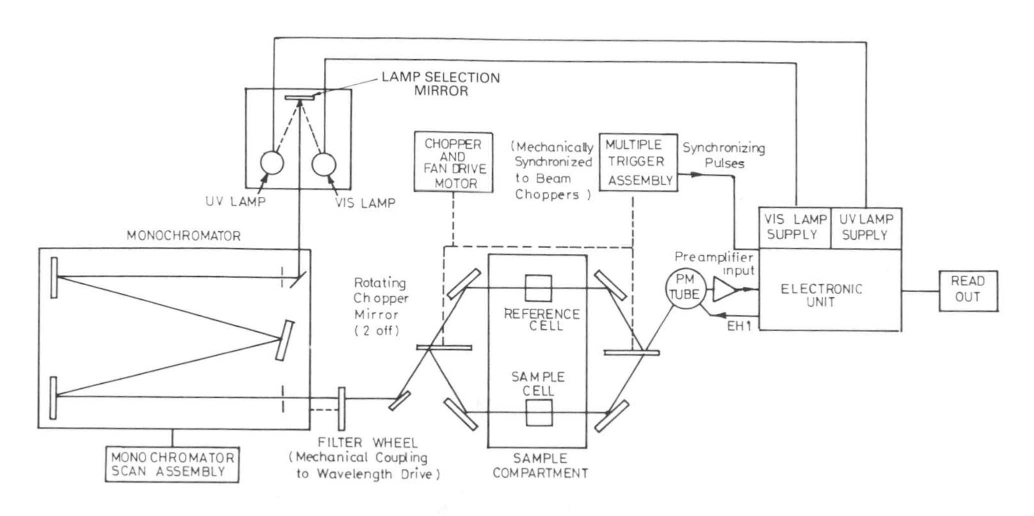


Fig. 7.11 — Schematic diagram of a variable-wavelength UV detector. Reproduced from Ref. 4 by courtesy of Elsevier Scientific Publishing Co.

to incidence by means of a grating or a monochromator so that the light entering the cell is of a given wavelength. The scanning over the range of wavelengths takes a certain time and requires stoppage of the flow of the mobile phase for that period of time. In the more recent diode-array detector [22], light comprising the complete range of wavelengths emitted by the source passes through the sample and is then dispersed over the array of diodes, each of which detects a discrete wavelength. The response of the individual diodes is then synthesized by a computer to give the normal UV spectrum. Thus, the full spectrum of an analyte from the column can be obtained in a few milliseconds 'on the fly' without having to stop the flow. It must be noted, however, that the incident light in this case is polychromatic and the light falling on the diode may or may not be entirely from the source; it may also be the light emitted by the sample by fluorescence. In other words, the spectrum obtained by the diode-array instrument may not be the true absorption spectrum. To obtain a genuine UV spectrum of the compound, the stop-flow dispersion spectrum is recommended, unless the solute is known to be totally non-fluorescent.

The main advantage of the diode-array system, however, is that, with the aid of the computer, the UV spectra of the eluant from the column can be continuously recorded for post-operative processing in a number of ways [23]. Fig. 7.12 shows the typical multidimensional, computer-synthesized output of a diode-array detector. From this, it is possible to optimize the output of the detector for each peak at the wavelength of its maximum absorption. Another advantage is the possibility of ascertaining the peak homogeneity and integrity. The peak homogeneity is established by extracting the spectra at different points on the peak; e.g. at the peak and at half-height on either slope of the peak, as shown in Fig. 7.13. The spectra from peak A (Fig. 7.13a) are all different, indicating that the peak represents more than one compound. On the other hand, the spectra from peak B (Fig. 7.13b) have the same

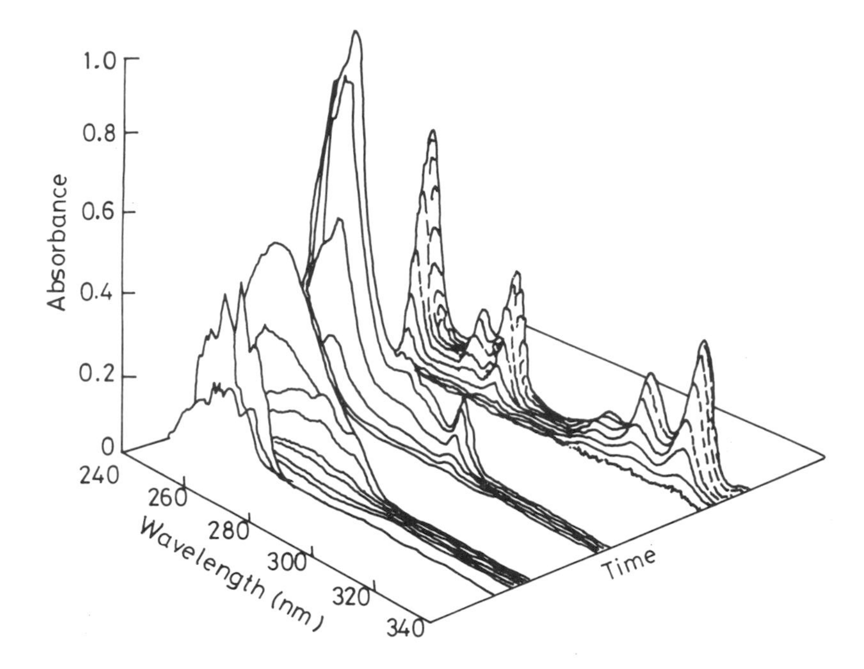


Fig. 7.12 — Typical output of diode-array detector. Reproduced from Ref. 23 the *Journal of Chromatographic Science* by permission of Preston Publications, Inc.

peak shapes and maxima, showing that the peak is homogeneous. It must be noted, however, that closely related compounds with the same chromophores (e.g. peptides) may have very similar UV spectra and this conclusion may be misleading if they also have identical retention times.

The photodiode-array detectors, though very versatile and powerful, are very expensive at the present time. They may thus be considered essentially research instruments for method development and other experimental activities, and not for routine applications.

#### *7.5.3.4 Fluorescence detectors*

Certain compounds (e.g. aflatoxins, riboflavins, polynuclear aromatics, porphyrins, etc.) absorb radiation of a particular wavelength (usually in the UV region) and upon excitation, emit radiation of a characteristic (usually longer) wavelength. This phenomenon, called fluorescence, obeys the equation:

$$F = I_0 \phi abc \tag{7.5}$$

where $\phi$ is the fluorescence yield and the other symbols represent the same parameters as in eq. (7.4). The intensity of fluorescence is often very high and the detection limit is thus much lower (by about 100 times) than the absorbance detectors described above. While the absorbance of radiation is monitored by measuring the

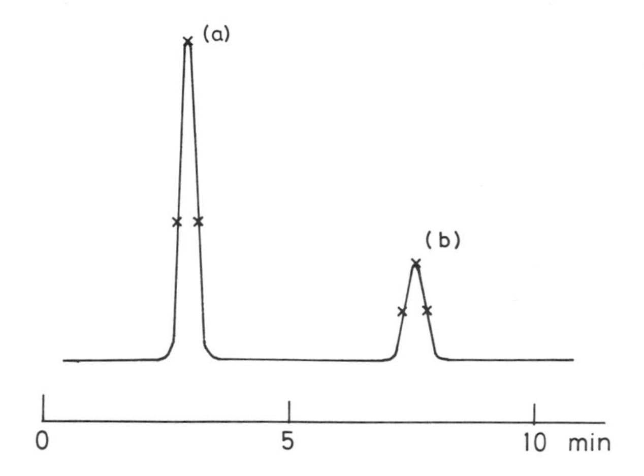


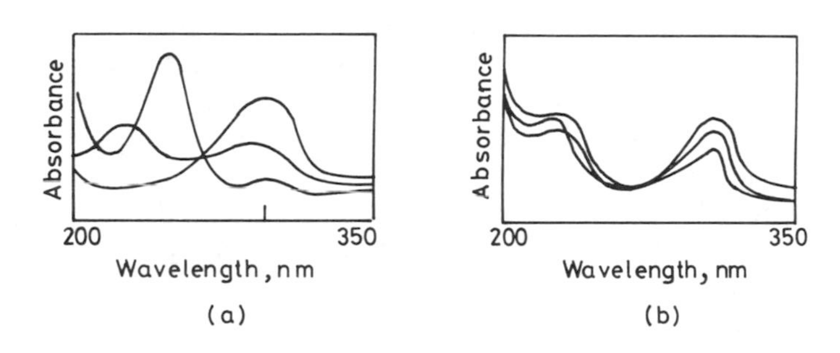


Fig. 7.13 — Detection of peak inhomogeneity by diode-array detector.

transmitted light on the opposite side of the detector cell, fluorescence is usually measured at an angle of 90°C to the direction of the incident radiation. It is thus possible to combine the two principles for simultaneous absorbance–fluorescence detection (Fig. 7.14). The advantage of the dual-detector system is that the fluorimetric mode may be used for low concentrations below the detection limit of the photometer and in the absorbance mode above the linear range of the fluorimeter. The fluorimetric detector has a linear dynamic range of $10^4$, but only at very low concentration; i.e. when the product of $a$, $b$ and $c$ in eq. (7.5) is less than 0.01.

A major drawback of the application of fluorimetry is that the phenomenon of fluorescence is exhibited by relatively few types of compounds, mainly those that are symmetrically conjugated and not strongly ionic. However, these include such important classes of compounds as drugs and pharmaceuticals, their metabolites, food additives, fossil fuels and environmental pollutants, which need to be detected at trace levels. Non-fluorescent compounds like amines and alcohols can be easily converted into fluorescent derivatives by treatment with reagents such as fluoresc-

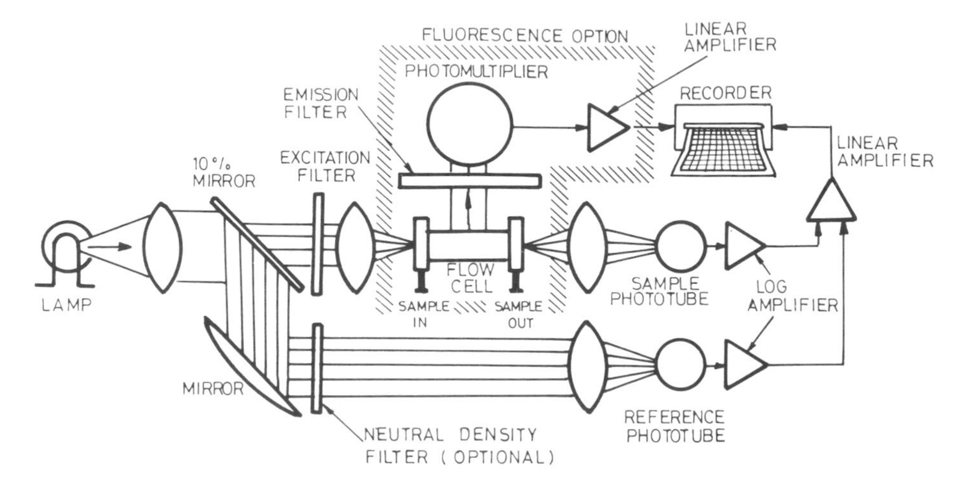


Fig. 7.14 — Schematic diagram of a combined UV-fluorescence detector. Illustration by courtesy of DuPont Instruments, Inc.

amine and dansyl chloride in order to take advantage of the high sensitivity of this technique (see section 7.5.5 on reaction detectors). The selectivity of fluorimetric detection may be enhanced using excitation and emission wavelengths that are characteristic of the solute. The sensitivity of fluorescence detection is also enhanced by increasing the intensity of the incident radiation; picogram-level detection is possible using laser-induced fluorescence detection. Amino-acid analysers based on post-column derivatization with *o*-phthalaldehyde, followed by fluorescence detection, are commercially available. The fluorimetric detectors make exceptionally high demands on the solvent purity as the phenomenon is very susceptible to interference and quenching from several sources; otherwise, they may be used for gradient LC also.

### *7.5.3.5 Infrared detectors*

As noted earlier, at the other end of the visible electromagnetic radiation is the infrared (IR) region (2–15 $\mu$m), exposure to which excites the vibrational, stretching, bending and other motions of chemical bonds. While the intensity of absorption of the radiation, and therefore the sensitivity of detection using this method, may be low, the infrared spectrum is a very characteristic feature of the molecule to the extent that it is called the fingerprint of the compound. It is thus useful in qualitative analysis for positive identification of the solutes. Apart from its low sensitivity, the two main features that prevented the routine use of IR spectroscopy in LC detection are the slow speed of scanning in most of the traditional IR spectrometers and the strong absorbance of most of the mobile phases (particularly water), without removal of which the IR spectrum of the solute could not be observed. Chlorinated solvents, because of their weak absorbance in the IR region were the most useful for LC–IR studies. Early experiments were conducted mostly by fraction collection and solvent evaporation. Transportation detectors (see later), in which the column effluent is conveyed on a wire or ribbon to a hot area to remove the solvent prior to

IR analysis was another alternative. Deuterated chloroform and a small-bore column also have been examined to minimize the interference from the solvent.

As with the GC–IR combination, the situation improved dramatically with the advent of the Fourier-transform infrared (FT–IR) spectrometers (also called the interferometers) and computers. With the help of these systems, IR spectra could be recorded 'on the fly' using very small quantities of the sample, either after removal of the solvent or by computer-assisted solvent-spectrum subtraction techniques. Obviously, this cannot be done when the solvent composition is continuously changing as in gradient elution.

The best approach to LC–IR appears to be that involving removal of the solvent [24]. In view of the large difference in the elution volumes of the peaks, the strategies adapted for the purpose are different for the conventional columns (elution volume about 1 ml) and the small-bore columns (elution volume around 50 $\mu$l). In the former case, a two-stage approach may be used. In the first stage, 80–90% of the solvent is removed by nebulization, and the remainder after depositing on a powder used as the matrix for IR spectroscopy (diffuse reflectance spectra). This technique furnishes the full IR spectrum with a sensitivity about two orders of magnitude higher than the flow-cell method. The matrices used most frequently for the purpose are alkali-metal halides, held in small cups. Carousels that can hold about 160 samples are available. This system again cannot be used with water-containing mobile phases since the alkali-metal halides are very hygroscopic and the moisture cannot be completely removed from the matrix. On the other hand, reversed-phase LC with aqueous mobile phases is so common that this limitation becomes very serious. The solution to the problem lies in either using a matrix that is insoluble in water (sulphur or diamond powder) or reacting the water with a chemical (e.g. 2,2-dimethoxypropane) that yields easily removable products (e.g. acetone and methanol in this case). Alternatively, the solute may be extracted from the aqueous phase by an organic solvent such as dichloromethane.

Use of small-bore columns, where the elution volumes are only 1–10% of the conventional HPLC columns, showed considerable promise initially. The effluents from 0.3 or 0.5 mm columns may be deposited on a moving plate that is transparent to IR radiation. The solvent evaporation (assisted by nitrogen draught) leaves spots of about 2 mm diameter which give satisfactory spectra when used with a 4 × beam condenser. The minimum detectable limit by this method is about 100 ng.

It must be noted that while several new approaches are still being examined, the best strategies appear to involve solvent removal. However, even in this, approaches described in the literature have proved satisfactory enough to be adapted on a routine basis.

### 7.5.4 Electrochemical detectors

Oxidation and reduction are defined electronically as the removal and addition of an electron to the molecule, respectively. Thus, in an electrochemical system, a compound may lose an electron to the anode and get oxidized, or may gain an electron from the cathode and undergo reduction. The electrochemical reaction, i.e. transfer of electrons from the solution to the electrode, and *vice versa*, causes a potential difference between the electrodes and the solution and this can be measured.

The electrochemical process is governed by the Faraday law which states that the electrochemical conversion of $m$ moles yields an amount of current equal to $mnF$ coulombs, where $n$ is the number of electrons transferred and **F** is the Faraday constant with a value of about $10^5$ coulomb/mole. Since measurement of picocoulombs of electricity is routinely possible, this indicates that a detection sensitivity of as low as $10^{-17}$ mole can be expected from this technique. In practice, however, the low ratio of actual conversion to the amount of sample present, the current yield, and the noise level need to be considered, which together bring the detection limit to about $10^{-15}$ mole/sec of a flowing solution; at a flow-rate of 10 $\mu$l/sec, this corresponds to a $10^{-10}$ M solution or, roughly, 100 pg/100 $\mu$l peak volume, which is still 100 times more sensitive than the highest achievable sensitivity in photometric detection.

The potential for high-sensitivity detection notwithstanding, the electrochemical technique has certain drawbacks which limit its application. One is that the mobile phase should be electrically conducting. This is easily achieved in reversed-phase chromatography by addition of a salt to the medium. Alternatively, post-column addition of a suitable high-dielectric constant solvent and supporting electrolyte is possible. Another requirement is that the solute of interest should be electroactive; this limitation can also be circumvented by pre- or post-column derivatization (see below). The requirement of susceptibility to electrochemical reaction can be used to advantage in selective detection. Selectivity can also be achieved by choosing the proper electrode potential so that only a specific functional group of interest is affected. Variation of detector response with electrode potential is another characteristic property which can be used for identification of a particular constituent.

The types of compounds that undergo electrochemical reduction include: aldehydes, ketones, oximes, conjugated acids, esters, nitriles and unsaturated compounds, aromatics, activated halides and heterocycles. Phenols, mercaptans, peroxides, hydroperoxides, aromatic amines, diamines, purines, dihydroxy compounds and heterocycles are subject to electrochemical oxidation.

There are several modes of electroanalytical chemistry. In potentiometry, generally the current is controlled at zero and the voltage measured; the voltage is normally related linearly to the logarithm of the electrochemical activity of the substrate. When the potential is controlled and the dependence of the current on it is measured, it is called voltametry. Voltametry using a dropping mercury electrode is known as polarography. Irrespective of whether a dropping mercury electrode is used, if the potential is controlled at a constant d.c. value and the current is simply measured, the technique is called amperometry. If the conditions are such that all of the active material reacts at the electrode, it is termed coulometry. Since the eluent continuously flows through the cell, the electrochemical conversion is seldom complete, and amperometric detection is virtually the only technique used routinely in LC.

A significant advantage of the amperometric detectors is the wide range of the cell and electrode designs as well as the materials that can be used. Flow cells with a hold-up volume of less than 1 $\mu$l can be constructed. There are normally three electrodes (auxiliary, reference and working electrodes), with the electrochemical reaction occurring at the working electrode and the reference electrode compensating for any change in the eluent conductivity. The electrode geometry can take a number of forms, the most popular being the thin-layer cells (Fig. 7.15a). Among the

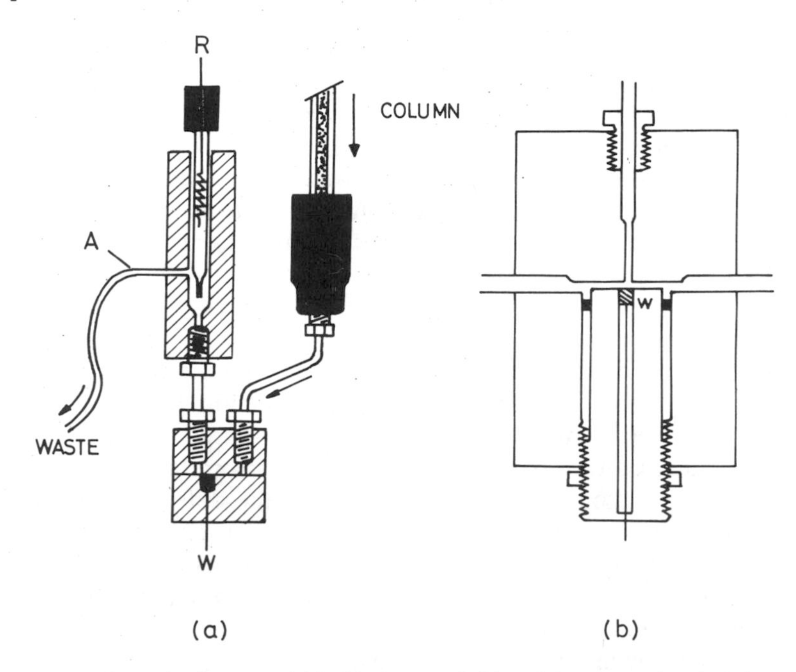


Fig. 7.15 — Schematic diagram of (a) thin-layer and (b) wall-jet electrochemical detectors; from Ref. 4 by courtesy of Elsevier Scientific Publishing Co.

alternatives, Fig. 7.15b shows a wall-jet electrode, where the eluent from the column is allowed to impinge directly on the working electrode. This arrangement increases the linear velocity of the solute at the electrode as well as the mass-transfer coefficient and both these effects increase the sensitivity of detection.

The best known electrode among the early instruments was the dropping mercury-type. Platinum electrodes have been popular for the determination of inorganic ions, but the problems of adsorption and filming make them less favourable for use with organic compounds, except in non-aqueous systems. Other metals (mercury, amalgamated gold, silver or nickel), carbon paste (a mixture of spectroscopic graphite powder and mineral oil or silicone grease) or graphite alone, have also been used. Owing to its long life and excellent results in the electrochemical reactions with several types of organic compounds, the carbon-paste was employed in many of the earlier designs. However, the life of the cell is relatively shortened when the mobile phase contains a high concentration of organic solvents. The more recent, glassy carbon electrode is a hard, amorphous material that can take a mirror-like polish and is resistant to organic solvents, including acetonitrile and methanol. Indeed, the glassy carbon electrode has been used even with non-aqueous mobile phases. For a detailed discussion on different electrode and cell designs and the principle of electrochemical detection in general, see Ref. 1.

A large variety of compounds are electrochemically active. Both reductive and oxidative modes have been used in electrochemical detection. The types of compounds that undergo electrochemical reduction include aldehydes, ketones, quinones, oximes, conjugated acids, esters, nitriles and unsaturated compounds, aromatic and aliphatic nitro compounds, *N*-oxides, activated halides, organometallics, peroxides, hydroperoxides, nitrosamines, azo compounds and heterocycles. Phenols, oximes, mercaptans, aromatic amines, diamines, purines, dihydroxy compounds, ascorbic acid and heterocycles are subject to electrochemical oxidation. The electrochemical detectors have been extensively used in the analysis of catecholamines, pharmaceuticals, enzymes and several other compounds of biomedical interest [25].

### 7.5.5 Reaction detectors

Post-column derivatization (i.e. treatment of the column effluent with a suitable reagent for improved selectivity and sensitivity of detection) has been widely used in traditional CLC. For example, classical amino-acid analysis involves treatment of the collected fractions with ninhydrin reagent to develop the characteristic colouration and determination of the optical density of the derivative at 570 nm. The reaction detectors in LC, in effect, use a post-column derivatization reaction to enhance the sensitivity of the above-discussed detectors. Reaction detection is particularly useful in trace analysis where the presence of large matrix effects may tend to interfere with the detection.

Adaptation of the principle of post-column derivatization to on-line derivatization in LC requires a very careful design of system. Suitable devices must be provided for storage and addition of the appropriate quantities of the reagents, mixing and incubation at different temperatures and other conditions required for effective and complete derivatization and, at the same time, minimize extracolumn band broadening. Fig. 7.16 shows the scheme of a suitable device for the purpose. As most LC

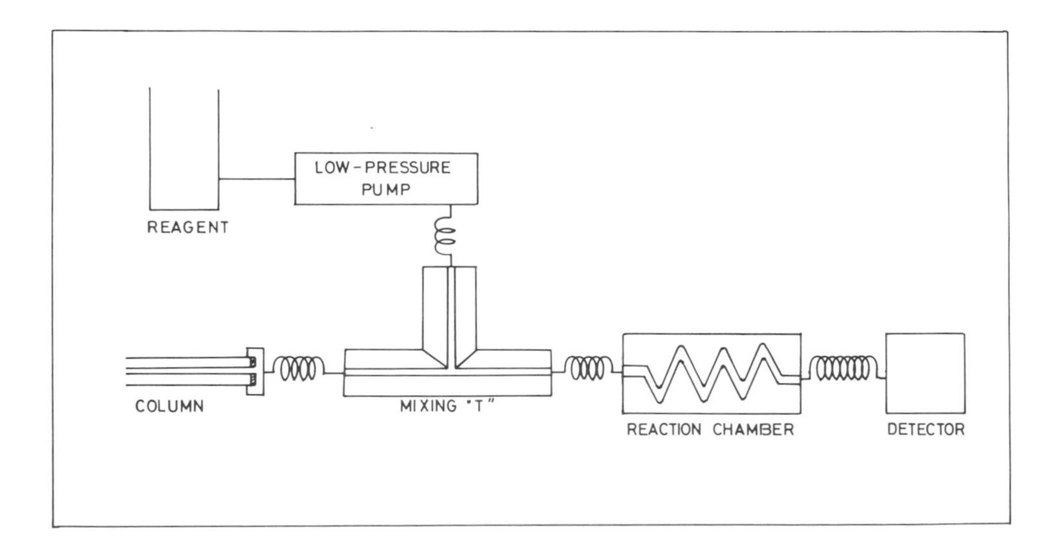


Fig. 7.16 — An arrangement for post-column reaction detection.

detectors are flow-sensitive, the pumping system for reagent addition should provide a pulse-free flow. A syringe pump is usually adequate. The mixing T-joint should provide effective mixing of the column eluent and the reagents without causing considerable dispersion. It typically consists of a 1.6 mm o.d., 0.18 mm i.d. tube, about 4 cm long, on the three arms [20]. Often the reactions are not instantaneous and the system should allow sufficient time between the mixing and the entry of the mixture into the detector. Also, certain reactions require heating or incubation for certain periods of time to force the reactions to completion. These are often achieved by means of delay tubes, which may be simple segments of tubes, packed tubes (e.g. when immobilized enzymes are used), coiled or serpentine tubes. Sometimes, a gas bubble is introduced into the post-column stream to generate segments that act as reaction compartments; a phase separator is needed to remove the segmentation agent, prior to the detection. A phase separator is also used when extraction of the derivative is needed to avoid interference from the reagents, by-products or the mobile phase.

Another limitation in on-line derivatization is the requirement of compatibility of the mobile phase with the conditions of derivatization. Since most of the common derivatives useful for detection are formed in aqueous medium, the reaction detectors are conveniently used in conjunction with ion-exchange and reversed-phase chromatography. However, rarely do optimal mobile-phase compositions offer ideal media for the derivatization reactions and detection. Thus, the electrochemical detectors can tolerate only a few water-miscible solvents in a narrow range of pH and ionic strength. Additionally, the reagents preferably should not be detectable by the system used. The reaction should also be fast enough to avoid unreasonably long analysis times. Two of the convenient and interesting reactions which are particularly useful for post-column derivatization are ion-pair formation and photochemical reactions. The principle of ion-pair formation has been discussed in the previous chapter. The technique is useful for derivatization of several antibiotics and tertiary amines that cannot be readily derivatized by other means. Photochemical decomposition or transformation to fluorescent compounds is another reaction that finds application in the detection of cannabinoids. The photochemical reaction products may also be further derivatized where necessary. The useful reagents for on-line, post-column derivatization, the reaction conditions and detection systems are given in Table 7.3 [2]. It may be suggested that the post-column derivatization be used as a last resort when the sensitivity of detection by the usual detectors proves inadequate because, however carefully the reaction detector is designed, band dispersion is inevitable and this would degrade the separation achieved after painstaking optimization of the other parameters.

### 7.5.6 Liquid chromatography–mass spectrometry (LC–MS)

The phenomenal success of GC–MS has naturally led to considerable effort to evolve similar on-line LC–MS systems. Indeed, the combination could be expected to be much more powerful in view of the wider applicability of LC, particularly in the field of biological sciences. While there is general agreement that LC–MS interfacing is practicable and many applications are on record, the technique is still to become as routine as GC–MS, owing mainly to the relatively high cost of the current instrumentation. Since both LC and MS are very well-developed methods, the main problem

**Table 7.3** — Reactions for post-column derivatization†

| Compounds detectable | Reagents employed | Reaction conditions | Remarks |
|---|---|---|---|
| Amino acids | Fluorescamine | Ambient | Fluorescence |
| | *o*-Phthaldehyde | Ambient | Fluorescence |
| | Ninhydrin | 140°C | 440 and 570 nm |
| Acids | *o*-Nitrophenol | Ambient | 432 nm |
| Carbonyls | 2,4-DNP | 3 min | 430 nm |
| Carbohydrates (reducing) | Ce(IV) | Variable | Fluorescence |
| | Ferricyanide | — | Electrochemical |
| | Neocuproin | 97°, 10 min | — |
| Steroids (3-oxo-4-ene) | Iso-nicotinyl hydrazine | 70°, 2 min | Fluorescence {360(I) and 450(II)} |
| Estrogens | Hydroquinone-H2SO4 | 120°, 15 min | Fluorescence {535(I) and 561(II)} |
| Guanidino compounds | 9,10-Phenanthrene-quinone | 60°, 2 min | Fluorescence {365(I) and 460(II)} |
| Catechol amines | Ethylenediamine/hexacyanoferrate | 75°, 5 min | Fluorescence {400(I) and 510(II)} |
| Thiols | 5,5′-Dithio-(2-nitrobenzoic acid) | Ambient, fast | 412 nm |
| Cannabinoids | Photolysis | Variable | Fluorescence |
| Nitrite(s), nitrosamides, nitrosocarbamates | Griess-reagent | Ambient, 3 min | 550 nm |

† Data from Ref. 2.

remains to be one of proper interfacing of the two, without reducing the separation efficiency of the former and the sensitivity of the latter.

It must be pointed out at the outset that the two methods, LC and MS, are basically incompatible. Thus, one system (the LC) operates at high pressures while the other works at very high vacuum. The fundamental requirement of MS is that the sample must be volatile (at least for the most common EI and CI modes of ionization) and the success of HPLC owes much to its applicability to non-volatile compounds. And, unlike in GC–MS, the carrier (the mobile phase) is not inert. Yet, there has been considerable success in the direct coupling of the two systems. The principles of mass spectrometry, various techniques of ionization of samples for mass spectrometry and the ways in which a mass spectrometer can be used for qualitative and quantitative analysis, have been discussed earlier in connection with GC–MS (section 5.9). Most of these are valid for LC–MS also. The present discussion is therefore limited to the main problem in LC–MS which is the removal of the solvent or the mobile phase from the column effluent. Most of LC analysis (particularly in the area of biological and biomedical applications) is currently being carried out in the reversed phase with aqueous mobile phases. The two approaches to the problem which are immediately apparent are: (a) the reduction of the solvent by splitting of the column effluent or by the use of small-bore columns and, (b) removal of the solvent by evaporation. The former reduces the sample availability to the mass spectrometer (and this could be a major drawback in the analysis of minor analytes) and the latter is fraught with problems of mechanics. A more recent approach is to use the mobile phase to aid ionization of the analyte in a way that is suitable for analysis by the mass spectrometer.

#### *7.5.6.1 Transport detector systems*

One of the earliest techniques of solvent removal is the moving-belt technique. This technique is ingenious in the sense that once the solvent is removed, the sample may be analysed or detected by any technique, including several of the universal and highly sensitive detectors which are very popular in GC; e.g. the flame-ionization and electron-capture detectors. In the so-called transportation detectors, the column effluent is deposited on a freshly cleaned, moving metallic wire, belt or chain which transports the material to a hot zone for removal of the solvent by nitrogen draft. In a combination of the moving belt and reaction detection, the sample is oxidized to carbon dioxide and then reduced to methane (by hydrogen, in presence of a nickel catalyst), which is detected by the FID [20].

The above principle of solvent removal could be used for LC–MS also [29], as shown schematically in Fig. 7.17. The column effluent, after partial evaporation by

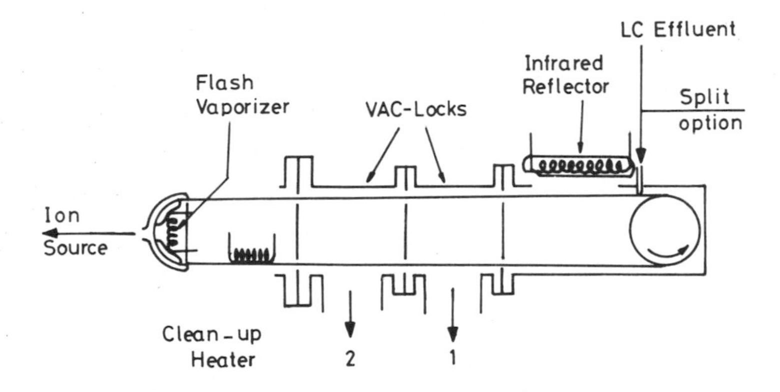


Fig. 7.17 — Schematic diagram of a moving-wire LC-MS interface; by courtesy of Finnigan Instruments, Inc.

an infrared heater or hot gas, is transported on a stainless-steel or high-temperature plastic ribbon (3.2 mm wide and 0.05 mm thick) through a series of vacuum locks. The vacuum pumps 1 and 2 help complete removal of the solvent after which, the sample is flash evaporated at or in the ionization chamber of the mass spectrometer. The sensitivity of the system is dependent on the minimization of the losses in both the sample deposition and the flash vaporization steps. The fidelity and efficiency of sample deposition is greatly improved by the use of a nebulizer and cold-gas spray technique [30]. The improvement of the performance is attributed to the reduced mixing of the analyte within the film deposited on the belt; the fact that the sample is evenly deposited on the surface in the form of small droplets also helps both solvent removal and sample vaporization. Still, when the mobile phase has a very high percentage of water, effluent splitting or use of a small-bore column may be needed to decrease the amount of the solvent that needs to be evaporated. The main advantage of the transport interface in LC–MS is that the solvent is completely removed prior to the ionization step so that any mobile phase can be used for separation and any ionization technique for detection.

#### *7.5.6.2 Direct liquid introduction (DLI)*

As the title indicates, this type of LC–MS involves introduction of the sample directly into the mass spectrometer. This may be achieved either by increasing the capacity of the mass spectrometer to tolerate large gas volumes (atmospheric pressure ionization) or by decreasing the quantity of the sample either by stream splitting or by the use of small-bore columns [15].

The principle of atmospheric pressure ionization (API) was briefly discussed in connection with GC–MS. Several API systems have also been designed for LC–MS. Ionization of the solvent vapour is accomplished either by $^{63}$Ni radiation or corona discharge and the solvent ions are used to ionize the sample molecules. The ions are then introduced into the mass spectrometer for analysis. However, several problems may be encountered when this technique is applied to LC–MS. The main drawback is that the sample has to be vaporized at atmospheric pressure and this is not easy because of the low volatility of many of the analytes in LC. Also, a prerequisite of API is that the molecule should have sufficient electron affinity. In any case, API yields only the $(M + 1)^+$ and $(M - 1)^+$ ions which severely restricts the structural information that a mass spectrum can give. Formation of polar solvent ion–molecule clusters is another problem that makes characterisation of low-molecular-weight compounds difficult. However, such problems may be overcome [28].

Column-effluent splitting to introduce only the amount of effluent tolerated by the mass spectrometer is probably the simplest approach to LC–MS. The standard electron-ionization mode can permit the use of only about 0.1% of the typical elution volume from a conventional LC column. The chemical ionization mode, with the mobile phase itself as the CI reagent, can normally tolerate up to 10–60 $\mu$l/min of the liquid, which is also a small fraction of the sample injected into the LC. Restriction of the mobile phases to those suitable for CI is another limitation. Nevertheless, the DLI–CI combination has been used for the analysis of several samples of biological interest [27].

Use of small-bore columns in LC–MS is a natural consequence of the limited elution volume that can be used in the DLI–MS system [15]. Because of the low elution volumes in the vicinity of 50 $\mu$l, the need for effluent splitting does not exist. Fig. 7.18 shows schematically a JASCO DLI micro LC/MS system, consisting of (1) a micro HPLC pump, (2) a 250 $\mu$l syringe, (3) 0.5 mm i.d. tubing, (4) sample inlet, (5) a small-bore column (0.5 mm i.d. × 7 cm), (6) variable UV detector, (7) a 0.3 $\mu$l flow cell, (8) a stainless-steel capillary, (9) PTFE tubing, (10) a glass-PTFE union, (11) glass capillary LC/MS interface probe, (12) direct insertion probe for the MS, and (13) the CI–mass spectrometer ion source. The capillary insertion probe ends in the ionization chamber in the form of a 5 $\mu$m hole in a stainless steel diaphragm. As the column effluent is forced through the pinhole, the micro-droplets are formed in the

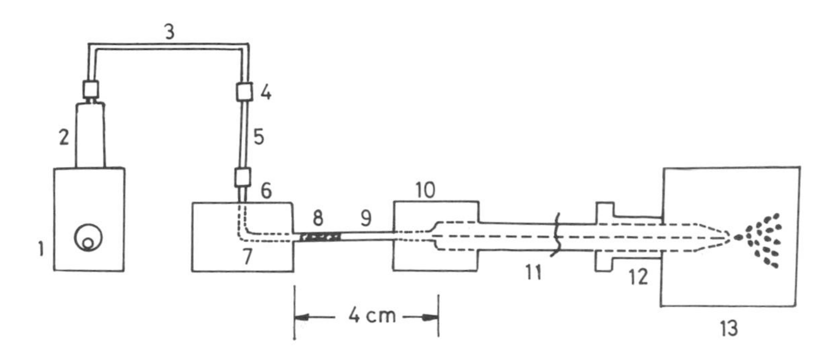


Fig. 7.18 — Schematic diagram of interface. Reproduced from JASCO DLI micro LC/MS/Ref. 15 by permission of Preston Publications, Inc.

ion source to facilitate chemical ionisation of the sample. The major drawback of the direct liquid introduction method of LC–MS is that only CI mode is possible. DLI has been recently reviewed [31].

#### *7.5.6.3 Thermospray ionization*

The thermospray interface is essentially a direct liquid introduction technique, with the difference that, in thermospray ionization, the ions are formed without the use of an external source of electrons [32]. The aqueous mobile phase containing an electrolyte (e.g. ammonium acetate) is introduced into an electrically heated stainless-steel capillary tube at flow-rates from 1 to 2 ml/min. The end of the capillary tube is situated close to a heated ion source, equipped with an auxiliary vacuum pump, opposite the capillary (Fig. 7.19); the pump serves to pump out the excess vapour from the source so that only part of it enters the skimmer leading to the ion source. The fluid from the capillary is released into the ion chamber in the form of a mist of fine droplets (or aerosol), at least some of which are electrically charged (positive and negative particles in equal number). As the size of the droplets reduce (because of continued evaporation of the liquid during its passage through the heated chamber), the charge or the electric field at the surface increases. When the charge is high enough, the ions in the drop are ejected and pass into the quadrupole analyser *via* the conical-shaped aperture.

While ammonium acetate has proved to be the best general-purpose electrolyte, other volatile salts, acids and bases may also be used. No additional electrolyte is necessary in the case of ionic samples. Since high water content in the medium is required to keep the electrolyte in solution and to enhance the sensitivity of the system, the technique is best suited for use with reversed-phase LC, ion-exchange chromatography and such other methods. It must be noted that the ionization process may sometimes fail even when the medium or the mobile phase contains volatile buffers. In such cases, an external filament may be used to generate conditions for chemical ionization; the technique is sometimes referred to as filament-on thermospray mode [28]. Alternatively, an electric discharge may be used

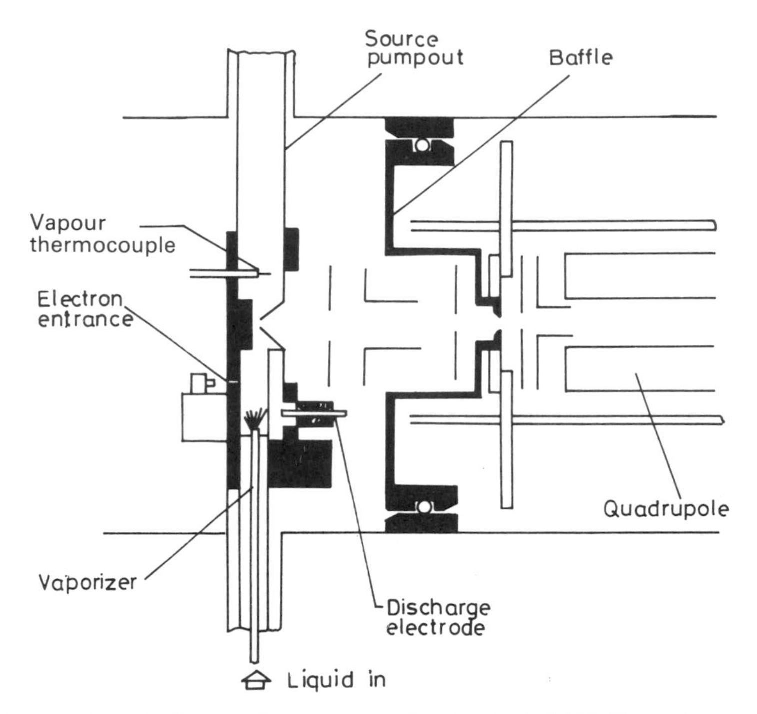


Fig. 7.19 — Schematic diagram of a thermospray interface for LC–MS. Illustration courtesy of Vestec Corp.

to generate reagent gas ions for chemical ionization. Thus, the thermospray system may be used to generate LC–MS data in the filament-off, filament-on, or discharge-ionization modes. The filament-on thermospray ionization mode is preferred when the mobile phase is non-aqueous as, for example, in the separation of enantiomers using chiral stationary phases.

Temperature control of the vaporizer tip, the ion-source block and the vapour can be critical in optimization of the thermospray ionization technique. The vaporizer temperature is determined by the mobile phase and its flow-rate, as well as the analyte composition and the diameter of the vaporizer. The last-mentioned parameter is subject to variation since non-volatile material may be deposited at the tip. Since change in the mobile-phase composition requires change of the vaporizer temperature, gradient elution would require temperature programming of the vaporizer also. Most commercial thermospray interfaces automatically control most of the parameters for optimal performance of the system.

Apart from the commercial availability and suitability for conventional (4 mm column) reversed-phase LC, one of the major advantages of the thermospray interface is its ability to ensure molecular-weight information. However, the ability of the soft-ionization techniques, in general, to give only the molecular weight (in the form of $[M^{+}+1]$ ions) is also a drawback since, unlike the electron-impact ionization spectra, these do not give much structural information on the molecule. The response of different analytes to thermospray ionization is also highly variable, making normalization and quantitation difficult.

#### 7.5.6.4 *MAGIC system*

All the above LC–MS systems, with the exception of the transport devices, are based on soft-ionization techniques that do not give much information about the structure of the molecule except for the molecular weight. The most recent approach to desolventization of the eluted sample is the monodisperse aerosol generator for introduction of (liquid) chromatographic effluents (better known by its acronym, MAGIC). In this system [33], the mobile phase is exposed to a nebulizing gas at an angle of 90° to yield highly uniform droplets or spray. A dispersing gas is used to prevent clustering of the droplets. The nebulized vapour passes through a desolventization chamber, maintained at atmospheric pressure and connected to the ion source of the mass spectrometer through an aerosol beam separator. Complete removal of the solvent is thus possible so that EI or CI spectra may be recorded as required. Flow-rates of 0.1–0.5 ml/min are found to be optimal. Demonstration of applicability of the system to less volatile compounds and ruggedness, as well as improvement of its sensitivity, may be needed before this promising technique finds wide usage.

### 7.5.7 Miscellaneous detectors

The above-discussed detectors, while being the most frequently used detectors in contemporary liquid chromatography, numerically form only a small fraction of the large variety of detectors that have been explored and that can be used advantageously. Among these, the detectors using the solute properties such as radioactivity (for compounds containing radioisotopes of the constituent elements), optical rotation (when the solutes are optically active), atomic absorption spectroscopy (for element-specific detection) and nuclear magnetic resonance (when structure determination is required) are of particular interest. These detectors are useful in practice and can be used as on-line detection systems. Their sensitivity may be poorer than some of the selective and better known detectors, but the structural information they provide is enormous. It is neither possible nor necessary to describe the principles and characteristics of these and several other detectors that find use in LC. Interested readers are referred to the monographs and advanced texts [1–5].

Among the lesser-used detectors which have been suggested for use in HPLC those based on photothermal (thermal lens), light scattering, photoacoustic, photoionization, photoconductivity, flame photometry, flame emission, phosphorescence and chemiluminiscence are worthy of mention, but several others still remain to be mentioned. Most of these are essentially of scientific curiosity and may never become available for routine use in the laboratory. Much of the effort in designing an LC detector is aimed at development of a universal detector, as in GC. One such promising and nearly universal detector is the light-scattering detector which is briefly described below.

The column effluent is nebulized and desolventized as described under LC–MS, using a high-velocity gas stream. The gas stream with the finely divided solute particles, is then passed through a photometer which measures the light scattered by the particles. The measured light is proportional to the amount of the sample present. Strictly, the light scattering depends on the number and size of the particles and these, in turn, depend on the properties of the mobile phase, its linear and volumetric flow-rates as well as those of the nebulizer gas. When these conditions are identical, the size and number of particles are proportional to the amount of the solute present. Since the light scattering is a property of any non-volatile substance present in the phototube, this detector may be considered nearly universal; the response of the detector to different solutes is also the same, irrespective of the chemical identity of the solute. This property finds application in the analysis of unknown samples whose response to different selective detectors (e.g. UV detector) is unknown. Some of the classes of compounds that can be determined by this technique are drugs, lipids, carbohydrates and polymers that do not absorb UV radiation. As discussed earlier, the refractive index also has similar applications, but with the drawbacks of instability to variation in temperature and solvent composition (e.g. as in gradient elution). An additional feature of the light-scattering detector is that it can be used for colloidal particles. It may be recalled that these materials can be chromatographed by techniques such as hydrodynamic chromatography and field-flow fractionation (sections 6.3.5 and 6.3.6).

### 7.5.8 Detector selection

When so many detectors are available, selection of an appropriate detector for a particular application could be problematic. The first choice would naturally go to the variable wavelength UV–visible absorption detector. Most compounds do absorb a finite amount of radiation above 190 nm or appropriate derivatives may be prepared using either the pre- or post-column derivatization procedures. The other advantages of the detector are its stability to a number of chromatographic variables like the flow-rate, temperature (within limits) and even mobile-phase composition (gradient elution). The other most popular general-purpose detector is the refractive index detector (with all its limitations of low sensitivity, and instability to changes in the temperature and the solvent). In view of their very high sensitivity and selectivity, the fluorescence and the electrochemical detectors are chosen wherever appropriate (i.e. when the solute exhibits fluorescence or is electrochemically active) and in trace analysis. The main drawback of these systems is the small number of compound classes that can be detected by these methods. Most often, post-column reaction detectors are used in conjunction with these detectors. Table 7.4 is a useful guide to determine the need for the less-common detectors which are also commercially available. Special-property detectors like atomic absorption or radioactivity are recommended only when element or isotope-specific detection is needed. Similarly LC–IR, LC–MS and LC–NMR may be used when structural information is required.

## 7.6 METHOD DEVELOPMENT IN LIQUID CHROMATOGRAPHY

### 7.6.1 Sample preparation and pre-column derivatization

As in gas chromatography, sample preparation prior to liquid chromatography may be crucial. The fact that sample volatility is not a prerequisite in LC is a mixed blessing; while it allows the analysis of a larger variety of samples, it also leads to considerable interference, overlapping and problems of resolution. Also, it is more important in LC not to introduce strongly retained compounds into the column since subsequent clean-up and reactivation takes a considerable time and, unlike in GC, it is not convenient to remove a portion of the damaged stationary phase and repack.

**Table 7.4** — Characteristics of some common LC detectors†

| Detector | Usability with gradients | Linear dynamic range | Sensitivity (g/ml) | Remarks selectivity, etc. |
|---|---|---|---|---|
| Refractive index | No | 104 | 10−7 | Universal |
| Conductivity | No | 2×104 | 10−8 | Universal |
| Absorbance (UV-visible) | Yes | 105 | 2×10−10 | Selective |
| Electro-chemical | No | 106 | 10−12 | Selective |
| Absorbance (Infrared) | Yes | 104 | 10−6 | Selective |
| Fluorimetric | Yes | 103 | 10−11 | Selective |
| Mass spectrometry | Yes | Wide | 10−9 | Universal |
| Light scattering | Yes | 103 | 5×10−4 | Universal |

† Data from Ref. 2.

The strategies employed for sample preparation depend mainly on the nature of the sample and the analytes, and are generally common to all chromatographic methods. These are discussed in detail in connection with the analytical applications of chromatography (Chapter 9).

Unlike in GC, derivatization is not normally necessary in LC to increase the mobility of the analyte along the column. However, derivatization in LC may be used to advantage to improve detectability, accuracy in quantitation and to obtain selectivity and improvement of resolution. Since detectability in LC is generally determined by the presence or absence of certain active groups (e.g. in photometric, fluorimetric and amperometric detection), derivatization is often necessary for achieving the the desired sensitivity. This aspect is particularly significant in view of the fact that general-purpose detectors in LC, namely, differential refractive index and infrared absorption detectors, are much less sensitive compared to, say, FID in GC. Derivatization is also helpful for achieving resolution of closely eluting but chemically dissimilar compounds and, occasionally, to prevent tailing of the peaks.

In practice, chemical derivatization in chromatography can be carried out in three ways; pre-column, on-column and post-column modes. The post-column derivatization methods using reaction detectors have been discussed above and on-column derivatization is rarely used in LC. Pre-column derivatization is convenient in many respects; e.g. unlimited reaction times, temperature and solvents, and the possibility of preliminary sample purification prior to injection. The additional manipulation and the time factor in derivatization may, however, be considered drawbacks. Another complication of derivatization is the possibility of a compound forming multiple derivatives with the same reagent, yielding more than one peak. The choice of the derivatization reaction should therefore be such that the solute of interest quantitatively forms a single, well-defined derivative in a short time. Complete conversion of the solute to the derivative is not essential even for quantitation, provided that the scheme is consistently reproducible.

Derivatization is normally a very simple organic chemical manipulation. A number of laboratory handbooks give the experimental procedures, which may be consulted. Many of the laboratory supply houses (e.g. Pierce Chemical Company, U.S.A.) not only supply the reagents, but also the appropriate labware and glassware as well as the optimized procedures for carrying out the derivatization reactions. The choice of a particular derivatization reagent and the procedure depend on a number of factors such as the chemical nature of the solute, interference from the accompanying compounds, ease of formation and the type of detector. Benzyl ethers and esters, phenacyl and benzoyl esters are easily prepared by treating the corresponding halide with an alcohol or a carboxylic acid; they serve the usual purpose of improved peak shape and detectability. Introduction of a bromo, methoxy or nitro group in the aromatic ring of the derivatizing agent considerably improves the sensitivity of detector response. The nitrobenzoyl derivatives can also be used in conjunction with electrochemical detectors. Use of 2,4-dinitrofluorobenzene and dansyl chloride for amino compounds and 2,4-dinitrophenylhydrazine for carbonyl compounds is well known. There are several books and reviews on derivatization for chromatography in general [34], and for LC in particular [35,36]. Some of the most commonly used derivatization reactions in LC are given in Table 7.5.

### 7.6.2 Method and column selection

The principles of different LC methods have been discussed in the previous chapter and the instrumentation available to facilitate their use is discussed in the foregoing sections of this chapter. The information should be adequate to determine the most suitable method for a given application. While this may be fairly simple, several strategies may have to be employed, within the selected system, to achieve and optimize separation of a complex mixture. Optimization is more easily achieved in instrumental LC than in conventional LC because of the speed of the analysis, manipulative facilities, near-instantaneous detector response and the convenience of computer-assisted data processing and optimization routines.

The choice of a particular LC method depends on the nature of the sample and the purpose of the analysis. There are several charts, schemes or guidelines (available from most chromatographic product supply houses); one such is given in Fig. 7.20. It must be emphasized that such schemes can only provide rough guidelines and it is virtually impossible to recommend simple procedures of column and method selection applicable under different circumstances. Such exercises are also probably unnecessary since, at the present time, it is not too difficult to locate literature procedures for separation of any given class of compounds and one rarely comes across a sample for analysis whose history and chemical nature is completely unknown. The major sources of information on the chromatographic systems for different classes of compounds are given under the reference sections of Chapters 1 and 9.

If the compounds to be separated differ from each other in their molecular size by 5–10% of the molecular weight, size-exclusion chromatography is an obvious choice. Within SEC, the choice of the column is dictated by that of the mobile phase, i.e. the solubility of the sample. Water-soluble solutes can be separated on controlled-pore silica, which can also be used with organic solvents and which, in other words, is a

**Table 7.5** — Some common precolumn-derivatization reactions†

| Compound class | Reagents used | Detection method |
|---|---|---|
| Carboxylic acids | Phenacyl bromide | UV-250 nm |
| | *p*-Bromophenacyl bromide | UV-260 nm |
| | Benzyl bromide | UV-254 nm |
| | *O*-*p*-Nitrobenzyl-*N*,*N*′-di-isopropylisourea | UV-265 nm |
| | 1-*p*-Nitrobenzyl-3-*p*-tolyltriazine | UV-265 nm |
| Alcohols/phenols (amines) | 3,5-Dinitrobenzoyl chloride | UV-254 nm |
| | Pyruvoyl chloride (2,6-nitrophenylhydrazone) | UV-254 nm |
| | *p*-Iodobenzenesulphonyl chloride | UV-254 nm |
| | Benzoyl chloride | UV-230 nm |
| | *p*-Nitrobenzoyl chloride | UV-254 nm |
| 1,2-, 1,3-, 1,4-Diols | Phenanthrene boronic acid | Fl-313/385 nm |
| Amines | *p*-Methoxybenzoyl chloride | UV-254 nm |
| | 2-Naphthacyl bromide | UV-248 nm |
| | *N*-Succinimidyl-*p*-nitro-2,4-dinitro-1-fluoro-benzene | UV-265 nm |
| Amines/amino acids | Dansyl chloride | Fl-360/510 nm |
| Amino acids | Pyridoxal | Fl-330/400 nm |
| Aldehydes/ketones | *p*-Nitrobenzyloxyamine | UV-265 nm |
| | 2,4-Dinitrophenyl-hydrazine | UV-430 nm |
| Isopcyanates | *p*-Nitrobenzyl methyl amine | UV-265 nm |

† Data from Refs. 2 and 4.

universal SEC packing. Organic-solvent-soluble solutes have been conveniently handled on cross-linked polystyrenes which are available in well-defined pore-sizes. The solubility of the solute also determines the choice of the mobile phase and, therefore, the stationary phase in partition chromatography, which is particularly recommended when the solutes differ significantly in their distribution coefficients between the two phases. Permanently ionic or ionizable compounds may be separated by ion (exchange) chromatographic methods and the choice between cationic- and anionic-exchangers is determined by the nature of the distinguishing ions in the sample. Either material may be used for separation of zwitterionic compounds, the choice depending on the isoelectric point and the range of pH in which the compounds are stable. Ionizable organic acids and bases may be separated either on the IEC packings, or on normal- or reversed-phase adsorption chromatography by ion suppression, or by using the ion-pairing techniques. Compounds that have very similar chemical structures but have distinctly different biospecificities are best handled by affinity chromatography.

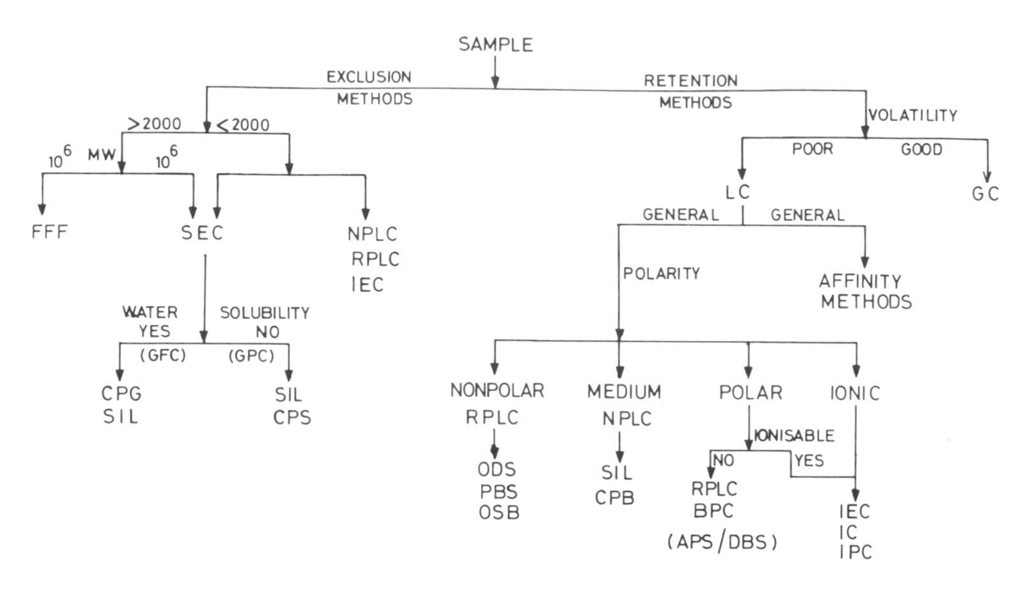


Fig. 7.20 — Column-selection chart; see list of abbreviations and Table 7.2.

Adsorption chromatography is a general-purpose technique using which, almost all classes of compounds have been separated either on normal or reversed phases. However, it is advantageous to use silica gel for medium polar compounds and reversed, hydrocarbonaceous (e.g. octadecylsilyl or ODS) bonded phases for both non-polar compounds, which are unretained or too weakly retained, and very polar compounds, which are too strongly retained on silica. Silica gel is preferable for the separation of isomers and for group separations, while the ODS packings show better selectivity for homologues within a group. Since both normal- and reversed-phase systems are of general applicability, the choice may also be dictated by the nature of application and convenience. Thus, biological samples, which are usually water-based, are conveniently analysed on the reversed phases, while the organic-solvent extracts of plants or synthetic reaction products are best separated on the normal-phase packings. The choice between the two modes is also often dictated by the detector; e.g. water-rich mobile phases and reversed-phase LC for electrochemical detection. The RPLC is also to be preferred where a large number of samples are to be routinely analysed as, for example, in quality control; the reason, obviously, is the low cost of the solvent. Similarly, the low-cost stationary phase and the facile solvent recovery favour the normal-phase silica gel in large-scale preparative separations. Between the two extremes of silica gel and ODS packings lie a series of bonded-phase packings of different polarities and selectivities. Certain types of highly polar compounds are too strongly retained on silica and completely unretained on the ODS packings. In such cases, amino-bonded (which is slightly less polar than silica) and phenyl-bonded (slightly more polar than the ODS packings) phases are useful. The former is particularly selective for carbohydrates, the latter for peptides. The cyanopropyl-bonded phases are very versatile and can be used both in the normal- and reversed-phase modes using appropriate mobile phases. It must be apparent that more than one column may suit a given type of sample; at the same time, it may not be possible to adequately separate all the components of a complex mixture on a single column even by varying the mobile phase. Multi-dimensional techniques may then have to be used, as described below.

### 7.6.3 Mobile-phase optimization

After choosing an appropriate LC method and the column for a given separation problem, the mobile phase is selected to achieve resolution of the components. The methods of choosing the mobile phase for different LC methods have been discussed under the respective sections of the previous chapter. The factors affecting resolution; namely, the number of theoretical plates ($n$), the capacity factor ($k$) and the separation factor ($\alpha$), were discussed in Chapter 2 (see eq. (2.40) and Fig. 2.3). It must be obvious that optimization does not always mean increasing the values of these three parameters. For example, when $\alpha=1$ there is no separation, no matter what the value of the other two parameters ($n$ and $k$) are. On the other hand, if $\alpha$ and $k$ are both large, the analysis time may be too long unless $n$ is reduced (usually by decreasing the column length, $L$). At the same time, large $k$ and small $n$ values would produce broad peaks, leading to inaccuracies in both qualitative and quantitative analysis. While it may be necessary to tolerate $k$ values up to 20 in case of complex mixtures, values between 1 and 5 may be considered as optimum in most cases, considering both resolution and the speed of analysis. The value of $k$ is easily manipulated by changing the selectivity of the stationary phase or the mobile phase, or both; once these are chosen, the value of $\alpha$ (i.e. the separation factor between two compounds) is fixed and improvement of resolution requires increasing the number of theoretical plates, $n$. As discussed in detail in Chapter 2, the value of $n$ is dependent on a number of factors such as the particle diameter, mobile-phase linear velocity, etc. Increasing the value of $n$ by increasing the column length may not be practical beyond a certain length owing to an increase in the pressure drop across the column; the extent of pressure drop for a given length of the column is determined by the diameter of the column-packing material. For a given column, the relation between the linear velocity ($u$) and the plate height ($h$) is given by the van Deemter plot (Fig. 2.8), the shape of which depends on the column. For any given column there is a minimum value of $h$ that can be attained, but the optimum value (considering the speed of analysis) is obtained at a slightly higher value of $u$ than the minimum indicated by the plot. The value of $u$ is obtained either by dividing the column length ($L$) by the time ($t_0$) taken by the mobile phase of volume $V_0$ (the void volume) to flow through the column (i.e. $L/t_o$) or by dividing the volumetric flow-rate ($F$) by the cross-sectional area of the column. In practical terms, this means the optimum flow-rate may be around 0.5 ml/min for a 25 cm×3.9 mm (i.d.) column. The value of $h$ is doubled (i.e. $n$ is halved) when the flow-rate is increased to 4.0 ml/min. For a detailed discussion on optimization of resolution by changing $u$, $n$ and $k$, see Ref. 2.

Change of $\alpha$ to optimize the resolution can be most effective and, at the same time, very difficult. The difficulty arises due to the unpredictability of $\alpha$ with changes in such parameters as the mobile-phase composition, pH, temperature and the stationary phase. As elaborated in the previous chapter (section 6.5.3), variation of the solvent strength of a binary mixture changes $k$, while replacing one of the components of the mixture by a solvent belonging to a different selectivity group can bring about a radical change in the separation factor. Selectivity in IEC may be achieved by changing the mobile-phase counter-ion or by varying the pH (in case of acids and bases). If these strategies do not give adequate separation, the stationary phase may have to be changed. This may involve a change in the method itself; for example, normal- to reversed-phase LC, adsorption to partition, ion-exchange to ion-pair LC, etc. Rise in temperature, while generally ineffective in adsorption chromatography (and impossible in partition chromatography), can be very useful in IEC, IPC and SEC. Separation of *cis* and *trans* olefins on reversed-phase columns using a mobile phase containing silver ions is an example of selectivity through chemical complexation; borate and bisulphite ions are similarly useful for separating diols and carbonyl compounds, respectively. However unpredictable, change in $\alpha$ is necessary if its value is initially 1, since, in that case, no other strategy would resolve the compounds chromatographically. A very large value of $\alpha$ is advantageous in preparative chromatography as this would enhance the loadability and throughput of the column. It is also very useful in developing a method for routine analysis (e.g. in quality control or clinical applications) since, in that case, the flow-rate and/or solvent strength can be increased to facilitate faster analysis.

The above discussion would indicate that method development in LC could take a considerable amount of time and effort. The goal is to achieve the highest resolution in the minimum analysis time. However, since enhanced resolution is often achieved at the cost of time (e.g. decreasing the flow-rate or increasing $k$), it may require a practical compromise between the various interdependent factors. A good understanding of the separation mechanisms and effects of different operating parameters is necessary for selecting and achieving rapid optimization of the method.

There are several manual and computer-assisted routines to help the decision on the method and column options. In recent years, statistical methods for optimization of resolution have been examined by several investigators and the approaches are dicussed in detail in several excellent books and reviews [37–43]. It may be noted, however, that statistical approaches can get very complicated when a number of interdependent factors control the result, as in chromatography. Another problem in this context is the difficulty in defining the criteria to establish the quality of separation, without which the statistical routine cannot even be initiated. For example, over-emphasis of the number of theoretical plates as representing the column efficiency should be avoided since, depending on chromatographic conditions, the plate count of a particular column may vary from 8,000 to 25,000.

A detailed treatment of the statistical methods of optimization of LC methods may be out of place here. Suffice it to say that these methods are currently under active development and a few microprocessor-based LC systems are on the market which offer on automated optimization facility, though none of them is very satisfactory. The sequential simplex or the multivariate method (as described briefly in connection with the optimization of GC conditions; see section 3.3) is probably the simplest. Optimization methods are built into several modern, microprocessor-based LC systems. Most of the routines are directed toward solvent optimization for adsorption chromatography in either the normal- or reversed-phase modes, often the latter. Occasionally, the column temperature, mobile-phase flow-rate, composition and pH and, if necessary, the gradient shape are also included.

Though the earliest to be introduced and now defunct, the Sentinel LC system of DuPont (U.S.A.) remains the most sophisticated and satisfactory among the commercial automated optimization facilities [39]. A four-solvent, seven-experiment method was used in this instrument. The four solvents may be hexane, methyl *tert*-butyl ether (MTBE), acetonitrile and methylene chloride, or MTBE, chloroform, methylene chloride and methanol for the normal-phase adsorption chromatography. For RPLC, the four solvents: water, acetonitrile, methanol and tetrahydrofuran may be chosen. A preliminary gradient run is performed to determine the solvent strength required to elute the analytes in a given time or range of $k$ values. Selectivity among the three solvent compositions (water–acetonitrile, water–methanol and water–tetrahydrofuran) is then evaluated using the solvent strength established in the previous run. Four experiments are needed to complete the simplex lattice design over the solvent selectivity triangle (see Fig. 3.6). The overlapping-resolution-mapping technique is used to determine the best composition. In this technique a triangle is constructed for every pair of peaks and the areas of the triangle giving less than the desired resolution (say, $R<1.5$) are shaded. Overlapping two or more (as the case may be) of these triangles gives the area in the triangle where all the pairs are separated with a resolution over 1.5. Seven runs are carried out to calculate the retention data and elution order of all the peaks, from which the optimum mobile-phase composition may be selected either by the computer or the chromatographer. The final run verifies the predicted composition. For a detailed description of the operation of the system, see Ref. 5.

### 7.6.4 Programmed liquid chromatography

#### *7.6.4.1 Gradient elution*

The general elution problem, where the peaks in a given stationary-phase–mobile-phase system have widely different $k$ values, and the principle of gradient elution to overcome this problem, have been briefly mentioned in the previous chapter. The modern equipment available for gradient elution has been described earlier in this chapter; these accessories make it possible to obtain, in addition to the linear and exponential solvent-strength gradients mentioned earlier, a number of complex, multicomponent gradients. These, and the possibility of automation in both selection and performance of the gradient elution methods, vastly increase the potential of this technique and, at the same time, make it routinely applicable. It may, however, be mentioned that a vast majority (about 80% as a rough estimate) of LC problems can be adequately solved by isocratic elution and about 80% of the remainder may be solved using a binary linear-solvent strength (LSS) gradient. The technique of gradient elution may, nevertheless, be profitably used even when isocratic elution is adequate, in view of certain intrinsic advantages. Thus, it often results in better or optimized resolution and faster analysis, enhancement of sensitivity (due to sharper peaks) for trace analysis, decrease in peak tailing and increase in peak capacity (i.e. the number of individual components that can. be separated in a single run). Since the elution is always terminated with a much stronger solvent, it also results in a simultaneous column clean-up.

The gradient elution is generally performed by choosing two solvent systems A and B, so that the most strongly retained components have a $k$ value of about 2–3 in the stronger solvent, B. The solvents are mixed before delivery to the column so that the percentage of B in the mixture increases with time and the last component has a $k$ value less than 10. The solvent systems A and B may be either pure solvents or mixtures of solvents and the programme may vary from 0–100% of B. When the polarities of the two solvents are very different, solvent demixing may occur; this phenomenon manifests itself in several ways such as inexplicable narrowing of certain bands in the middle of the chromatogram, poor resolution owing to clustering of peaks, etc. The cause of demixing could be that the polar solvent first gets adsorbed on the column (in the normal phase) and, when the column is saturated, there is a sudden increase in its proportion in the mobile phase. In such cases, three (or, even four) solvents may have to be chosen so that the gradient could be continued from A through B to C, etc. Multisolvent gradients, however, are very rarely used.

The gradient-elution profile is described by three characteristics, i.e. the identity of the two solvents A and B, the gradient shape and the gradient steepness. The gradient shape is the percentage change in the proportion of B as given by

$$\%\mathrm{B} = KT^n \tag{7.6}$$

where $K$ is the slope of the profile, $T$ is the time of the total analysis and $n$ is the exponent value, which may vary from 0 to 9.9. Fig. 7.21 shows the shapes of gradient profiles with different values of $n$. Most commercial gradient formers provide a larger number of options (10 to 12) for the value of $n$; more complex gradients are possible in some microprocessor-controlled equipment. When the value of $n$ is one, the gradient is linear and its steepness is directly given by the change in %B with time; in other cases, it may be expressed as the average value obtained by dividing 100 by the time in minutes taken for completion of the gradient. To the extent possible, the slope of the gradient is chosen to obtain all the peaks nearly equally placed and of equal width as in Fig. 6.2e. If a linear gradient gives a pattern shown in Fig. 6.2b or 6.2c, a concave gradient ($n > 1$) will have to be used; on the other hand, a convex gradient ($n < 1$) is recommended if the later eluting peaks are too closely placed. Sometimes, when the peaks with lower $k$ values are too close even with a concave gradient, it may be necessary to introduce a gradient delay so that the first few peaks are eluted isocratically with a solvent A; similarly, pumping the solvent may be continued at the end of the gradient for all the remaining peaks to be eluted by a solvent of composition B.

The change in $k$ of a given component during the gradient elution can be calculated and, for LSS gradient, it is given by:

$$\log k_{\mathrm{x}} = \log k_{\mathrm{o}} - b\left(\frac{t}{t_{\mathrm{o}}}\right) \tag{7.7}$$

where $k_{\mathrm{x}}$ and $k_{\mathrm{o}}$ are the $k$ values at time $t$ and $t_{\mathrm{o}}$, respectively. The parameter $b$ is

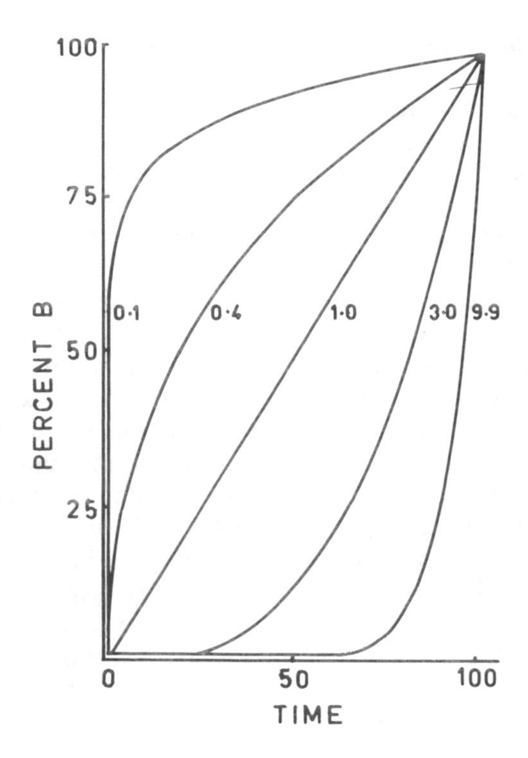


Fig. 7.21 — Solvent-gradient profiles for different values of $n$.

nearly constant for all the sample components and has an optimum value of 0.2–0.3. The value of $b$ also determines the steepness and the time of gradient. Changing the value of $b$ is similar to changing the solvent strength in isocratic elution in the sense that an increase in $b$ increases the speed of elution and sensitivity while decreasing the resolution; the optimum value is determined by the complexity of the mixture and the purpose of analysis.

If the gradient shape and steepness are optimum for a given combination of the solvents A and B, and if the resultant chromatogram is still unsatisfactory, variation of $n$ (by changing the length or flow-rate) or $\alpha$ (by changing one of the solvents or adding a third solvent) may be examined as discussed in the preceding section. At the conclusion of gradient elution, the systems will have to brought back to the initial conditions which may be achieved by washing the column with a large quantity of solvent A or, more conveniently and rapidly, by using a reverse gradient of the same profile. Since the regeneration depends on the volume of the solvent used for washing and not the time of column conditioning, the highest possible flow-rate (or gradient steepness) can be used.

The above is certainly an oversimplified treatment of the rather complex subject of gradient elution LC, which has attracted considerable attention in recent years. The comprehensive monograph by Jandera and Churacek [6] may be consulted for an up-to-date account of the technique and its applications.

#### *7.6.4.2 Programmed-temperature liquid chromatography*

While programmed-temperature GC has been very useful and extensively used, the utility of this technique in LC has been less dramatic, its place having been taken by gradient elution (or programmed-solvent composition LC). A rise in temperature usually increases the sample solubility (in both the stationary and the mobile phases) and mobile-phase diffusivity and viscosity. While these changes improve the resolution marginally in most chromatographic methods, their effect on SEC could be significant; however, no change in the elution time can be expected in SEC as long as the flow-rate is constant. Increased temperature can displace water and alcohol (which normally deactivate the stationary phase in adsorption chromatography). This may be expected to increase retention, but the situation is somewhat complicated since the displaced deactivating groups are now in the mobile phase and can compete with the solute molecules for the active sites. On bonded phases, increase in temperature normally causes a decrease in the retention and this feature may be used where necessary. Selectivity changes in other forms of chromatography are minimal, but, in IEC, dramatic changes can be expected. Rise in temperature invariably increases the ionization of weakly ionized compounds and thus may be expected to improve the retention. Here again, the increased solubility of the solute at higher temperatures could have the opposite effect. Thus, the retention characteristics of the column can be unpredictable in all forms of LC, except that the improvement in the mass-transfer properties generally make the chromatography more efficient.

The upper temperature limit with most solvents is about 20°C below the boiling point of the solvent. Some biological samples are also unstable at higher temperatures and cannot be analysed at high temperatures. There are also several examples of improved resolution at lower temperatures. However, decrease in the mobile-phase viscosity would require much higher head pressures to maintain the flow-rate. In any case, change of temperature during chromatography is seldom employed in LC.

#### *7.6.4.3 Flow-programmed liquid chromatography*

Change of mobile-phase flow-rate is effected manually by resetting the pump when required. Continuous change in the flow-rate is more conveniently effected either electronically or by using a two-pump gradient former, and switching off the pump A. Since these programmes operate by increasing the pumping rate of the pump B, any of the different programmes of Fig. 7.21 can be obtained. It is well known that any change in the flow-rate from the optimum only decreases the efficiency and the method can only be used to speed up elution of some late eluting, but well-separated compounds.

### 7.6.5 Multi-dimensional and related techniques

The methods and techniques of LC discussed above should be adequate for solving most of the analytical problems one might normally come across. However, in certain rare cases, one might use a combination of two or more methods to improve the efficiency of analysis. This can simply be done by collecting the fractions represented by the unresolved peaks from a particular method (stationary-phase–mobile-phase combination) and reinjecting into another system with a different stationary phase or mobile phase or both. Since the solvent can be removed in such a case, any combination of the LC methods discussed in Chapter 6 can be used. The need for fraction collection, concentration and the use of another chromatograph is obviated if the eluent from one column can be automatically directed into another column, to be eluted with the same or a different solvent. Such multi-dimensional chromatographic techniques have been used extensively in planar chromatography (see Chapter 8, below), conveniently adapted to GC and, inevitably, attempted in CLC. There is, however, a basic difference between the first two and the last methods. In the first case the stationary phase is often the same; the sample remains on the same plate and, after removing the mobile phase, the chromatographic bed is irrigated with a different mobile phase in a different direction. In GC, the mobile phase remains unchanged, and the eluant from the column, along with the sample, also remain in the vapour state, so that it can be easily introduced into a second column. On-line multi-dimensional CLC, on the other hand, is fraught with several problems of mutual incompatibility of methods and of mobile phases, as well as the difficulty in high-pressure sample transfer from one column to the other. Nevertheless, notable success has been achieved in recent years in the methodology of multi-dimensional column programming, particularly with the help of microprocessor-based external event controllers. As will be seen below, apart from increasing the resolution, a major application of the multidimensional technique is in shortening the analysis time and thus increasing the throughput of samples. Such methods are extremely useful in quality control and environmental control laboratories where a large number of analyses are routinely performed and where only a few of the components in a complex mixture may be of interest.

Several types of multi-column LC can be performed without any additional gadgets. Use of guard columns for preliminary clean-up and coupling of two or more columns of the same type in series to increase the theoretical-plate number have been mentioned before. Bimodal columns have been used for increasing the molecular weight range in SEC and columns of cation- and anion-exchange resins have been combined to separate different ions simultaneously. Ion-suppressor columns to facilitate conductometric detection of IEC effluents has also been elaborated in the previous chapter. Also mentioned earlier is the fact that most LC instruments are equipped with a recycling facility where the effluents from the column may be returned to the column through a closed loop for further separation. This technique can be used only for the separation of the last-eluting peak as, otherwise, mixing-up of the retained components with the recycling components would occur. Another drawback of recyling is the considerable amount of band broadening with each recycle. Fig. 7.22 shows an alternative to the closed loop recycling using a 6-port valve and two columns of the same type [44].

Arrangements similar to those in Fig. 7.22 can also be used for multi-dimensional LC by suitably choosing the columns and appropriately switching the valves. In two of the simpler examples, the two columns are packed with the same stationary phase but column 1 is about one-half to one-fifth of column 2. When the sample contains a number of components with $k$ values over a wide range (say, 1– 40), the smaller column may not resolve compounds with $k$ values of 1–5, but may rapidly and efficiently separate the later-eluting components. The eluent containing the first few peaks is, therefore, diverted to the longer column on which they can be adequately separated. Alternatively, the two columns may be packed with the same stationary

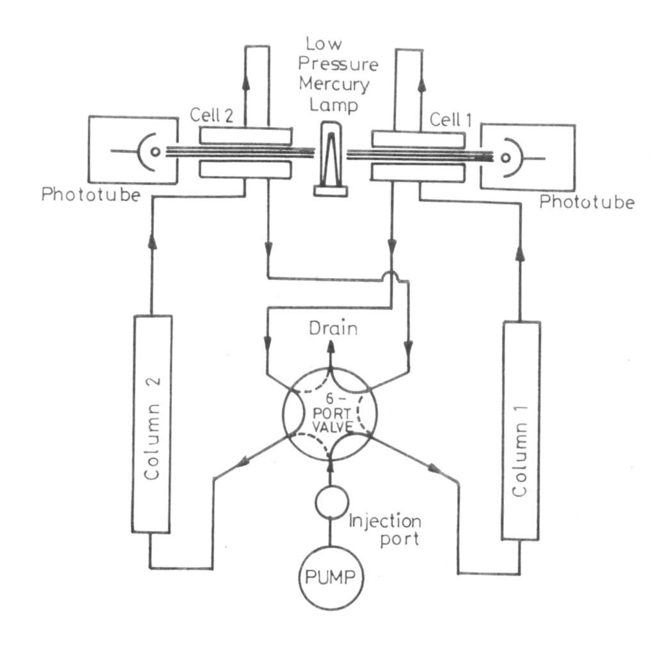


Fig. 7.22 — Two-column coupling with a six-port valve for multidimensional LC. Reproduced from, Ref. 44, by permission of Preston Publications, Inc.

phase but the phase ratio may be changed by packing column 1 with a pellicular packing, and column 2 with a porous one. In the chromatography of biological samples on RPLC, the compounds of interest are often retained ($k=2$–$10$), while much of the accompanying material elutes in the void volume. The void volume portion containing the unretained material can then be vented out and the remainder diverted to the second column.

Two coupled columns of differing capacity have been used in an ingenious way in the so-called boxcar chromatography [45] to increase the sample throughput. Advantage is taken of the fact that when samples with a narrow range of $k$ values are travelling through a high-capacity column, much of the column is unused even though the mobile phase is passing through it. Thus, if a sample has components with $k$ values in the range of 1–10 but only those in the range of $k=2$–$4$ are to be routinely monitored, the sample may be first passed through a low-capacity column and the portion containing the analytes of interest ($k=2$–$4$) is diverted to the second column; the remainder of the effluent from the first column is sent to the drain. After a lapse of a few minutes, and even before the analytes are eluted from the second column, a second sample of nearly the same composition may be injected into the first column and the above column-switching operation repeated. Fig. 7.23 shows an optimized use of the 'boxcar' technique for the analysis of drugs using two columns, the first of 6 cm length and the second of 15 cm. Eighteen samples could thus be analysed in 20 min. The above technique is certainly more convenient than gradient elution for at least two reasons: (a) the arrangement is inexpensive and does not need a gradient-

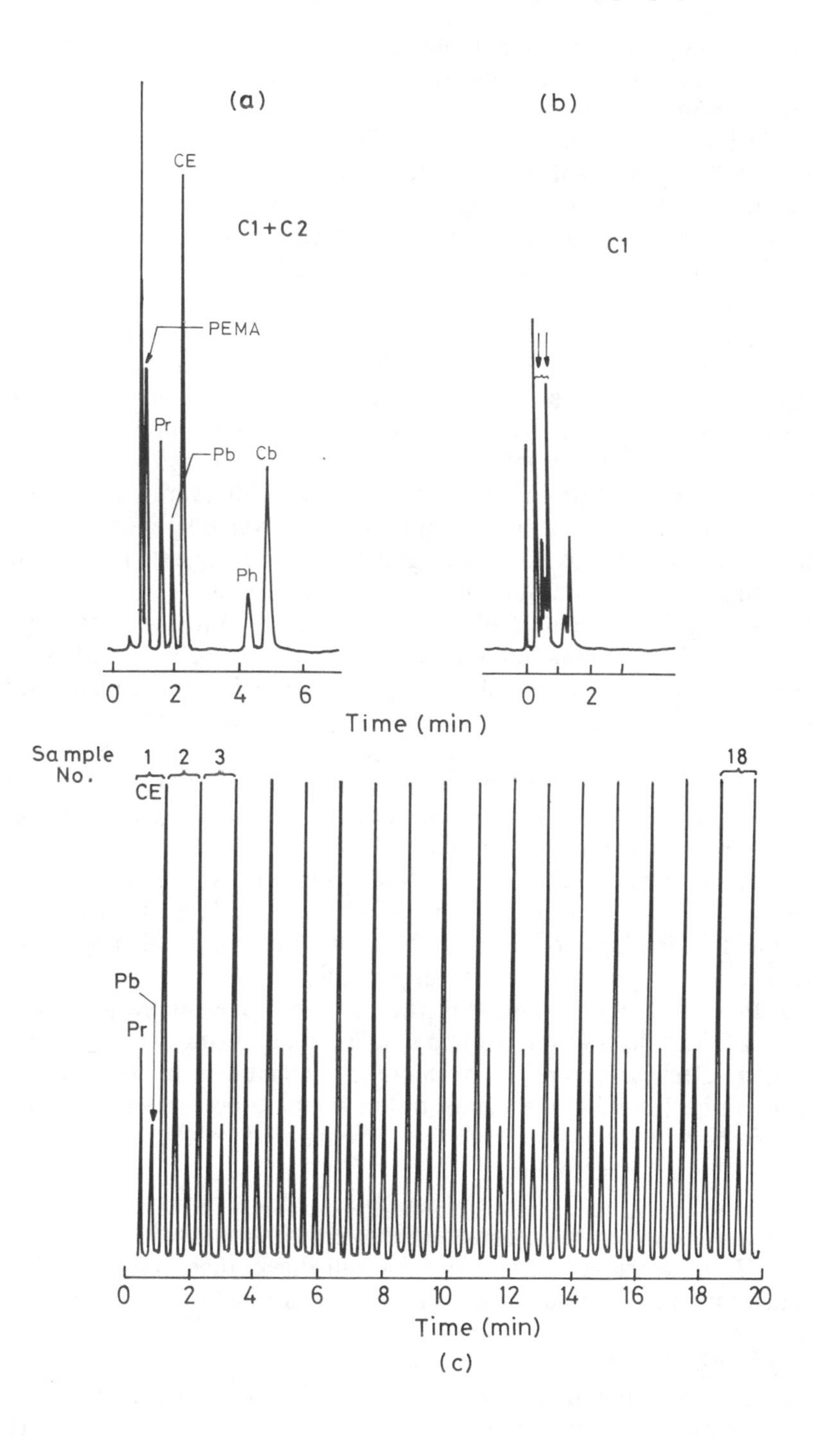


Fig. 7.23 — Illustration of boxcar chromatography. Reproduced from Ref. 45 by permission of Preston Publications, Inc.

elution device, and (b) since the same solvent is used, re-equilibration of the column and flushing of the whole LC system between samples is not necessary. Use of a single mobile phase also brings down the cost of the analysis and permits use of detectors which are not compatible with gradient elution.

The stationary phase in columns 1 and 2 can be different if the mobile phase for both the columns is the same; more than two columns can also be connected by a similar arrangement. The mobile phase can also be different and, using different solvents and stationary phases, it is possible to analyse the samples using different LC methods, keeping in view the requirements of solvent miscibility. Thus, adsorption chromatography on silica, normal bonded-phase LC and SEC in organic solvents can be coupled; similarly, SEC in aqueous medium, reversed-phase LC, and IEC are mutually compatible. However, coupling of normal-phase LC using water-immiscible organic solvents and RPLC using aqueous mobile phase are difficult to combine; addition of a third solvent such as tetrahydrofuran (THF), which is miscible with both the mobile phases, may be helpful. However, it should be noted that THF is a powerful solvent in both normal- and reversed-phase LC, and large injection volumes would cause partial migration of the samples down the column, thereby limiting the resolution. It is often advantageous to use SEC in the appropriate media as the primary mode (column 1) followed by any one of the other LC methods (column 2). Combination of RPLC and IEC has also been widely used for analysis of biological samples. In all cases of multi-dimensional chromatography, precise control of the timing and period of switching is necessary to avoid mix-up of the fractions as well as to prevent incomplete transfer that would cause errors in quantitation. This may sometimes be difficult since the peaks in HPLC are usually sharp, but reproducible transfers are easily achieved if carried out using automated, microprocessor-controlled switching valves.

By suitably modifying the valve injection system, it has also been possible to couple LC with GC [46]. As in GC, backflushing the column for clean-up after elution of the first few components may be considered yet another type of multi-dimensional chromatography; backflushing in LC, however, is not recommended because of the possibility of disturbing the column. Other applications of multi-dimensional LC include attachment of up to five columns using two 6-port valves for rapid column selection, sample enrichment by injecting a large volume into a precolumn and backflushing the concentrated sample into the analytical column, regeneration of one column while the other is being used for analysis, and stop-flow scanning of UV or IR spectra, etc.

High-pressure switching valves and pneumatic actuators for multi-dimensional LC are available from several chromatography supply houses, some of them (e.g. Rheodyne, U.S.A. and Valco, U.S.A.) specializing in these products. Fig. 7.24 shows some of the multi-port valves and multi-dimensional arrangements.

### 7.6.6 Preparative liquid chromatography

The preparative chromatography (either GC or LC) is distinguished from analytical chromatography which has been the subject of discussion in the foregoing part of the book. The term is generally used to signify isolation of the separated components as against on-line manipulations aimed at qualitative and quantitative analysis (section 9.2.1). The reasons for isolation of the samples may be several, including off-line

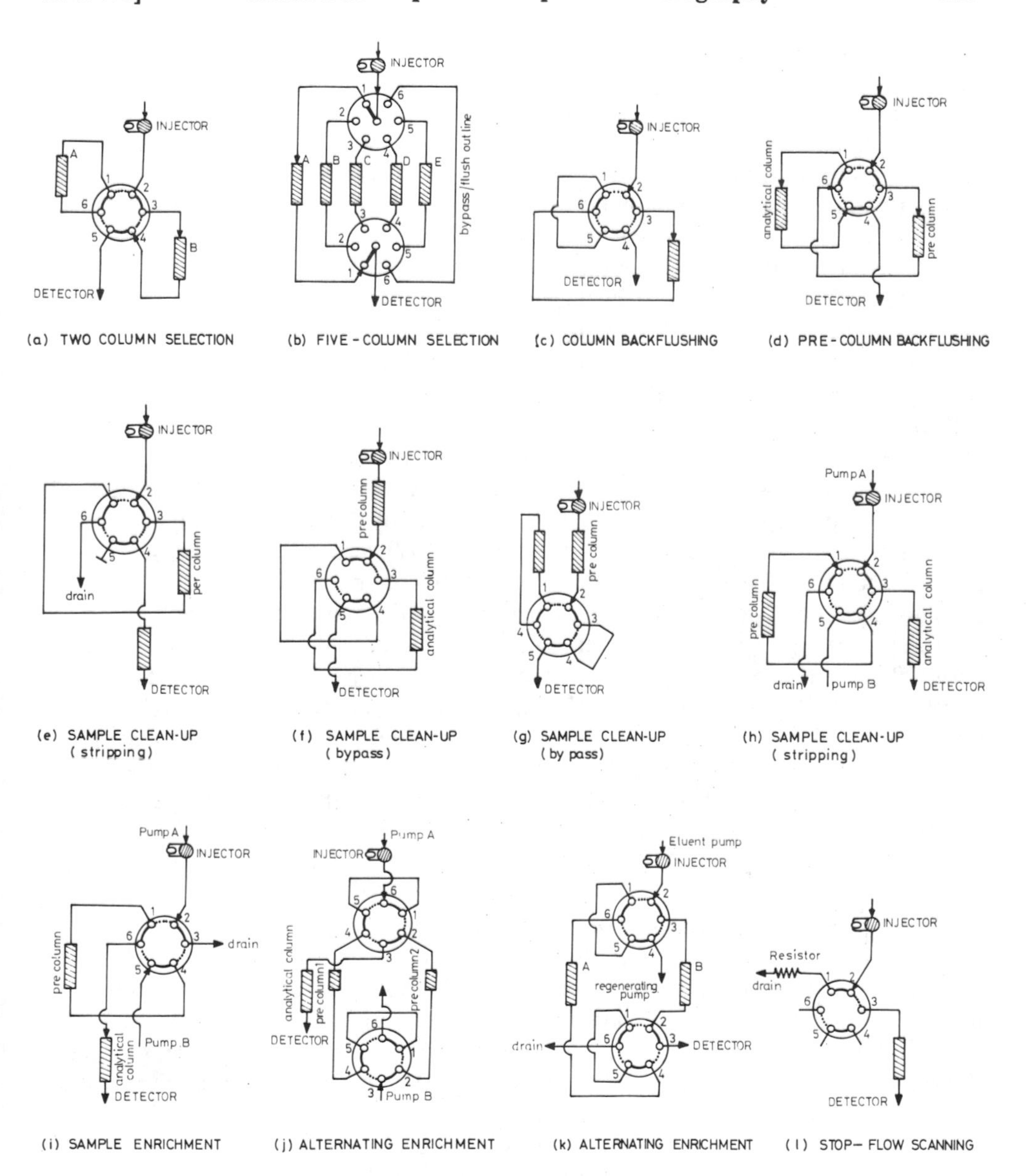


Fig. 7.24 — Switching valves for different multidimensional LC arrangements; illustration by courtesy of Rheodyne Inc.

spectroscopy (IR, MS, NMR, etc.), purification of a natural product or a synthetic intermediate for structural studies, degradative studies, biological evaluation or for further synthesis, or purely industrial where process-scale separation may be required in the pharmaceutical, biotechnological or perfumery industries. Thus, the quantity of the material required to be handled may vary from a few milligrams for spectroscopic studies through a few grams in synthesis to kilograms and tonnes for process-scale separations.

Traditionally, the development of LC was closely associated with the requirement of organic chemists and biological chemists to isolate sufficient quantities of pure compounds for structure and activity studies. With the advent of GC instrumentation, the potential of chromatographic techniques for qualitative, quantitative and trace analysis proved so attractive that much of the effort during the past two decades was directed toward improvement in resolution and detectability. Nevertheless, preparative chromatography, particularly in LC remained important. The principles of preparative chromatography are the same for both GC and LC to the extent that they are directed toward enhancing the throughput through the column with adequate resolution. The difference is mainly in the area of effluent (fraction) collection which is very convenient in LC owing to ambient temperature of operation and collection. An automatic fraction collector is a convenient device for column-effluent collection which can also be done manually. Collection of the effluent from GC, on the other hand, is fraught with problems of cryogenic trapping, aerosol formation, etc.; nevertheless, laboratory-scale [47] and even process-scale [48] preparative GC is feasible using simple devices for the collection.

Three distinct types of preparative chromatography may be considered, based on the sample size. They are: semi-preparative (0.5 mg to 0.5 g), preparative (0.5 g to about 50 g) and process-scale chromatography. Semi-preparative-scale chromatography is normally performed using the same chromatographic instrumental support (injection ports, pumps, detectors, etc.); the stationary and the mobile phases are also generally the same. Some of the strategies of method development for large-scale chromatography are discussed below. Assuming that the separation has been optimized on an analytical scale, the primary object of the scale-up is to enhance the throughput to the desired level while maintaining adequate, if not the best resolution.

The present understanding of the principles of chromatography permits a systematic approach to enhancement of throughput while retaining several of the advantages of the analytical instrumentation. As noted earlier, the diameter of a packed column does not appear in the resolution equations. Thus, for a well-packed column the diameter can be increased without affecting the performance, provided the other conditions, including the mobile-phase linear velocity, remain the same. The sample size and the flow-rate can, therefore, be increased in proportion to the cross-sectional area. The implications of this approach may be interesting. A typical analytical HPLC column of dimensions 2 mm × 25 cm can take a load of 1–3 mg and has an optimum flow-rate of, say, 0.5 ml/min. If the diameter is increased to 8 mm, the figures respectively are 10–100 mg and 8 ml/min. The loadability can be increased to 500 mg with a column of the same length, but a diameter of 24 mm would make the flow-rate 72 ml/min. It may be noticed that as the column diameter is increased, the flow-rates that could be used exceed the capability of several solvent delivery systems now on the market. Also, much of the plumbing, couplings and detector are based on low to moderate flow-rates. These may have to be suitably changed if the diameter exceeds 8 mm.

For a given diameter of the column, the sample capacity is dictated by the amount of the stationary phase, in other words, the length of the column. The column is said to be overloaded when the capacity factor ($k$) decreases by more than 10%; but, in preparative chromatography, overloading must be the point where the required separation does not occur. The significance of this approach is that, if the compounds are initially (on an analytical scale) well separated, the load may be increased beyond the traditional overload limit of 10% reduction in $k$. Even when the compounds elute rather closely, increase in the sample size up to the point where the resolution drops by 50% may be tolerated. The effect of increasing the column length is to proportionally increase the number of theoretical plates. However, the resolution is a function of the square root of the plate number, which means that, to offset the loss of resolution of 50%, the column length must be increased four-fold. This would, in turn, increase the analysis time four-fold at the given mobile-phase flow-rate.

Considering the stationary and the mobile phases, the large quantity of the stationary phase limits the choice to some of the more common and inexpensive stationary phases. Thus, between normal-phase silica and the reversed-phase ODS packings the choice may be the former, but the cost of the solvent may be much lower in the reversed-phase mode as it is then based essentially on water. On the other hand, removal of large quantities of water (for recovering the solute) may be expensive. Considering laboratory safety and other matters, the reversed-phase systems may have an edge unless the sample can be readily separated using a single-composition solvent. Gradient elution is not normally preferred in preparative chromatography since considerable time may lapse in re-equilibration between cycles and solvent recovery and re-use is difficult.

The stationary-phase particle size also is of prime importance. Lately, the 4–5 $\mu$m particles have become very popular in analytical LC. Though a low flow-rate is desirable for high efficiency, higher throughput for a given column size is achieved by increasing the flow-rate or the sample load. The flow-rate can only be marginally increased for the small-particle columns without substantially increasing the pressure drop. 8–10 $\mu$m particles are now widely used in preparative LC, so that faster analysis is possible, though at the cost of some efficiency.

To summarize the choice of the separation system, it must be based on the isocratic mode, either in the normal or reversed phase (ion-exchange and size-exclusion methods are also used). Particles of 8–10 $\mu$m are an ideal compromise between resolution and speed for sample loads in the semi-preparative range. The length of the column is chosen to give an adequate plate number and then the load and/or the mobile-phase flow-rate are increased until the resolution falls by about 50%. The diameter can then be increased as required, keeping in mind that the flow-rate and sample throughput increase in direct proportion to the cross-sectional area of the column.

The volume and mode of sample introduction must be carefully considered. The six-port valve injection is common. For a given quantity of the sample, it is preferable to use a large volume of a dilute solution than a small volume of a concentrated sample. The total volume of the sample injected may be maintained to be between 25–30% of the elution volume of the first peak; up to about 10 ml may be injected in the large diameter columns. Most of the common detectors of analytical chromatography may be too sensitive for large-scale chromatography. Low sensitivity universal detectors, like the refractive index detector, may be used. Alternatively, the path length of the UV detector, which is normally 10 mm, may be reduced to 1 mm. The detector signal being within a certain range is important if it is used to trigger the tube movement in an automatic fraction collector.

Several commercial instruments are available which all use much larger packing particles (30–75 μm) compared to the 8–10 μm particles discussed above; two distinct examples may be mentioned. One is from Jobin-Yvon (France) and the other is from Waters (U.S.A.). The French system consists of a 60 × 8 cm column which is filled with the adsorbent and longitudinally compressed with a piston. The Waters instrument employs a radial compression system, based on the principle discussed earlier and using 30 × 5.7 cm polypropylene cartridges. Each cartridge can take about 375 g of the adsorbent; 20 g of a binary mixture with a separation factor of two can be separated in a single run.

Columns for industrial-scale chromatography are generally custom-built and may have diameters up to 20 cm. The technology and engineering could be quite complex in view of the large size of the columns and the consequent difficulties in packing them. As noted above in connection with boxcar chromatography, much of the stationary phase is not used at any given time. Large throughputs are achieved by multiple moving-spot injections, resulting in a virtual countercurrent flow of the solvent and the sorbent, though the sorbent is stationary. A number of short and wide-bore columns may be operated in parallel. The sample is introduced in pulses into each column in turn using a rotating multiport valve. The frequency of injection is adjusted so that the last peak of the previous sample elutes just before the first peak of the present sample. The throughput then is the maximum possible with no portion of the bed remaining unused. The rate of rotation of the injection device must be synchronized with the fraction collection device, so that a nearly continuous system can be obtained. Several thousand tonnes of *p*-xylene (from ethylbenzene and other isomers), and over a million tonnes of fructose (from molasses) have been separated chromatographically [49]. High selectivity of the stationary phase is necessary in such cases and, in the case of the fructose example, this is achieved by strong cation exchangers in the $Ca^{2+}$ form. A two-volume monograph [50] describes the theoretical and technological aspects of large-scale chromatography.

## REFERENCES


[1] J. F. K. Huber (ed.), *Instrumentation for High-Performance Liquid Chromatography*, Elsevier, Amsterdam, 1978.

[2] L. R. Snyder and J. J. Kirkland, *Modern Liquid Chromatography*, 2nd edn., Wiley, New York, 1979.

[3] H. Engelhardt, (ed.), *Practice of High-Performance Liquid Chromatography: Application, Equipment and Qualitative Analysis*, Springer-Verlag, Berlin, 1985.

[4] C. F. Poole and S. A. Schuette, *Contemporary Practice of Chromatography*, Elsevier, Amsterdam, 1984.

[5] N. A. Parris, *Instrumental Liquid Chromatography*, Elsevier, Amsterdam, 1984.

[6] P. Jandera and J. Churacek, *Gradient Elution in Column Liquid Chromatography*, Elsevier, Amsterdam, 1985.

[7] J. J. Kirkland, W. W. Yau, H. J. Stocklosa, and C. H. Dilks, *J. Chromatogr. Sci.*, **15**, 303 (1977).

[8] H. D. Christensen and R. Isernhagen, in *Biological and Biomedical Applications of Liquid Chromatography* (G. L. Hawk, ed.) Vol. III, Marcel Dekker, New York, 1981.

[9] R. P. W. Scott, *Adv. Chromatogr.*, **22**, 247 (1983).

[10] R. P. W. Scott, *Small-Bore Chromatography Columns: Their Properties and Uses*, Wiley, New York, 1984.

[11] M. Novotny (ed.), *Microcolumn Separations*, Elsevier, Amsterdam, 1984.

[12] P. Kucera (ed.), *Microcolumn High-Performance Liquid Chromatography*, Elsevier, Amsterdam, 1984.
[13] F. M. Rabel, *J. Chromatogr. Sci.*, **23**, 247 (1985).
[14] N. Sagliano, Jr., H. Shih-Hsien, T. R. Floyd, T. V. Raglione, and R. A. Hartwick, *J. Chromatogr. Sci.*, **23**, 238 (1985).
[15] E. D. Lee and J. D. Henion, *J. Chromatogr. Sci.*, **23**, 253 (1985).
[16] L. T. Taylor, *J. Chromatogr. Sci.*, **23**, 165 (1985).
[17] K. Slais, *J. Chromatogr. Sci.*, **23**, 321 (1985).
[18] T. M. Vickery, *Liquid Chromatographic Detectors*, Marcel Dekker, New York, 1983.
[19] E. S. Yeung (ed.), *Detectors for Liquid Chromatography*, Wiley, New York, 1986.
[20] R. P. W. Scott, *Liquid Chromatography Detectors*, Elsevier, Amsterdam, 1986.
[21] I. S. Krull, *Reaction Detection in Liquid Chromatography*, Marcel Dekker, New York, 1986.
[22] D. G. Jones, *Anal. Chem.*, **57**, 1057A (1985); **57**, 1207A (1985).
[23] L. N. Klatt, *J. Chromatogr. Sci.*, **17**, 225 (1979).
[24] P. R. Griffiths and C. M. Conroy, *Adv. Chromatogr.*, **25**, 105 (1986).
[25] I. N. Mefford, *Meth. Biochem. Anal.*, **31**, 221 (1985).
[26] D. M. Desiderio and G. H. Fridland, *J. Liquid Chromatogr.*, **7**, 317 (1984).
[27] C. Eckers and J. Henion, in *Therapeutic Drug Monitoring and Toxicology by Liquid Chromatography*, S. H. Y. Wong (ed.), Marcel Dekker, New York, 1984.
[28] T. R. Covey, E. D. Lee, A. P. Bruins, and J. D. Henion, *Anal. Chem.*, **58**, 1451A (1986).
[29] N. J. Alcock, C. Eckers, D. E. Games, M. P. L. Games, M. S. Lant, M. A. McDowall, M. Rossiter, R. W. Smith, S. A. Westwood, and H.-Y. Wong, *J. Chromatogr.*, **251**, 165 (1982).
[30] M. J. Hayes, E. P. Lankmayer, P. Vouros, B. L. Karger, and J. M. McGuire, *Anal. Chem.*, **55**, 1745 (1983).
[31] W. M. A. Niessen, *Chromatographia*, **21**, 277 (1986), **21**, 342 (1986).
[32] M. L. Vestal, *Anal. Chem. Symp. Ser.*, **24**, 99 (1985).
[33] R. C. Willoughby and R. F. Browner, *Anal. Chem.*, **56**, 2626 (1984).
[34] R. W. Frei and J. F. Lawrence, *Chemical Derivatisation in Analytical Chemistry, Vol. 1: Chromatography*, Plenum, New York, 1981.
[35] J. F. Lawrence and R. W. Frei, *Chemical Derivatisation in Liquid Chromatography*, Elsevier, Amsterdam, 1976,
[36] K. Imai, *Adv. Chromatogr.*, **27**, 215 (1987).
[37] G. Guiochon, in *High-Performance Liquid Chromatography: Advances and Perspectives*, C. Horvath (ed.), Vol. 2, Academic Press, New York, 1980.
[38] R. E. Kaiser and E. Oelrich, *Optimisation in HPLC*, Huethig, Heidelberg, 1981.
[39] H. J. G. Debets, *J. Liquid Chromatogr.*, **8**, 2725 (1985).
[40] S. N. Deming, J. G. Bower, and K. D. Bower, *Adv. Chromatogr.*, **24**, 35 (1985).
[41] H. Issaq, *Adv. Chromatogr.*, **24**, 55 (1984).
[42] L. de Galan and H. A. H. Billiet, *Adv. Chromatogr.*, **25**, 63 (1986).
[43] J. C. Berridge, *Techniques for the Automated Optimisation of HPLC Separations*, Wiley, New York, 1986.
[44] R. A. Henry, S. H. Byrne, and D. R. Hudson, *J. Chromatogr. Sci.*, **12**, 197 (1974).
[45] L. R. Snyder, J. W. Dolan, and Sj. van der Wal, *J. Chromatogr.*, **203**, 3 (1981).
[46] K. Grob Jr, D. Frolich, B. Schilling, H. P. Neukom and P. Nageli, *J. Chromatogr.*, **295,** 55 (1984).
[47] R. Teranishi, in *Flavor Research*, R. Teranishi, R. A. Flath, and H. Sugisawa (eds.), Marcel Dekker, New York, 1981.
[48] P. E. Barker, in *Developments in Chromatography–I*, C. E. H. Knapman (ed.), Applied Science, London, 1978.
[49] M. Verzele and E. Geeraert, *J. Chromatogr. Sci.*, **18**, 559 (1980).
[50] P. C. Wankat, *Large-Scale Adsorption and Chromatography*, Vols. I and II, CRC Press, Boca Raton, Florida, 1986.

# 8

# Planar chromatography

## 8.1 INTRODUCTION AND THEORY

The term planar chromatography is currently being used to refer to both paper and thin-layer chromatography of different forms. Except for the historical reasons of being the oldest chromatographic technique (in the form of Kapillaranalyse) and the forerunner of partition chromatography, paper chromatography may now be considered all but extinct. Nevertheless, for the same reasons as just mentioned, a brief reference to certain aspects of paper chromatography is made in this chapter. The reasons for the unpopularity of the otherwise inexpensive and convenient technique are not hard to see. Paper chromatography is much less efficient with respect to the analysis time and resolution compared to TLC and the more recent high-performance thin-layer chromatography (HPTLC). Therefore, wherever the stationary phase (cellulose) has the desired selectivity, paper chromatography can be profitably replaced by the cellulose TLC, though the results are not strictly identical. Commercial availability of pre-coated cellulose TLC plates, glass-fibre sheets and loaded papers (see later) have certainly contributed to the popularity of cellulose TLC over the conventional paper chromatography. In any case, the two techniques have many operational features in common; e.g., sample application, methods of development, detection procedures and techniques of qualitative and quantitative analysis.

The theoretical principles of liquid chromatography discussed in Chapter 6 are generally applicable to both the column and the planar techniques, the main difference being in the form of the stationary phase and the technique of development and detection. In the simplest form of TLC, the sample is dissolved in a volatile solvent and a small volume (1–100 $\mu$l) is placed on the TLC plate, about 0.5–2 cm from one edge. The plate is placed in a jar containing the mobile phase at a level below that of the applied sample. The mobile phase then moves up the plate by capillary suction, carrying and separating, in the process, at least some of the components. When the mobile phase reaches the top, the plate is removed from the tank and the mobile phase allowed to evaporate. The separated spots that remain on the plate are visualized by exposing the plate to, for example, UV light, iodine vapours or sulphuric acid spray (followed by heating to 100°C for a few minutes). There are several variations of each of the different aspects of TLC, namely, (1) the stationary phase, (2) the plate, (3) the technique of sample application, (4) mobile-phase application, (5) development and (6) visualization. These are discussed in the following sections. Several excellent texts have appeared on the subject [1–10].

The differences between the CLC and TLC methods are quite obvious. In CLC, the sample application is sequential, one at a time, while in TLC, it is multiple and concurrent; up to 60 samples can be spotted on a 10×10 cm HPTLC plate. Thus, while the separation efficiency of HPLC (with a plate height of 2–5 $\mu$m by 30 $\mu$m for TLC or 12 $\mu$m for HPTLC) is larger, the sample throughput in routine analysis is much higher in the planar techniques owing to the possibility of multiple and simultaneous analysis of a large number of samples. On the other hand, HPLC has unquestioned superiority in its ability to separate complex samples, the separation number (or the number of components that can be separated in a given sample in a single run) being 20 to 40 while that in TLC is 7 to 10 and in HPTLC, 10 to 20. The comparative separation efficiency can be gauged from the total number of theoretical plates that can be obtained on a system of typical dimensions; the figures are 5,000 to 10,000 in the case of HPLC, about 5,000 for HPTLC and <600 for conventional TLC. Separation and detection is by elution from the stationary-phase bed in the former and developmental in the latter. Also, the detection and quantitation in the former is dynamic or on-line, while in TLC, it is static; this aspect of TLC can be exploited to advantage in a number of ways. Resolution can be optimized for a few compounds at a time, and the choice of the mobile phase is not limited by considerations such as refractive index or light absorption of the solvent, since it is removed from the plate prior to detection and evaluation. Reactions may be carried out *in situ* with different reagents on the same plate. With the aid of scanning densitometers and computers, manipulations such as optimization of the detector signal by signal averaging, digital filtering, background subtraction or cross-correlation can be easily handled [11].

A major drawback of TLC, however, is the difficulty in automation of all the various steps such as sample application, placing the plates in the development chamber, their removal, application of spray reagents and further treatment for visualization, transfer to densitometer, and quantitation; some of these operations have been automated and these are mentioned at appropriate places.

Regardless of the relative merits of CLC and TLC, the two techniques should be considered complementary to each other. Thus, while HPLC may have high efficiency for the separation of individual samples, TLC, because of the inexpensive equipment, ease of operation and speed, is routinely used for scouting solvent systems for CLC and for monitoring its effluents. TLC is also preferable when a large number of easily separable samples are to be routinely analysed. Also, compounds of high $R_F$ value appear rather broad and diffused on TLC, but being of low $k$ value, yield sharp peaks in HPLC. In contrast, compounds which are strongly retained appear as broad peaks in HPLC but give compact spots on TLC (see Fig. 2.4).

In spite of several attempts, satisfactory theoretical treatment of TLC remains elusive [12,13]. Much of the difficulty arises from two factors. Firstly, the exact composition of the stationary as well as the mobile phases remains ill-defined as they exist in a constantly changing environment owing to the presence of solvent vapour of indefinite composition in the development chamber. As soon as the TLC plate comes into contact with the mobile phase and its vapour, the stationary phase (e.g. in the normal phase) tends to preferentially adsorb some of the more polar constituents of the solvent system, resulting in an undesigned solvent gradient both in its composition and velocity. The adsorptive properties of the stationary phase may be under continued modification during the run, because of the changes in its porosity and the apparent velocity of the mobile phase. The flow profile of the mobile phase also is subject to inconsistency owing to its uneven penetration through the bed; the capillary forces being stronger in the narrow interparticle channels, these are filled first.

The second factor responsible for the inaccuracies in the theoretical treatment is that the experimental parameters of TLC, as it is normally practiced, are not easy to control or measure. For example, the mobile-phase flow-rate ($Z_f$, defined as the migration distance of the mobile-phase front in cm after $t$ seconds) is given by eq. (8.1), where $x$ is the mobile-phase velocity constant ($cm^2/s$). The value of $x$ is a function of several factors such as the permeability constant of the thin layer ($K_o$), the mean particle diameter ($d_p$), surface tension of the mobile phase ($\gamma$), the mobile-phase viscosity ($\eta$) and contact angle ($\theta$), as shown in eq. (8.2):

$$Z_f^2 = xt \tag{8.1}$$

$$x = 2K_o d_p \frac{\gamma}{\eta} \cos\theta \tag{8.2}$$

The permeability constant, again, is a complex, dimensionless variable and is a function of the porosity and pore-size distribution profile of the stationary-phase particles, as well as the ratio of the bulk mobile-phase velocity to the solvent-front velocity. The value of $K_o$ is about 8000 for HPTLC silica plates (see below). The contact angle for most organic solvents on silica plates is normally zero, so that $\cos\theta = 1$. On the reversed-phase plate, the contact angle of the aqueous organic phases is much higher than zero owing to wettability problems and, therefore, the rate of flow of the mobile phase on these plates is very slow. Equation (8.2) shows that the rate of solvent flow is considerably improved by using coarser particles (i.e. $d_p$ is large, as in the conventional TLC plates), but the more uniform and tightly packed layers used in HPTLC yield larger $K_o$ values, thus compensating for the slow mobility of the solvent.

As in the case of the other chromatographic systems, the efficiency of separation (or resolution) on TLC is dependent on the mobile-phase linear velocity, the diffusivity of the solute in the stationary and the mobile phases, the resistance to mass transfer of the solute in the two phases, the mean particle diameter and the pore-size distribution. As can be expected, the separations on the small-particle HPTLC are much superior to that observed in the conventional TLC plates and are characterized by near-symmetrical (Gaussian distribution profile) and compact spots. The number of theoretical plates ($n$) is given by:

$$n = 16\left(\frac{Z_s}{w_b}\right)^2 = 16\left(\frac{R_F Z_f}{w_b}\right)^2 \tag{8.3}$$

where $Z_s$ is the migration distance of the spot $s$, and $w_b$ is the spot diameter. As defined earlier, $R_F$ of a substance in TLC is given by the ratio of $Z_s$ to $Z_f$. In contrast to GC or CLC, where all solutes elute from the column and thus travel the same distance (i.e. the length of the column) but in different times (i.e. the retention times of the individual components), the different components of a sample migrate to different extents in TLC, but in a given time. Thus, separation in TLC is a function of distance rather than time and a uniform plate number for the bed cannot be defined. This feature (i.e. the freedom from the time constraint) is exploited in a number of ways using rather ingenious development methods to be elaborated later. Since the efficiency is a function of the migratory distance of the spot, the definition of the number of theoretical plates should also specify the $R_F$ value and this figure is conventionally 0.5 or 1.0. The height equivalent to a theoretical plate is then given by the simplified equation:

$$h = a\left(\frac{Z_f^{2/3} - Z_o^{2/3}}{Z_f - Z_o}\right) + b(Z_f + Z_o) \tag{8.4}$$

where,

$$a = \frac{3}{2} A\, d_p \left(\frac{d_p}{2D}\right)^{1/3} \tag{8.5}$$

and,

$$b = \frac{2\gamma D}{x R_F} \tag{8.6}$$

In the above equations, $D$ is the diffusivity of the solute (assumed to be the same in both the phases), $A$ is a constant (approximately equal to 1), $Z_o$ is the distance between the initial position or the sample application point and the solvent level in the tank and $\gamma$ is the tortuosity factor (see Chapter 2).

The efficiency of conventional TLC generally increases with the distance of the mobile-phase flow. However, since the rate of flow of the solvent on HPTLC plates falls steeply as the migration length increases, the separation efficiency is counteracted by the diffusional spreading of the separated zones and a run of only about 5 cm is normal. Both the efficiency of separation and the reproducibility of TLC data can be greatly improved by minimizing the fluctuations in the vapour–solid equilibrium by using a sandwich chamber, in which a glass plate is held close to the stationary phase bed by means of a gasket and spring clips (see later). In the so-called over-pressured or forced-flow TLC, the glass cover plate is replaced by a flexible plastic membrane that is held in intimate contact with the bed by the use of hydraulic pressure on the other side [14]. This would render the capillary forces inadequate to cause the solvent flow and an external metering pump is required for the purpose. Use of a constant-volume pump permits complete control of the mobile-phase velocity independently of the plate length. The relationship between the theoretical-plate height ($h$) and the distance of mobile-phase flow for different types of TLC are given in Fig. 8.1. As discussed above, the efficiency of HPTLC is much superior ($h$ is

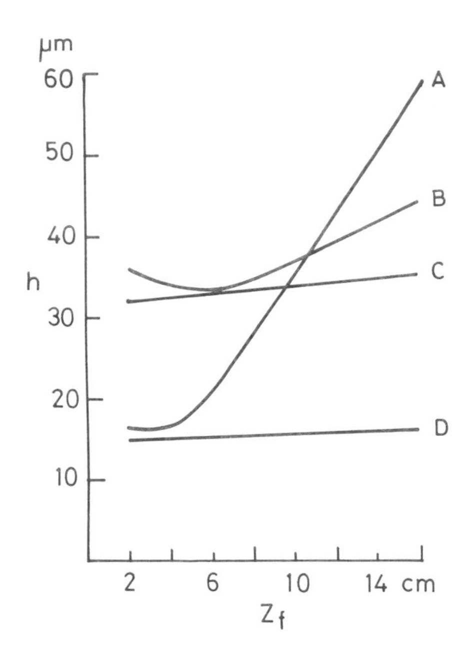


Fig. 8.1 — Variation of theoretical-plate height ($h$) with solvent-migration distance, $Z_f$. (A) HPTLC, normal development; (B) conventional TLC, normal development; (C) conventional TLC, over-pressured development; (D) HPTLC, over-pressured development. Reproduced from Ref. 13 by courtesy of Elsevier Scientific Publishing Co.

much lower, see curve A) to conventional TLC (curve B) at lower migratory distances of the solvent front ($Z_f$), but it sharply declines ($h$ sharply increases) as $Z_f$ crosses about 10 cm. In the over-pressured (OPTLC) mode, on the other hand, the plate height remains nearly constant for both conventional and high-performance TLC plates (curves C and D) and the theoretical plate number ($n$) increases linearly with the distance of solvent flow. As Fig. 8.1 indicates, the value of $h$ in HPTLC is only a third (about 12 $\mu$m) of that in the coarse-particle TLC. Indeed, a plate height as low as 8 $\mu$m was achieved when the flow was optimized so that a plate count of as high as 31,000 was possible for a substance moving by 25 cm. If the particle-size and bed homogeneity, as well as optimal solvent flow, are assured, a plate count of 50,000 seems feasible [13].

The $R_F$ value in TLC can be related to the capacity factor of CLC by the following equation:

$$R_F = \frac{1}{(1+k)} \quad \text{or} \quad k = \frac{1-R_F}{R_F} \tag{8.7}$$

The resolution equation for TLC can thus be written analogously to that in LC (see eq. (2.40)) as given below, using the same terms; namely, $n$, $k$ and $\alpha$:

$$R_s = (\alpha - 1)\left(\frac{n}{1+k}\right)^{1/2}\left(\frac{k}{1+k}\right) \tag{8.8}$$

The first term in the above equation reflects the selectivity of the system, while the second, the quality of the layer and the third, the relative position of the spots on the chromatogram. The plot of the resolution of two closely placed spots as a function of the migratory distance (Fig. 8.2) shows that the curve is bell-shaped, with the

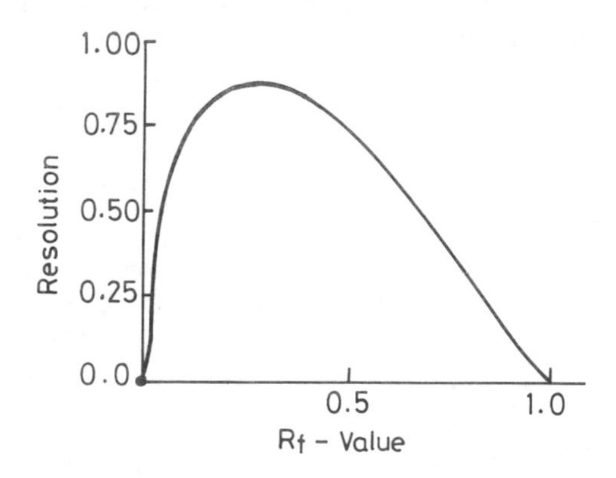


Fig. 8.2 — Variation of resolution in TLC as a function of solute migratory distance ($R_F$-value). Reproduced from Ref. 13 by courtesy of Elsevier Scientific Publishing Co.

maximum resolution at an $R_F$ of about 0.3, or essentially between 0.2 and 0.5. Under the forced-flow conditions, however, there is no maximum attainable resolution as a function of the $R_F$ value since the resolution continues to increase with the migratory distance. Thus, while the peak capacity (the number of compounds that can be separated in a single run) in the conventional TLC is only about 30, that in OPTLC may be of the order of 150–200.

## 8.2 STATIONARY PHASES

### 8.2.1 Paper and cellulose

The main advantages of paper chromatography have been the ready availability of the stationary phase and the low cost. Whatman Inc., U.S.A., and Schleicher & Schuell, F.R.G., have long been the chief suppliers of the material and, as such, uniform quality and reproducibility of results can be realized. Specially selected cotton linters are used for the manufacture of the papers. Cellulose, the predominant constituent of the paper, is a 1,4-β-linked naturally occurring polymer of glucose. The polysaccharide chains, held together by hydrogen bonding, form an ordered structure, parallel to the machine direction (usually marked on the packing). The paper contains interconnected capillaries and pores of varying size and shape and holds a quantity of strongly bound water which is not available for dissolving the solute. Unless thoroughly dried, it also contains loosely bound multilayers of water and for a long time it was believed that paper chromatography was only a partition process between the water and the organic mobile phase, cellulose being the solid support for the aqueous stationary phase. However, the fact that separation can be achieved using water-miscible solvents, or even water alone, showed that other mechanisms are also operative. Separation can be due to the differential adsorption of the solute molecules on cellulose or cellulose–water complex. The carboxylic acid groups present in the cellulose structure can also act as ion exchanger; the ion-exchange capacity of Whatman # 1 paper is about $3\text{–}8 \times 10^{-3}$ meq/g of dry paper. Possibly all three mechanisms, namely, partition, adsorption and ion exchange, operate in paper chromatography.

Generally, ordinary laboratory filter paper, cut into the desired shape and size, is adequate for most purposes. For more demanding applications and reproducible results, special chromatographic papers may be used; a large variety of papers, varying in thickness and porosity, are commercially available. Inorganic impurities, which are often troublesome (causing irregular zoning, etc.), may be removed by successive washing with an aqueous solution of a metal-complexing agent (e.g. ethylenediaminetetraacetic acid, EDTA), dilute hydrochloric acid and water, followed by drying. Organic impurities may be washed off with ether or with the mobile phase to be used. The practical aspects of paper chromatography have been discussed in detail in a monograph by Sherma and Zweig [2]. The technique has been used extensively for the analysis of a variety of compounds, notably, polyhydric alcohols, carbohydrates, amino acids, alkaloids, heterocycles, flavonoids, anthocyanins and oxo compounds [14]. As noted above, the inconveniences of using large papers, tanks, and the slow development, are overcome by using cellulose thin-layer plates.

### 8.2.2 Modified papers

Papers impregnated with oils are used for the separation of hydrophobic substances and those coated with polar liquids are used to facilitate separation of certain polar compounds. Silicone-treated papers of various grades are available commercially. Separation of inorganic ions is achieved using papers coated with liquid ion-exchangers such as high molecular weight amines and acids; impregnating the papers with complexing and precipitating agents is also useful. Among the chemically modified papers, the acetylated papers and oxidized papers containing up to 5% COOH groups are more common.

*Loaded papers* consist of up to 40% of silica gel, Kieselguhr or alumina powders dispersed on cellulose paper. They are mostly used for separation of non-polar compounds in much the same way as the respective TLC adsorbents. A variety of these have been introduced commercially and their applications have been reviewed

[15]. Papers loaded with either cation-exchange (e.g. carboxymethylcellulose and cellulose phosphate in monoammonium form) or anion-exchange (DEAE-cellulose and ECTEOLA-cellulose, see section 6.6.2.2) materials are also commercially available.

Glass-fibre papers have the advantage of being stable to corrosive detecting agents containing chromic acid or sulphuric acid [16]. When sprayed with a liquid of appropriate refractive index, (e.g. methyl salicylate or ethyl benzoate), they become transparent, making them convenient for quantitative analysis by transmittance densitometry. Other advantages include easier and complete elution of the material chromatographed, high loadability, and temperature stability.

### 8.2.3 Silica gel

Silica gel is by far the most frequently used stationary phase in TLC. Though chemically similar, silica gel, silicic acid, Kieselguhr, diatomaceous earth, celite and powdered glass are some of the various forms of silica used as the stationary phase in chromatography. Preparation and properties of the chromatographic grade of silica were briefly described in section 6.5.2. Different types of materials, with a wide range of specifications, i.e. particle size (6–66 $\mu$m) and pore size (60–150 Å) and surface area (300–600 $cm^2/g$) are generally available commercially and it is seldom necessary to prepare them in the laboratory. However, for some special requirements, the gel may be washed successively with 50% aqueous hydrochloric acid, then with an abundance of water, a 0.1% aqueous solution of EDTA, ethanol, chloroform and hexane, and finally dried at 120°C for 4 h. Both inorganic and organic impurities are removed this way.

Specifically adsorbing silica gels with a marked affinity for a particular compound have been prepared by coagulating (precipitating the silica gel in the presence of the compound, particularly organic bases, and then removing the compound from the gel by extraction with a solvent). Silica gel prepared this way shows enhanced selectivity for the particular compound in comparison to closely related compounds. Presumably, these compounds occupy the sites vacated by the sensitizing molecules of the same structure. In view of the stereoselectivity of the active sites, these adsorbents have potential application in separation of optical isomers, determination of relative configuration and study of biological models [17].

Almost all types of compounds have been analysed on silica-gel layers. However, it is sometimes advantageous to modify the chromatographic properties by impregnating the gels with certain metal ions or liquid phases, or by chemical bonding (see below).

### 8.2.4 Bonded phases

The spectacular success of bonded-phase silica in HPLC naturally led to the development and use of similar stationary phases in TLC. Bonded silicas carrying alkyl chains of 1, 2, 6, 8, 12 and 18 carbon atoms have been prepared with varying degrees of selectivity [18]. The resolution power of these phases for non-polar compounds, stability, selectivity and capacity (or loadability) appear to increase with the length of the hydrocarbon chain. The octadecylsilyl-bonded phases are commercially available as pre-coated plates (see below). Non-polar compounds, which move too fast on the normal-phase silica (e.g. hydrocarbons, fatty-acid esters, carotenoids, etc.), as well as those that are retained strongly (polyphenols, carboxylic acids, glycosides, etc.), are conveniently separated on the reversed-phase layers. As in the case of HPLC, the mobile phase is generally water- or alcohol-based and the substances move in the decreasing order of polarity, the least polar solutes being retained strongly. The techniques such as sample application, visualization and quantitation are similar to those used in the normal-phase silica except that aggressive charring by reagents like sulphuric acid may lead to background discolouration. An introductory account on the subject of RPTLC is available from Whatman Chemical Separations, U.S.A. [19]; see also Ref. 20.

### 8.2.5 Miscellaneous adsorbents

With the ready availability of silica gels of a wide variety of activity, selectivity and form, other adsorbents, in effect, are only of academic interest. Alumina, which was at one time a serious contender for the first place, is relegated to a distant second position. Different grades of alumina (acidic, neutral or basic) may be obtained commercially, or easily prepared by heating precipitated aluminium hydroxide to 400°C. Alumina, prepared this way is usually basic. Neutral alumina plates are obtained by dipping the plate in ethanol containing 1.5% acetic acid, and acidic alumina plates are obtained by dipping them in acetate-buffered ethanol at pH 5–6. As with silica gel, a variety of hydrocarbons, carbonyl compounds, alkaloids, etc., have been analysed on alumina. Its usage has considerably decreased following the realization that a variety of transformations take place on the active surface of alumina. Structural changes in organic molecules occur on silica gel also but much less frequently [21]. A number of oxides, carbonates, silicates and other insoluble salts of various metals (particularly those belonging to group IIA of the periodic table) have been suggested as adsorbents for specific separations, but none has found general application. Phosphates and arsenates of cerium and zirconium have been examined for their ion-exchange properties. Almost all the adsorbents employed in CLC (Chapter 6) have been examined for use in TLC.

Among the organic adsorbents, polyamides rank next only to cellulose (and paper). Though the early promise of polyamides (polycaprolactam, perlon, polyacrylontrile, polyvinylpyrrolidone, etc., and mixtures thereof) could not be realized, they remained important in the limited area of polyphenols and are occasionally used to resolve other proton-active substances such as acids, alcohols and amines. The sorption mechanism is considered to be reversible hydrogen-bonding of the solute molecule to the amide moiety of the sorbent, but resolution of non-hydroxylic compounds like esters, imidazoles and phenylhydrazones on polyamide indicates that other mechanisms could also be operative [22]. Dextran layers have also been used for size-exclusion chromatography on thin layers but the procedures are certainly more tedious than in CLC.

## 8.3 THE PLATES

### 8.3.1 Laboratory preparation

Preparation of uniform TLC plates of reproducible performance is an art that can be easily mastered by practice. Choice of the plate, use of binder, preparation of the slurry, spreading the layer on the plate, drying, activation and impregnation of the layers with different reagents (when necessary) are all carried out rather routinely in most laboratories. However, considerable variation is possible in each of the manipulations and individuals may have their own preferences.

Glass, in view of its relative inertness and ready availability at low cost, has been the most popular support for the layers. Smooth as well as ground, or sandblasted, surfaces may be used. Stainless steel, aluminium and plastic sheets have also been used. Plates of different sizes and shapes have been suggested for different types of application. Rectangular and square plates (particularly for two-dimensional development) are the most common; the sizes vary from 2.5×7.5 cm (microslides) to 100×200 cm. The larger plates are used mainly for preparative work; 5×20, 10×10 and 20×20 cm plates are common for everyday use.

Earlier workers prepared the thin layers without a binder, resulting in much softer plates, which were rather difficult to handle. Cellulose, because of its fibrous structure, does not require a binder but it is convenient to use one in the case of silica gel or alumina. Several binders have been suggested, the most common being plaster of Paris (5–20%) and starch (2–5%), in that order. The popular brands of silica gel G and alumina G contain plaster of Paris (gypsum) as the binder. Starch, which gives harder layers with a high rate of solvent mobility, may be preferable except that the plates may darken when treated with corrosive spray reagents such as those containing sulphuric acid.

For the preparation of the plates, a slurry of the adsorbent in a suitable medium is spread as a thin layer on the chosen backing. The nature and the proportion of the dispersing liquid depend on the adsorbent and the method of spreading; this information is usually furnished by the supplier of the adsorbent. Water-based slurries are preferred for silica gel, alumina and other inorganic adsorbents, particularly when a spreader is used. The ratio of the adsorbent to water may vary (normally 1:1 for silica gel and 1:2 for alumina), depending on the thickness of the desired layer; the object is to obtain a somewhat thick, but fluid, slurry. Cellulose and ion-exchange layers are also slurried in water while methanol is used for polyamide.

The simplest method of preparing the larger plates is to pour the slurry on to the plate and vibrating the plate either mechanically or by hand on a horizontal surface. Preparative plates, which are relatively thicker, are conveniently prepared this way. The layer may also be spread with a glass or a metal rod using the finger tips or rubber sleeves at the ends as a guide, to give a uniform thickness.

The slurry is more conveniently and uniformly spread on the plates using a spreader, several types of which are commercially available. Basically, they all contain a rectangular reservoir to hold the slurry and either a fixed or adjustable slot for the slurry to flow on to the plate. A very inexpensive device of the former type can be easily constructed using four glass plates and an adhesive [23]. In the CAMAG (Switzerland) machine, the plates are moved manually or electrically over a conveyor, under the slurry reservoir. In a variation designed by Stahl, the spreader containing the slurry is moved over a series of plates, side by side, on a horizontal surface; the slurry is dispensed on to the plates through a slot of adjustable thickness. More recently, the spreader has been modified to allow preparation of gradient layers [24]; these consist of two adsorbents, the proportion of which may change from 0 to 100% from one end to the other. These 'movable hopper' spreaders can also be moved electrically, eliminating the problem of uneven coating. One of the problems associated with the movable spreaders is the possible disturbance in their movement resulting in a non-uniform thickness of the plate. This is overcome by using a plate leveller; the plates are placed over a series of rollers held in place between two guide rails and the air-bag underneath the rollers is then inflated to bring the upper surface of all the plates to the same level.

Ordinary microscope slides ($2.5 \times 7.5$ cm, microslides) are very convenient for routine use in the laboratory for scouting suitable solvent systems, monitoring column effluents, following the progress of a reaction, etc. The slurry is prepared in an organic solvent (acetone, ethyl acetate, methanol, chloroform, or a mixture of two or more of these); chloroform, because of its high density, gives a better dispersed gel and a more uniform layer. The suggested procedure for preparation of the plates is to hold two plates together, back to back, dip them into the slurry and withdraw slowly. The following procedure is very convenient. A slurry of silica gel (25 g) (e.g. Merck 60) in chloroform (100 ml) is prepared in a 10 cm high wide-mouthed, glass-stoppered bottle. The bottle is held in a slightly inclined position and the slide is dipped into the slurry with the help of forceps and slowly drawn out, touching the sides of the slide to the inner wall of the mouth of the bottle. A uniform layer of 50–100 $\mu$m thickness is formed on the upper side. The thinner layer formed on the lower side is scraped out and added back into the bottle for re-use. Each plate prepared in this fashion has only about 50 mg of the adsorbent.

The plates, prepared by any of the above methods or obtained ready-made, need to be activated before use by removing part of the moisture. Silica gel layers are usually dried in air for about an hour and then heated at 110–120°C for 30 min to 3 h, depending on the thickness of the layers. Air-dried plates are preferred for certain applications.

### 8.3.2 Commercial products

While home-made plates are still being used in many laboratories, they are being fast replaced by commercial pre-coated plates, which are available in various sizes, shapes and thicknesses, and coated on glass, aluminium or plastic sheets. Plastic sheets are particularly convenient as they can be cut easily into the desired size and shape. They are also convenient for quantitative analysis using transmission densitometry or the inexpensive cut-and-weigh method (see later). Silica gel, alumina, cellulose and reversed-phase layers with or without a fluorescent indicator, are readily available. It is also possible to obtain layers impregnated with silver nitrate or ammonium bisulphate; the latter material decomposes to sulphuric acid on heating thus obviating the need to spray the corrosive reagent for visualization (see later). A more recent innovation is the permanently coated plate, which is described in greater detail below (sintered layers). Apart from saving considerable time, the advantages of pre-coated plates include: (1) better resolution, (2) reproducibility, and (3) increased accuracy in densitometric measurements, all due to the very uniform thickness.

### 8.3.3 Impregnated layers

Impregnation of the adsorbent is carried out to enhance the selectivity of the stationary phase. The property of the silver ion to complex with unsaturated compounds (particularly, to *cis* disubstituted olefins), or that of the borate ion to 1,2-diols, is exploited by preparing silver-nitrate- or borax-impregnated plates. Similarly, selectivity for certain metal ions is obtained on tri-isooctylamine-impregnated silica gel. The principles of partition chromatography are readily applied to TLC by impregnating the plate with one of the two immiscible liquids and using the other liquid as the mobile phase. Liquid ion exchangers are widely used for impregnation.

Water-soluble impregnating agents are conveniently added to the slurry during the preparation of the plates. Organic-solvent-soluble materials may be dissolved in a volatile solvent such as ether, the requisite quantity of silica gel added and the solvent allowed to evaporate in a draft of hot air. Silver-ion impregnated silica can also be prepared in a dry form [25]. For convenient preparation of silver-ion impregnated layers on microslides, silver nitrate (1 g) in methanol (25 ml) is added to silica gel (25 g) and the mixture diluted with chloroform (75 ml) with vigorous shaking. The plates are made as described earlier.

Impregnation of the plates after preparation can also be carried out in a number of ways. Hard and firmly bound layers can be dipped into a solution of the impregnating agent in a volatile, preferably non-polar, solvent. Another method is to allow the solution to flow on the plate as in normal TLC experiments. Though they can be sprayed on to the layer using an atomizer, reproducibility of results is doubtful in view of the difficulty in obtaining uniform impregnation. Impregnation of the layers after preparation of the plates may be used to advantage in two-dimensional chromatography by developing first on the normal layer, impregnating with the reagent and then developing in the second direction (see later). The technique of impregnation has been extensively used in paper chromatography. Chemical modification (acetylation) of polyamide layers has also been achieved by dipping the plate in a mixture of acetic anhydride and pyridine (20:7) for 5 min [26].

A wide range of impregnating agents has been used for specific separations. These include inorganic salts such as acetates, borates, formates, halides, molybdates, nitrates, phosphates and tungstates of cadmium, copper, iron, lead, manganese, sodium, thallium, etc., and organic substances such as carbowax, decalin, dimethlyformamide, methoxyethanol, nitromethane, paraffin oil, propylene glycol and vegetable oils [3].

### 8.3.4 Sintered layers

An interesting innovation in TLC is that of Okumura [27], who prepared sintered glass plates with adsorbents such as silica, alumina, magnesia, titania and zinc oxide fixed on them. These plates are highly porous (facilitating rapid development) and being free from organic binders, they are also acid-resistant and can be repeatedly regenerated and used even after spraying with corrosive reagents such as sulphuric acid or chromic acid. Polyolefin-sintered layers could also be prepared using cellulose and ion-exchange and size-exclusion materials. Chromatographic behaviour is comparable and sometimes even superior to the normal TLC.

The sintered glass plates or rods are prepared by mixing one part of silica gel (TLC grade) with 2–5 parts of finely powdered glass (1–10 $\mu$m), slurried in an organic solvent (acetone) and spread on a soda-lime glass plate. After drying, the plate is heated at 470–770°C for 2–7 min. It has been demonstrated that the properties of silica critical to chromatographic behaviour, such as specific surface area and specific pore volume, are unaffected by the treatment. The sintered glass powder acts as the binder. Sample application, development and detection are carried out as in traditional TLC. Regeneration of the plate is easily achieved by cleaning with organic solvents, chromic acid or nitric acid and water, followed by reactivation at 110°C for 30–60 min.

Taking advantage of the thermal stability of the sintered glass layers, a novel method of detection and quantitiation based on flame ionization could be developed. A commercial device (IATROSCAN) using this principle is available from Iatron Laboratories, Inc., Tokyo. The 'chromarods' used in this equipment are of 152 mm length and 0.9 mm diameter with a usable length of 120 mm. They carry silica or alumina layers of about 75 μm thickness. The sample (1–20 μg) is applied on one end and developed in a chamber in the usual manner. After development, the rods are exposed to the flame in a specially designed flame-ionization detector (Fig. 8.3). The

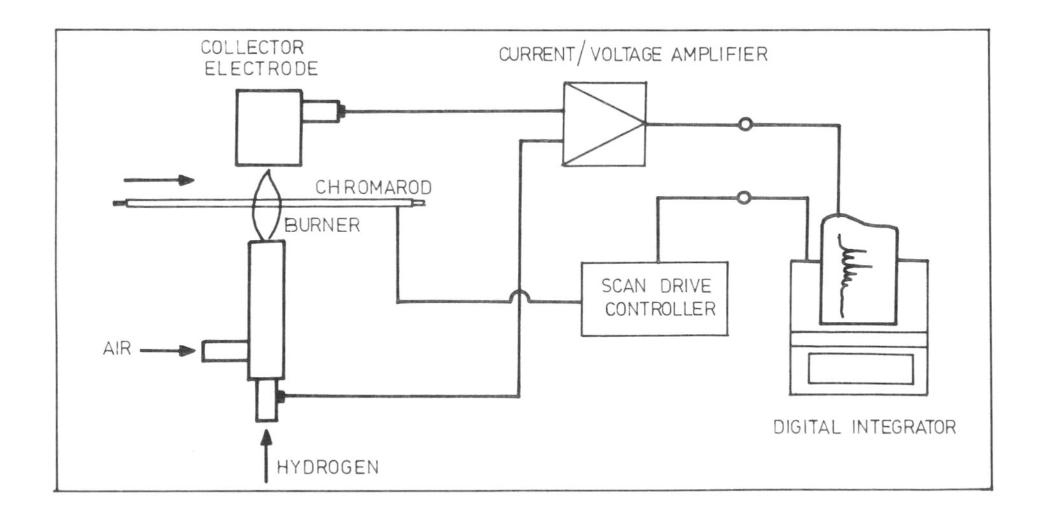


Fig. 8.3 — Schematic diagram of Iatroscan. Illustration by courtesy of Iatron Laboratories, Inc.

response is recorded and integrated for quantitation as in the other techniques of chromatography (Chapter 9). Excellent resolution, sensitivity (0.1 μg) and re-usability (100 times!) are claimed.

A very detailed review by the innovator gives information on the different methods of preparation of the plates with inorganic and organic adsorbents, mechanism of binding, comparison of performance of these plates relative to the classical plates, and their applicability to various classes of compounds [27].

### 8.3.5 High-performance thin-layer chromatography (HPTLC) plates

Strictly speaking, HPTLC is not different in principle or application from conventional TLC; but because of improvements in materials and manipulations, considerable enhancement in the effectiveness of the method could be realized. The technique, with a potential to operate at femtogram level, certainly deserves special mention. As discussed earlier, the performance of a sorbent is influenced by such factors as the mean particle size, particle-size distribution, pore system, surface texture of the skeletal substance, and the amount of absorbed water. Selected silica gels with a narrow pore-size distribution of 2–5 μm (compared to 6 to 66 μm in the conventional TLC plates) are used to prepare the plates of 200 μm thickness. Because of the sophistication in the technology of preparation of plates with very smooth surfaces required for HPTLC, these are not prepared in the laboratory; commercial HPTLC plates are available from several supply houses. Solvent flow-rates on these layers are much slower and smaller plates are used. Excellent results are obtained using 5 × 5 cm and 10 × 10 cm plates. Because of the tendency of the layers to get overloaded, great care should be taken in sample application; a sample smaller than 10 μg with a spot diameter of less than 1 mm should be used (see the following section). Performance characteristics of HPTLC plates, methods of development, quantitation and special equipment necessary are all well documented [8,9]. A comparison of the operational parameters of the conventional TLC and HPTLC are given in Table 8.1.

**Table 8.1** — Comparison of the operational parameters of conventional TLC and HPTLC†

| Parameter | Conventional TLC | HPTLC |
|---|---|---|
| Plate size | 20×20 cm | 10×10 cm<br>10×20 cm |
| Layer thickness | 100–250 μm | 200 μm |
| Particles size | | |
| average | 20 μm | 5–15 μm |
| distribution | 10–60 μm | narrow |
| Sample volume | 1–5 μl | 0.1–0.2 μl |
| Starting spot diameter | 3–6 mm | 1.0–1.5 mm |
| Separated spot diameter | 6–15 mm | 2–6 mm |
| Solvent migration distance | 10–15 cm | 3–6 cm |
| Time for development | 30–200 min | 3–20 min |
| Detection limits | | |
| Absorbance | 1–5 ng | 0.1–0.5 ng |
| Fluorescence | 0.05–0.1 ng | 0.005-0.01 ng |
| Tracks per plate | 10 | 18 or 36 |

† Reproduced from Ref. 13 by courtesy of Elsevier Scientific Publishers, Amsterdam.

## 8.4 SAMPLE APPLICATION

Assuming that the mobile phase, stationary phase and the developmental methods are optimum, the results of the TLC experiment depend to a large extent on the manner in which the sample is applied on to the plate. Prior to application of the sample a preliminary clean-up of biological samples and plant extracts is advantageous; the procedures naturally depend on the purpose of analysis, the nature of the substance to be analysed and the nature of the impurities to be removed. The procedures, which are common to all chromatographic methods, are discussed in the following chapter.

For spotting the sample on the plate, the sample should be dissolved in a volatile solvent, preferably of low polarity, and applied on the plate to obtain as small a spot as possible. The diameter of the spot may be up to 5 mm on paper and 1–2 mm in the case of conventional TLC plates; HPTLC plates require less than 1 mm spots for the resolution. In the simplest procedure, sample application is carried out by drawing a small volume of the solution into a glass capillary tube and gently touching the tip of the tube to the surface of the layer, taking care to see that the adsorbent surface is not disturbed; an ordinary melting-point tube is adequate for qualitative analysis. The amount of sample to be applied depends on the detectability of the methods, to be discussed later (section 8.7). Ideally, the smallest detectable concentration of the minor constituent should be applied. Calibrated micropipettes (1–10 $\mu$l) are available commercially. If the $R_F$ values of different samples are to be compared on the same plate, they are spotted along a line parallel to the edge of the plate. Spotting guides of different designs, usually made of plexiglass and with holes or notches for receiving the pipettes, are also available. To prevent undue atmospheric exposure, which could affect the $R_F$ value and even resolution, the remainder of the plate may be kept covered with glass plate during sample application.

The main advantage of automatic sample applicators is that they deliver repetitive, reproducible and accurate sample volumes on to the plate. They are convenient (a) when large quantities of a sample are to be applied for preparative work, (b) when a large number of samples are to be analysed routinely, (c) for accurate quantitation and (d) when HPTLC plates are to be used. Sophisticated spotters, either of the contact or the spray-on type, can accurately deliver from 50 nl for HPTLC to about 500 $\mu$l for conventional TLC. The sample may be applied either as a round spot or as narrow streak of variable length.

In most applicators, the sample solution is held in a syringe whose plunger is depressed either by a guide rod on its head or by the turning of a knurled head as it moves along the bar. The sample may also be delivered using air or gas pressure. A contact-spotting procedure is described [28] in which a specially treated fluoropolymer film is placed on a perforated plate on a device that can apply either suction or pressure on the film. The sample drop is placed on the film and a small depression is created under it by mild suction to prevent the spot from spreading. The plate is gently heated by a hot nitrogen draft on the sample. After evaporation of the solvent the residue is transferred on to a TLC plate (placed on it with its face down) by applying a slight pressure; spots of less than 1 mm diameter have been obtained using 0.1 ml of the solution. Several samples can be simultaneously applied at precise locations on the TLC plate. The procedure is illustrated with a single-point application scheme in Fig. 8.4. A possible drawback of the method is a non-quantitative transfer of the sample, particularly when the sample is highly crystalline. In such cases, a non-volatile solvent (e.g. methyl myristate) may be added to the solution.

A unique method of sample application, somewhat analogous to pyrolysis GC, is the TAS (thermomicro application–separation) technique or thermofragmentography. It involves the use of the Tasomat (from Desaga, F.R.G.). This equipment is useful for studying the products of pyrolysis of high molecular weight compounds as a function of temperature, or for cleaning-up of a volatile sample from a non-volatile

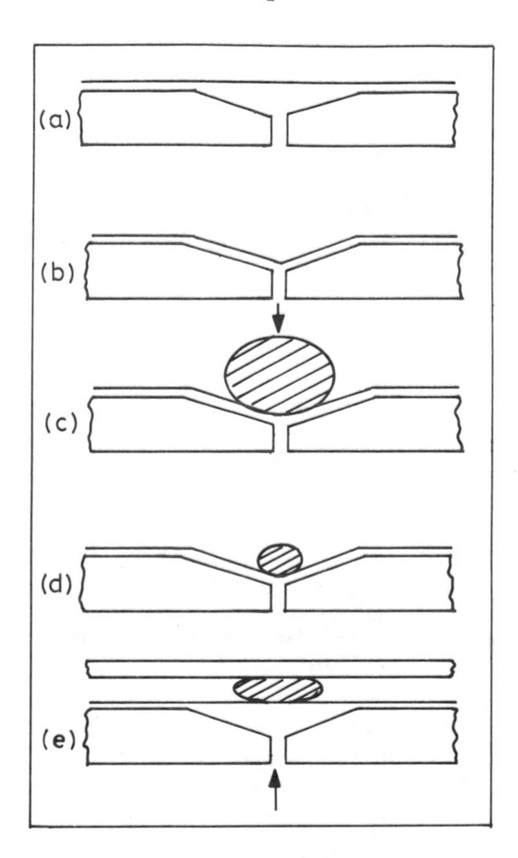


Fig. 8.4 — Contact-spotting procedure. Reprinted with permission from Ref. 11. Copyright (1981) American Chemical Society.

bulk. It consists of a narrow, cylindrical, programmable tube furnace which holds the sample. The exit end of the tube is drawn into a nozzle and a TLC plate is held vertically on a movable trolley and perpendicular to the axis of the cylindrical furnace. The gap between the cylinder orifice and the TLC layer is about 1 mm. An inert gas is passed through the other end of the cylinder and the furnace temperature is raised while the TLC plate is moved across the exit capillary. The escaping volatile compounds are carried by the gas and deposited on the TLC layer in the form of a streak, in the order of decreasing volatility; Fig. 8.5 shows the result of a chromatogram obtained in this manner [29].

## 8.5 MOBILE PHASES

The mobile phase in TLC may normally consist of one to four components depending upon the nature of the solute and the stationary phase. As in other types of chromatography, compatibility among the solute, solvent and stationary phase is a prerequisite for effective separation of the solute components. Though a wide

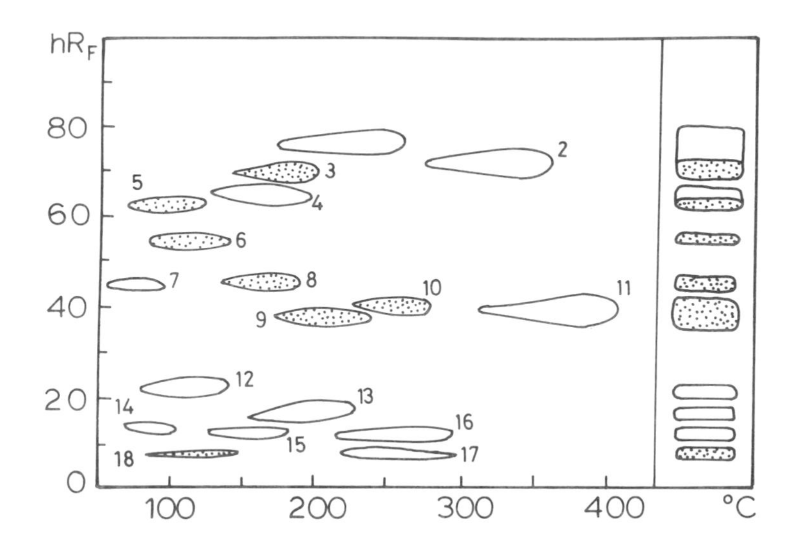


Fig. 8.5 — A typical separation using TAS technique. Substances are separated along the abscissa according to their volatility and on the ordinate according to their $R_F$ value. On the right is the chromatogram of the sample developed unidimensionally using the same mobile phase. Reproduced from Ref. 29 by courtesy of Elsevier Scientific Publishing Co.

variety of stationary phases have been suggested, silica gel, ODS-silica and, to some extent, cellulose are the most common, convenient and readily available adsorbents on which a majority of chemical classes of compounds have been separated. The quality of separation therefore depends, to a large extent, on the mobile phase. The properties of the mobile phases and their role in controlling the selectivity and mobility of the solutes have been discussed in detail in section 6.5.3. The observations made in that connection are all valid for TLC also. Nevertheless, a few general comments may be recalled. It may be noted that there is a wider choice of mobile phases in TLC since the requirement of detector compatibility becomes irrelevant.

On silica gel, the mobile phase strength and the mobility of the solute increases with the polarity of the solvent from hydrocarbons to hydroxylic solvents. This statement is generally valid for medium polar compounds for which the silica gel is ideal. For very non-polar solutes, which move very fast even in hydrocarbon solvents, and for very polar compounds which are strongly retained, reversed-phase adsorbents may be used. In the case of non-polar compounds which are strongly retained on the reversed phases, mobility can be increased by decreasing the polarity of the solvent. The mobility of polar compounds on a reversed phase is largely determined by the solubility of the compound in the mobile phase. Water is generally a component of the mobile phase in reversed-phase chromatography, employed to help (a) retention of non-polar compounds and (b) mobility of polar compounds. However, high water content causes problems of wettability on reversed phases, and alcohols or acetonitrile may be used as diluents; tetrahydrofuran, dioxan, etc., are sometimes added as modifiers. Silver salts (to separate *cis* and *trans* double-bond isomers) and counter ions (ion-pair chromatography) are conveniently added to the mobile phase in reversed-phase chromatography.

Though method development was discussed in detail in Chapter 6, in connection with CLC, the appropriate mobile phase is often sought using the much faster and inexpensive TLC. The ideal range of $R_F$ values, 0.25 to 0.66, corresponds to $k$ values of 0.5 to 3.0 in CLC (eq. (8.7)); similarly, the desired range of $k$ values, i.e. 2 to 9, on CLC corresponds to $R_F$ values of 0.1 to 0.33 on TLC. Thus, the method developed by Saunders [30] for determining the $E_3$ values (i.e., the solvent strength required to give a $k$ value of 3) for different compounds can be conveniently used for TLC. About 5% difference may be expected between the value calculated from eq. (8.7) and that observed in practice, because of the difference in the degree of equilibration between the mobile phase and the stationary phase in the two techniques.

Though the discussion on the above relationships and solvent properties can sometimes be used to determine the composition of the mobile phase, quite often one relies on one's own experience and that of others for choosing the proper system. Fortunately, almost all classes of compounds have been analysed and the systems are well documented. Statistical analysis to determine the ideal solvent system was found to be useful in correcting rather than developing a new system [31]. Numerical taxonomic techniques in chromatography have been reviewed [10,32].

## 8.6 DEVELOPMENT OF THE CHROMATOGRAM

The basic difference between closed column (CLC or GC) and planar chromatography is that while, in the former, the mobile phase is forced through the column under a head pressure in order to *elute* the sample components out of the stationary-phase bed, the mobile phase is *allowed* to flow over the stationary phase in the latter by capillary suction. Thus, the rate of flow of the liquid in TLC is determined by the capillary structure of the adsorbent and the viscosity of the mobile phase. Resolution in both cases occurs on the bed, but while detection and quantitation of the solutes occurs outside the bed (i.e. after elution) in closed-column systems, it is performed with the compounds still on the plate in TLC. Another major difference is that the amount of the mobile phase that flows over the TLC bed is limited by the size of the plate. For this reason, several strategies or techniques have been employed to optimize the resolution in TLC mainly through different developmental modes.

### 8.6.1 Linear development

Normally, TLC is developed in the linear mode, either in the ascending, descending or horizontal fashion. The choice is mostly one of convenience, ascending development being generally preferred for TLC and descending for paper chromatography. $R_F$ values are generally higher for a given system in the descending mode. By normal development it is understood that the plate is developed in the linear mode with a single run of the solvent across the length of the plate.

#### *8.6.1.1 Ascending development*

Ascending development is understandably popular for its simplicity of apparatus and procedure. The plate, with the sample applied in the appropriate manner, is placed in a closed chamber containing the mobile phase. Any chamber that can hold the plate in a near-vertical position is suitable. For a microslide, a 100 ml laboratory beaker, without the spout and covered with a glass plate (or a test tube with a bark cork for thin chromatostrips) is adequate. For larger plates, suitable jars (cylindrical or rectangular) covered with a ground-glass lid may be used. Several sizes and shapes of developing chambers, some of which are capable of holding 2–10 plates at a time, are available. Fig. 8.6 shows the uses of a commercial twin-trough chamber (CAMAG, Switzerland).

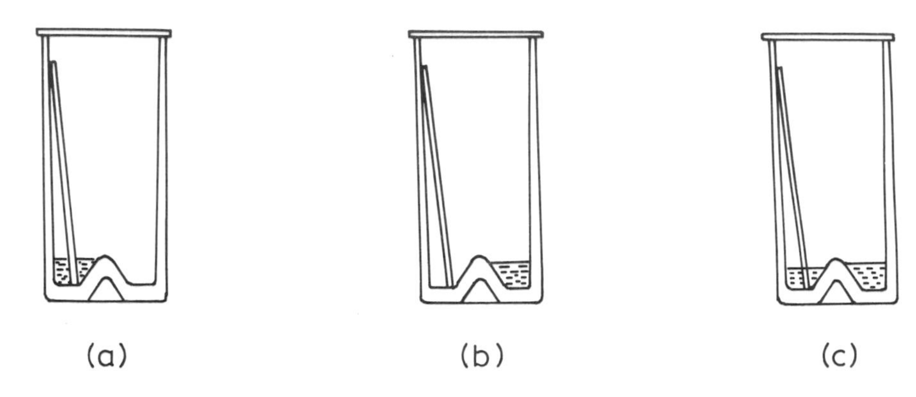


Fig. 8.6 — A commercial twin-trough chamber, showing (a) normal development; (b) pre-equilibration of the layer and (c) gas-phase equilibration. Illustration by courtesy of CAMAG AG, Switzerland.

For best results, the development of the plate is carried out in a chamber saturated with the vapours of the mobile phase. Otherwise, the solvent tends to evaporate from the edges of the plate, causing a concave solvent front and thus distorting the final result. This can be prevented to some extent by trimming the adsorbent along the three edges excepting the one that dips into the solvent. Chamber saturation is achieved by lining the walls of the chamber with a fast-flow filter paper that dips into the solvent. A twin-trough chamber (Fig. 8.6) is convenient if saturation with a different solvent is desired.

'Sandwich plates' probably have the minimum chamber volume. In this technique, the adsorbent is removed from three edges (about 5 mm wide) of the plate; the sample is applied along the fourth edge. A U-shaped gasket, made of glass, metal or plastic, is placed to cover the area; the thickness of the gasket should be slightly larger than that of the adsorbent layer. A second glass plate is then placed on the gasket and two plates with the gasket and the adsorbent layer in the middle are held together by spring clips. For development, the open end is dipped in a narrow trough containing the mobile phase (Fig. 8.7). The spots developed on sandwich plates are generally sharper and smaller than in chamber development.

#### *8.6.1.2 Descending development*

In the descending mode of development, the solvent is allowed to flow down the layer (or paper) by the combined action of capillary suction and gravity. The solvent

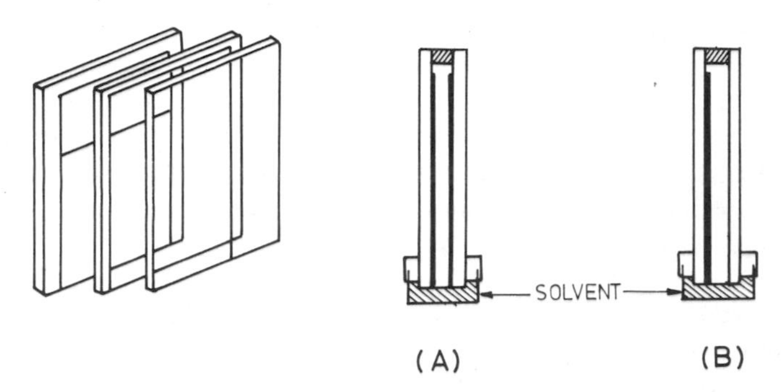


Fig. 8.7 — A sandwich chamber; (A) saturated by using a counterplate with the same adsorbent soaked in the same solvent and (B) unsaturated, using a glass cover plate. Illustration by courtesy of CAMAG AG, Switzerland.

flow is faster and is therefore preferred in paper chromatography, which is otherwise very slow. Moreover, the papers being relatively large (usually 35–40 cm length is required for good resolution compared to 10–12 cm in TLC), handling is inconvenient in ascending chromatography; also, the solvent slows down considerably in ascending development after about 25 cm of solvent run. Suitably large rectangular development chambers (about 60×20×60 cm) that can hold up to four solvent boats at the upper portion of the trough are commercially available for use with the larger papers. For descending development on TLC, a thick filter paper of the same width as the plate can serve as a wick for transferring the solvent from the boat to the plate which is placed on it, face down. Alternatively, as is usual for loose layers, the plate is held in an inclined (near horizontal) position, face up, and the solvent is fed by a paper wick.

### *8.6.1.3 Horizontal development*

Better reproducibility is claimed in horizontal linear development than either ascending or descending modes. Horizontal development can also be carried out using paper supported on glass rods or held between glass or metal plates. Fig. 8.8

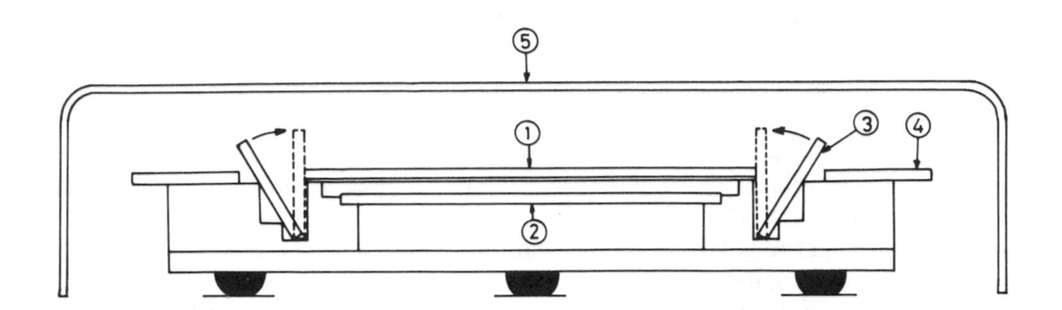


Fig. 8.8 — Cross-sectional view of a horizontal development chamber. Illustration by courtesy CAMAG AG, Switzerland. For explanation of the operation and key for the numbers, see text.

shows the cross-sectional view of CAMAG's horizontal development chamber used in HPTLC. Development can be either in one direction using the entire length of the plate, or in opposite directions using only half the length of the plate and thus doubling the number of samples that can be separated on a plate. Development is performed in the sandwich configuration and the amount of solvent needed is minimal. With reference to the diagram, the HPTLC plate (1) is placed face down on the chamber and glass strips (3), either or both, are tilted inward to start the development. The counter plate (2) has a 0.5 mm clearance from the plate for sandwich configuration and (4) and (5) are cover plates for the solvent reservoirs and the whole system respectively.

### 8.6.2 Radial development

Variations of horizontal linear development, using circular papers or plates, are described below.

#### *8.6.2.1 Circular development*

Circular development was widely used in the early days in all forms of planar chromatography but was replaced by the more convenient linear techniques. It is now receiving renewed attention with the development of the U-chamber by CAMAG (see below). In its simplest form, the circular chromatography with paper is carried out using two petri dishes of equal diameter and a filter paper of slightly larger size. Two parallel cuts are made 2–5 mm apart from the edge of the paper to the centre and bent down to form a wick that dips into the solvent placed in the lower petri dish. The sample is spotted at the centre and the second dish is used to cover the paper [2]. Separated components appear as concentric rings. Several samples can be simultaneously analysed on the same disc by drawing a small circle (about 2 cm diameter) at the centre and spotting the samples on it. The separated components would then form radial segments and the technique is therefore called radial chromatography. There is a concentration effect as the annular zones are formed; the zones increase in length but decrease in width, thereby aiding resolution. Circular development on TLC plates is performed in a similar manner [22]. For transportation of the solvent from the petri dish on to the layer, a wick is made by stuffing with cotton a small Teflon tube of diameter about 1 mm, and length a little less than the inner height of the dish. For a given plate, the speed of development is related to the diameter of the Teflon tube. The spotted TLC plate is placed face down with the centre of the plate touching the wick.

For gradient development, or while working with opaque metal sheets, it is advantageous to keep the plate on a horizontal surface with the coated side up, and supply the solvent from above. As in sandwich plates, a Teflon or metal gasket (2–3 mm thick and 90 mm diameter for a $10 \times 10$ cm plate) is placed on the plate and covered with another glass sheet with a tiny hole at the centre. A Teflon tube is held in place through the hole and its other end is connected to the solvent reservoir.

Circular development under highly controlled conditions is carried out in the CAMAG's U-chamber (Fig. 8.9). A $50 \times 50$ mm HPTLC plate (1) is placed face down in the holder ring of the body (2). The mobile phase is fed to the centre of the plate through a platinum–iridium capillary (3) and its flow is controlled electronically. The layer may be conditioned by gas or vapour fed through a circular channel (4) and exited through a bore (5); the direction of the gas flow can be reversed. By slight modification, the U-chamber can be used for simultaneous development using

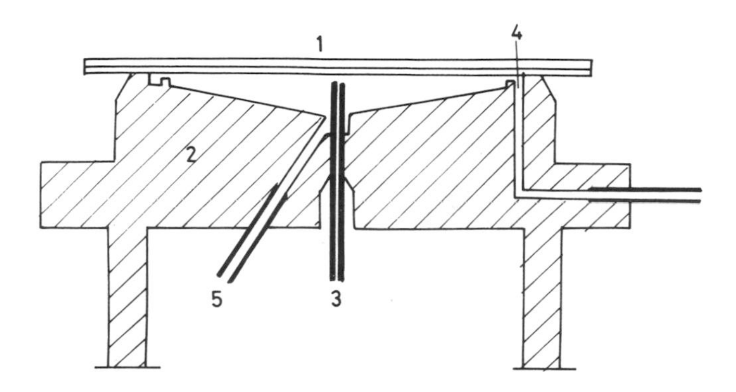


Fig. 8.9 — Schematic diagram of CAMAG's U-chamber. Illustration by courtesy of CAMAG AG, Switzerland.

up to four different solvents for rapid optimization of the mobile phase. Protagonists of the circular-development technique claim many advantages such as increased separating region and thus the separation number (i.e. the number of substances completely separable in the development region), increased resolution and reproducibility, increased sensitivity and accuracy in quantitation and enhanced speed of analysis [8].

The $R$ value in radial chromatography ($RR_F$) is given by the ratio of the distance travelled by the substance to the distance travelled by the solvent. It is related to the $R_F$ value in linear chromatography by the relation, $R_F = RR_F^2$. The practical significance of this relation is the enhanced separating region in the case of radial chromatography. The optimum separating region in linear chromatography is from about 0.25 to 0.65; below the lower limit, the components may be too close for complete separation and above the upper limit the band width and the consequent dilution of the substance increases, affecting both resolution and detectability. The 0–0.25 region of linear chromatography doubles to 0–0.5 in circular chromatography and the region 0.65–1.0 contracts 60% to 0.8–1.0; in comparison the 0.25–0.64 area suffers a contraction of only 25%.

### *8.6.2.2 Centrifugal development*

In this variation of circular chromatography, the sample is applied eccentrically on a circle of small diameter at the centre of the plate (or paper), held nearly horizontal on a rotatable disc. The mobile phase is delivered from a tube held just above the centre of the disc, while the disc is rotated at about 1000 r.p.m. (the speed variable between 250–2000 r.p.m.). A number of developments have been made in the assembly of the apparatus. A commercial apparatus (Chromatotron) for preparative application has become available (Harrison Research, U.S.A.). In this system, the solvent delivery is continued after the front reaches the edge of the chromatoplate so that the separated bands are eluted and spun-off into a collection system. Stahl and Mueller [33] have evaluated the performance of the system with respect to layer thickness (1–2 mm optimum), solvent flow-rate (3–6 ml/min/mm of layer) and rotational speed

(750 r.p.m.) and concluded that while there are certain definite advantages of speed, cost, sample size and re-usability in this system over the conventional preparative TLC and HPLC, there is room for improvement; see also section 8.9 (below) and Ref. 34.

#### *8.6.2.3 Anticircular development*

In the so-called anticircular development, the samples are applied on a circle of diameter slightly less than that of the solvent feeding line on the periphery of the circular plate and the solvent is allowed to flow to the centre. A technique analogous to the centrifugal chromatography but employing anticircular movement is centripetal TLC, in which the plate is placed on a turntable and the solvent is delivered on to a felt ribbon surrounding the plate using a stationary syringe. Advantages claimed are speed and more compact bands for substances in the high $R_F$ region. Similar effects are more conveniently obtained in triangular TLC by applying the sample at the base and allowing the solvent to flow to the apex. Varying degrees of success have been reported with other 'wedge-shaped' layers and papers [35].

### 8.6.3 Extended development

#### *8.6.3.1 Continuous development*

Continuous development simulates CLC and is useful for separation of strongly retained solutes without changing the solvent or removing the plate from the chamber. One of the greatest advantages of the descending mode of development is its easy adaptability to continuous development. Continuous development can also be carried out in ascending or horizontal mode by allowing the solvent to evaporate from the edge of the TLC plate. This is achieved by using a development chamber with a slotted lid and a plate longer than the chamber. To facilitate easy evaporation of the solvent, hot air or heat is applied on the edge.

#### *8.6.3.2 Multiple development*

In multiple development, the plate is removed from the chamber after each run, allowed to dry and redeveloped in the same solvent. Slow moving compounds of very close $R_F$ values can often be resolved in this way. The number of solvent passes required to effect the separation is determined by the $R_F$ difference between the two compounds and their absolute $R_F$ values. The number of developments needed to separate compounds of different relative and absolute mobilities have been tabulated [36]. Understandably, its utility is limited only to certain combinations of mobilities. Thus, it is not very useful for compounds with an $R_F$ larger than about 0.3 as the length of available space above the spot decreases considerably with each successive development. Also, as the lower spot has a longer development distance, it races closer to the faster-moving spot as the number of passes increases, and this problem is more acute if the two spots are very close. Multiple (or stepwise) development is also possible with different solvents and is generally performed with mobile phases in the increasing order of their strength. It is advantageous to develop the chromatogram to different heights with the length of development increasing with successive developments.

#### *8.6.3.3 Programmed multiple development*

This technique is essentially the same as described in the foregoing section except that a programmer is used to control both the distance and the number of developments. After each development, the solvent is evaporated by heat from an infrared source and/or an inert gas. The spots are usually elliptical, the narrow zone being in the direction of the development [35].

The possibility of chemical transformation of the solutes by the heat-assisted evaporation of the solvent is avoided in the so-called automated multiple development which is claimed to be a superior version of the programmed multiple development [10]. As in PMD, the chromatogram is developed repeatedly in the same direction, with each successive run being longer than the previous one. Most often, the elution strength of the mobile phase is reduced after each successive run (e.g. starting with methanol and progressively increasing the proportion of chloroform and then benzene). Almost all classes of compounds, including inorganic ions, carboxylic and amino acids, carbohydrates, phenols, etc. have been successfully separated by this technique. Unlike in the PMD, however, the solvent is removed after each successive run, the solvent from the plate evaporated under vacuum and the solvent changed. The combined focusing and gradient effects result in extremely narrow bands of about 1 mm width, so that as many as 40 compounds may be separated on the normal final run of 8 cm.

Figure 8.10 shows the arrangement for the automated multiple development unit (from CAMAG). It consists of a developing chamber (1) that can accommodate a 20

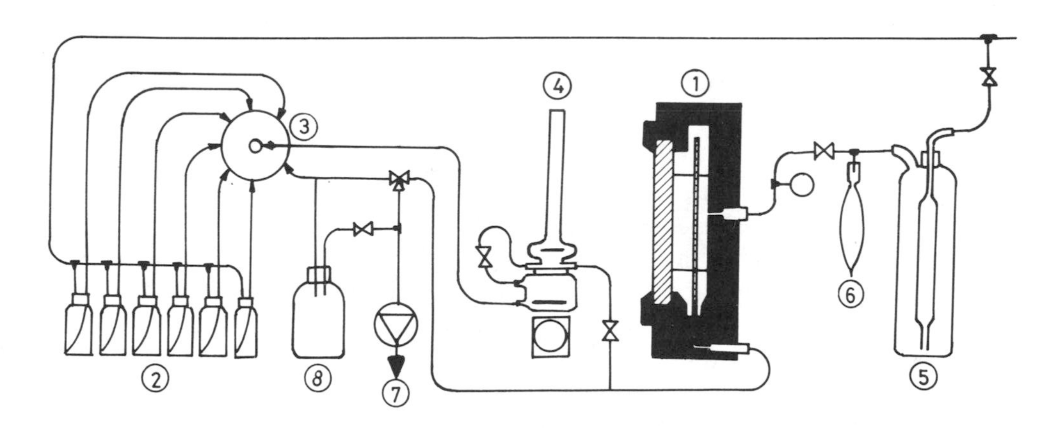


Fig. 8.10 — Schematic diagram of an automated multiple development unit. Illustration by courtesy of CAMAG AG, Switzerland.

× 10 cm plate, six solvent-reservoirs (2), a seven-port motor-driven valve (3), a two-step gradient mixer (4), wash bottle (5) for the external preparation of the gas phase, gas-phase reservoir (6) and the vacuum-tight tube system with control valves. (7) and (8) are the vacuum pump and the waste-collection bottle, respectively. The first solvent (about 8 ml) is fed to the gradient mixer and the run begun immediately.

After a certain amount of time, as determined by the programmed distance of travel by the mobile phase, the solvent is drained, the plate is then dried using the vacuum pump and, simultaneously, the mixer (4) is filled with the second solvent. The plate is reconditioned by feeding gas from the reservoir (6).

### *8.6.3.4 Two-dimensional development*

In a complex mixture of compounds, it is possible that certain components separate well on a particular system and others on another. Two-dimensional chromatography may be used to advantage in such cases. In the simplest form, where the adsorbent and the method are the same, the sample is spotted at the corner (about 1–2 cm from either edge) of a square plate and developed in the first solvent. The plate is removed from the chamber, the solvent allowed to evaporate, and the second development carried out at right angles to the first. The technique is very versatile and can be modified in a number of ways using different stationary phases and separation techniques. Electrophoresis, where ionic compounds are separated by differential migration under the influence of an applied electric field, has been extensively used in combination with paper chromatography for separation of amino acids, peptides and proteins; Fig. 8.11 shows 61 such compounds separated by this

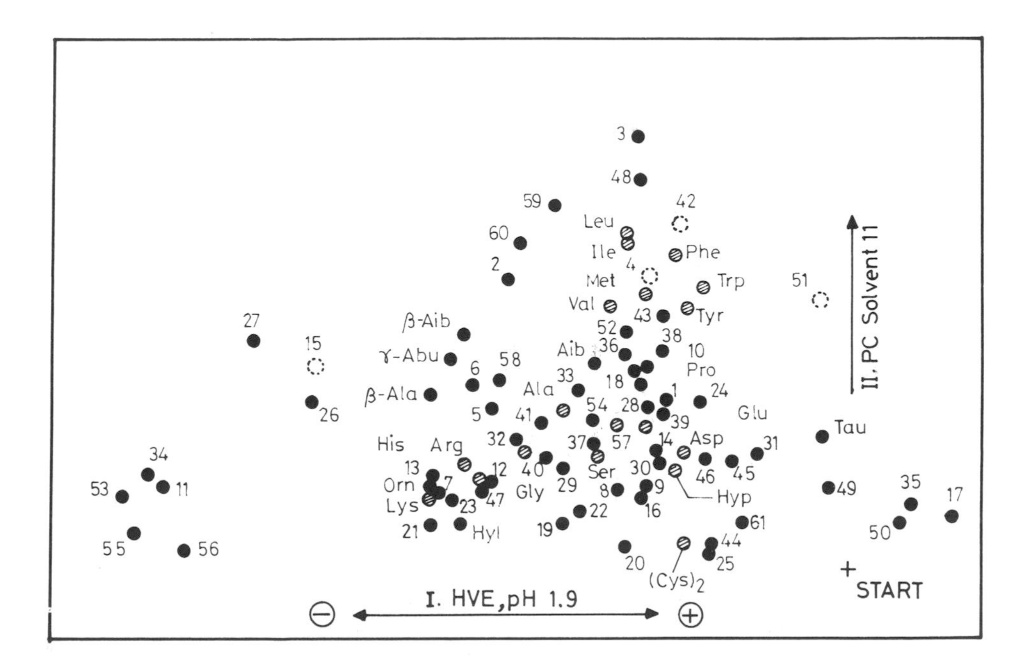


Fig. 8.11 — Illustration of separation of 61 amino acids and peptides by 2-dimensional development using high-voltage electrophoresis at pH 1.9 (horizontal direction) and chromatography on butanol:acetic acid:water (4:1:1) (vertical direction). Reproduced from Ref. 37 by courtesy of Van Nostrand Reinhold Publishing Co.

method [37]. Electrophoresis–adsorption, adsorption–steric-exclusion and steric-exclusion–electrophoresis combinations have also been used. The stationary phase, after the first development, may be modified by impregnating with silver nitrate or with a non-polar liquid; modification of the solute by chemical reaction on the plate has been employed in some cases. Two different adsorbents, silica gel for one direction and charcoal, magnesium silicate or alumina for the other have also been used on the same plate. Multilayers with octadecylsilyl-bonded phase and normal-phase silica gel are commercially available; they may have a 3 × 20 cm strip of the reversed phase, the remainder of the 20 × 20 cm plate being silica. The sample is applied on the bonded phase and the first development is carried out on this layer. The plate is dried and developed in the perpendicular direction by dipping the bonded-phase side in the organic mobile phase (Fig. 8.12).

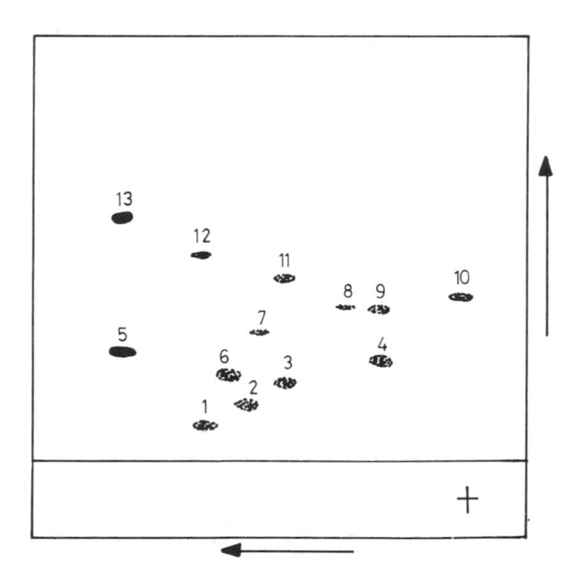


Fig. 8.12 — Illustration of separation of some suphonamides on 2-dimensional development on multilayers using normal and reversed-phase adsorbents. Reproduced from Ref. 19 by courtesy of Whatman Chemical Separations, Inc.

### 8.6.4 Gradient techniques

Gradient techniques in TLC may be considered a carry-over from CLC, but with the advantage that not only the mobile phase but also the stationary phase can be changed continuously or in steps.

#### *8.6.4.1 Vapour programming*

Normal TLC development almost always has a gradient effect; the plate above the solvent front is exposed to the solvent vapours, the components of which have different affinities towards the adsorbent. It is for this reason that tank saturation and equilibration of the plate with solvent vapours was studied to eliminate inconsistencies and to facilitate resolution by controlled gradients. In the Vario-KS chamber of CAMAG (Fig. 8.13), the conditioning tray consists of 3, 5, 10 or 25 sub-divisions

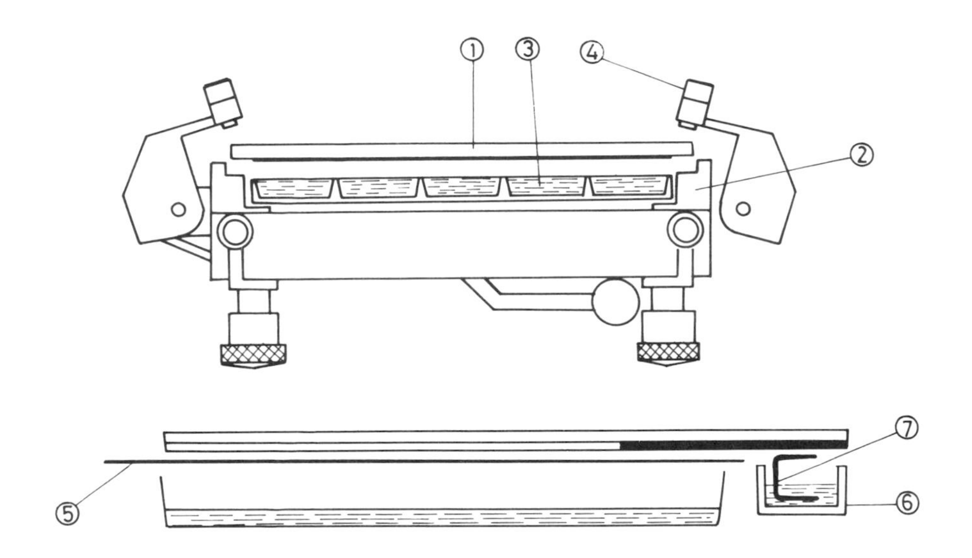


Fig. 8.13 — Cross-sectional view of CAMAG's Vario-KS chamber. Illustration by courtesy of CAMAG AG, Switzerland.

which may be filled with the same solvent (for uniform preloading) or with solvent mixtures of different composition and polarity (to obtain a gradient). The TLC plate with the samples is set on the tray with the sorbent layer facing the conditioning tray and allowed to equilibrate for 5–30 min. A sandwich plate is then inserted between the layer and the tray to prevent mix-up of the loaded vapours with the mobile phase vapours. The mobile phase in the solvent trough is transported to the TLC plate by means of a wick. The conditioning liquid and the mobile phase need not be the same.

### *8.6.4.2 Adsorbent gradient*

Gradient layers may consist of two adsorbents, A and B, ranging from 100% A on one side to 100% B on the other. Development can be along or perpendicular to the gradient. Gradation in stationary-phase characteristics can also be achieved by varying the amount of moisture on the plate, by impregnation, pH variation, or by using complexing agent. A spreader for the preparation of gradient layers was mentioned earlier.

### *8.6.4.3 Solvent gradient*

In the chamber developed by Wieland and Determann [38], the solvent mixture is stirred by a magnetic stirrer while the solvent is being delivered through a tube and the mixture withdrawn through the other. The TLC plate is supported on a perforated plate above the magnetic bar. Several variations have been discussed for obtaining a solvent gradient and delivering it to the plate [39].

### 8.6.5 Over-pressured (forced-flow) thin-layer chromatography (OPTLC)

The solvent vapour to which the adsorbent is exposed during the development has considerable influence on the mobility of the solvent as well as that of the solute. Tank saturation is, therefore, recommended to avoid inconsistent results caused by the variation in the vapour pressure of the solvent. Since the plate above the solvent front is always exposed to the solvent vapours, true adsorption chromatography cannot be easily achieved. Sandwich plates may be used for rapid saturation. The interstitial space is further reduced in over-pressured TLC [40]. It is claimed that the system simulates CLC, but whereas in CLC the stationary phase is in equilibrium with the mobile phase, in OPTLC, the layer is not exposed to the mobile phase except when the solvent is pumped through it. The theoretical considerations of the OPTLC (also known as the forced-flow thin-layer chromatography), have been briefly discussed in section 8.1. The features of this technique have been elaborately reviewed very recently [41]. The TLC plate is covered with a flexible plastic sheet (membrane) and subjected to air pressure over the membrane. The solvent is forced through the stationary phase using a pump at a pressure lower than the air pressure on the membrane. The operation needs a special device that can be easily constructed. Both conventional and HPTLC plates have been used. Owing to the low interstitial volume and the high flow-rates achieved by the pumping system, diffusion is minimal and, consequently, smaller and well-defined spots suitable for quantitative evaluation can be obtained. The time taken for development is 5–20 times faster than normal TLC. Apparatus have been designed both for linear and circular development.

### 8.6.6 Temperature-controlled development

The effect of temperature on chromatographic development and resolution has been investigated by a number of workers. As the vapour pressure increases with temperature, pre-loading of the sorbent layer by the solvent increases and only the remainder of the pore volume of the sorbent is available; this causes a decline in the resolution. Conversely, sub-ambient development sometimes helps resolution. Thus, conformational isomers of some heterocyclic nitrosamines have been resolved by developing the plate in a cryostatic jar at −77°C [42]. Varying the temperature during development has also been attempted. Sparingly soluble and thermally stable compounds have been chromatographed at high temperatures (150 to 160°C) using high-boiling solvents. An eluotropic series was established for high-boiling solvents using indigo as the test sample. They are, as expected, in the order: halocarbons < aromatic hydrocarbons < ethers < esters < aromatic nitro compounds < *N*-heterocycles [43].

## 8.7 VISUALIZATION

Unlike GC and CLC, the components after separation remain on the adsorbent in TLC and need to be detected or visualized *in situ*. Coloured compounds naturally appear as spots or bands without any treatment, but colourless compounds require special treatment before they can be located and quantitated. For convenience, the visualization procedures may be classified into non-destructive methods, and methods involving chemical change. About 450 reagents of varying degree of applicability are on record [14]. Only some of these are of general utility and widely used, most others being applicable to specific classes of compounds and sometimes only to a specific compound. Application of the visualizing agent to the chromatogram is done in a number of ways. They include: (i) exposure of the chromatogram to its vapours, (ii) dipping the plate in the reagent solution, (iii) allowing the solution to develop as in the normal TLC procedure, or (iv) spraying or impregnating the plate with the reagent. Occasionally, the reagent (or a chemical that produces the reagent on some simple treatment like heating) may also be applied prior to sample application and development. Spraying the reagent is probably the most common. Glass sprayers or atomizers are available from most suppliers of glassware and TLC products. Atomization requires air pressure and this is delivered either by a rubber bulb, a spray gun, an aerosol spray can, or compressed air. The methods used for visualization are usually determined by the nature of the compounds, availability of reagents and apparatus, requirements of sensitivity and quantitation, or simply convenience.

### 8.7.1 Non-destructive methods

#### *8.7.1.1 UV light*

A large number of compounds containing aromatic, heterocyclic or conjugated chromophores absorb or fluoresce in the ultraviolet (UV) region and can be easily detected by viewing the chromatogram under UV light. Commercial UV-view cabinets and portable UV lamps emit light of two selectable wavelengths: 254 and 366 nm. Enhanced sensitivity and contrast is achieved by use of fluorescent materials, either incorporated into the layer (e.g. zinc silicate and zinc cadmium sulphide) or sprayed on to the plate (e.g. dichloro- or dibromo-fluorescein in ethanol). The fluorescence is quenched by UV-absorbing materials which appear as dark zones on a light background. Adsorbents as well as pre-coated plates containing the fluorescent indicator are commercially available.

#### *8.7.1.2 Iodine vapour and water*

Like the UV method, exposure to iodine vapours is also a generally non-destructive procedure applicable to a wide variety of compounds, particularly those containing unsaturated functional groups. Though the iodine may sometimes react with the sample spot, the colour generated on the plate is often reversible; under favourable conditions, the sensitivity could be very high; e.g. 0.1 $\mu$g for lipids and alkaloids. Exposure to iodine vapours followed by UV light is often more sensitive. The iodine may be sprayed on to the plate in the form of dilute solution in a volatile non-polar solvent or, more conveniently, a few crystals of iodine are placed in a jar which gets filled with the volatile iodine vapours and the chromatogram (after removal of the solvent) is simply placed in the jar for up to 5 min.

By spraying water heavily on the plate, it is often possible to detect hydrophobic components as opaque zones against a translucent background.

#### *8.7.1.3 pH indicators*

Indicators commonly used in acid–base titration can be conveniently used for their detection on chromatograms. An aqueous or aqueous alcoholic solution of the indicator, such as bromophenol blue, bromocresol green or methyl red, is sprayed on to the plate after adjusting the pH close to the end-point using citric acid or sodium carbonate. If necessary, the indicator can be washed off the plate to recover the test compound.

### 8.7.2 Reagents causing irreversible reactions

The most common reagent belonging to this class is sulphuric acid. Aqueous sulphuric acid, 5–50%, without or with additives such as nitric acid, chromic acid or vanillin (for less reactive compounds), is uniformly sprayed on to the chromatogram and the plate is heated to about 110°C in an oven or on a controlled hot plate for 2–20 min. Acetic anhydride and concentrated sulphuric acid (Liebermann–Burchard reagent) gives characteristic colours with different steroids. Impregnation of the plate with sulphuric acid prior to sample application is also known, but there is the risk of the compound undergoing acid-catalysed changes. Incorporation of salts such as ammonium bisulphate which decomposes on heating to sulphuric acid has the advantage of avoiding corrosive sprays. Most organic compounds get charred by sulphuric acid on heating and thus give dark spots on a light background after the above treatment.

Antimony trichloride or pentachloride and phosphomolybdic acid are also reagents of wide applicability. Unsaturated compounds and alcohols can be detected by spraying with a solution of potassium permanganate in aqueous acetone, when the compounds appear as yellow zones on a dark pink background. Dragendorff reagent (acidified bismuth subnitrate + potassium iodide solution) for alkaloids, 2,4-dinitrophenylhydrazine for carbonyl compounds, and ninhydrin for amino compounds are well known. Other spray reagents and their applications are found in a number of good manuals on TLC [1–7,14].

### 8.7.3 Miscellaneous detection methods

Where detection is desired at a particular wavelength other than the common 254 nm and 366 nm, the chromatogram may be divided into a number of horizontal bands of 1–5 mm width; each band is scraped, shaken with a suitable solvent and the absorbance of the solution measured using a spectrophotometer. Flame-ionization detectors have been used for detection of volatile products formed by pyrolysis of the resolved compounds directly on the chromatogram. The combination of FID and sintered glass rods has been mentioned earlier. Chromatograms containing radioactive compounds can be photographed by direct contact, or examined by use of a Geiger, scintillation or gas-flow counter. Antibiotics can be detected by contacting the chromatogram with an agar layer inoculated with the appropriate micro-organism and containing a suitable indicator, such as the light-yellow triphenyltetrazolium chloride which is reduced by the micro-organisms to reddish brown triphenylformzan. Several other biological and enzymatic methods have been suggested [3].

## 8.8 ANALYTICAL METHODS

All the discussion in the foregoing sections on the choice of the stationary phase, the mobile phase and the various developmental techniques, is aimed at improving the separation or resolution of the sample. Higher resolution is desirable because the main purpose of chromatography, in general, is to determine the number of compounds present in the sample, their identity and quantity. High resolution yields round, sharp spots, which helps both qualitative and quantitative analysis. Diffuse spots make it difficult to determine the exact position of the spot for determination of the correct $R_F$ value for identification, as well as to obtain accurate densitometric data for quantitation. While the methods of qualitative and quantitative analysis are very similar in GC and HPLC (and, therefore, treated together in the following chapter), they are somewhat different on TLC. The differences are elaborated here, but the principles are essentially the same. As will be seen later, the modern densitometric evaluation of the planar chromatogram converts it into the typical *X–Y* trace (see Fig. 2.4) similar to the elution chromatogram (Fig. 1.3). This data is accepted by the modern integrators and data processors and the retention and quantitation data is obtained exactly in the same manner as for GC and HPLC and is used in the same manner (see section 9.2.1.3).

### 8.8.1 Qualitative analysis

The paper chromatography during the earlier period and TLC in recent times are widely used for qualitative analysis of samples to determine the number of compounds present in the sample and, if possible, to identify them by comparison with the standards. The planar chromatographic techniques are the most convenient for the purpose for three reasons: (1) the retention data (i.e. the $R_F$ values on planar chromatography) are very easy to calculate on a visualized plate; (2) the precision of the retention data depends on the precision with which the stationary phase and the mobile phase are reproducible and this occurs as a matter of routine in TLC since the sample and the standard (as well as a mixture of the two) can be spotted on the same plate (co-chromatography), so that the stationary phase, the mobile-phase composition, mobile-phase flow-rate and the temperature are exactly the same; and (3) a number of reactions can be easily carried out on the developed plate to determine the chemical nature of the analyte.

The reactions that can be conveniently carried out on the TLC plate include oxidation (aerial, ozonolysis, hydrogen peroxide, chromic acid and potassium permanganate), reduction (lithium aluminium hydride, aluminium isopropoxide, sodium borohydride and catalytic hydrogenation over colloidal palladium), dehydration and hydrolysis (using concentrated hydrochloric acid, sulphuric acid and hydrogen peroxide, sodium or potassium hydroxide, hydriodic acid or acetic acid), esterification (boron trifluoride-methanol, trifluoroacetic anhydride or acetic anhydride-pyridine), halogenation (bromine or iodine), nitration (nitric acid vapours or inorganic nitrate and hydrochloric acid), derivatization (2,4-dinitrophenylhydrazine), and several others [3]. The number of ways in which these reactions can be used for identification (or determination of the chemical nature) of the compound is limited only by the ingenuity of the analyst. As an example, a number of spots of the same sample are applied on a TLC plate. Prior to development, each spot is treated with a different reagent. The TLC plate is then developed in the normal manner and the reactivity of the sample (as revealed by its absence or shift) reflects its chemical nature.

Multi-dimensional techniques like TLC–GC and TLC–HPLC have the advantage of combining preliminary clean-up and high resolution and, at the same time, use the retention data on both the systems for qualitative analysis. Elution of the spots with a suitable volatile solvent, and transferring the solute to a potassium bromide pellet for recording the infrared spectrum, is feasible on microscale. The eluent can also be used for recording the UV and mass spectra. Direct introduction of the adsorbent (along with the material in the spot) into the spectrometer has also been used. Photoaccoustic spectroscopy permits recording of the UV spectrum directly from the TLC plate [44]; this technique can also be used for three-dimensional quantitation.

### 8.8.2 Quantitative analysis

Quantitation by TLC is attractive in view of the very large number of samples that can be simultaneously run on a given plate. The simplest method, of course, is a visual comparison of the developed and visualized spots with a series of spots (usually run on the same plate) of different concentration. But this method can only, at best, give a rough idea of the amount that may be present, the error being in the range of 10–30% depending on the experience of the analyst. Where possible (e.g. paper, or plastic- or aluminium-backed TLC), the spots may also be cut out and weighed. Eluting the visualized bands (or spots) with a suitable solvent, centrifugation and measurement of the intensity of the colour by a spectrophotometer at the appropriate wavelength is commonly practised in most laboratories as a semiquantitative method. Alternatively, the bands or spots are visualized using a non-destructive method (e.g. exposure to UV) and eluted from the plate; a suitable chromogenic agent is then added for photometric determination. Commercial devices are available that can suck the adsorbent layer containing the spot and elute the compound using a suitable solvent. These methods are probably satisfactory in the research laboratory for occasional use, but not for routine analysis in an industrial set-up where the 'one-at-a-time' approach neutralizes one of the main advantages of TLC, i.e. the simultaneous analysis of a large number of samples. Scanning densitometers which have been in the market for nearly two decades have been improved sufficiently for use as a viable quantitative technique for TLC.

In principle, the densitometer measures the intensity of the light absorption, transmission, scattering or fluorescence relative to the background just as in a photometer. The densitometer consists of a light source (UV–visible), a detector (photomultiplier) and a recorder. Normally, the position of the light sources and the area of focus are stationary so that a device to move the TLC plate, either linearly or in a radial fashion (for circular TLC), is incorporated into the instrument. As already noted, the function of the densitometer is to convert the thin-layer chromatogram into a form useful for quantitation by the 'area-under-the-peak' measurements as in GC and HPLC, so that the integration and quantitation by chromatographic data processors can be carried out, as discussed in greater detail later (see Fig. 2.4 and section 9.2.1.3).

It is well known that the absorbance of radiation by a compound is proportional to its concentration in solution and the path length (Beer–Lambert law). Applicability of this law to substances adsorbed on solid surfaces, where scattering of the light interferes, has been examined by Kubelka and Munk (1931); most of the current densitometers operate on the relationship developed by them [45,46]. The Kubelka–Munk equation may be represented as:

$$C = \frac{(1-R)^2}{2R} \cdot \frac{S}{\varepsilon} \tag{8.9}$$

where $C$ is the concentration of the spot, $R$ is the reflectance, $\varepsilon$ the absorbance, and $S$ is the coefficient of scattering. Since the densitometric measurements are made using a narrow slit-shaped aperture, the calibration curves are not strictly linear, but parabolic. Use of an internal standard with similar optical properties helps to improve the accuracy of quantitation.

Densitometric measurements can be made in a number of ways. These include absorbance, transmission, reflectance, simultaneous transmittance-reflectance, fluorescence or, when a fluorescent indicator is used, fluorescence quenching. Absorbance or transmittance measurement requires the light source and the photomultiplier to be on opposite sides of the plate and the plate to be transparent to light. Since glass absorbs a considerable portion of the UV radiation, this mode of measurement is limited to the visible range (above 320 nm). On the other hand, reflectance measurements are possible over the whole range of 185–2500 nm. Baseline noise or fluctuation is considerably compensated by simultaneous reflectance and transmittance measurement on the same spot and using a weighting factor in the calculation. When the compound to be determined has fluorescence in the UV or visible region, this mode of measurement is preferred in view of the low background noise and high sensitivity of up to two orders of magnitude in favourable cases. Reproducibility of densitometric measurements, either in the fluorescence or absorbance mode, is in the range of 1–3%.

Both radial and peripheral scanning is possible in the evaluation of radially (circularly) developed chromatograms. Radial scanning (which is normally preferred over the alternative) involves scanning of the chromatogram in the direction of the development, while peripheral scanning signifies scanning along the band, perpendicular to the direction of mobile-phase flow.

Since the scanning by the densitometers is usually in a given direction, this method of evaluation is not normally suitable for two-dimensional TLC, which may be required for resolving certain complex samples. However, improvement in resolution may be achieved in a post-development operation using a video camera and computerized image-processing techniques. The technique involves storing the TLC image made by the video camera on a $512 \times 512 \times 8$ bit computer memory and initiating manipulations like image subtraction, filtering, thresholding and use of false colour to enhance the quality of the original picture [10].

In some densitometers, it is also possible to record the UV or fluorescence spectrum of the compound in the spot. This may be obtained by scanning the plate repetitively and recording the absorbance or fluorescence at different wavelengths. The intensity of the peaks is different at different wavelengths and this depends on the absorbance (or fluorescence) of the compound at that wavelength. Joining the peak maxima gives the characteristic absorption band of the compounds. Spectrodensitometers also permit choice of the appropriate wavelength for each spot to maximize the response and sensitivity.

Several other methods have been suggested for quantitative evaluation of the thin-layer chromatograms. A radiochromatogram may be obtained (when the compounds are radioactive) by exposing the TLC plate to photographic plate and processed as above using a densitometer; alternatively, a scintillation counter may be used. Use of a flame-ionization detector in combination with sintered TLC plates or rods has been mentioned earlier.

## 8.9 PREPARATIVE PLANAR CHROMATOGRAPHY

As in the case of the closed column chromatography, the term preparative planar chromatography signifies the technique of carrying out the operation on a relatively large scale with a view to obtaining sufficient quantities of the separated compounds for further studies on the isolated compounds. To improve loadability, considerably thicker plates or papers are used in preparative planar chromatography. A preliminary clean-up or coarse separation on column chromatography is useful for increasing the amount of the sample that can be applied on the layers. Resolution on the preparative plate is generally poorer but the technique is useful when small quantities of a few well-separated compounds are required for structural, spectroscopic, synthetic or biological studies.

Thick chromatographic papers (e.g. Whatman 3MM) are available, which are very convenient for separation of a few hundred milligrams of the sample. These papers have been extensively used in natural-products chemistry for the isolation of several plant pigments during the 1940s, 50s and 60s. While many fewer reports now appear, the convenience of their use (where applicable) is that the bands after separation can be easily cut and eluted without the problems normally associated with scraping and elution of small-particle adsorbents from TLC plates. Preparative TLC plates may have thicknesses from 0.5–10 $\mu$m, but 1–2 mm plates are more common. Because of the increased thickness, care must be taken to avoid cracking or flaking during the preparation of the plates, and in drying and activation. The coated plates are allowed to dry at ambient temperature for 1–2 days and then activated by heating to 110°C for 3–4 h. Sorbents for preparative separations are specially formulated to contain particles of different sizes to improve adhesion. Sample size naturally depends on the size and thickness of the plate: about 5–25 mg per 1 mm thickness of the layer on a 20$\times$20 cm plate. Kirchner [47] prepared 3–12 mm plates of silica gel using 20% gypsum. The plates were held in a steel frame with parallel and diagonal internal wires for support; as there was no back plate, samples could be applied on both sides.

Several methods are used for sample application in large quantities to obtain a thin band at the origin. Automatic application using any of the devices described earlier is convenient. Manual application as a streak may need some practice. Alternatively, a 1–2 mm portion of the layer across the thickness of the plate may be removed and the groove filled with the sample which has been adsorbed on thc adsorbent in a separate operation as described earlier.

After the separation by development of the chromatogram in a suitably large tank, any of the non-destructive methods may be used to visualize the bands (e.g.

UV light). If no suitable method is available, a major portion is covered with a glass plate and the exposed portion sprayed with a chromogenic agent that does not require heating of the plate. The located zones are scraped off the plate, transferred into a Soxhlet apparatus or a filter funnel, or a column for extraction (or elution) with an appropriate solvent. If silica gel is used, a few drops of water may be added to the scraped silica gel to deactivate it and facilitate elution. Several small devices, like inverted sintered-glass columns or funnels, are available commercially to facilitate collection of the zones from the chromatogram. They are all based on the principle of vacuum suction and can be easily constructed in the laboratory.

Fast preparative planar chromatography as in rotating-disk thin-layer chromatography (centrifugal TLC, see section 8.6.2.2) has been recently reviewed [34]. The RDTLC system (e.g. Model CLC-5 from Hitachi, Japan) consists of a circular plate chamber, a motor, a flow-through detector and a fraction collector. The turntable (about 25 cm in diameter, Fig. 8.14) is placed horizontally in the plate chamber and

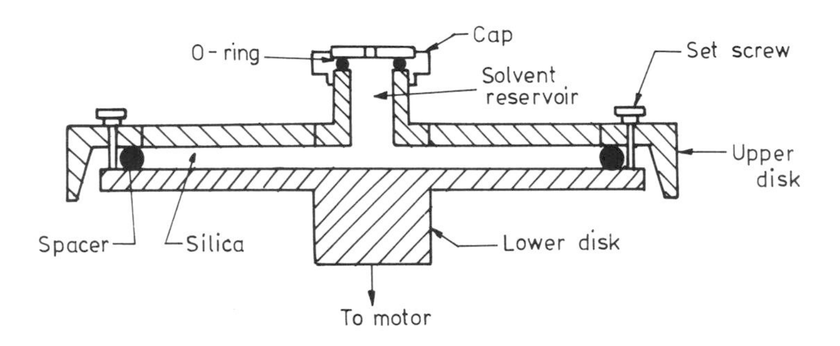


Fig. 8.14 — Schematic diagram of the turntable for RDTLC. Reproduced from Ref. 34 by courtesy of CRC Press, Inc.

covered with a transparent plastic or a tempered glass sheet; a stainless-steel spacer between the plate and cover determines the thickness of the adsorbent layer, which may be 1–2 mm. The layer is prepared by introducing a slurry of activated silica gel in dichloromethane (through a receptor at the centre of the plate which also holds the solvent reservoir) while rotating the plate at around 400 rpm. The speed is increased to 1000 rpm when the slurry fills the space up to the edge of the solvent reservoir; when the adsorbent gets compacted, more slurry is added. The process is continued until a well-packed layer of the desired thickness forms. The sample is introduced at the bottom of the reservoir, while rotating the plate at 300–500 rpm, using a syringe and rinsed down with the help of some solvent. A 1–2 mm ring of the sample is formed at the centre. The reservoir is then filled with the mobile phase to initiate chromatography (the solvent may also be fed using a peristaltic pump). The flow-rate of the mobile phase (10–35 ml/min) is determined by the speed of the rotation, which is normally around 1000 rpm, and the solvent. The flow-rate increases linearly with the speed of rotation. When the solvent reaches the outer edge of the plate, it is funnelled into the fraction collector, if necessary, through a detector. The sample capacity depends on a number of factors, including the thickness of the plate, the separation factor or the complexity of the sample and the dilution. As a rule of thumb, 2–5 mg of the sample per 1 mm thickness of the plate may be used; 10 mg/mm could also be used under favourable cases.

The RDTLC is claimed to be useful for simultaneous trace-component concentration and 'analytical-cum-preparative' separations. It may be mentioned, however, that with the ready availability of HPLC systems, the popularity of preparative planar chromatography has considerably declined.

## REFERENCES


[1]. E. Stahl, *Thin-Layer Chromatography: A Laboratory Handbook*, Springer-Verlag, Berlin, 1969.
[2] J. Sherma and G. Zweig, *Paper Chromatography and Electrophoresis: Vol. II, Paper Chromatography*, Academic Press, New York, 1971.
[3] J. G. Kirchner, *Thin-Layer Chromatography*, 2nd edn., Wiley, New York, 1978.
[4] K. Randerath, *Thin-Layer Chromatography*, 3rd edn., Verlag-Chemie, Wienheim, 1980.
[5] J. C. Touchstone and M. F. Dobbins, *Practice of Thin-Layer Chromatography*, 2nd edn., Wiley, New York, 1983.
[6] B. Fried and J. Sherma, *Thin-Layer Chromatography: Techniques and Applications*, 2nd edn., Marcel Dekker, New York, 1986.
[7] L. R. Trieber, *Quantitative TLC and Its Industrial Applications*, Marcel Dekker, New York, 1986.
[8] A. Zlatkis and R. E. Kaiser (eds.), *HPTLC: High-Performance Thin-Layer Chromatography*, Elsevier, Amsterdam, 1977.
[9] W. Bertsch, S. Hara, R. E. Kaiser, and A. Zlatkis (eds.), *Instrumental HPTLC*, Huethig, Heidelberg, 1980.
[10] R. E. Kaiser (ed.), *Planar Chromatography*, Vol. 1, Huethig, Heidelberg, 1986.
[11] D. C. Fenimore and C. M. Davis, *Anal. Chem.*, **53**, 252A (1981).
[12] G. Guiochon, L. A. Beaver, M. F. Gonnord, A. M. Siouffi, and M. Zakaria, *J. Chromatogr.*, **255**, 415 (1983).
[13] C. F. Poole, *Trends Anal. Chem.*, **4**, 209 (1985).
[14] G. Zweig and J. Sherma, *Handbook of Chromatography*, CRC Press, Cleveland, 1972.
[15] C. Michalec, in *Stationary Phase in Paper and Thin-Layer Chromatography*, K. Macek and I. M. Hais (eds.), Elsevier, Amsterdam, 1965.
[16] F. C. Haer, *An Introduction to Chromatography on Impregnated Glass Fiber*, Ann Arbor-Humphry, Ann Arbor, Michigan, 1969.
[17] H. Bartels and B. Prijs, *Adv. Chromatogr.*, **11**, 115 (1974).
[18] R. K. Gilpin and W. R. Sisco, *J. Chromatogr.*, **124**, 257 (1976).
[19] J. Sherma, *Practice and Applications of Thin-Layer Chromatography on Whatman $KC_{18}$ Reversed Phase Plates*, Whatman Chemical Separations, Clifton, NJ, 1987.
[20] W. Jost and H.E. Hauck, *Adv. Chromatogr.*, **27**, 129 (1987).
[21] S. Dev, *J. Sci. Industr. Res.*, **31**, 60 (1972).
[22] K.-T. Wang, Y.-T. Lin, and I. S. Y. Wang, *Adv. Chromatogr.*, **11**, 73 (1974).
[23] S. Anandaramn, N. B. Shankaracharya, C. P. Natarajan, and N. P. Damodaran, *Lab. Pract.*, **26**, 870 (1977).
[24] E. Stahl and J. Mueller, *J. Chromatogr.*, **189**, 293 (1980).
[25] A. S. Gupta and S. Dev, *J. Chromatogr.*, **12**, 189 (1963).
[26] K.-T. Wang, P.-H. Wu, and T. B. Shih, *J. Chromatogr.*, **44**, 635 (1969).
[27] T. Okumura, *J. Chromatogr.*, **184**, 37 (1980).
[28] D. C. Fenimore and C. J. Meyer, *J. Chromatogr.*, **186**, 555 (1980).
[29] E. Stahl, *J. Chromatogr.*, **165**, 59 (1979).
[30] D. L. Saunders, *Anal. Chem.*, **46**, 470 (1974).
[31] S. Turina, M. Trbojevic, and M. Kastelan-Mecan, *Anal. Chem.*, **46**, 988 (1974).
[32] D. L. Massart and H. L. O. De Clercq, *Adv. Chromatogr.*, **16**, 75 (1978).
[33] E. Stahl and J. Mueller, *Chromatographia*, **15**, 493 (1982).
[34] R. J. Laub and D. L. Zink, in *Recent Developments in Separation Science*, N. N. Li and J. M. Calo (eds.), Vol. 9, CRC Press, Boca Raton, Florida, 1986.
[35] T. H. Jupille and J. A. Perry, *J. Chromatogr. Sci.*, **13**, 163 (1975).

[36] J. A. Thoma, *J. Chromatogr.*, **12**, 441 (1963).
[37] A. Niederwieser, in *Chromatography*, 2nd edn., E. Heftmann (ed.), Van Nostrand Reinhold, New York, 1975.
[38] T. Wieland and H. Determann, *Experientia,* **18**, 431 (1962).
[39] A. Niederwieser and C. C. Honegger, *Adv. Chromatogr.*, **2**, 123 (1966).
[40] E. Tyhak, E. Mincsovics, and H. Kalasz, *J. Chromatogr.*, **174**, 75 (1979).
[41] Z. Witkiewicz and J. Bladek, *J. Chromatogr.*, **373**, 111 (1986).
[42] H. J. Isaaq, M. N. Mangino, G. M. Singer, D. J. Wilbur, and N. H. Risser, *Anal. Chem.*, **51**, 2157 (1979).
[43] G. Szekely and P. Baumgartner, *J. Chromatogr.*, **186**, 575 (1980).
[44] A. Rosencwaig and S. S. Hall, *Anal. Chem.*, **47**, 548 (1975).
[45] V. Pollak, *Adv. Chromatogr.*, **17**, 1 (1979).
[46] R. J. Hurtubise, *Solid Surface Luminescence Analysis*, Marcel Dekker, New York, 1981.
[47] J. G. Kirchner, *J. Chromatogr.*, **63**, 45 (1963).

# Part IV
# Applications

# 9

# Applications of chromatography

## 9.1 INTRODUCTION

Applications of chromatography are too numerous and varied to be adequately covered in a general text of the present type. As noted earlier, the applications of chromatography encompass disciplines ranging from chemistry and biology on one side to manufacturing industries and environmental control on the other. The size of the analyte molecules may vary from small metal ions to high polymers with molecular weights of over a million. The scale of operation may also vary from ultratrace analysis with attogram-level ($10^{-18}$ g) measurements to large-scale operations with throughputs in the range of tons per hour. Classification of the applications may be based on (1) the major disciplines of study and research (chemistry, biology, etc.), (2) major disciplines of use (clinical medicine, toxicology, etc.), (3) areas of application (food, drugs, fuels, etc.), (4) chemical classes of compounds (hydrocarbons, alcohols, acids, etc.), (5) the chemical or biological activity (antioxidants, vitamins, etc.), (6) the major chromatographic classes (GC, LC, SFC, TLC, etc.), (7) the methods used (affinity chromatography, size-exclusion chromatography, ion-pair chromatography), and so on. None of the above is really satisfactory since, under any of the above, considerable overlap and repetition cannot be avoided. For example, carotenoids can be covered under either terpenoids, plant pigments or vitamins; prostaglandins, which at one time challenged the organic chemists' ingenuity in separation and structure determination, are now of immense interest to biologists.

The principles and methodologies of the different chromatographic techniques and their specific applications have been discussed in the foregoing chapters. The present account is, therefore, limited to an overview of the different areas of applications of chromatography in general, and no attempt is made to describe in detail the various systems and their use in the chromatographic analysis of the different classes of compounds. The applications of chromatography have been very broadly classified into three categories: (1) chemical, (2) biological and (3) industrial and miscellaneous applications. Chemical applications are discussed in some detail; other applications are more or less based on the principles of chemical analysis. The chemical applications begin with general methods of sample preparation and analytical techniques which are common to all applications of chromatography; this is followed by what may be considered essentially chemical applications (physicochemical measurements) and continue through the separation of natural products, which may be of interest to both chemists and biologists.

Certain common features of chromatographic applications may be mentioned here. Though basically a separation method, the combination of chromatography with different detection and spectroscopic techniques has made it a very powerful analytical tool for both qualitative and quantitative analysis. The possibility of obtaining spectral data on the chemical entities by the use of on-line spectroscopic instruments is a great convenience that helps in their identification, or in confirmation of their identity made by other methods.

As an analytical tool, chromatography offers a unique advantage. Both in qualitative and quantitative analysis, the result of the chromatographic experiment (i.e. retention data and/or the detector response) are largely unaffected by the presence of the other compounds since these are separated during chromatography. However, a preliminary clean-up and concentration procedure could be advantageous in many ways. For example, unnecessarily large sample size can be avoided, thereby increasing the efficiency of the separation and, at the same time, ensure longer column life and maintenance of the detector performance. It is also helpful to concentrate the components to be analysed in materials such as air and water as the hazardous constituents may be present in trace quantities. Lately, it has been realized that the chromatographic analysis itself could be much faster than the sample preparation and much attention is being paid to systematize and automate this aspect. The methods of sample preparation (and concentration), strategies to enhance the detectability of the analytes, and treatment of the data for qualitative, quantitative and trace analysis are generally common to all the chromatographic methods. These are discussed in the first part of this chapter.

The nature of the sample and the purpose of analysis usually determines the applicability of a particular chromatographic method. In view of its convenience, rapidity of analysis, high separation power, inexpensive instrumentation and ready availability of high-purity mobile phases, GC is the most frequently used when the material to be analysed (or, sometimes, its derivative) is sufficiently volatile. However, since only about 20% of the compounds can be conveniently handled by GC, LC methods are indispensable. LC is also more versatile and offers a much wider choice of methods, mechanisms of separation, stationary phase, mobile phase combinations and techniques. A large proportion of the analytes is amenable to separation by the reversed-phase LC methods. They are especially convenient for the analysis of the water-based samples, which include a variety of biological fluids. The choice among the large variety of LC methods can be difficult in the absence of previous experience. The chart in Fig. 7.20 may be used as a guide. Almost all the LC methods can also be conveniently carried out using the planar chromatographic techniques.

## 9.2 CHEMICAL APPLICATIONS

### 9.2.1 Analytical chemistry

The importance of chromatography as an analytical technique need not be re-emphasized. Almost all the topics to be discussed in this chapter fall under this classification. This section elaborates those aspects of analytical methodologies which are common to all the chromatographic techniques and which enhance the utility of chromatography as an analytical technique.

#### *9.2.1.1 Sample preparation*

As noted earlier, one of the major advantages of the chromatographic methods of analysis is the non-interference of the accompanying substances, assuming that the compounds being analysed are separated from the rest by the chromatographic bed. Indeed, in most laboratory situations, it is possible to introduce the sample directly into the chromatographic systems for satisfactory analysis. Nevertheless, it is generally profitable to perform certain preliminary sampling exercises for the following reasons: (1) the chromatographic efficiency drops steeply beyond a certain sample load and overloading invariably causes band broadening which is detrimental to the analytical precision; (2) some of the accompanying compounds may overlap on the peak being detected; (3) some of the components of the sample may be retained strongly and their elution may delay the analysis; (4) certain compounds may also be so strongly retained by the stationary phase that the bed may eventually deteriorate and require replacement; (5) the compounds of interest may be present in very minute quantities to be determined without pre-concentration; (6) quite frequently, the sample may not be in a state suitable for direct injection (e.g. an aqueous sample cannot be injected into normal-phase LC or most GC columns and a solid sample often needs to be dissolved in a suitable solvent); and (7) on occasion, conversion of the compound of interest into a suitable derivative helps detection.

The method of sample preparation depends on (a) the nature of the sample, (b) the compounds of interest, (c) the purpose of analysis and (d) the chromatographic method. Methods like headspace collection and pyrolysis GC, as well as some of the concentration techniques using precolumns in multi-dimensional chromatography, which have been discussed earlier (Chapters 3 and 7) may also be considered sampling techniques. Some of the general methods are discussed here. Again, a few of the very specific procedures used in certain speciality applications (e.g. environmental monitoring) may be treated elsewhere as appropriate. For a recent exhaustive review of the sampling methods, see Ref 1.

The samples for chromatographic analysis may be gaseous, liquid or solid. They may be simple or complex mixtures. The compounds of interest may be either the major or trace constituents of the sample. In addition they may be non-volatile or volatile in relation to the accompanying bulk; they may be ionic or non-ionic, acidic, basic or neutral, and so on. The purpose of analysis may also vary from assaying the major constituents or trace impurities in quality-control analysis to detection of adulterants in foods or toxins in the environment. Whereas in qualitative analysis the number and identity of the constituents of the sample are important, quantitation usually requires that the components being analysed are well separated and isolated without loss. Trace analysis almost always requires a pre-concentration step.

The authenticity of the analytical data depends on the representative character of the sample. Sample collection and storage must therefore ensure that the sample truly represents the composition of the analyte since errors occurring at this stage cannot be corrected later. Sampling from a bulk is always problematic (except for gaseous samples in a closed container) since the composition of the sample at one part of the bulk may be different from the other. The material must then be homogenized by any or several of the techniques such as mixing, shaking, stirring, grinding, etc. Where homogenization of the sample is not possible owing to its bulk, a statistically reliable number of the representative samples may be drawn from different portions of the bulk. If the material to be analysed is in more than one phase, samples must be drawn from each of the phases. Changes in sample composition may occur during storage because of the chemical reactivity of the constituents under such conditions as exposure to air, light, temperature, metals (e.g. cans), microbial action, or even simple adsorption on the surface of the packing material. When rather long periods of storage are required, the samples must be stored in clean glass vials, tightly stoppered with the exclusion of air and stored at sub-zero temperatures in the dark. Homogenization of food and biological products could liberate certain enzymes that could change the composition of the sample; they are, therefore, best stored whole, at sub-zero temperature.

Sample preparation methods may be broadly divided into two groups; isolation and concentration. However, either of these may involve the other. The former is particularly necessary when, for example, volatile compounds present in essentially non-volatile material are to be analysed by GC. Concentration may occur during the above process of isolation but deliberate concentration (invariably involving isolation) is necessary in certain cases. Alternatively, the methods may be classified by the nature of the actual process involved; e.g. distillation, adsorption, solvent extraction, etc.

*9.2.1.1.1 Temperature-assisted methods*

Distillation is the most convenient method of isolating and concentrating volatile constituents of any material from the bulk of the non-volatile matter provided; of course, the components are thermally stable. Alternatives to the conventional atmospheric distillation are vacuum distillation, steam distillation or sweep co-distillation, which are all common practices in a chemical laboratory. Several designs of distillation apparatus are available from the suppliers of laboratory glassware; most of them are also available in miniature versions for handling very small quantities of the samples. The effectiveness of separation by distillation depends on the relative volatilities of the desired and unwanted components. When the differences in the volatilities are not large enough, fractional distillation using a packed column or spinning band column may have to be used. Rapid evaporation and condensation in molecular distillation or in rotatory-, climbing- or falling-film evaporators help to minimize artefact formation. Details of the practical aspects of methods are available in most laboratory manuals in chemistry.

Essential oils and several other aroma compounds which are not ordinarily distillable at atmospheric pressure can often be distilled with steam or water. When the quantity of the distillate is large enough, phase separation of the oil and water occurs and the two can be readily separated. On a smaller scale, an organic solvent is needed to extract the organic compounds from the aqueous distillate. In one of the more widely used apparatus, called the Likens–Nickerson apparatus, a small quantity of the water-insoluble organic solvent (usually pentane or dichloromethane) is taken in the 'receiver-cum-phase separator'; the design permits the aqueous portion (after extraction with the organic solvent) to return to the still containing the raw material. The volatiles are concentrated in the small volume of the organic phase in the receiver, ready for introduction into the chromatograph. Several improved versions, suitable for small-scale operation have been designed, notable among them being those designed by Peters [2] and Verzele [3].

In sweep co-distillation, the volatiles are forced out of the warm or boiling water medium by a stream of nitrogen and/or a low-boiling organic solvent like Freon (b.p. −30°C) and trapped in a suitable organic solvent like pentane. Sublimation is rarely used as a sample-preparation method because of the possible decomposition of some of the components by heat.

Freeze-drying (lyophilization) can be used to remove large quantities of water under high vacuum from the frozen aqueous samples, provided the samples are non-volatile. Freeze concentration is another mild procedure but part of the sample may be trapped or adsorbed by the ice formed.

*9.2.1.1.2 Solvent extraction*

Most samples are introduced into the chromatographic bed in the liquid or solution form; exceptions are the use of solid- or gas-sampling devices in GC. Extraction of the sample with a suitable solvent is, therefore, the most frequently used technique and is often the first step in sample preparation. It may be used for gaseous, solid or even liquid samples or solutions in immiscible solvents. The object is to obtain a sufficiently concentrated solution of the analyte in a solvent compatible with the chromatographic method that is to be used. Frequently, it is a volatile solvent for GC so that the solvent moves much ahead of the sample components. The solvent compatibility is more critical in LC. It is preferably the same solvent that is being used for elution or, if that is not possible, it must be of lower strength for the given system; e.g. water for the reversed-phase LC, and hexane or other non-polar solvent for the normal-phase LC.

The factors affecting the solubility of compounds and the principles of distribution of the solutes between different solvents have been discussed in Chapter 1. These principles may be used to advantage in developing solvent-assisted methods for isolation and concentration of a variety of samples. Gaseous samples may be simply bubbled through the solvent in a gas wash-bottle. The extractability depends on the relative affinity of the solute in the solvent relative to that in the gas, in other words, volatility. The solubility of the compounds in the solvent may thus be improved by cooling or by the use of a complexing agent in the solvent. Changing the solvent and the complexing agent can bring about considerable selectivity in retaining desirable compounds.

As mentioned earlier, occasionally the sample may be present as a solution in a liquid that is incompatible with the chromatographic method under consideration. When required, the analyte in one liquid phase can be transferred to another liquid by liquid–liquid extraction, provided (a) the two liquids are mutually immiscible, and (b) the distribution constant is favourable. Under these conditions, the operation is very simple and requires equipment no more complicated than a separatory funnel. A favourable distribution constant can sometimes be enforced by using a complexation reagent in the appropriate phase. When the analyte is an acid or a base, it can be brought into the aqueous medium simply by modulating the pH (greater than 7 for acidic solutes and less than 7 for basic compounds). The same principle may be used when the desired substance is neutral and the accompanying substances are acidic or basic. If prolonged extraction is required, the appropriately designed continuous liquid–liquid extractor may be used. Countercurrent distribution, being tedious and time-consuming, is not recommended as a routine sample-preparation method.

Thick liquids, semi-solids, and solids can be extracted with the appropriate solvents simply by triturating them in a test-tube with a glass rod or by means of a magnetic stirrer. For prolonged extraction, a Soxhlet extractor may be used. In this device, the low-boiling organic solvent is refluxed with an arrangement for the condensed vapours to fall into a container of the pulverized raw material to be extracted; when the container is full with the solvent, it automatically siphons back to the solvent flask for repetitive reflux. Laboratory supply houses offer Soxhlet apparatus of different sizes (from a capacity of, say, 1 g to 1 kg or more) and multiple Soxhlet trains for handling a large number of samples simultaneously.

The number and variety of solvents available for use in the above procedures is very wide and the choice depends on the solubility and the requirements of the chromatographic procedure. A typical scheme for isolating a plant constituent is given below; this can generally be applied to any raw material with appropriate modifications. Several alternative schemes are given in Ref. 1. The plant material (wood chips, root, bark, leaves, flowers or mixtures of these) is air-dried and disintegrated into a coarse powder in a blender. The quantity of the material required may vary from 10 g to over 1 kg depending on the concentration of the compounds to be isolated and analysed. The plant material is extracted successively with hexane (or, better, pentane), chloroform (or dichloromethane) and methanol. Extraction with the hydrocarbon solvent may yield a variety of non-polar constituents like hydrocarbons, terpenoids, essential or fatty oils, steroids and other lipids, including fatty acids and esters. The haloalkane solvents are very efficient in extracting a large number of medium polar compounds including alcohols, carbonyl compounds, esters, alkaloids, carboxylic acids, phenolics and the like. The methanol extract may contain glycosides and some carbohydrates, in addition to the classes of compounds noted above. Each of the above extracts, which contain a large number of compounds, may be too complex for direct chromatographic analysis. Solvent-extraction procedures can be used for further fractionation using any two of the large number of possible combinations of immiscible solvents (or mixtures of solvents); e.g. methanol–water–hexane (4:1:5), methanol–hexane, acetonitrile–hexane, butanol–water, and so on. Alternatively, or following the above, each of the fractions may be refractionated to yield basic, neutral and acidic compounds. The dichloromethane (or diethyl ether) solution of the material is successively extracted with 2–10% HCl, water, 2–10% aqueous sodium carbonate, 0.1–1% aqueous sodium hydroxide and water. The aqueous HCl fraction is neutralized to a pH around 7 using aqueous sodium carbonate and re-extracted with dichloromethane to yield basic compounds (e.g. alkaloids). The sodium carbonate and sodium hydroxide extracts, when similarly neutralized (with dilute HCl) and re-extracted, yield carboxylic acids and phenolics, respectively. Occasionally, the material may be extracted directly with a base (e.g. very dilute aqueous sodium hydroxide or sodium carbonate for the isolation of phenolics and carboxylic acids alone) or an acid (e.g. aqueous citric acid or 1% aqueous HCl for the isolation of alkaloids); in the latter case, some oxygen heterocycles (capable of forming oxonium salts, e.g. anthocyanins) may also be extracted. Anthocyanins, deep red to violet and blue pigments of flowers, are generally extracted directly from the plant material with 1% methanolic HCl.

While the above procedures are very simple and can usually be performed without special skills, occasionally problems which are not so easily surmountable may arise. One of the frequent complications is the formation of strong emulsions that do not break on standing. Often, they may be broken by filtration through glass wool, centrifugation or addition of a small quantity of methanol or salt. Some, of the compounds of interest may be present in the biological tissue bound to a macromolecule (e.g. protein) or strongly encapsulated in the matrix. The macromolecule may be broken down by hydrolysis with acids, bases or enzymes. Disruption of the encapsulating material by treatments like osmotic shock, heat or ultrasonic agitation is possible.

#### *9.2.1.1.3 Adsorptive methods*

Several of the principles of adsorption chromatography can be conveniently used in the low-resolution, low-pressure mode; the multi-dimensional GC, HPLC and TLC methods described in the previous chapters, or the use of guard columns, may be considered as instrumental variations of these principles. It must be remembered that the purpose of the manipulation at this stage is essentially sample clean-up and preliminary fractionation and not separation of individual components. This distinction is made only to adapt certain convenient and very simple procedures for rapid sample preparation, avoiding certain time-consuming precautions normally to be taken in analytical chromatography.

Generally, the procedure involves the use of an appropriate adsorbent, loosely packed in a small, pencil-sized column for adsorbing the crude sample. The adsorbent may be silica gel, alumina, Celite or Fluorisil (magnesium silicate) for general purpose; carbon or carbon-polystyrene-polyacrylamide (macroreticular resins) for adsorption of organics in water; hydroxyapatite (a form of calcium phosphate) for fractionation of medium and high molecular weight biopolymers; polyamides for compounds with hydrogen-bonding functional groups, cellulose and several of the size-exclusion gels belonging to the dextran, agarose, polyacrylamide or cross-linked polystyrene types. The cross-linked polystyrenes (Amberlite XAD-type) and reversed-phase, chemically bonded (ODS) silica are also used for selective adsorption of organics from water and for clean-up of aqueous samples. Silica gel (normal and ODS-bonded) is the most popular.

Taking the example of silica gel as the adsorbent, the sample may be dissolved in the weakest possible solvent; i.e. a solvent of low polarity like pentane and mixtures of pentane and dichloromethane. Most samples (particularly biological and environmental samples) may contain water which may deactivate the silica gel. The sample solution, therefore, is passed through a small bed of anhydrous sodium sulphate to remove water prior to adsorption on the silica gel column. The adsorbed material is eluted with two, or in the case of more complex mixtures, three solvents of widely different strength. Elution with the weaker solvent (hexane for the normal-phase or water for the reversed-phase adsorbent) may be used either to elute the weakly adsorbed analyte or the undesired impurities as the case may be. Where three solvents are employed, the first solvent usually removes the weakly adsorbed impurities, the second elutes the components of interest to the analysis (leaving strongly adsorbed material on the column) and, finally, the third solvent cleans up the column, if needed for re-use. The three-solvent procedure is also frequently useful for preliminary fractionation of the sample into three portions, each of which may be separately chromatographed on different systems, optimized for the respective fractions. Where sample concentration is the objective, the sample (dissolved in a weak solvent) is adsorbed on the column and washed with a large quantity of the non-eluting or the weak solvent to remove as many of the accompanying impurities as possible. The column is then washed with a strong solvent so that the remainder is eluted as a thin, concentrated band. By repetition and fine tuning of the solvent systems, this procedure can quantitatively yield a clean sample.

Concentration of trace constituents by trapping on beds of high-efficiency adsorbents is a particularly useful technique for the analysis of materials such as environmental air and water, where the components to be analysed are at ppm or ppb levels. Large quantities of the raw materials therefore need to be sampled for satisfactory analysis. As the number of samples increase, storage and transportation of the sample, from the collection centre to the laboratory (which may not always be in the same city) becomes highly problematical. A number of procedures, generally known as the solid-phase extraction, are now available for the purpose. The most popular entrapping sorbents for organics in water, and sometimes air, are the so-called macroreticular resins based on cross-linked polystyrenes such as the Amberlite XAD-type, manufactured by the Rohm & Haas Co., U.S.A. Other common adsorbents are charcoal, polyurethane foam, and bonded reversed-phase silica packings. Ion-exchange resins are also occasionally used when the analysis of only acidic or basic constituents is required. Comparative evaluation of different adsorbents for a given analysis is possible with the knowledge of the capacity of the resins and the breakthrough volumes (i.e. volume of the sample that can be treated before the substance under examination begins to elute). These may be determined using model compounds but may not accurately reflect their efficiency unless the model compound is closely related to the analyte; the procedures to determine the capacities and the breakthrough volumes are also time-consuming and tedious. In any case, general information on most of the common adsorbents is available which may be taken as valid for most of the common organics in the absence of information to the contrary. It may be understood that the larger the breakthrough volume, the greater is the amount of the bulk raw material that can be treated and the greater is the concentration factor.

The chemical nature and the methods of preparation of most of the common solid-phase extraction packings have been described in section 6.5.2. A majority of these are commercially available, in the ready-to-use form, from the suppliers of chromatographic materials. As noted above, the Amberlite-type macroreticular resins and charcoals of different grades are commonly used for trapping trace organics from large quantities of water. While the capacity of activated carbon to absorb organics from water is reasonably high (about 600 bed volumes), the rate of absorption is low (optimum flow-rate is about 3 bed volumes/h). On the other hand, Amberlite XAD-2 can absorb organics from water at the rate of 300 bed volumes/h with a breakthrough volume ranging from 100–10,000 bed volumes depending on the compound. There is also a considerable selectivity difference between the two adsorbents. For example, the Amberlite resins have pronounced affinity for phenols and other aromatic compounds which are poorly adsorbed by charcoal. On the other hand, charcoal can trap aliphatic hydrocarbons and their halogenated derivatives more effectively than the macroreticular resins. Ionic and ionizable compounds are poorly retained by both the materials. Reversed-phase LC packings (particularly the ODS-bonded silica) are also used for adsorbing non-polar organic compounds from water. These packings are similar in selectivity to the RPLC packings described earlier except that the particle size is around 40 $\mu$m to facilitate faster throughput.

In addition to the above adsorbents, molecular sieves (zeolites) are also used to trap organics from air by pumping it through a column packed with them. Several hundred volumes of air may be pumped through the columns. The trapped organics can be released from the adsorbents by extraction with a volatile organic solvent (e.g. dichloromethane) or by pumping nitrogen at high temperatures (around 250°C). The volatile organics so purged are collected by cryogenic trapping methods or injected directly into a cold GC column. The gas chromatogram is obtained in the normal manner by raising the temperature of the column. Care must be taken to avoid exposure of the adsorbents (particularly the organic resins) and the eluants to oxygen in view of their instability to high temperatures and oxygen.

The above purge-and-trap techniques can also be used for isolating volatiles from aqueous samples (see section 3.6.1). Commercial devices consisting of sample vials for introduction of an inert (purge) gas and traps (typically, 2.5 mm i.d.×250 mm length, packed with methylsilicone-gum-coated GC supports, Tenax, charcoal and silica gel) are also available (e.g. Tekmar Co., U.S.A.). The inert gas (helium, nitrogen or argon) may be introduced into the vial containing the sample (20–100 ml) at the rate of about 40 ml/min for not less than 10 min, while holding the trap at room temperature. The trapped volatiles are released by purging the trap with the inert gas at an elevated temperature (200°C).

Ion-exchange resins are particularly effective in trapping ionic analytes specifically (e.g. phenols in water at a pH at least two units above the $pK_a$ value of the phenol). The adsorption and recovery of these materials on the porous polymers and charcoal is often variable and incomplete. They may also be used to remove ionic contaminants from non-ionic analytes. These ion-exchange packings, as well as the silica and ODS-bonded phases are available in convenient, disposable polyethylene columns that hold about 100–500 mg of the adsorbent.

Since chromatography is mainly used for quantitative analysis, sample loss and contamination during isolation and clean-up should be minimized. A few elementary precautions (such as using clean glassware, repeatedly rinsing the glassware, and other precautions normally given in undergraduate analytical chemistry programmes) need to be taken for the purpose. Even when these precautions are taken, inadvertent loss of sample is possible and, to compensate for such loss, use of an internal standard is advisable. The internal standard is added at the earliest possible stage of the sampling procedures. An ideal internal standard should be close to the analyte in polarity and solubility so that it is not even partially removed during the clean-up protocol; at the same time it should not overlap with the analyte or any other peak in the chromatogram. The last-mentioned feature, however, is irrelevant in the case of radioisotope-labelled solutes where the radioactivity is used for detection and determination. To the extent possible, the internal standard should respond to the detector in a similar fashion (with nearly the same response factor; see later) and should be added in nearly the same concentration. When the sample contains a large number of components for simultaneous analysis (as in, for example, environmental analysis), more than one internal standard may be necessary to match the chromatographic properties and concentrations of the different sets of compounds. The preferred internal standards are structural analogues, homologues, isomers, enantiomers (when a chiral separation system is used), or isotopically labelled compounds (when a radioactivity detector or a mass spectrometer is used).

*9.2.1.1.4 Automated methods*

It may be noticed that although the materials and methods described above for sample preparation are several, they all fall into a few simple laboratory manipulations like solid–gas, solid–liquid or liquid–liquid extraction, phase separation and concentration. These methods are routine and repetitive, but they consume much more time than the chromatographic analysis itself, given the high-speed analytical systems and data processors now available. In other words, the limiting factor in the speed of analysis is no longer the analysis itself but the sample processing, which is often sequential unless a large number of analysts are employed. It is thus attractive to automate as many unit processes as possible so that the most routine part of the analytical programme can be run unattended. Following the successful application of the robotics in chemical, electronic and machine-manufacturing industries, the technology is being increasingly used in the laboratory [4,5].

The robots are electromechanical devices that can be programmed to carry out many of the routine functions of the laboratory technician and often have attachments that are referred to as 'hands, arms and wrists', with 'touch' and 'tactile' senses. Unlike the machine-manufacturing robots which are designed to perform specific functions (mainly, the 'pick and place' mass-movement type activities), the laboratory robots may be 'trained' or reprogrammed to carry out different manipulations or sequences of manipulations as the situation demands. Thus, they may be used to weigh out or withdraw predetermined quantities (volumes) of samples, dilute with solvents, dissolve by shaking, stirring and heating to a certain temperature, add internal standards, extract with different solvents or solid-phase extractors, carry out phase separations, reconcentrate, suitably derivatize, load the samples in a magazine for autosampling, and the like. While the currently available laboratory robots are all designed to use the readily available glassware, shakers and other laboratory appliances, one limitation could be the axes of mobility. Since these robots have a fixed base or can only move in fixed directions (in straight lines or circles on fixed rails), the laboratory lay-out may have to be redesigned to keep the glassware, balances, etc., within the 'arm-reach' of the robots. The 'working envelope' of the robots is often radial (i.e. all round the base). In some systems (e.g. the 'Zymate' of Zymark Corp., U.S.A.), the 'working envelope' of the robot may be enlarged by using two or more robots (joined by a single controlling computer) which can coordinate the activities and pass samples around the laboratory. Often, the same computer can also control the chromatograph, acquire the data and process it to make a completely automated analytical system. Add 'artificial intelligence', and the scope for laboratory automation would seem unlimited.

#### *9.2.1.2 Qualitative analysis*

The purpose of qualitative analysis is to determine the number of compounds present in a given sample and identify them. A prerequisite for this application of chromatography is separation or resolution and the possibility of peak overlap is always an uncertainty. Improvement of resolution would thus continue to be pursued to its limits. Lower than the optimum resolution would also adversely affect the precision in determination of the retention parameters since, as shown in Fig. 2.2, the peak apex does not represent the true peak centre unless the resolution ($R_S$) is at least 0.8. If each peak in the chromatogram represents a single compound, determination of the number of compounds is fairly simple. The retention data in GC and LC are widely used for identifying the known compounds.

At the outset, two terms, which are often used incorrectly and interchangeably, need to be clarified; they are: accuracy and precision. If the amount of an analyte in a sample is 1.000 units and an analytical method gives a value 0.999, it is accurate, but the method may not be precise if it is not reproducible with the same accuracy. On the other hand, another method, which on three experiments gives values 0.972, 0.969 and 0.970, is very precise but not accurate. Both accuracy and precision are desired in analysis, either qualitative or quantitative.

##### *9.2.1.2.1 Retention parameters*

Comparison of retention data remains the simplest and the quickest method of identification and is normally the first technique to be employed if the purpose of analysis is to determine if a particular compound is present in the given sample. The accuracy of the identity by retention data understandably depends on the reproducibility of the chromatographic conditions and precision in the measurement of the data. Reproducibility of the chromatographic conditions is more easily achieved in the same laboratory if the same column (or TLC plate) is used with the mobile phase composition, flow-rate, temperature, etc.

The precision of the retention data on a given stationary-phase–mobile-phase combination depends on the reproducibility of the chromatographic conditions, i.e. the temperature and flow-rate. As noted earlier, a fall in temperature by 30°C would double the retention time and to maintain a reproducibility of 1% in the retention parameter, the temperature should not vary by more than 0.3%. At a constant temperature, the retention measurement is as accurate as the precision in the flow-rate control; i.e. a 1% change in the latter is reflected by a 1% change in the former.

In the technique of co-chromatography, the sample and the standard are injected into the column, under exactly the same conditions, often followed by chromatography of the mixture of the sample and the standard. If the standard is present in the sample, the peak due to it is more intense. Direct comparison of retention data by co-chromatographic methods has two limitations: (1) the need for availability of all the standards and (2) the time-consuming requirement of repetitive analysis. Indirect comparison methods use retention data from published literature but with the drawback that reproducibility of the exact conditions is uncertain. In any case, the identification by comparison of retention data is only indicative and not confirmative. The probability of peak overlap and wrong identity increases with the number of components in the sample. It is recommended that the data be compared on more than one chromatographic system (stationary-phase–mobile-phase combination), preferably differing in their selectivity. It is well recognized that comparison of retention data alone is insufficient for positive identification. Negative identification (i.e. the conclusion of non-identity of two compounds that differ in their retention values on any chromatographic system), however, is certain.

To compensate for the differences in packing density of the bed and the flow-rate, adjusted retention times ($t'_R = t_R - t_M$) and capacity factor ($k = t'_R/t_M$) must be used. Relative retention ($t'_{R(A)}/t'_{R(B)}$) of compound A to that of a standard compound B is often a recommended parameter for comparison, as it may nullify variations in such 'not-so-precisely' reproducible parameters as column temperature and the mobile-phase flow-rate. However, there is no accepted standard for all compounds, or even for each class of compounds. Any commonly available compound, preferably belonging to the same chemical class (e.g. fatty-acid esters) may be used as standard.

Uncertainty in retention-time measurement can also arise from sample overloading. Unsymmetrical, 'leading' peaks that arise when too large a sample is injected would exhibit a larger apparent retention time compared to the actual. The retention-time repeatability may be ensured by decreasing the sample size by half. If the retention time changes, the operation is repeated by further reducing the sample size by half, and so on.

In view of the uncertainties associated with the reproducibility of the retention times, a number of retention indices (both linear and logarithmic) have been suggested for the comparison of retention data on GC [6]; among these, the Kovats retention index is the best known (see section 4.2.3.2). It is based on the fact that the logarithm of the adjusted retention time ($t'_R$) for $n$-alkanes varies linearly with the carbon number for a given column and a given temperature. Recapitulating eq. (4.1), the Kovats retention index, $I$, for a compound ($x$) eluting at an adjusted retention time $t'_{R(x)}$ is given by:

$$I = 100z + 100\left[\frac{\log t'_{R(x)} - \log t'_{R(z)}}{\log t'_{R(z+1)} - \log t'_{R(z)}}\right] \tag{9.1}$$

where $t'_{R(z)}$ and $t'_{R(z+1)}$ represent the adjusted retention times of two $n$-alkanes with $z$ and $(z+1)$ carbon atoms, so that $t'_{R(z+1)} > t'_{R(x)} > t'_{R(z)}$. Signifying the fact that the retention index varies with the stationary phase and the temperature, these two parameters are often given as the superscript and subscript, respectively, following the letter $I$. At a given temperature (isothermal GC), the Kovats index is fairly independent of the other chromatographic parameters and comparison with those determined elsewhere and from the literature is possible. Kovats retention index libraries and library search programmes are commercially available on computer diskettes [7]. The retention index may vary within 1 unit for non-polar stationary phases and 2 units for polar phases.

Owing to the non-reproducibility of several parameters, retention times are not very accurately comparable on programmed-temperature GC and, therefore, the log $t'_R$ of eq. (9.1) may be replaced by the elution temperature ($T_R$) to give the temperature-programmed retention index, $I_{TP}$, as in eq. (9.2).

$$I_{TP} = 100z + 100\left[\frac{T_{R(x)} - T_{R(z)}}{T_{R(z+1)} - T_{R(z)}}\right] \tag{9.2}$$

Retention indices on open-tubular GC and temperature-programmed GC have received considerable attention lately [8]. The $I_{TP}$ values vary considerably with the temperature-programme rate ($r$), the average carrier-gas flow-rate ($F$), the internal diameter of the column, and the amount of the stationary-phase ($w$). The ratio $rw/F$ is a characteristic parameter in linear temperature-programmed GC and, so long as this ratio and the initial column temperature are constant, the $I_{TP}$ value is reproducible within 1 unit on non-polar stationary phases; there may, however, be moderate slight changes with change in the column diameter and film thickness.

The reduced retention of *n*-alkanes on polar stationary phases makes them unsuitable as universal standards for the Kovats indices since it would require comparison of low-boiling solutes with much higher-boiling *n*-alkanes. Also, the hydrocarbons do not respond adequately to certain selective detectors (e.g. the electron-capture detector). Several alternatives (based on alcohols, carbonyl compounds, esters, etc.) have been suggested. Standards based on similarity of structure (e.g. equivalent carbon number for fatty acids (CN), steroid number (SN), etc.) are also used so that better correlation between structure and retention behaviour is possible [9]. The retention index is calculated as for the alkanes eq. (9.1).

*9.2.1.2.2 Subtractive methods*

This general term is used to indicate all methods in which peak identification is based on removal of some of the compounds in the sample either by chemical reaction or physical adsorption and comparison of the chromatogram with that of the original sample. In the simplest form, this involves use of a precolumn containing a reagent supported on a solid, to selectively remove compounds containing a specific functional group [10]; e.g. β-diketonate chelates can be used for selective retention of nucleophilic materials. After eluting the weakly retained hydrocarbons and their halogen derivatives, the precolumn is suitably treated (e.g. heated to a higher temperature in GC or eluted with a solvent containing a more powerful nucleophile in LC) for detecting the nucleophiles. Fig. 9.1 demonstrates the use of subtractive chromatography using a mixture of alcohols, carbonyl compounds, esters and hydrocarbons [11]. The trace (a) is obtained with the total mixture of 17 compounds on programmed-temperature GC using a Carbowax 20M-coated PLOT column. Trace (b) was recorded after introducing a 25 mm precolumn of 3-nitrophthalic anhydride (40% w/w), and trace (c) with a similar precolumn containing semicarbazide (40% w/w). The former column retained all the alcohols represented by the peaks 1, 2, 5, 7, 8, 12, 14 and 17, while the second precolumn removed all the carbonyl compounds (peaks 3, 6, 10, 11, 15 and 16).

Though not strictly subtractive, catalytic hydrogenation using palladium or

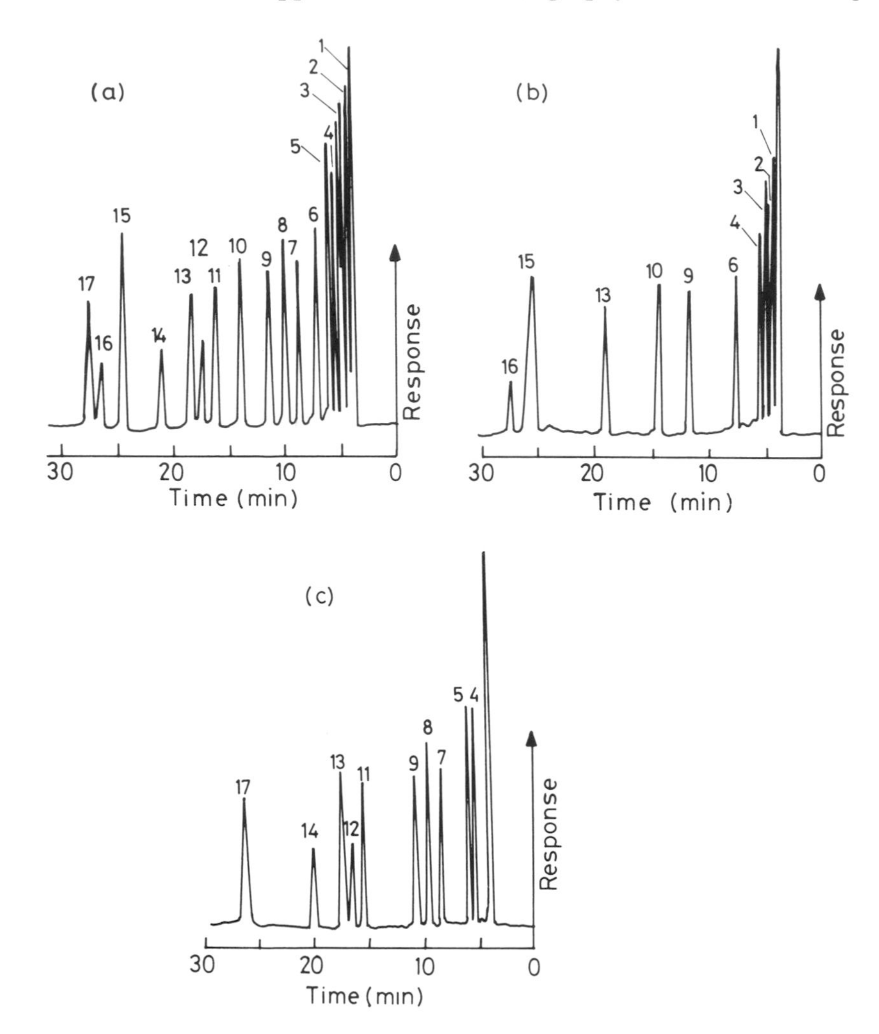


Fig. 9.1 — PTGC analysis of a synthetic mixture of volatiles using subtraction chromatographic technique; (a) no pre-column, (b) a 2.5 cm pre-column of 3-nitrophthalic anhydride (40% w/w) operated around 190°C, and (c) a 2.5 cm pre-column of semicarbazide (40% w/w) maintained around 115°C. Analysis conditions: 0.5 mm i.d. × 37 m PLOT column coated with Carbowax 20M (0.5% w/v); carrier-gas, nitrogen at 4 ml/min; initial temperature, 70°C; programme rate, 2°C/min. Composition of the sample mixture: 5 μg of each of (1) isobutanal, (2) 2-methylbutanal, (3) ethyl methyl ketone, (4) *n*-propylacetate, (5) *n*-propanol, (6) *n*-hexanal, (7) 3-buten-1-ol, (8) 2-methylbutanol, (9) *n*-butyl butyrate, (10) 2-octanone, (11) *n*-hexanal, (12) *cis*-3-hexen-1-ol, (13) tridecane, (14) 1-octen-3-ol, (15) benzaldehyde, (16) 2-nonenal, and (17) *n*-octanol. Reproduced from Ref. 11 by courtesy of Elsevier Scientific Publishing Co.

platinum on a finely divided porous support in the precolumn, and hydrogen as the carrier-gas, has been used for reducing several functional groups to yield hydrocarbons in the so-called carbon-skeleton chromatography [12]. The types of compounds that can be used in this type of analysis include unsaturated hydrocarbons, carbonyl compounds, esters, amines, halides, nitriles, epoxides, sulphides, fatty acids and others. Fatty acids undergo decarboxylation and give hydrocarbons containing one carbon atom less. Pyrolysis and other precolumn derivatization reactions discussed earlier can all be used in a similar manner with appropriate modification.

#### *9.2.1.2.3 Miscellaneous methods*

Use of the so-called hyphenated methods, coupling the chromatography to UV, IR, NMR, AAS and MS, are very powerful techniques of identification of compounds in complex samples. The MS, diode-array (UV) and FT-IR techniques, coupled with the spectral libraries and library-search programmes available on computers, have made chemical characterization of the unknown solutes relatively easy. Post-column chemical reactions and selective detectors in multi-channel mode using effluent splitters are also possible. Where the compounds are new and their isolation is required for structure determination by the classical methods (degradative and spectroscopic studies that cannot be carried out on-line), preparative chromatography is used.

### ***9.2.1.3 Quantitative analysis***

Though chromatography is basically a separation process, it is well recognized that the capability of instantaneous, on-line detection and the quantitative nature of the response of most detectors are what have made the technique an indispensable tool in analytical chemistry. It must be emphasized that it is this latter feature which contributed to the phenomenal growth of the subject. There are several aspects of quantitation in chromatography that need to be carefully considered. The accuracy and precision depends on the methods of sampling and injection, sample size, efficiency of the column, linearity and time constant of the detector, and recording and processing of the data. There are several detailed reviews [13,14] and monographs [15] on this very important application of chromatography.

#### *9.2.1.3.1 Sources of error*

There are three main sources of error in quantitation using chromatography. They are (a) sampling and sample introduction, (b) design of the chromatograph, and (c) peak-size measurement. As in any analytical method, the representativeness of the sample used for analysis is a limiting factor of the final result. Since only a very small portion of the bulk is used, non-homogeneity could cause serious errors; these can be minimized by multiple sampling and repetitive analysis. The problem of sample loss during storage and transportation by phenomena such as evaporation, adsorption and decomposition must be considered and, to the extent possible, the time gap between sampling and analysis should be minimal. Trial runs with standard samples under simulated conditions of storage and handling might establish the stability of the compounds to the conditions of sampling, storage and pre-treatment. Use of internal standards (see below) is recommended for minimizing the errors from these sources. The methods of sample preparation have been discussed earlier in this chapter.

The accuracy with which the sample size introduced into the chromatograph could be measured is limited by the technique and equipment used. The most common method of sample measurement is by use of a microsyringe, particularly in GC and TLC. Fixed-volume sample loops are more common in LC; the loops are more precise provided the sample injected is more than the loop capacity. There are several commercial sources of excellent microsyringes, specially designed for chromatography. They are generally of 10 $\mu$l capacity. As the efficiency of chromatography increases, the sample size decreases (as in, for example, open-tubular GC) and as the sample size decreases, the errors in measurement of the sample size increase. For practical reasons, the volume of the sample introduced is normally not less than 1 $\mu$l and the amount of the analyte required is about 1% of this value. This would occasionally require dilution of the analyte with a suitable solvent, which could be another source of error, arising out of the dilution technique, its measurement and the impurities in the solvent that might overlap on the peaks of interest. The latter problem can be easily overcome by ensuring that the solvent is pure and using a blank run to ascertain the absence of such impurities.

The errors arising out of reading a certain volume of the liquid have been briefly mentioned earlier (Chapter 3). The error in manual reading of the meniscus in the barrel of the syringe may be 2–5%. Added to this is the possibility of partial evaporation of the sample remaining in the needle at the time of its introduction into the hot injection port (when a heated injection port is used in GC). The 'plunger-in-the-needle' type of syringes are more precise (within 1% error) than the more common 'plunger-in-the-barrel' type; but the former type of syringes are much more expensive. Again, the use of an internal standard can minimize the errors from these sources, provided the amount of internal standard used is precisely measured.

The gas-tight syringes, which are normally used for introducing gaseous samples into a GC, have a better reading accuracy as the volumes measured are three orders of magnitude greater (about 1 ml) compared to the liquid samples. However, since the pressure in the column is generally higher (about 3 atm), the possibility of leakage of the sample out of the syringe is high. Here again, the sampling valves are more accurate and are therefore recommended. The leakage and bleeding of the septum also adds to the error in sample introduction unless good-quality septa are used and are frequently replaced to ensure satisfactory performance.

Fast and slug injection is recommended in all cases of dynamic chromatography (GC or LC). The injection time is particularly critical for fast-eluting compounds ($k < 5$). Crevices and unswept areas in the injection port are the other sources of error in sample introduction. It may be assumed for the purpose of the present discussion that there are no chemical changes occurring during the passage of the sample through the injection port, the column and the detector, but this could be a real problem in GC where the temperatures are high. It is also assumed that the system is optimized for the given analysis and that peaks are nearly Gaussian in shape. Unresolved and unsymmetrical peaks can cause serious errors in quantitation by peak-size measurement.

The detector cell volume and response time constant, both of which should be as small as practicable in all cases, can also be the source of errors in the case of high-efficiency columns using very small sample sizes, particularly in the case of the fast eluting (needle-sharp) peaks. The other detector factors of importance to quantitation; i.e. sensitivity, linear dynamic range, selectivity and stability, have been discussed earlier with reference to the specific GC and LC detectors.

*9.2.1.3.2 Peak-size measurement*

The basic premises in the use of chromatography for quantitative analysis are that (a) the response of the detector to the sample is proportional to the quantity (at least over the range in which the sample is present), (b) the response fed to the appropriate device is recorded with high fidelity in the form of a peak (either graphically or digitally), (c) the size of the peak can be measured accurately, and (d) the peak size can be converted or related to the amount of the sample with a reasonable degree of accuracy. While there are limitations to the realization of these assumptions, the state of the present instrumentation (see below) meets the requirements to a very large extent.

The size of the peak is traditionally measured in two ways, by the peak area, and by the peak height. The relation of the Gaussian peak to the height, standard deviation (peak width at different heights), the area and the concentration are shown in Fig. 2.1. The peak size is measured in a number of ways [14], which include determination of the peak height (i.e. the distance of the apex from the baseline) or the peak area (height × half the peak width at the base, or height × peak width at half height). In addition to calculating the area of the triangle formed by the baseline and the tangents to the peak, the peak area may be directly measured by means of a planimeter, or a disc integrator or simply by cutting and weighing the chart paper covered by the peak. With the advent of modern digital integrators which can process the signal directly from the detector, and their widespread use in most laboratories, much of the discussion on the relative merits of the different methods of peak-size measurement appear irrelevant. Nevertheless, a brief discussion on the subject may be useful in choosing the right type of integrator, several kinds of which are commercially available.

The peak height (the length of the perpendicular line connecting the apex of the peak to the baseline drawn by joining the beginning and end of the peak) is the easiest to measure. If the peak width is constant, the height is proportional to the concentration of the analyte. It is generally known that the peak width increases with the retention time and this method cannot, therefore, be used for comparing the concentrations of two peaks in the same chromatogram. Even when an external standard calibration curve is used, quantitation by the peak-height method requires that the temperature (or solvent strength in LC) be kept constant. Since the retention time decreases by 3% for every 1°C rise in temperature and the peak width is proportional to the retention time, the error in quantitation also varies accordingly. It is therefore advisable to maintain a temperature stability of within 0.1% for a quantitation accuracy of over 99%.

The retention time is also directly affected by changes in the mobile-phase flow-rate. Some of the detectors (notably the thermal conductivity and the photo-ionization detectors in GC, and the refractive index and electrochemical detectors in LC) are also flow-sensitive in the sense that the electrical output of the detector for a given amount of the sample changes with the flow-rate. Thus, the flow sensitivity is reflected in any method of peak measurement, whether by height or area. The peak height of the early-eluting peaks is also greatly affected by the injection time, a fraction of a second difference in which could change the peak height two-fold.

Strongly retained peaks are rather broad and, when adsorption on the solid support occurs, there can be significant tailing. The effect of tailing is more pronounced when the concentration (and hence the peak height) is very small because, in this case, the relationship between the peak height and the amount of the material is not linear. The non-linearity is also observed when the column is overloaded because of the distortion and broadening of the peaks. In short, there are several factors which affect the accuracy of quantitation by peak-height measurement and the method is generally not recommended for quantitation except in the case of repetitive analysis under well-controlled conditions.

While the peak height and width change with the chromatographic conditions, the product of the two parameters is nearly the same. In other words, within the linear dynamic range of the detector (irrespective of whether it is concentration-sensitive or mass-sensitive), the peak area is proportional to the amount of the sample and is, therefore, the method of choice under most circumstances. As noted above, the area under the peak can be calculated in a number of ways, both manually, mechanically and electronically [14]. Several of the mechanical devices, which have been widely used earlier, are now almost completely replaced by the electronic instruments and are therefore not discussed here. The area of the triangle formed by the apex of the peak and the baseline (intercepted by the tangents drawn along the two sides of the peak) is given by the height of peak multiplied by half the basewidth. However, since the exact positions of the peak start and peak end are difficult to pin-point, the peak width at half height is usually taken for the calculations. If the baseline is not horizontal but sloping, the calculation of the area becomes somewhat more complicated. Peak-area measurement is also difficult when the peaks are not well resolved. The electronic integrators, which are at least an order of magnitude more accurate than the manual and mechanical devices, handle the peak-asymmetry, poor-resolution and unsteady-baseline and baseline-drift problems in a number of ways, depending on their capabilities. These are usually described in the manufacturers' literature.

#### *9.2.1.3.3 Normalization and standardization*

The terms normalization and standardization are used here to describe the methods employed to convert the peak-size measurements into the amount of the analyte. In both cases, the quantitation is based on comparing the peak size of the sample to that of a standard. In all types of chromatographic quantitation, the standard, which is also invariably chromatographed, must be structurally as close to the unknown sample as possible, so that the retention time is not very different from the sample, and response factors of the detector for the two samples and the peak shapes are also very similar. To the extent possible, the quantity of the standard is also adjusted to be close to the concentration of the sample. Preferably, the standard is added to the same matrix so that the recovery values are comparable. In the choice of the standard, attention should also be paid to the purity and stability of the compound. Several standards, belonging to different classes of compounds and application areas (lipids, amino-acid derivatives, pesticides, environmental pollutants, etc.) are available commercially in easily dispensable form (e.g. in 1 ml vials with rubber-septum caps so that a measured quantity can be withdrawn using a microsyringe with a hypodermic needle).

In the internal normalization method, the area percentage of all the peaks is required even if some of the peaks are not of interest to the given application. If it is assumed that all the components of the sample respond equally (i.e. they have identical response factors), the area-percentage values directly give the relative concentration of the components in the sample. But, this situation is rarely found in practice even among compounds of the same functionality. Therefore, the next step involves preparation of a mixture of these compounds in known proportions. The calculation method for obtaining the concentration of the individual compounds is shown in Table 9.1, using a hypothetical mixture of compounds A, B, C and D [14].

**Table 9.1** — Internal normalization†

| *Standard* | | | | | | |
|---|---|---|---|---|---|---|
| Component | Weights taken (g) | Weight % | Peak area | Area % | Weight % area | Response factor, $F$ |
| A | 0.3786 | 21.74 | 4231 | 22.41 | 0.005138 | 1.000 |
| B | 0.4692 | 26.94 | 5087 | 26.94 | 0.005296 | 1.031 |
| C | 0.5291 | 30.38 | 5691 | 30.14 | 0.005328 | 1.039 |
| D | 0.3648 | 20.94 | 3872 | 20.51 | 0.005408 | 1.053 |
| Total | 1.7417 | 100.00 | 18881 | 100.00 | | |

| *Unknown* | | | | | |
|---|---|---|---|---|---|
| Component | Peak area | Weight % | Normalized weight % | Area $\times F$ | Weight % |
| A | 3862 | 19.84 | 19.67 | 3862 | 19.66 |
| B | 5841 | 30.93 | 30.66 | 6022 | 30.66 |
| C | 4926 | 26.29 | 26.06 | 5118 | 26.06 |
| D | 4406 | 23.82 | 23.61 | 4639 | 23.62 |
| Total | 19035 | 100.88 | 100.00 | 19641 | 100.00 |

†Reprinted with permission from Ref. 14; copyright (1985) John Wiley & Sons, New York.

First, the weight-percent and the peak-area values (in any convenient units) for the individual compounds are tabulated. By dividing the weight percent by the area, the concentration per unit area is obtained. Arbitrarily assigning the response of the compound A to the detector as 1.000, the response factors ($F$) of the other components are calculated. The corresponding areas under the peaks for the test sample are also similarly tabulated. Multiplication of the areas by the concentration per unit area calculated above, gives the weight percent of the individual compounds. However, the total may not be exactly 100 (in this example, the figure is 100.88) for the sample sizes may not be identical. However, this can be easily adjusted by normalizing the weight percent values to the total of 100 (i.e. by multiplying all the weight-percent values by 100/100.88). It may be noticed that the normalized weight percent is equal to the response-corrected area percent.

Some of the disadvantages of the internal normalization method are quite obvious. Unless the detector responds equally to all the compounds, the identity of all the compounds present must be known and the standard samples available in the laboratory for calculating the response factors of them. Errors could still arise if the response of some of the compounds is very poor and the others very high. Thus, it cannot be used with the specific detectors like, for example, the electron-capture detector in GC. It also requires that the response of the detector be linear for all compounds and in all concentration ranges in the sample. When the sample contains widely differing concentrations of certain constituents, it is advisable to dilute both the sample and the standard with a non-interfering solvent so that the relative concentration in the total sample does not vary significantly. Since the absolute quantities are not taken into the calculations, dilution is not a factor and the response due to the solvent can be completely ignored. Determination of the response factors for all the compounds in a complex sample may also be quite tedious. Nevertheless, internal normalization is routinely used in natural-gas analysis, since mostly the same components are present (with known response factors) and complete analysis is required for determining the heating value of the sample [14].

The external standardization procedure is typical of any analytical method where a particular parameter is related to the concentration (e.g. colorimetric determination of a compound using a chromogenic agent). The method involves drawing a standard calibration curve (concentration or quantity *vs.* the detector response) using the same compound(s) in the same matrix and in the same concentration range as may be present in the analytical sample. Ideally, the calibration curves are linear, passing through the origin. This situation may be realized if the column and the detector are not overloaded and the response of the detector and the electronics is linear. As in the other analytical methods, this technique can also be used with confidence only in the linear dynamic range of the detector for the given compound. Outside the linear range, the calibration curve has to be well defined with a large number of standards, and each time the analysis is carried out with a different sample.

Obviously, the use of external standardization requires that the amount of the sample and the standards injected be accurately known. Therefore, all the sources of error discussed earlier in connection with the sampling by syringe techniques are of importance. The sample size in high-resolution chromatography means not only the absolute quantity of the solute but the total volume of the sample injected. This would mean that to draw the calibration curve, one must use different concentrations of the standards and not different volumes of a single-concentration solution. Naturally, it is to be expected that the injection volume of the unknown sample should also be the same. Thus, the precision of the external standardization method is likely to be much higher if sampling valves rather than syringe injection techniques are used. For reasonable accuracy, at least five different concentrations of the standard are required. Once a linear standard curve is obtained, the concentration of the sample is easily calculated from the peak area for the sample of identical injection volume. Preparation of standard solutions of different concentrations, is fairly simple. It is convenient to prepare a master solution of the compound (1% wt/vol) in a suitable solvent. Progressively increasing volumes (say, 10–100 $\mu$l) of this solution are added to the required number of volumetric flasks of the same volume (say, 10 ml) and diluted with the same solvent to the mark. Preparation of gaseous standards is more difficult and would require special equipment and techniques [14].

The internal standardization method, in view of its convenience and improved accuracy, is the most popular quantitation method in chromatography. It is akin to the internal normalization method, described above, to the extent that the calculation part is similar. However, the internal standard to be used is not expected to be present in the sample but added to it early in the process of sample preparation. The characteristics of an ideal internal standard have also been described earlier (section 9.2.1.1).

The principle of internal standardization is best understood with reference to Table 9.1, assuming that compound A is not originally present in the sample but added as the internal standard. As before, a standard mixture containing compound A and any one or all of the analytes B, C and D is prepared, chromatographed and the response factors determined. Using the same figures as in Table 9.1, the weight ratio of compound B to A is 1.2393 and the area ratio is 1.2023. The response factor ($F$, which here is the reciprocal of that in the internal normalization method) is then given by the area ratio divided by the weight ratio, which is, 0.9701. The response factors for the other components of interest in the sample are similarly determined. A series of different weight ratios of the standard to the analyte may be examined to ascertain the linearity of the response. A known quantity of the standard (compound A) is then added to the sample and the mixture chromatographed. From the relative response factors of the analytes, their concentrations relative to the known concentration (or weight) of the internal standard can be easily calculated using eq. (9.3):

$$\frac{W_B}{W_A} = \frac{A_B}{A_A} F \tag{9.3}$$

where $W_A$ and $W_B$ are the weights, and $A_A$ and $A_B$ are the areas of the peaks for compounds A and B, respectively.

Since chromatography of the standard mixtures and the analytical samples are carried out under the same conditions and since the ratios, and not absolute responses, are used in the calculation, the more convenient peak-height measurements can be used instead of the peak areas. While all the advantages of the internal normalization technique are valid for the internal standardization method, the main disadvantage of the former, is that the requirement of measuring all the components does not apply and only the components of interest need be followed. Also, quantity of the sample injected or the extent of dilution need not be accurately known. However, it is advisable to maintain a certain degree of equivalence or similarity in concentrations.

*9.2.1.3.4 Recorders and data processors*

The most commonly used device for recording the chromatogram (the two-dimensional plot of the retention time or volume *vs.* detector response) has long been the potentiometric recorder. From the chromatogram, the data required for qualitative (retention times and volumes) and quantitative analysis (peak size) can be readily obtained. The output of the detector, which is in the form of an electrical signal, is fed to the potentiometer either directly from the conductivity-type detectors or through an electrometer-amplifier. There is also an attenuator (voltage divider) between the electrometer and the potentiometer to decrease the signal amplitude in steps of 1, 2, 4, . . . 1024; i.e. the actual size of the peak is given by the observed size multiplied by 1, 2, 4, . . . 1024.

The potentiometer records the chromatogram on a strip-chart. The chart speed may be selected from 0.5–10 cm/min and the sensitivity is determined by the potentiometer range which can be set from 0.1 to 10 millivolts; the lower the range, the higher is the sensitivity of response of the recorder. The difference between the upper and the lower limits of the range is called the span, which is normally set at 1 millivolt. The step response time of the recorder is the time required by it to respond to an electrical signal input (say, of 1 millivolt); this period is usually 1 second in most strip-chart recorders. Another characteristic of the recorder is the deadband, which is the amplitude of the signal that can vary without initiating a response on the recorder; this value should not normally exceed 0.1% full scale, or some of the smaller peaks may be missed. The linearity of the recorder, which is the ratio of the full-scale peak to that of the smallest detectable peak, is normally about 0.1%; the linear range is effectively extendable by attenuating the input signal.

The dynamic properties of the recorder (i.e. the response of the recorder as a function of the signal) can be critical when the signal changes occur rapidly as in capillary or high-resolution chromatography. While the above-described functional properties of the potentiometric recorders (i.e. the range, the step response time, deadband and linearity) are generally adequate for the packed-column GC and LC, they are not suitable for high-resolution capillary chromatography. The large number of very sharp peaks that are generated in open-tubular GC also makes it difficult to accurately determine the retention times and peak areas manually. In the modern chromatographic data processors, the electrical (analog) signal is converted into a digital signal and fed to a microprocessor; for a description of the different types of analog-to-digital convertors (ADCs), see Ref. 16. Using the ADCs, a large number of chromatographs can be interfaced to a desktop computer or even a mainframe computer for data management in several ways (see below). Intermediate between the strip-chart recorder and the microprocessor-based chromatographic data systems are the low-cost reporting integrators.

A number of reporting integrators are available which automatically plot the chromatogram and print out the retention times and the peak areas (or heights) in both absolute units and as percentages of the total. They can also be programmed to compute the quantitative data by either internal normalization or relative to an internal or external standard; the response factors can be fed to the integrators or generated by them from the chromatograms of the standards. These features of the integrators are indeed so common that they are normally referred to as the standard calculations.

The current popularity of the reporting integrators and the more sophisticated data processors has made the potentiometric recorders nearly obsolete. As the data is processed completely digitally, it is no longer required to record the chromatogram in the traditional *X–Y* pattern, except, probably, to meet the habitual needs of the chromatographer. The data systems can also furnish a reconstructed two-dimensional chromatogram. Even so, some of the information that can be obtained by examining the chromatogram at a glance (e.g. peak skew or distortion which indicates non-linear isotherms and/or column deterioration) may be lost unless the microprocessor is programmed to give such specific information. As will be discussed later, the actual peak shape reflects several physicochemical parameters of the solute, stationary phase and the mobile phase.

Using the modern microprocessors, a large amount of chromatographic data can be conveniently stored in the computer memory or discs and retrieved on demand in a form suitable for particular application. The data processors are characterized by three main features; data acquisition and storage, data manipulation, and reporting (report generation). The data acquisition, again, has two main parameters; the sampling (or acquisition) rate, and the number of channels. In general, the higher the sampling rate the better is the raw data; sampling rates of 10 to 100 Hz (cycles) per second per channel are common. The number of channels that a computer can simultaneously acquire data from may also vary widely; up to 75 (or 100) different instruments can be hooked to a single computer. The data-storage capacity is seemingly unlimited since it can be expanded at will. Presentation of the chromatogram by the data processors can be either in real time, or can be manipulated in many ways at the end of the run (or at convenience). The chromatogram may be obtained in the standard form, or a baseline-compensated, or a normalized form. An important and very useful facility in the modern chromatographic data processors is the display of the 'soft chromatogram' on a cathode-ray tube (CRT). A hard copy of the processed or the raw chromatogram can be obtained using a high-speed, microprocessor-controlled, printer–plotter.

The baseline-compensated chromatogram not only makes the picture more presentable with a horizontal baseline, but also improves the apparent resolution and quantitative accuracy. The baseline compensation is achieved by first obtaining a blank run using the same column, mobile phase and temperature solvent programme. The blank baseline is then stored in the computer and subtracted from the chromatogram of the sample. In this manner, the drift in the baseline, arising out of the possible column bleed or from impurities in the mobile-phase components, is compensated. Similarly, two chromatograms can be compared and the differences highlighted by subtracting one from the other.

Two chromatograms can also be compared by computer-assisted ratioing. When a CRT display is used, two or more chromatograms can be compared by splitting the screen into windows and the different chromatograms displayed simultaneously for visual comparison. A particularly useful function is the presentation of the data from several chromatograms in a 'three-dimensional' format for easy comparison. Fig. 9.2 shows a portion of the gas chromatograms of the extracts from 17 varieties of black pepper [17].

In the normalized form of the chromatogram, the highest peak is automatically brought within the scale (full-scale or 100%) and the other peaks proportionally adjusted; it is useful for gaining an overall picture of the composition of the sample. Other types of axis-scaling are also possible so that different aspects of the chromatogram can be highlighted or suitably enlarged (zooming). For example, regions of the chromatogram where the peaks are too closely placed (cramped) or are too small, can be expanded with respect to the *X*- or *Y*-axis, or both, and replotted. Logarithmic scaling permits the closely placed early eluting peaks and the well-separated later-eluting peaks to be nearly uniformly placed.

One of the main conveniences of the chromatographic data systems is that the data can be acquired unattended and the entire elution profile stored in a digitized form. The raw data stored in the computer can be subsequently processed in a number of ways to obtain the maximum qualitative and quantitative information on the sample analysed. To obtain the retention data, the computer needs to receive a signal to denote the instant of sample injection. This can be done either manually or

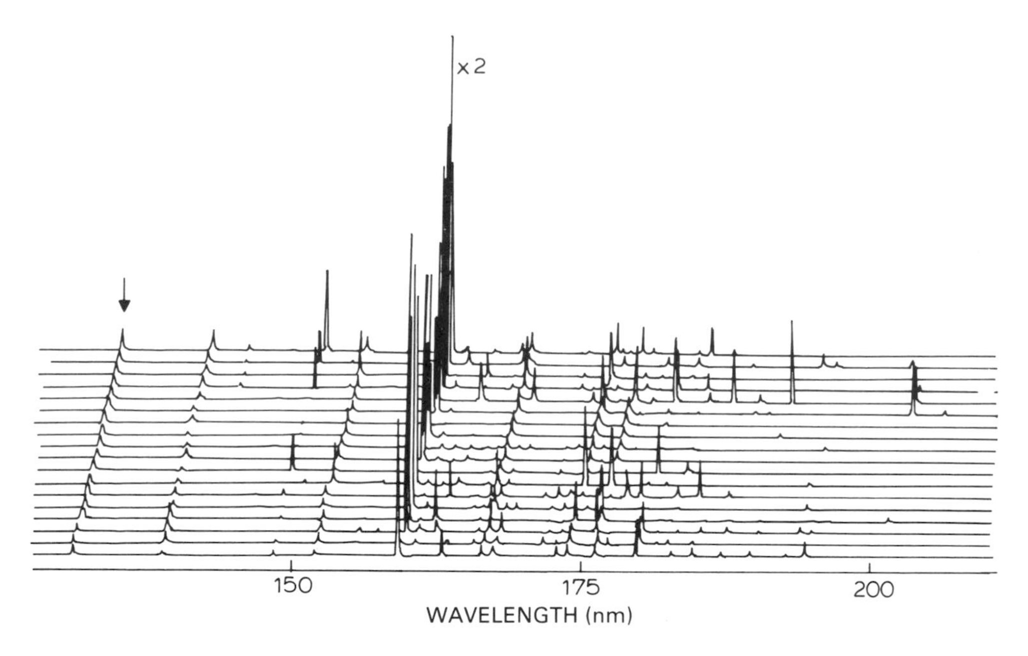


Fig. 9.2 — Open-tubular gas chromatography of the extracts of 17 varieties of black pepper. The illustration shows portions of the chromatograms (sesquiterpenes and volatiles), replotted by a computer for better comparison. The first peak on the left is that of the *flag* or the internal standard (n-tridecane). Reproduced from the *Journal of Chromatographic Science*, Ref. 17, by permission of Preston Publications, Inc.

electronically if an autosampler is used. The electronic internal (quartz) clock of the computer takes note of it for the later presentation of the precise retention times. With the help of standard runs, the corrected retention times, relative retention times, the Kovats' indices, etc., may be obtained. Comparison of the retention index data with the stored information helps sample identification. Kovats' retention-index databases on several classes of compounds are commercially available.

Hyphenated techniques, coupling GC or LC to UV, IR, NMR, AAS and MS, are the most powerful methods for sample identification (qualitative analysis). However, these techniques require such high-speed data acquisition, storage and processing capability that they cannot be profitably used without the aid of computers. This is particularly true of on-line FT-IR and MS, which require fairly large data-storage capability for the spectral libraries and a search, retrieval and comparison facility. The currently available mass spectral library contains about 43,000 spectra and FT-IR libraries contain about 6,000. UV spectral data on individual compounds can be reconstructed after multi-wavelength detection; to these may be added spectra recorded in the laboratory also. Again, the use of the very powerful diode-array detectors that can continously acquire UV spectral data of the column effluents for subsequent retrieval and manipulation requires the microprocessor facility for the data management.

Most microprocessors now available for use in chromatography have sufficient software-handling capacity that they can be used to control several instrument parameters. Thus, they can activate autosamplers and sampling valves, control oven temperature and temperature programmes, solvent delivery and composition, detectors and external devices. They may also be used for method development and HPLC solvent optimization with the appropriate software. Size-exclusion chromatography software (for molecular-weight calibration and averaging) are available from some suppliers. The data systems may also be used for statistical evaluation of the columns and their performance. They may be programmed to aid in column selection and to alert the operator when the performance of a given column deteriorates.

Indeed, with the increasing data handling and storage capabilities of the modern microprocessors, the amount of data that can be stored, and the number of ways in which the data can be processed, appear unlimited. Coupled with the laboratory robotics mentioned earlier, the recently introduced laboratory information management systems may be expected to increase the output of analytical laboratories multifold. These systems can be programmed to suit the particular requirements of a laboratory and are particularly useful in large, analytical quality-control laboratories.

Typically, a laboratory information management system (or LIMS, as it is popularly known), has all the features needed for a large quality-control laboratory. Quality control is an essential activity for any manufacturing industry that wishes to keep its products to certain specifications of quality and performance and would not compromise on this matter in its own long-term interests. However, the analysis protocol and the number of samples (and the people who handle them) may be so large that keeping track of the samples and the analytical reports would seem too complicated. Normally, a situation may exist where samples are to be drawn from a number of units and to each of the units, the raw materials are supplied from different sources. In a formulation industry (e.g. pharmaceuticals, foods and beverages), the specifications and quality of the ingredients directly relate to the quality of the finished product. This would require quality assurance on the incoming raw materials on one hand and the finished products on the other. Given below is the working of a typical quality-control laboratory that can be completely and conveniently handled by the commercially available LIMS.

As the samples are received for analysis, they are logged, coded and the tests to be carried out on the sample noted against the sample and the code. The specifications the sample should meet, as well as the analytical procedures and the method of calculation are also stored for retrieval when desired. Occasionally, the same product may have different specifications for different applications. These are highlighted at the time of work allocation. Worksheets are prepared for each sample or for each batch (if the tests are different), to be distributed to the analysts. Modifications to the specification, when needed, must be made only by the authorized person and there must be a security system to avoid tampering by others. The analyses are carried out by the analysts (or, possibly, by robots) usually on automated instruments hooked to a computer. The LIMS can simultaneously receive and store data from as many as 100 analytical instruments (e.g. GC or HPLC). They may also be entered *via* a keyboard by the analyst. The data is processed according to the pre-programmed methods. The results are compared with the specifications and any results which are outside the accepted limits of variation are immediately highlighted, and exception reports are made. Otherwise, the reports are made according to the format for internal use or for an outside agency. When a sample is passed, a certificate of analysis is prepared, to be signed by the lab supervisor. Midway during the analysis, the sample may be tracked and an *ad hoc* report generated. It may be necessary to reschedule and reallocate the samples, if for some reason, the analyst is not available or the work is held-up. Graphic reports can be prepared and the long-term trends or periodical changes, if any, can also be obtained using the archived analytical data.

#### *9.2.1.4 Trace analysis*

Analysis of trace compounds has attained such great importance and the methods are so unique, that the subject is developing into a discipline by itself. Considering the potential toxicity of several trace compounds that we come into contact with in everyday life, the importance of the subject cannot be over-emphasized. Examples of such toxic compounds that frequently accompany bulk materials of common use include: pesticide residues in the soil, polychlorinated hydrocarbons in water, *N*-nitrosamines and polycyclic aromatic hydrocarbons in cigarette smoke, industrial and automobile emissions in air, aflatoxins in peanut products, metal contaminants in canned foods and toxic contaminants and isomers in pharmaceuticals. In addition, several important and biologically active constituents occur naturally in the plant and animal kingdom. With the availability of several high-sensitivity specific detectors, the chromatographic methods are an obvious choice for the analysis of such compounds.

The proper definition of the word 'trace' is still elusive. Obviously, it is a relative term. About three decades ago, determination of compounds present in a concentration of about 0.1% would have been considered trace analysis, but now this is routine [18]. For the purpose of this discussion, analysis of materials present in a concentration around 1 part per million (1 ppm, 1 $\mu$g/g or $10^{-4}$%) in a matrix may be considered trace analysis. The concentration of the trace is usually expressed as *a* parts per $10^x$ parts or simply *a* pp $10^x$.

The driving force in the development of the methods of trace analysis is essentially legal in the sense that governmental enforcement agencies have an obligation to protect man and animals from exposure to toxic chemicals in the environment. Thus, in the U.S. (as in several other countries), the Foods and Drugs Administration (FDA) and the Environmental Protection Agency (EPA) set standards of maximum allowable contaminants and these must be followed by all agencies whose activities might contribute to their occurrence in the foods, drugs or environment. The other areas of application of trace analysis include clinical chemistry, pharmaceutical industry, food industry, geology, forensic chemistry, biological chemistry and the like.

The very idea of analysing such a wide variety and number of compounds (present in trace concentrations) could be daunting. In addition, the compounds to be analysed may be closely related (separation problems), unstable (sample preparation problems) and unknown (identification problems). There is also the contamination problem during sample preparation and analysis. In view of the very small concentration levels being monitored, care must be taken to prevent any contamination from the glassware, stirring paddles, solvents and other materials that may come into contact with the analyte. Good laboratory practices for organic trace analysis have been described in detail [18].

It is conceivable that the matrix (or the bulk of the material which needs to be analysed for the trace constituent) could be the source of the problem in trace analysis. Since the concentration of the trace is often well below the detection threshold for direct analysis by most analytical methods, isolation of the trace from the bulk is often indicated. While several chemical and physical separation methods for isolation of minor compounds and sample preparation methods have been discussed earlier, one formidable problem in trace analysis is that the chemical nature of the analyte may be unknown. However, this is not true in the analysis of some of the pharmaceutical formulations, where the active constituent could be in trace quantities. In any case, the isolation procedure should ensure a reproducible and, to the extent possible, quantitative yield of the trace compound. To avoid any possible loss of the trace compound by such mechanisms as adsorption and catalytic reactions, the procedures should involve the minimum number of very simple steps. The method of sample preparation should be such that it is accompanied by concentration of the trace compound. Solvent extraction methods are seldom used for the isolation because multiple extractions may be needed for complete extraction, and the isolated fraction is often more dilute than the original sample. It may be noted, however, that most of the common methods of isolation in trace analysis involve phase transfers of some sort.

The procedures employed for the concentration of the trace constituents depend on several factors, particularly the nature of the matrix and the trace constituent (if known) and the purpose of analysis. It is also important to know the approximate concentration of the trace that may be present, so that the degree of concentration and the type of detector can be selected. This information is necessary because the range of concentrations encountered in trace analysis is very wide (1 ppb to 100 ppm). For the same reason, it is also essential to know the minimum limit of concentration below which the presence of the trace is not of consequence. For example, if the allowed concentration of a compound in a particular sample is 1 ppm, it may be wasteful to attempt to detect the compound at less than this level. This knowledge thus permits a wider choice of detectors and sample preparation methods.

As noted earlier, the types of material for trace analysis are quite variable. Environmental samples like soil, water and air are very common. But, less well known are the body fluids and tissues, which may have several widely different trace compounds, foods (for analysis of flavours, minerals and vitamins), additives in polymers, and residual solvents in industrial products, etc.

All precautions discussed earlier in connection with the sample preparation in general are valid in trace analysis, but some of them acquire special significance. For example, it is very difficult to ensure homogeneous distribution of a trace constituent even in fluid samples like water and air. Multiple sampling at different points in the bulk must therefore be used for the analysis. Solid and semi-solid samples must be thoroughly homogenized before the sample is used for the analysis. Water and air sampling methods depend on the source and purpose. In public water systems, stagnant water near the pipes and valves should be avoided. Sampling of flowing water (streams) is even more complicated. It is desirable to collect the water from different parts of the cross-section of the stream. The composition of waste water streams of industrial effluents can change dramatically with time. Again, multiple sampling is the solution.

Air sampling can be tricky. Two extreme types may be encountered. One is the industrial chimney, the sample from which can be collected at the mouth of the chimney or by placing the sampler inside the stack. Secondly, sampling of air in a city environment at any given time is not representative since it is normally moving. The samples should therefore be collected at different points and at different times. Three types of container may be used. An evacuated jar (or tube) with a Teflon or rubber-septum stopper, a collapsible balloon in an evacuable box, or a large syringe may be used [19]. These methods are suitable if the airborne analytes are gaseous or very volatile and the chromatographic analysis could be carried out by GC. Suspended solids are trapped using a low-porosity filter through which a known volume of air is pumped. Acids or other liquid droplets suspended in air may be trapped using an alkaline solution or any suitable liquid.

Concentration of the collected sample usually involves transferring the trace component from the original matrix to the new phase. As already noted, the strategy is not to dilute it more than the original concentration and liquid–liquid partition is not carried out under the normal circumstances unless the distribution constant is very high as in extraction of a very non-polar compound from water by a non-polar organic solvent. For volatile trace compounds in solution, the headspace collection techniques are very convenient. The liquid sample is purged with helium and the gas (as well as the air sample) is passed through a column packed with active carbon, Tenax, Porapak or Amberlite XAD-2, to be desorbed later. These adsorbents can also be used to trap organics directly from water. If the water contains non-volatile trace constituents, the bulk of the water can be conveniently removed by freeze drying. Organics from water are preferably trapped by pumping it through an HPLC column of ODS-bound silica. Use of Sep-Pak cartridges for the purpose was mentioned earlier. If necessary, the concentrated sample is broadly fractionated over columns of alumina, silica or Fluorisil prior to separation by high-resolution chromatographic techniques like open-tubular GC and HPLC. Derivatization of the sample prior to the analysis is sometimes suggested to enhance the sensitivity and to improve resolution.

Unlike the air and water samples discussed above, biological samples (urine, serum or tissues) are not available in unlimited quantities. This puts severe restrictions on the methods of analysis and the types of detectors that can be used for the high-sensitivity analysis. The tissue samples are generally homogenized along with a solvent and filtered. Alternatively, Soxhlet extraction may be used. Fluid samples are passed through columns of ODS-silica or Amberlite XAD-2. To retain drugs of abuse like amphetamine or phenobarbitone from urine, the pH has to be adjusted to around 9.

It has been mentioned earlier that the nature of the trace constituents may be unknown. The identity may be established by one of the methods of qualitative analysis discussed earlier. The most frequently used are the retention data on standardized chromatograms. When the class of compounds is known (e.g. pesticides), the partition ratio ($p$) of the compounds is obtained simply by dissolving the sample in a suitable solvent (hexane or iso-octane) and equilibrating it with an equal volume of water or water–acetonitrile mixture. The gas chromatogram is recorded (using an electron-capture detector) before and after equilibration and the ratio of the peak heights is the *p* value. The value is compared with that recorded in the literature for over 100 pesticides [19]. GC–MS techniques are extremely useful for both qualitative and quantitative analysis using the selected-ion monitoring methods (section 5.9.5.3). Quantitative analysis is possible using any of the standardization methods described earlier, provided the calibration curves are made using standards in the same range and in the same matrix. Specific applications of trace analysis are mentioned in the later sections.

More recently, detection and determination of ultratrace quantities ($10^{-12}$ to $10^{-18}$ g) of substances has been explored [20]. Ultimately, the detectable levels may be down to the theoretical limit of $10^{-23}$ g (i.e. just one molecule). The principles of sample preparation and the precautions to prevent contamination are the same as in the methods of trace analysis, but far more rigorous. Detectability at the ultratrace levels depends largely on the sensitivity and stability of the detector since even a marginal fluctuation in the stability can translate into very imprecise results. Flame-emission spectroscopy shows sub-picogram sensitivity to several metal ions (e.g. alkali metal ions). Similar sensitivities are also realized in GC using a hydrogen-atmosphere flame-ionization detector ($1.7 \times 10^{-14}$ g for Mn). *N,N*-Dipentafluorobenzoyl pentafluoroaniline was detectable at $90 \times 10^{-18}$ g (attogram) level using an electron-capture detector. Femtogram ($20 \times 10^{-15}$ g) levels of fluoranthracene are detectable in LC with krypton-ion laser detection. Picogram sensitivity is feasible with fluorimetric or electrochemical detection under appropriate conditions. Using electron-capture negative-ion mass spectrometry, 10 fg of amphetamine is detectable.

### 9.2.2 Physical and inorganic chemistry

#### *9.2.2.1 Physicochemical measurements*

Chromatography is essentially a separation technique and, most often, the goal of a chromatographic experiment is to separate a complex mixture into the individual components for further qualitative and quantitative analysis. However, as discussed in Chapter 2, the chromatographic behaviour of a given compound in a given system is dependent on a number of physicochemical properties of not only the solute but also of the stationary and the mobile phases. Dependence of the retention characteristics of the stationary phase on the surface area and functionality and, if liquid, on its thickness and polarity are well known. The relationship between distribution constants and capacity factors has also been discussed earlier. It has also been noted that linear adsorption isotherms lead to Gaussian peak shapes; conversely, non-Gaussian peaks indicate non-linear isotherms. Dependence of peak widths on stationary phase and mobile phase diffusivities has also been discussed. In addition to these, chromatographic behaviour is quantitatively related to several other physicochemical properties of the stationary phase, mobile phase and the solute. It is thus possible to directly or indirectly measure these properties by chromatography.

There are several obvious advantages of the chromatographic methods for determination of thc physicochemical properties. These methods use relatively small quantities of the compounds and are generally very rapid. Analytical chromatography is usually conducted under nearly infinite dilution conditions and, for such studies, it is probably more accurate than the gravimetric methods as no extrapolations are required. A single chromatogram can give several different types of physicochemical information. Chromatography being a separation process, initial purity of the sample is not essential in most cases and, with appropriate experimental design, several physicochemical measurements can be simultaneously made on the different components in a mixture. A wide temperature range is available for such studies, particularly in GC. Further, GC experiments are conducted in an inert atmosphere and LC can also be carried out in a closed system.

Certain limitations of chromatographic methods for physicochemical determinations should also be mentioned. The most advantageous aspect of chromatography is also the most serious drawback; thus, while several measurements can be made simultaneously, it is possible that interference from the other physicochemical properties of the materials may render the results inaccurate. A GLC system contains at least five phases, namely, gas, liquid, solid and the two interfaces; measurements may thus reflect interactions in more than one phase. Another important limitation is the stationary-phase volatility in GC and mutual miscibility of the phases in LLC in the study of solution properties; even a partial miscibility can produce a whole new system. Thus, the choice of the stationary phase is restricted and, for a given stationary phase, the operating temperature for GC and and the mobile phase for LC are also restricted. Most measurements assume Gaussian peak shapes, which are often not achieved in practice owing to either improper sample size (or sample introduction technique) or adsorption at the solid or liquid interface. Surface activities can, of course, be considerably reduced by silanization or by using a non-polar polymeric support. Errors are also caused by non-uniformity of the coating of the liquid phase on the support. While measurements can be easily made at infinite dilution by GC, lack of universal and high-sensitivity detectors is still a limitation in LC. Chromatography being a non-equilibrium process, peak-maximum retention volumes may differ significantly from hypothetical equilibrium values. Nevertheless, chromatographically determined thermodynamic quantities are generally in good agreement with those obtained by statistical methods.

A prerequisite for measurement of solution thermodynamics by GC is the knowledge of the precise amount of liquid stationary phase in the column; in practice, this quantity can be determined with an accuracy of only $\pm 0.2\%$ at best. Other sources of inaccuracies in measurements include the flow-rate, column temperature, temperature gradient, head pressure, etc. As an example, the retention volume, $V_R$, is actually given by $t_R F_c j$, where $t_R$ is the retention time, $F_c$ the mobile phase flow rate, and $j$ is the mobile-phase compressibility factor at the given column head pressure. The flow rate is usually measured at ambient temperature with a soap-bubble flow meter rather than at the column temperature; correction should also be made for the vapour pressure of the soap solution at ambient temperature. It has been shown that in order to achieve a precision of $\pm 0.01\%$ in $t_R$, one must stabilize the pressure drop to $\pm 0.006\%$ and temperature to $\pm 0.001\%$. Clearly, an unmodified, ordinary chromatograph is not suitable for such purposes. Special, computerized systems are available for precise measurement and control of all the operating parameters.

A variety of physicochemical measurements have been made by GC, HPLC and TLC. GLC and LLC are widely used in the study of solution thermodynamics and solubility properties. The activity coefficients at infinite dilution have been determined by GC for a wide range of solute–solvent systems and the values compare favourably with those obtained by traditional methods. GC has also been used to study vapour-phase interactions of the solute and mobile phase and the method is particularly suited to study the interactions in the supercritical state. Physical properties of polymers have been determined by using them as stationary phases and volatile compounds as molecular probes. Adsorption propensity of polymeric stationary phases for low-molecular-weight solvents can be readily determined by GC; use of GC for measuring the properties of liquid crystals and SEC for molecular-weight determinations have been discussed earlier. Techniques such as donor–acceptor complex chromatography, ligand-exchange chromatography and metal-chelate affinity chromatography involve formation of co-ordinate complexes of certain compounds with either the stationary phase or the mobile phase. These complexations impart a high degree of selectivity to the chromatographic system and the chromatographic data can be used to calculate the stability constants and related parameters of the complexes.

LC is particularly suited for studies on the correlation of physicochemical properties, biological activity and chromatographic behaviour. The hydrophobic (i.e. the lipophilic) character and aqueous solubility of a substance often determines its behaviour in different biological systems. Several studies relate biological and pharmaceutical activities of compounds to their distribution constants between water and an organic solvent, usually 1-octanol. It has been established during the early stages of investigations on the mechanism of retention in partition LC that there exists a linear relationship between the capacity factor, $k$ (more precisely, $\log k$) and the distribution constant, $K$ ($\log K$) in the above binary system for a variety of compound classes; octadecylsilica is the stationary phase of choice for such correlations. A major disadvantage of the use of bonded phases, however, is their instability beyond a limited pH range (1.5–8.0), making them suitable for only neutrals, and acids with a $pK_a > 3$, and bases with $pK_a < 6.5$. Several compounds which fall outside these limits, and which are of considerable importance in the pharmaceutical field, require modification of the mobile phase by using an appropriate ion-pairing agent.

The electronic properties of a substance are usually described by a single solute parameter; namely, the acidity constant, $K_a$, often expressed as the logarithm of its reciprocal value, $pK_a$. Since retention of an ionizable solute in reversed-phase LC is directly related to its $pK_a$ value, RPLC systems could be used to determine the $pK_a$ values of the solutes. Octadecylsilica and polymeric stationary phases of the type of Amberlite XAD-2 have been extensively used for this purpose. The retention of an ionic compound has been correlated with the difference between the mobile phase pH and the solute $pK_a$. The capacity factor and the mobile phase pH have a sigmoidal relationship. Excellent agreement has been demonstrated between the $pK_a$ values determined chromatographically and those obtained by conventional potentiometric and spectrophotometric methods.

Assuming that the solute–stationary-phase interactions for similar solutes is essentially solute dependent, it is possible to correlate and determine the solubility of the compound in the mobile phase. Thus, the product of the capacity factor ($k$) and the molar solubility ($S$) for a given solute using reversed-phase stationary and aqueous organic mobile phases is approximately constant for a given solvent–water combination.

It is impossible to adequately describe all the applications of chromatography to physicochemical measurements. Studies on solution and gas-phase thermodynamics, vapour-phase interactions, polymer characterization, complexation phenomena, interfacial effects, surface properties of solids and catalysts, transport properties and reaction kinetics, as well as determination of diffusion coefficients, virial coefficients, latent heat of vaporization, boiling point, vapour pressure and freezing point and transition temperature, stationary phase molecular weight, molecular structure determination, etc. by GC have been described [21–23]. Determination of hydrophobic parameters and complexation constants, etc., by LC have been reviewed [24].

### *9.2.2.2 Inorganic analysis*

There are several sensitive, non-chromatographic methods of analysis of inorganic elements; for example, polarography and atomic absorption spectroscopy. However, with the advent of HPLC and, more recently, ion chromatography, the study of the structural and analytical aspects of inorganic species by chromatographic methods has been on the increase [25–28]. At first sight, chromatography of inorganic compounds may appear to be a simple matter of separating the metal ions (cations) and the acid ions (anions) on appropriate ion exchangers. Indeed, ion exchange is the most frequently used method of separation of inorganic compounds; the stationary phases include inorganic heteropoly acids, metal oxides and polymeric and liquid ion exchangers. It must be mentioned, however, that the actual composition of the inorganic ions and, therefore, their mobilities, are not always well defined. For example, the pink ferric nitrate hydrate, on dissolution in water, yields a pale brown solution which contains several species; namely, $Fe(H_2O)_6^{3+}$ (unstable), $Fe(H_2O)_5^{2+}$, $Fe(H_2O)_4^{+}$ and $Fe(H_2O)_3$ (polymers). In HCl, the same substance gives a yellow to brown solution, which may contain $Fe(H_2O)_6^{3+}$, $FeCl^{2+}$, $FeCl_2^{+}$, $FeCl_3$ and $H^+FeCl_4^-$; in about 6N HCl, the last two species predominate.

The differences in the forms in which different metal ions exist in solution may be taken advantage of for their separation from other metals, which may occur in a different form under the same conditions. Thus, in strong HCl, while $Fe^{3+}$ is in the strongly hydrophobic anionic form noted above, $Ni^{2+}$ exists as $NiCl^+$, which means that the two can be readily separated by either ion exchange or partition. Since the separation factors can be readily changed by varying the acid concentration, high plate numbers are not very crucial in inorganic chromatography. In view of the high concentration of the acid, however, metal columns and plumbing of the kind that are normally used in HPLC cannot be used. It must also be mentioned that several metal ions (e.g. $Rh^{3+}$, $Ru^{3+}$ and $Cr^{3+}$) which form complexes do so rather slowly, and a rapid change in the acid concentration may lead to mixtures of different species of the same ion. Some of the ions, such as those of the last row of the periodic table, are too strongly retained on anion exchange resins (presumably in the form of, say, $AuCl_4^-$) and cannot be eluted by HCl at any concentration but are readily desorbed by oxygenated anions such as $ClO_4^-$ or $NO_3^-$; retention in these cases is believed to be hydrophobic in nature.

In addition to ion exchange, size separation and partition methods have also been used in inorganic chromatography. Thus, it has been demonstrated by exclusion on cation exchange resins that $Zr^{4+}$ in dilute acid solution forms a polymer of ionic weight of about 20,000; several other polymeric hydrolytic species of ruthenium, rhodium and soluble ferrocyanides have been separated on cross-linked dextran gels. The development of rigid packings based on styrene and silica allowed rapid separation of several unstable organometallic complexes such as those of aralkyl esters of phosphorus and nickel as well as aluminosilicate sols. Size-exclusion methods have also been used to study the interaction between metal ions and macromolecules such as proteins and polyphosphates.

The aqueous solubility (and relative insolubility in organic phases) of most inorganic compounds makes reversed-phase partition chromatography very attractive for their analysis. RPLC also allows the use of a variety of detectors for inorganic analysis. Thus, inductively coupled plasma detection allowed determination of Hg(I), Hg(II), Cr(III) and Cr(VI) ions at ppb levels in environmental samples. Coupling of LC with an atomic absorption spectrometer (AAS) has also been reported; unlike the stand-alone AAS, this technique allowed monitoring of different compounds and different oxidation states of the same metal. Metal complexes of PAN (1-(2-pyridylazo)-naphthol) can be detected by light absorption in the visible region; however, only the complexes of Cu, Ni and Co were separable by RPLC. Several anions (e.g. bromide, iodide, iodate and nitrate) absorb in the ultraviolet region and can be detected in the presence of a large excess of non-absorbing ions such as chloride and sulphate. The method could thus be applied to the analyses of sulphuric acid, brine and silage. Flame photometry allows specific detection of phosphorus- and sulphur-containing species in the presence of a large excess of other materials. Electrochemical or amperometric detection was used to simultaneously determine Cu(II), Ni(II), Co(II), Cr(III) and Cr(VI) as dithiocarbamates after RPLC separation. It is possible to combine the principles of ion exchange and reversed-phase partition to achieve very high efficiencies by the ion-pairing techniques. Alkylsulphonates and tetra-alkylammonium ions, containing a large number of carbon atoms, are effectively retained by the octadecylsilica columns which then act as high-performance ion exchangers. By the use of appropriate stationary ions, baseline separation of Cu(II), Co(II) and Mn(II) occurred in 200 seconds and that of $IO_3^-$, $S_2O_3^{2-}$, $NO_2^-$, $NO_3^-$ and $I^-$ in 400 seconds.

Ion chromatography, which, in principle, involves mobile-phase ion suppression followed by conductometric detection of the eluate ions, has recently become a method of choice for the analysis of anions and, less frequently, for cations [29,30]. Several hundred ions have been chromatographed using this technique. The specificity of interaction between crown ethers (large, cyclic polyethers, usually containing four to six oxygen atoms) for alkali and alkaline earth metal ions has been used for the separation of these ions; the stationary phase was silica gel, covalently bonded to a crown ether. Optical isomers of metal complexes have been separated either on chiral stationary phases or on normal ion-exchange resins using optically active eluents.

Paper chromatography and, more recently, TLC have been extensively used for the analysis of inorganic compounds. Alumina, organic and inorganic ion exchangers, cellulosic ion exchangers (sulphoethyl, carboxymethyl, ECTEOLA and phosphate esters) have all been used for inorganic TLC. Silica gel G is used most often but, in recent years, organic-bound, reversed phase silica layers have been more popular. Mobile phases for inorganic TLC range from aqueous acids, bases and buffers for ionic species to essentially organic solvents for the non-ionic entities; often, mixtures of aqueous and organic solvents are used. Detection is typically by use of chromogenic or fluorogenic agents such as potassium iodide, hydrogen sulphide, ammonium sulphide, quercetin, PAN and 8-hydroxyquinoline solutions.

For obvious reasons, gas chromatography is not generally associated with inorganic analysis. The common perception is that inorganic compounds are inherently not volatilizable, and hence, incompatible with most of the conventional GC detectors. However, there are several classes of inorganic compounds that can be readily separated by GC, either directly or indirectly by derivatization techniques. They are gases and vapours, certain metal halides, hydrides and oxides, organometallics and metalloids such as alkyl, aryl and carbonyl derivatives, and chelated metal complexes containing N, O, S, P, etc., as ligand atoms.

Historically, GC separations of inorganic compounds were based mainly on adsorption and partition processes. Chemical reactivity of the solutes required rigorous control of stationary phases and columns to prevent undesirable interactions and decomposition. Adsorption GC is widely used for the separation of permanent gases; alumina, silica and molecular sieves have been in use for a long time. Cross-linked polystyrene resins are favoured for more reactive and polar compounds such as hydrides, volatile organometallics, ammonia and water.

The choice of suitable liquid stationary phases for partition GC has been somewhat difficult owing to elution problems and non-ideal behaviour. Methyl silicones and, less frequently, phenyl, cyano and fluoroalkyl silicones found general application. Glass columns are preferred for most inorganic separations since the presence of metals in the column material is often responsible for on-column catalytic decompositions. The fused-silica capillary columns, with their extremely low residual trace metal concentrations in the silica matrix, show promise of relative inertness towards reactive inorganic volatiles. A variety of metal hydrides, halides, alkyls, alkoxides, carbonyls and chelates have been separated. Of particular interest may be the trace analysis (at nanogram and picogram levels) of lead and mercury compounds in environmental samples, using appropriate detectors. Depending on the metals and the substituents, TCD, FID, ECD and MIPE (microwave induced plasma emission) detectors have been generally used. As an element-specific detector, atomic absorption spectroscopy (AAS) has enormous potential as an on-line detector for both GC and LC [31]. Analysis of inorganic compounds by GC has been thoroughly reviewed by Uden [28,32]. HPLC of metal complexes (i.e. organometallic compounds and metal co-ordination compounds) using the principles of normal-phase, reversed-phase, ion-pair, ion-exchange and size-exclusion techniques, see Ref. 33. A voluminous handbook on inorganic chromatographic analysis has been recently published [34].

### 9.2.3 Organic chemistry

#### *9.2.3.1 Elemental analysis*

Determination of the elemental composition and, thence, the molecular formula of a compound has long been the first step in classical organic structure determination procedures and continues to be important in the characterization of several types of materials [35]. Elemental analysis is routinely performed as a confirmatory step in the characterization of the different products formed in synthetic reactions. A vast majority of organic compounds contain carbon, hydrogen, oxygen and nitrogen, in that order, and some contain sulphur, phosphorus and halogens; other elements occur much less frequently. Traditonally, organic elemental analysis (generally called microanalysis) consists of heating a few milligrams of the compound in the presence of a metal oxide and oxygen to convert carbon into carbon dioxide, hydrogen into water, and nitrogen into a mixture of oxides of nitrogen. The carbon dioxide and water formed are trapped in suitable adsorbents and weighed. The oxides of nitrogen are reduced to nitrogen and the volume measured. Two separate experiments are required, one for carbon and hydrogen and one for nitrogen; oxygen is usually calculated by difference. The procedures are tedious, time-consuming and require considerable skill on the part of the operator.

Since the products of oxidation (that is, the oxides of carbon, hydrogen, nitrogen and sulphur) and reduction (methane, ammonia, phosphine, hydrogen sulphide, etc.) are all volatile, their separation and determination by GC appeared attractive and considerable effort has been expended to achieve satisfactory results. A large number of commercial apparatus, which use these methods are available. Briefly, the principle of chromatographic elemental analysis involves mineralization (combusting) the sample ($<1$ mg) in the presence of a metal (Cu, Ag, Mn, Mg, V, W or Co) oxide to form carbon dioxide, water and nitrogen oxides. The nitrogen oxides are selectively reduced to nitrogen on heated copper and the three products; carbon dioxide, water and nitrogen, are separated by GSC on molecular sieves. Helium is generally used as the carrier-gas and a thermal conductivity meter as the detector; the use of the TCD is preferred in view of its concentration-dependent response to all the products of the combustion analysis. Usually the same equipment can be used, after slight modification, for the determination of oxygen and sulphur in the form of carbon monoxide and sulphur dioxide. The advantages of these methods are that they are very sensitive and rapid, and the analysis can be carried out unattended; about 50 analyses per day can be performed using less than 1 mg of each sample. The rapid, microscale manipulation makes it possible to couple a gas chromatograph to an elemental analyser, so that the GC effluents can be analysed on-line [36].

#### *9.2.3.2 Synthetic organic chemistry*

Synthetic organic chemistry pervades several areas of activity in the chemical laboratory and industry. Synthetic organic chemicals in every day use include drugs, pesticides, dyes, food additives, plastics, synthetic rubber products, etc. Laboratory activity in synthetic organic chemistry involves preparation of these compounds by different methods and analysis of the products, intermediates and by-products. Synthesis of these products is often a multistep process and an analytical monitoring of the course of the process at each step is normally necessary to avoid complications at later stages. Analysis of the products is essential to claim or confirm conformity to the specifications required for a particular application; e.g. certain impurities in drug preparations could be fatal. Similarly, analysis and identification of the by-products often leads to their better use or, preferably, readjustment of the reaction conditions to prevent formation of the by-products and the undesirable side reactions. More than one compound is frequently formed in a reaction, and the formation of the desired product in a larger proportion depends on the experimental conditions. Thus, a number of different conditions need to be tried and the products analysed in each case before the optimum conditions are arrived at. Being a very convenient analytical tool, chromatography plays an important role in all these activities as also in the determination of the reaction mechanisms.

The study of the reaction mechanisms is important not only from the theoretical point of view but also to help optimization of the reaction conditions to promote the desired course of the reaction and to improve the yield of the required products. Among other things, such studies involve separation and identification of all the products formed, kinetics of disappearance of the reactants or of formation of the products, trapping and identification of the intermediates, etc. As a rapid and sensitive technique, chromatography finds increasing use in this field.

A majority of the organic compounds handled in an organic chemical laboratory belong to one or other of the various classes of compounds which will be discussed in the following sections and the same systems may be used for the analysis of the synthetic products. Chromatography of synthetic products, however, does not require very sophisticated systems and methodologies since the nature of the compounds is normally well defined. Being composed of mostly the reactants and products, the materials for analysis are generally very simple mixtures unlike in natural products chemistry or biochemistry. However, whereas the plant and animal products are generally produced by nature in optically pure form, synthetic compounds are normally obtained in the racemic form unless the steps involve enantiospecific reactions. The chiral separation methods (see below) are thus most useful in synthetic organic chemistry.

Depending on the volatility of the compounds, GC or TLC is frequently used for monitoring the progress of a reaction; the conventional column liquid chromatography is routinely used for the preparative separation and purification of the products formed in the reaction; in a multistep sequence, it may often be necessary to purify the products at each step.

#### *9.2.3.3 Chiral separations*

The term chiral separations refers to the technique of separation of racemic mixtures of optical isomers into their individual enantiomers. The enantiomerism (or optical isomerism) stems from the non-superimposability (chirality) of the mirror images of the structures of compounds, which contain an element of asymmetry. Since the chemical structures of these enantiomers are otherwise identical, their physical and chemical properties are also identical except for the fact that the two enantiomers rotate the plane of polarized light in opposite directions (optical activity). Thus, separation of enantiomers has always been a formidable problem. However, such separations are very important in synthetic organic chemistry (particularly in the total synthesis of biologically active compounds), study of the mechanisms and kinetics of reactions, catalysis (enzymatic or chemical), geology (or geochronology), biochemistry, pharmacology and medicine.

The most common approach to the separation of enantiomers has been to react the racemate with an optically active compound so that the derivative will be a mixture of diastereoisomers, which can then be separated by one of the conventional methods such as fractional crystallization. These methods are often tedious and time-consuming and seldom clear-cut; in other words, total separation is generally not achieved. Also, the separated derivatives are required to be decomposed to the original compounds without racemisation. An alternative approach is to convert one of the isomers of the racemate into another chemical entity by an enzymatic reaction; in which case the isomer is lost unless it can be easily reconverted back to the original compound after separation. The principle of biological recognition of asymmetry has been extensively used in affinity chromatography wherein an immobilized enzyme or receptor could specifically retain (or react with) one of the optical isomers.

The chromatographic approach to enantiomer separation naturally involves selective and reversible interaction of the isomers with chiral stationary or mobile phases or both. Chiral stationary phases available for use in GC (section 4.2.5.5) and LC (section 6.5.2.3.2) have been discussed earlier. It can be easily recognized that, for effective enantioselection, the molecular interaction between the solute and the stationary (or mobile) phase should occur at least at three points in the vicinity of the asymmetric centre. These interactions may be of either the hydrogen-bond or dipolar type. Even when this condition is fulfilled, the separation factor of enantiomers has seldom been much larger than unity, so that very high-efficiency columns are often required.

Open-tubular GC, because of the high plate numbers that can be readily achieved, has been extensively used in much of the early work on chiral separations [37]. However, the requirements of temperature stability and non-volatility severely restrict the choice of the chiral stationary phases that can be used. In view of the ready availability of enantiomerically pure compounds, the use of several derivatives of L-amino acids and peptides as stationary phases has been extensively reported in the early literature. More recently, cross-linked or polymeric siloxane phases carrying a covalently bonded amino acid derivative have become commercially available (e.g. Chirasil-Val, which is *N*-acyl-L-valine *tert*-butylamide bonded to the polysiloxane skeleton). The relatively high column temperature that needs to be normally used in GC also makes the intermolecular interactions between the solute and the stationary phase comparatively weak and the separation of the enantiomers minimal. High column temperatures may also cause racemisation of both the stationary phase and the solute. Another limitation, which is common to all GC experiments, is the requirement of sample volatility. Also large-scale (preparative) separations are almost impossible using the capillary GC and the separation, even when achievable, is largely useful only for analytical work.

Liquid chromatography, on the other hand, offers several advantages over GC for the separation of enantiomers. Apart from the fact that most of the above noted drawbacks are invalid, the possibility of using chiral additives to the mobile phase in LC is a tremendous advantage. Amino acid derivatives, in view of their importance, availability and ease of monitoring, have been natural candidates for investigation in this field. Liquid chromatographic separation of enantiomeric amino acids by use of chiral derivatization reagents (e.g. *O*-acetylated sugar isothiocyanates), chiral stationary phases and chiral mobile phases (containing, for example, *N*-tosylamino acid–copper complex) have been attempted. Addition of copper ion to the mobile phase permits the formation of a ternary complex between the chiral amino acid attached to the stationary phase and those in the solute (ligand-exchange chromatography). Several variations are possible in the nature of the solid matrix, the bound amino acid and the metal ion, and this technique remains one of the most popular for the separation of underivatized amino acids.

A large number of chiral stationary phases have been synthesized and several of them are commercially available [38]. Amino acid derivatives, again, are the natural choice for immobilization and used as centres of chirality in the stationary phase. When an *N*-acyl amino acid such as (*R*)-*N*-(3,5-dinitrobenzoyl)phenylglycine is linked to aminopropylsilica, enantioselectivities (separation factors) of the order of 100 could be realized [39]. Although these stationary phases can be used with polar mobile phases also, greater enantioselectivity is attainable with non-polar mobile phases such as hexane–isopropanol.

A large number of aryl sulphoxides, alkyl and acyl amines, catecholamines, amino acids and their derivatives, alcohols, thiols, hydroxylacids, lactams, epoxides, cyclopropanes, hydantoins, acyloins, allenes, helicenes, ferrocenes, phosphonates, organometallics, binaphthols, terpenoids, steroids, alkaloids, pyrethroids, drugs and herbicides have been separated using the chiral stationary phases [40]. Covalently bound chiral, binaphthyl crown ethers have been used to resolve asymmetric compounds containing a primary amino group. Cyclodextrins have been similarly immobilized and used for the separation of the racemates of several drugs, alkaloids, steroids, metallocenes and even polycyclic aromatic hydrocarbons. Natural chiral polymers based on proteins and carbohydrates and certain synthetic polymers derived from chiral monomers (e.g. polytriphenylmethyl methacrylate) have also been useful for the separation of several amino acid derivatives, aromatic sulphoxides, insecticides, barbiturates, drugs, organophosphorus compounds, etc. The most popular among these stationary phases is based on bovine serum albumin, bound to microparticulate silica; retention and selectivity is controlled by varying the pH, the ionic strength of the aqueous mobile phase, and by the addition of small amounts of organic modifiers. Certain other chiral stationary phases are based on hydrogen-bonding (e.g. tartaric acid), charge-transfer complexes and insoluble transition metal complexes.

Enantioselectivity can also be obtained on the conventional, inexpensive, achiral stationary phases by using appropriate chiral mobile phases or mobile-phase additives. The main disadvantage of this approach is that the additives in the mobile phase often interfere with the detection unless removed in a post-column operation. Removal of the additive from the eluant is also necessary to obtain the pure substances in preparative applications. However, this approach, particularly in the ligand-exchange mode, has been attractive in view of the fact that the readily available octadecylsilica columns can be used [29]. A chiral bidentate ligand, consisting of a hydrophobic moiety is added to the mobile phase along with a transition metal ion such as Cu(II), Zn(II) or Cd(II). The ligand associates itself with the reversed-phase medium because of its hydrophobic nature. Amino acids form ternary complexes with the ligand and the metal ion and are retained selectively by the stationary phase. Retention of ionic species by the reversed phase is also possible as neutral diastereoisomeric pairs if an appropriately charged chiral couter-ion is added to the mobile phase; e.g. quinine, (+)-10-camphorsulphonic acid and tartaric acid derivatives. Cyclodextrins have also been used as mobile phase additives, in which case part of the cyclodextrin is also adsorbed on the stationary phase; mandelic acid, dansylamino acid and piperazine derivatives have been resolved using this technique.

Optimization of chiral separation is somewhat more complicated than the conventional LC because chiral recognition and resolution requires specific and simultaneous interactions and demands stringent control of the mobile-phase composition, flow-rate, temperature, etc. Some of the competing retention mechanisms may not be enantioselective; therefore, a greater understanding of the molecular interactions is necessary to eliminate the non-enantioselective forces. These problems notwithstanding, chiral separations are very popular as evident from (1) the very large number of reports of chiral separations appearing lately, and (2) the wide variety of commercial stationary phases now available. As the techniques become more common, the demand for analogous preparative systems is likely to grow. Indeed large-scale chiral separations are very valuable since many of the synthetic drugs are readily obtainable in the racemic form but the physiologically active material is often only one of the enantiomers.

#### *9.2.3.4 Natural products*

While several classes of substances, from elements to enzymes, can be considered as natural products, the organic chemist's perception of the term signifies solvent-extractable, small to medium-sized organic substances present in the plant and animal kingdom. Historically, the technique of chromatography was developed in connection with the separation of natural products; for example, adsorption chromatography of plant pigments by Tswett, and partition chromatography of amino acid derivatives by Martin and Synge. Following the early developments, the growth of the subject, particularly in relation to the separation of natural products, has been phenomenal. A very large number of compounds could be isolated from plant and animal tissues from which only a few were earlier isolated using the classical methods. The contribution of this development to the understanding of chemistry and biology cannot be overemphasized. While there are several spectroscopic techniques (for example, UV, IR, NMR and MS) in the organic chemist's arsenal for the structure determination of the organic compounds, their utility is limited without prior separation of the natural product into the individual compounds. Application of chromatographic methods to natural product chemistry eventually led to the opening up of new fields of research of incalculable economic and social value. Thus, it may be said with confidence that but for the chromatographic methods, it would not have been possible to isolate and characterize such physiologically active and economically useful classes of compounds as prostaglandins and pheromones, which occur in very minute quantities and often as complex mixtures of closely related compounds.

The natural products may be classified either on the basis of their chemical functionality (for example, alcohols, carbonyl compounds, carboxylic acids, etc.) or on the basis of their co-occurrence and utility (lipids, pigments, antibiotics, etc.). Neither classification is entirely satisfactory since each group in a particular class (say, terpenoids) encompasses several groups of the other. Apparently for this reason, the *Journal of Chromatography* uses both the classifications in their bibliography section, which is a veritable source of information on the application of chromatographic techniques to the separation of different classes of compounds. Thus, after four sub-sections of general topics, each section (LC, GC and PC) of the bibliography is sub-divided into: 5) hydrocarbons and halogen derivatives, 6) alcohols, 7) phenols, 8) substances containing heterocyclic oxygen, 9) oxo compounds, ethers and epoxides, 10) carbohydrates, 11) organic acids and lipids, 12) organic peroxides, 13) steroids, 14) steroid glycosides and saponins, 15) terpenes and other volatile aromatic compounds, 16) nitro and nitroso compounds, 17) amines, amides and related nitrogen compounds, 18) amino acids and peptides, 19) proteins, 20) enzymes, 21) purines, pyrimidines, nucleic acids and their constituents, 22) alkaloids, 23) other substances containing nitrogen, 24) organic sulphur compounds, 25) organic phosphorus compounds, 26) organometallic and related compounds, 27) vitamins and various growth regulators, 28) antibiotics, 29) insecticides, pesticides and other agrochemicals, 30) synthetic and natural dyes, 31) plastics and their intermediates, 32) pharmaceuticals and biomedical applications, 33) inorganic compounds, 34) radioactive and other isotope products, 35) some technical products and complex mixtures, 36) cells and cellular particles, and 37) environmental analysis.

For convenience and because of the accent on chromatographic applications, the classification based on utility of the compounds is used in the present discussion. The major problem in covering this area lies in the fact that several chromatographic methods are used for each class of compounds; for example, metal ions and enzymes have been analysed by GC, and volatile flavour chemicals by LC. Fortunately, a few methods may be identified as preferable for each class of compounds. Again, under each method for a given class of compounds, there can be any number of stationary phases, mobile phases and methods of operation and detection. There are, however, a number of recent reviews and books which treat the subject in detail. Worthy of special mention is the recent manual edited by Heftmann [41]; the work contains articles that nearly comprehensively cover the large number of chromatographic methods used for different classes of compounds.

For reasons of brevity, the treatment of the subject could not be elaborate enough; however, an attempt has been made to describe the various ways in which chromatographic methods could be used to analyse the different classes of compounds present in nature. The compound classes have been arranged approximately in the order of their structural complexity that would require increasingly complex chromatographic systems for their analysis. No distinction is sought to be made among the compounds that could be essentially of biological, industrial or academic interest.

#### *9.2.3.4.1 Hydrocarbons and halogen derivatives*

Hydrocarbons and their halogen derivatives have very similar chromatographic properties. Separation of hydrocarbons has an important place in the development of the technique of chromatography, particularly gas chromatography. In addition to the commercial importance of the petroleum products, their complexity on the one hand and their chemical inertness on the other required a highly efficient physical method for their separation. It may also be mentioned here that the relative inertness of the hydrocarbons made them less demanding with respect to the instrument design and materials. Thus, it should be interesting to note that while on-column injection was nearly universal in the early days of GC, the independently heated flash-vaporizing injection ports became quite popular during the 1960s, in spite of their obvious disadvantages. This was probably because many of the users were from the petroleum industry and the heated injection ports did not cause decomposition of the hydrocarbons.

For the purpose of convenience, the hydrocarbons may be broadly divided on the basis of their boiling ranges which, in turn, are related to their molecular weights and polarity (degree of unsaturation, aromaticity, etc.). Except for the very high-boiling derivatives, GC has long been the method of choice for the analysis of hydrocarbons and the halogenated compounds. Though gas–solid chromatography was widely used in the beginning for the analysis of low-boiling hydrocarbons, over the years gas–liquid chromatography has overtaken it mainly because of the wide range of stationary phases that became available, and its reproducibility. With the increasing use of capillary columns with very large plate numbers and the consequent high separating power, the number of stationary phases required is considerably reduced. Porous polymers (e.g. Porapak Q or Chromosorb century series), which are chemically the co-polymers of styrene and divinylbenzene, are very useful for the separation of very low-boiling hydrocarbons (containing, for example, one to eight carbon atoms) and permanent gases such as oxygen, nitrogen and carbon dioxide. Certain stationary phases selectively remove specific classes of the mixtures so that the remaining, less complex mixture can be more easily analysed. For example, *n*-paraffins are removed by 5Å molecular sieves, which are essentially zeolites of specific pore size; on the other hand, olefins and aromatic hydrocarbons are completely and irreversibly removed by chemical reaction with mercuric perchlorate.

The flame-ionization detector is very sensitive for hydrocarbons and is almost invariably used. However, it does not respond to water, permanent gases and other compounds that do not contain carbon. The thermal conductivity detector, in this respect, is more universal, albeit at lower sensitivity. Very small quantities of sulphur and nitrogen compounds normally accompany hydrocarbons from petroleum. Certain selective detectors can be used for their detection in trace proportions relative to the hydrocarbons. Thus, the flame photometric detector is selective for sulphur and phosphorus compounds while the thermo-ionic detector is used for the nitrogen compounds. The electron-capture detector is extremely sensitive for polyhalogenated and polynuclear aromatic hydrocarbons. GC–MS is, of course, the most powerful and nearly universal technique and is particularly useful for the identification of the compounds in very complex mixtures such as the petroleum fractions.

In the analysis of natural gas, the complication arises mainly from the presence of permanent gases in addition to the hydrocarbon gases of one to five carbon atoms. While all the compounds are separable on Porapak R, two detectors may be required (that is, TCD for permanent gases and hydrocarbons of one to three carbon atoms, and FID for all the hydrocarbons); the response of the two detectors to ethane or propane may be used to superimpose the chromatograms to obtain the total picture.

Gasoline is a mixture of hydrocarbons with boiling points ranging from −42°C (propane) to 216°C (*n*-dodecane) and may typically contain paraffins, olefins and aromatics. Naphtha and olefin feed-stocks for petrochemical manufacture and several industrial solvents (e.g. benzene, toluene, xylenes) also fall in the above boiling range. Long capillary columns of 50 m or longer are typically used for the separation of the very complex mixtures of this group. Squalane, in spite of its serious limitation of very low usable maximum temperature (100°C), has been the most popular stationary phase due mainly to its superior separating power for low-boiling hydrocarbons and to the enormous retention data on record [42], facilitating identification of the peaks. The main disadvantage of this particular stationary phase (with its maximum allowable temperature of 100°C) is that compounds continue to be eluted even after 60 min and some higher boiling aromatic hydrocarbons take about 100 min for elution. Alternatively, multiple columns with different stationary phases (such as molecular sieves, porous polymers, non-polar and polar stationary phases with selectivity for different types of hydrocarbons) and appropriate switching valves may be used. Routine analysis and data manipulation of complex hydrocarbon mixtures on such systems require specialized, computer-controlled instruments which are commercially available.

Kerosine and gas oils may have a boiling range from 100 to about 380°C and may contain hydrocarbons from seven to 24 carbon atoms. Some of these (e.g. pristane and phytane) are isoprenoid hydrocarbons, whose relative amounts vary considerably in oils of different origins; this information can thus be used to identify oil spills. GC of these materials is best carried out using capillary columns coated with a non-polar liquid phase such as dimethylpolysiloxane (e.g. OV-1 or SP-2100) and temperature programme starting at 0 and on to 275°C. While all the components of the fractions may not be resolved under these conditions, it is possible by use of a computer facility to determine the carbon-number distribution or the approximate *n*-paraffin content if required for a given application.

Analysis of polynuclear aromatic hydrocarbons (PAHs) and polychlorinated biphenyls (PCBs) has attracted considerable attention in recent years in view of their potent carcinogenic properties and their wider environmental implications. The PAHs may be detected in the exhaust gases of automobiles, kerosine, lubricating oils, cigarette smoke, marine organisms and sediments. The PCBs are present as residues when polychlorinated pesticides are used. Since these compounds often occur as complex mixtures and in very small quantities, great care needs to be taken in their pre-concentration and chromatography. The raw material is usually distributed between aqueous organic systems such as water–dimethylformamide (or acetonitrile, nitromethane, dimethylsuphoxide, etc.) and cyclohexane; the cyclohexane layer is then subjected to silica gel chromatography to separate the hydrocarbons from the accompanying materials. The hydrocarbon portion is further fractionated by chromatography on lipophilic Sephadex (LH-20) using dimethylformamide–hexane to separate the PAHs. Analysis of the PAH fraction may be carried out on capillary GC using a non-polar stationary phase (e.g. dimethylpolysiloxanes) or reversed-phase liquid chromatography. RPLC of PAHs have received much attention lately in view of the convenience, and easy and sensitive detectability by UV and fluorescence techniques, and the high resolution and selectivity of the octadecylsilyl silica packings [43]. The column selectivity of the RP packings depends on the physical and chemical properties of the silica substrate, the conditions of preparation of the packings and surface coverage with the octadecylsilyl units. Polymeric phases have better surface coverage and retention capacity than the monomeric phases. Best separations of isomeric PAHs are obtainable on heavily loaded polymeric phases, bonded to wide-pore silica substrates; about 80% aqueous acetonitrile or acetonitrile itself may be used as the mobile phase.

LC permits better control of separation of even hydrocarbons since selectivity can be improved by appropriate choice of the stationary and mobile phases. However, except for the PAHs discussed above, their detectability by conventional detectors is very poor. The sensitivity of the refractive index detector can be improved by using a fluorocarbon solvent (e.g. Fluorinert FC-78 from the 3M company) as the mobile phase. With a refractive index as low as 1.25, these solvents allow detection of even lower saturated hydrocarbons. TLC, whose resolution cannot compete with that of GC or HPLC, has not been very useful for the analysis of closely related hydrocarbons.

#### *9.2.3.4.2 Lipids*

Lipids are a major class of natural products; they are usually understood to comprise long-chain fatty acids and their derivatives but, in a broader definition, may include several other non-polar compounds; for example, steroids. Chemically, they may be classified as free fatty acids, glycerides (mono-, di- and triacyl derivatives of glycerol with fatty acids), phospholipids (glycerides or fatty acids containing phosphate moiety) and glycolipids (derivatives containing carbohydrates, mostly glucose and galactose). Prostaglandins are substituted long-chain (20 carbon atom) fatty acids and are very important, physiologically active compounds.

Lipids are usually extracted from animal or plant tissue by a chloroform:methanol (2:1) mixture. Prostaglandins and other oxygenated fatty acids may be extracted with ether after acidification with dilute formic acid or, preferably citric acid. Preliminary separation of the lipid fraction from the extract by chromatography on a dextran gel is often employed. The non-lipid substances can also be avoided by pre-extraction of the tissue with aqueous acetic acid. Fractionation of the lipid portion into different classes is also possible by silica gel chromatography. Neutral lipids are eluted by hexane or chloroform and glycolipids by acetone or acetone–methanol. Further separation of neutral lipids into individual classes is conveniently achieved by chromatography on Florisil or silicic acid. A linear gradient of ether in hexane successively elutes hydrocarbons (0% ether), cholesteryl esters and waxes (1% ether), triacylglycerols (5%), free fatty acids (8%), cholesterol and diacylglycerols (15%), and monoacylglycerols (100% ether). Prostaglandins can be separated on the basis of their oxygen content by chromatography on acid-washed silicic acid using increasing concentration of ethyl acetate in hexane, cyclohexane or benzene; more polar compounds are eluted by methanol–ethyl acetate. Phospholipids are fractionated on a silicic acid column using chloroform:methanol (4:1) and increasing amounts of water (1–2%). Isolation of glycolipids requires more polar mobile phase combinations of chloroform, methanol and water. Ion-exchange chromatography on DEAE-cellulose is an alternative to adsorption chromatography for the separation of complex lipids. Chloroform elutes most of the simple lipids, while chloroform:-methanol (9:1) elutes choline phospholipids, and chloroform:methanol (1:1) elutes ethanolamine phosphatides and certain glycolipids.

Among the various classes of lipids, chromatography of fatty acids has attracted considerable attention, they being the most common constituents and their identification being a prerequisite for the identification of the natural lipids. GC using polar liquid phases, TLC on silver-ion-impregnated layers, and HPLC on reversed-phase columns, appear to be the most effective methods for fatty acid analysis. Esterification of short-chain fatty acids greatly improves their chromatographic properties without affecting the order of elution. GC of fatty acid methyl esters and trimethylsilyl esters both on polar (DEGS or, preferably, Silar-10C) and non-polar (Apiezon L or SE–30) continues to be popular in fatty acid analysis. The structure-retention relationships and protocols for systematic separation and identification of long-chain fatty acids have been reviewed [44]. In addition to FID, coupling of a capillary GC with mass spectrometry proved very valuable in qualitative and quantitative analysis of unknown samples. TLC on silica gel using a non-polar mobile phase (for example, hexane and ether) and HPLC on reversed phase using aqueous ethanol or acetonitrile as mobile phase are frequently used for the separation of fatty acids and their derivatives. In both cases resolution is improved by modifying either the stationary phase (TLC) or the mobile phase (HPLC) with borate (for 1,2- or 1,3-dihydroxy compounds) and silver ions (for *cis* and *trans* double-bonded isomers). Charring by sulphuric acid followed by densitometry is widely used for quantitation by TLC; more recently, TLC on sintered quartz rods, which can be passed through a hydrogen FID for quantitative detection, proved very convenient. Refractive index is almost universally used for detection of lipids in HPLC, because of the inadequate light absorption by most lipids in the UV region. The detection sensitivity can be improved by preparation of UV-absorbing derivatives such as phenacyl, nitrobenzyl and 2-naphthacyl esters. Coupling of LC with MS or with a mass detector (based on light scattering) are viable alternatives. For more detailed information on the chromatography of lipids, including prostaglandins, see Refs 41, 45 and 46.

#### *9.2.3.4.3 Terpenoids and steroids*

These compounds form a large group of very important natural products, comprising flavours and perfumery chemicals, plant growth regulators (gibberellins), hormones and vitamins (A, D, E and K group). Structurally, they are formed by head-to-tail linking of isoprene units and are called mono-, sesqui-, di-, sester- and triterpenoids when the molecules contain 2, 3, 4, 5 and 6 isoprene units; those containing 8 units (strictly, dimeric diterpenes) form tetraterpenes, the best known group among them being the orange-yellow carotenoids. Steroids are related to the terpenes to the extent that they are also biogenetically derived from the isoprenoid precursors but are structurally distinguished by their perhydrocyclopentanophenanthrene skeleton. Within the broad class of isoprenoids, the skeletal and functional variations are almost unlimited, ranging from hydrocarbons and epoxides to alcohols, carbonyl compounds, carboxylic acids and esters.

In view of their volatility, mono-terpenoids are most conveniently analysed by GC. SE-30 and Carbowax 20M are preferred stationary phases; Tenax GC and graphitized carbon black are also frequently used for the purpose. Essential oils of most aromatic plants and flavour concentrates of several foods and beverages are often complex mixtures requiring high plate numbers for their separation. For this reason, capillary gas chromatography is almost always employed for their analysis. Relative retention times and Kovats indices are often used for the identification of known mono-terpenoids. The combination of GC, MS and data systems provides a very powerful tool for the separation and identification of the individual constituents of these complex materials. Several compendia provide extensive retention and mass spectral data on the aroma chemicals; e.g. Ref. 47. The mono-terpenoids are often conveniently isolated from the matrix by steam distillation, their well-known thermal instability notwithstanding; solvent extraction is a preferred alternative. Degradation (e.g. isomerization, dehydration and polymerization) may also occur during introduction into the heated injection port or during the GC analysis itself. These transformations may be minimized by using direct, on-column injection, minimum possible column temperature, a glass column, silanized support, etc.

Silica gel is generally used for TLC of terpenoids, with hexane–ethyl acetate (0–10%) as the mobile phase. To prevent acid-catalysed degradation of the terpenoids, the plates may be prepared in dilute alkali instead of water. Impregnation with silver nitrate is useful for the resolution of isomeric unsaturated compounds. In HPLC, the silver salt is usually replaced by thallium nitrate in view of the latter's lower solubility, permitting use of more polar solvents. Because of the convenience, high resolution and sensitivity of GC, HPLC of monoterpenoids is very infrequently used and often restricted to those derivatives which contain a UV-absorbing chromophore. Normal, reversed-phase and size-exclusion modes may be employed for the HPLC.

Sesquiterpenoids have chromatographic properties very similar to those of monoterpenoids and, often, the same systems are useful. Di- and higher terpenoids would need derivatization to render them volatile enough for GC. LC on both normal, reversed-phase and size-exclusion modes is common [41]. Worthy of special mention among diterpenoids are the gibberellins, which are of economic importance as plant growth regulators. A large number of them (over 50 of which are known) co-occur and their separation and analysis can be challenging. TLC and HPLC on silver-ion impregnated silica and reversed phases using *p*-nitrobenzyl and *p*-bromophenacyl esters have been used for their analysis [48]. Carotenoids, being polyunsaturated compounds, are unstable to heat and cannot be analysed by GC except after hydrogenation. In addition to LC on normal and reversed phases, chromatography of carotenoids [41,49] and steroids [41,50,51] on DEAE-cellulose and lipophilic Sephadex are known.

Chromatography of steroids is of immense value in the study of hormonal imbalance and metabolism. HPLC on silica with isocratic, gradient and recycling techniques have great resolving power for sterols, epoxides and carbonyl compounds, acetates, etc. Dichloromethane, hexane and ethyl acetate (e.g. in the ratio 94:5:1) is a commonly used mobile phase. Non-aqueous reversed phase LC using octadecylsilica and hexane containing 0.5% isopropanol or hexane–methanol–acetone (18:1:1) is also popular; alternatively, 35% aqueous acetonitrile may be used. GC of steroids and their TMS derivatives have been carried out on a number of stationary phases and their retention data tabulated [41]. High-resolution GC using capillary columns and mass fragmentography with a coupled mass spectrometer yields highly specific data for positive identification and determination of steroid derivatives in biological fluids.

Like lipids, terpenoids and steroids are also extractable from the tissues by dichloramethane or ether. Columns of Amberlite XAD-2 or octadecylsilyl silica effectively retain these compounds which can later be eluted with methanol with very good recoveries. Sephadex LH-20 has been extensively used for fractionating steroids; dichloromethane–methanol or cyclohexane–ethanol have been the solvent systems of choice. The solvent systems for RPLC of steroids generally consist of water admixed with methanol, acetonitrile or dioxan.

*9.2.3.4.4 Pigments*

Natural colouring matters fall into several structurally distinct groups. Chlorophyll, the ubiquitous green pigment of plants, haem, a constituent of the red pigment of blood, and bilirubin, the open-chain bile pigment, belong to a group called porphyrins, made up of four pyrrole units. They have important biological functions (e.g. oxygen transport) and any change in their production, function and metabolism can impair the normal health. Sharing the term chloroplast pigments with the chlorophylls are the carotenoids, which are conjugated polyenes derived from eight isoprene units. Carotenoids are orange-yellow pigments occurring in higher plants, fungi, photosynthetic bacteria and algae. Another widely distributed class of natural pigments is called the flavonoids. They are polyphenols containing two aromatic rings joined by a three-carbon chain. Depending on the oxidation state of the carbon chain, several closely related structural subclasses are possible; e.g. flavones, flavonols, anthocyanidins, chalcones, aurones, catechins, etc. In addition to these, there are simple phenolic compounds, phenolic acids, coumarins, xanthones, etc., which are biogenetically related to the flavonoids and often co-occur with them. The best known among the phenolic pigments are the anthocyanins (glycosides of anthocyanidins), which impart bright colours to flowers and the red colour to wine; they have attracted renewed attention in recent years for their potential use as natural food colourants. A common feature of all the above classes of pigments is their non-volatility, making GC not routinely applicable.

Dolphin [41], Petryka [52] and Roy [53] have recently reviewed the various systems used for the chromatography of porphyrins. Structural variations among the porphyrins involve both the peripheral groups and the central metal ions; these variations impart different chemical and chromatographic properties which, in turn, determine the systems to be used. For this purpose, the porphyrins may be broadly divided into two groups; hydrophilic and hydrophobic. The porphyrins are generally unstable to light, and oxygen; and the metal-free porphyrins have a great affinity to metal ions. Chromatography should therefore be performed in the absence of light and with highly purified solvents, free of peroxides and metal ions. Porphyrins are highly coloured with strong absorption in the whole range of the UV and visible spectrum and this feature makes their detection in LC and on TLC plates extremely easy even at submicrogram levels. Even greater sensitivity (to picogram levels) can be achieved, using the fluorescence detector, in the case of most metal-free porphyrins, bile pigments and chlorophylls which fluoresce. A large number of systems are available for column and planar chromatography of these compounds on cellulose, polyamide, silica and reversed phases. Historically, these pigments (along with the carotenoids) were the first type of compounds to be chromatographed and the use of several other adsorbents, such as talc, alumina, Celite and oxides, hydroxides or carbonates of calcium, magnesium and zinc, is on record.

Hydrophilic porphyrins, like those containing carboxylic acid groups, are best converted into their non-polar derivatives (esters) for efficient chromatography. The acids themselves have been chromatographed extensively on paper or silica plates using polar mobile phases, usually containing water, an alcohol and a nitrogen-containing base (e.g. 2,6-lutidine or ammonia). HPLC on anion exchange resins using gradients of methanol–acetic acid and reversed-phase chromatography using aqueous alcohols containing a phase-transfer agent (e.g. tetrabutylammonium dihydrogen phosphate) is also convenient with carboxylic acid derivatives of porphyrins. Metallated porphyrins readily exchange the ligands with the metal ions which are generally present in the aqueous mobile phases. While this property may not be of much consequence in analytical chromatography, preparative separations need to be followed by 're-ligandation' with the appropriate metal ion. Silica, cellulose and polyamide have been used for the planar chromatography of metallated porphyrins; the mobile phase generally is a combination of a hydrocarbon (benzene or hexane), an alcohol (methanol or ethanol) and an acid (acetic or formic). Hydrophobic porphyrins, e.g. methyl esters of porphyrin carboxylic acids, are readily separated on silica gel using a variety of solvents. Acetonitrile–water mixtures may be used for RPLC of these derivatives. Both silica gel and alumina have been extensively used for preparative separation of hydrophobic porphyrins. Powdered sucrose is an excellent adsorbent for the chromatography of chlorophylls, the mobile phase being 0.5–2% propanol in light petroleum; about 3% starch may be added to the adsorbent to prevent caking.

Carotenoids are readily separable on silica-gel columns using hexane and propanol. RPLC on octadecylsilyl columns using 95% aqueous methanol or acetonitrile is also effective. Sephadex LH-20 is particularly effective in the separation of carotenoids from steroids. A mixture of silica gel and Hyflo Super-Cel has been recommended for rapid analysis of carotenoids. Hyflo Super-Cel, admixed with magnesium oxide has been used for TLC of these pigments using light petroleum ether as the mobile phase. Octadecylsilyl sililca layers could separate several carotenoids when methanol–acetone–water (20:4:3) was used as the mobile phase. GC is of limited value in the analysis of carotenoids in view of their thermal instability. However, if *cis/trans* isomerism is ignored, they can be conveniently analysed by GC following hydrogenation. Carotenoids containing hydroxyl groups would require derivatization using trimethylsilylation or acetylation, and those containing carboxylic acid groups need to be esterified (with diazomethane). SE-52 and OV-17 are the stationary phases from which the retention data for a number of carotenoid derivatives have been recorded both under isothermal (275°C) and programmed conditions [41,49].

Paper chromatography has long been used for the chromatography of the flavonoids and still retains its importance probably because of the vast amount of retention data on record, which helps in their easy identification [54]. The mobile phases most often used are *n*-butanol–acetic-acid–water (6:1:2, or the upper layer of the mixture in the ratio of 4:1:5), acetic-acid–conc. HCl–water (30:3:10, the Forestal solvent) and phenol, saturated with water. Flavonoids answer a variety of colour reactions which enable easy detection on paper or TLC; exposure to ammonia, ferric chloride, ammoniacal silver nitrate, sodium borohydride-ethanolic HCl are some examples. In addition, several of these compounds are visible to the naked eye or strongly absorb or fluoresce under the UV light. The structural variation in the naturally occurring flavonoids involve the number and position of the hydroxyl, methyl, methoxyl, isoprenyl groups, and of different sugar groups. The well-defined structural variations and, correspondingly, the chromatographic behaviour is useful for discerning the structure–retention relationships. In fact, the relationships are so regular that they have been used for structure determination of unknown compounds. The easy identifiability of flavonoids by paper chromatography makes them convenient markers for classification of plants (chemotaxonomy).

As in the other fields, TLC and HPLC are also being increasingly used in the analysis of polyphenols. Cellulose, silica gel and polyamide are the most preferred stationary phases for TLC. A large variety of mobile phases have been in use for different types of flavonoids. Highly polar compounds (e.g. anthocyanins and other glycosides) are more conveniently analysed on reversed-phase LC using aqueous methanol or acetonitrile, occasionally modified with acetic or phosphoric acid. Droplet countercurrent chromatography (using chloroform–methanol–water, 7:13:8) is very effective for the separation of flavonoid glycosides [41].

#### *9.2.3.4.5 Carbohydrates*

Carbohydrates are important constituents of food and, being the source of energy in living systems, are of immense biochemical importance. Starch, cellulose and some other polysaccharides (e.g. plant gums) are also of considerable industrial value. Carbohydrates may be distinguished as monosaccharides, consisting of the well-known sugars such as glucose, xylose, etc., oligosaccharides (e.g. sucrose and other soluble carbohydrates) and polysaccharides such as starch and cellulose. In addition, there are several other classes of carbohydrates that occur in nature; e.g. saponins (steroid and triterpene glycosides), anthocyanins and related compounds (glycosides of polyphenols), nucleosides, glycolipids and glycoproteins. As the number of sugar moieties increase, their chromatographic properties may be expected to resemble those of the free carbohydrates.

Paper chromatography was extensively used in carbohydrate analysis in the early decades of partition chromatography, but is now largely replaced by TLC on cellulose or silica. Impregnation of the silica layers with different salts and buffers (e.g. boric acid, sodium acetate or phosphate and molybdic or phosphotungstic acid) may be necessary for satisfactory resolution. Different combinations of alcohols (e.g. butanol), esters (ethyl acetate), acids (acetic or formic acid), bases (pyridine or ammonia) are used as mobile phases. Recently, thin-layer, ligand-exchange chromatography using a Cu(II)-loaded stationary phase and water as the mobile phase has been shown to separate mixtures of carbohydrates which are not resolved by other methods. Lead tetraacetate-2,4-dichlorofluoresceine spray reagent causes carbohydrate spots to fluoresce so that quantitation by densitometry is convenient. HPTLC is capable of achieving much higher speed and resolution. HPTLC on silica gel layers at elevated temperatures (40–60°C) using continuous development with mixtures of acetone, alcohol and water effectively resolves oligosaccharides.

Though carbohydrates are non-volatile, they can be readily converted into volatile derivatives suitable for GC. Methyl and trimethylsilyl ethers, acetates and trifluoroacetates of the sugars or their sodium borohydride reduction products (alditols) are the most frequently used derivatives. Both polar (e.g. OV-17 or OV-225) and non-polar (SE-30) stationary phases may be used; each carbohydrate may give two peaks due to the two anomers formed during derivatization. GC, in combination with MS, is particularly useful in the structural studies of carbohydrates.

GC and TLC are now completely over-shadowed by HPLC in view of the latter's convenience and excellent results. Microparticulate silica is very effective for the separation of derivatized (e.g. benzoylated, methylated or acetylated) carbohydrates; hexane, containing ether, ethyl acetate or acetone may be used as the mobile phase. Benzoylated (230 nm) or, preferably, *p*-nitrobenzoylated (260 nm) derivatives are conveniently detected by photometric detectors, which cannot otherwise be used for carbohydrates. These derivatives can also be separated on RPLC using octadecylsilyl silica columns and aqueous acetonitrile. Bonded phases carrying an aminopropyl or cyanopropyl functionality are quite commonly used for mono- and oligosaccharides with a degree of polymerization (DP) of up to five monosaccharide units; the mobile phase is usually 75–85% aqueous acetonitrile. By use of a gradient from about 70 to 55% aqueous acetonitrile, oligomers with a DP of up to 20 could be analysed. The main drawback of the aminopropyl silica columns is that they deteriorate rather rapidly owing to the high reactivity of the amino functional group. Simple separations can be more inexpensively carried out by addition of a polyfunctional amine (about 0.1%) to the mobile phase before the sample introduction; the proportion of the amine may be reduced to about 0.01% after injection. The amine forms a coating on the silica surface which then mimics the amino-bonded phase.

Ligand-exchange chromatography using cation-exchange resins in Li, K, Ca or Ba ionic form may be used to separate sample mixtures of mono- and oligosaccharides (with a DP of 6 or 7); the Ca ion form is the most popular. The mobile phase generally is plain water, but the column efficiency may be improved by addition of triethylamine. The column is normally maintained at an elevated temperature of 50–80°C to prevent peak distortion due to partial separation of anomers. The mechanism of separation is believed to be ligand exchange of the solute molecules with the molecules of water in the hydration sphere of the metal ion. Members of homologous series are eluted in the order of decreasing molecular weight, indicating steric-exclusion effects in the case of oligosaccharides. Cross-linked dextran, agarose and polyacrylamide gels are used for size separation in the traditional manner and porous silica is for the same purpose in HPLC. Hyaluronidase-digested hyaluronic acid (a carbohydrate polymer) may be used as a molecular weight marker; the marker fractions have molecular weights of 4000, 3000, 1900 and 1150. Anion-exchange resins carrying a sulphate ion, or cation-exchange resins in the lithium form are used with aqueous ethanolic mobile phases for the separation of mono- and oligosaccharides. Anion-exchange resins in borate form are used with borate buffers for separating sugars. Addition of ethylenediamine to the effluent and heating the mixture yields fluorescent derivatives of carbohydrates for convenient and sensitive detection. Affinity chromatography is becoming increasingly popular for the isolation of polysaccharides and glycoproteins following the discovery of an increasing number of lectins having appropriate specificity.

In spite of its low sensitivity, the differential refractive index measurement has hitherto been used for detection of carbohydrates after HPLC for want of a better method. Short wavelength (210 nm) UV, polarimetry and the more recent mass detector are the alternative detectors. The ECD with copper bis(phenanthroline) as mediator has also been employed. Recently, however, post-column labelling is often employed when a sensitive detection method is required. Reaction of the eluant with phenol, orcinol or anthrone in sulphuric acid, tetrazolium blue in alkali, and 2-cyanoacetamide are used for photometric detection, and 2-cyanoacetamide, ethylenediamine, ethanolamine, taurine and argenine are used for fluorimetric detection. A variety of derivatives carrying chromophores may be conveniently prepared prior to chromatography to facilitate resolution as well as detection; e.g. alditol benzoates, *p*-nitrobenzyloximes, phenylisocyanates and dansylhydrazones. Chromatography of carbohydrates has been recently reviewed by Churms [41], who has also edited a handbook on the subject [55]. For a more recent compilation of the chromatographic systems for carbohydrates, see Refs. 56–58.

*9.2.3.4.6 Amino acids, peptides and proteins*

Amino acids and their derivatives were one of the first types of compounds to be separated by paper partition chromatography. The methods of planar chromatography are still used for qualitative analysis of amino acids in several laboratories, particularly when a large number of analyses are to be routinely carried out. In addition to paper, cellulose, silica, charcoal, polyamide, polyacrylamide, Sephadex, ion-exchange resins and, more recently, octadecylsilyl thin layers have been used. Depending on the stationary phase, a number of mobile phases, generally containing water, an alcohol (methanol, ethanol, propanol or butanol) and a carboxylic acid (formic or acetic acid), phenol or ammonia, may be used. Two-dimensional planar chromatographic techniques are used when a large number of amino acids are present in the sample. A large number of derivatives are also used for the separation of amino acids, generally on silica-gel layers; these include 2,4-dinitrophenyl, 1-dimethylamino-5-naphthalenesulphonyl (dansyl), carbobenzoxy, *tert*-butyloxycarbonyl, dinitropyridyl and nitropyrimdyl derivatives and phenylthiohydantoins.

The classical amino acid analysis on ion-exchange resins is still being used quite commonly; cation, anion and dual-ion exchangers, based either on polystyrene or cellulose, are generally employed. Cation exchangers with a gradient of acidic buffers are the most common systems used for the separation of free amino acids; the more acidic amino acids elute first while those with more than one primary amino group, or possessing a guanidyl residue, appear later in the chromatogram. These methods may be used either in traditional CLC or the modern HPLC modes. Except in the case of aromatic amino acids and certain other derivatives, these compounds cannot be detected by spectrophotometry unless very short wavelengths are used. They are therefore converted into suitable derivatives either before or after chromatographic separation. Colourimetry using post-column derivatization with ninhydrin or, when higher sensitivity is required, fluorimetry after derivatization with *o*-phthalaldehyde, is frequently used for detection of free amino acids. Use of 4-fluoro-7-nitrobenzo-2,1,3-oxadiazole yields very highly fluorescent derivatives with a detection limit of 5–10 pmol depending on the nature of the amino acid. HPLC of free amino acids on reversed phases, using ion-pairing techniques is possible; but, HPLC of methyl or phenylthiohydantoin, dansyl (dimethylaminonaphthalenesulphonyl), 2,4-dinitrophenyl, 2,4,6-trinitrobenzenesulphonyl, fluorescamine and *o*-phthalaldehyde derivatives of amino acids, with UV or fluorimetric detection at the appropriate wavelength, is more common. The mobile phase is usually a phosphate buffer with 20 to 80% acetonitrile, methanol, tetrahydrofuran or other organic modifier. Trimethylsilyl and other derivatives of amino acids have been chromatographed by GC but their use is now largely limited to GC–MS combinations. The use of on-line LC–MS techniques with the thermospray ionization approach is becoming increasingly popular for the identification of certain amino acids. For excellent, state-of-the-art reviews on the LC methods of analysis of amino acids, see Refs. 41, 59 and 60.

Oligopeptides may also be analysed by ion exchange or by the more recent reversed-phase LC methods. Acidic, basic or hydroxylated peptides are poorly retained on the reversed-phase columns. Therefore, alkanoic acids containing five to eight carbon atoms have been used as surfactants. The most useful technique of RPLC of peptides involves the addition of tetra-alkylammonium salts to the mobile phase at low pH. Chromatography of peptides is greatly facilitated by predicting the retention based on the contribution of the various amino acids and end groups; a high degree of correlation is claimed [61].

Chromatographic separation of proteins has had immense application in modern biochemistry over the past quarter century. The labile nature of the proteins, particularly the tertiary and quaternary structures, requires careful control of the pH, and the ionic strength of the medium. The solubility properties of the proteins and the denaturing effect of several organic solvents limit the choice of the mobile phase to aqueous buffers. However, certain detergents and denaturing agents (e.g. mercaptoethanol) are occasionally used to denature and solubilize the proteins. In view of the large size, ionic character and the biological activities of the proteins, size exclusion, ion exchange and affinity methods are most frequently used for the chromatographic separation of proteins.

Size-exclusion chromatography on dextran-based gels is often the first step in the purification of proteins. Resolution of this method being low, multistage procedures using gels of progressively narrower pore-size need to be used for effective separation. The size-exclusion techniques are generally used for separation of the proteins from smaller species such as inorganic ions of the buffers, low molecular weight contaminants, and the denaturing agents, and for the molecular-weight determination of the purified proteins.

Ion-exchange chromatography of proteins on hydrophilic matrices has been extensively employed; DEAE-Sepharose or DEAE-cellulose have been used more frequently than CM-cellulose. The electrostatic interactions depend on both the pH and the ionic strength of the medium. Acetate or citrate buffers may be used for chromatography on the cation exchangers and tris or phosphate buffers for separation on anion exchangers. If a pH gradient is used, the pH is increased with cation exchangers and decreased with anion exchangers. Owing to the large number of ionic centres on both the solute and the stationary phase, the protein is essentially immobile on the column until the ionic strength of the mobile phase is increased to the appropriate level. For the same reason, the elution bands of the larger proteins are usually broad unless a steep salt gradient is used. Sodium or potassium chloride may be added to the buffer to increase the ionic strength.

Affinity chromatographic techniques are often employed when proteins of specific biological activity need to be isolated. Covalently bound inhibitors or cofactors may be used for retention of enzymes; and similarly, antigens for antibodies, receptors for hormones, lectins for glycoproteins and appropriate small molecules for the corresponding transport and binding proteins. In recent years, the high performance instrumental methods have come to be increasingly used in all the above methods; namely, size exclusion, ion exchange and bioaffinity. There are several excellent books and reviews which deal with protein separation methods; [41,62–64].

*9.2.3.4.7 Nucleic acids and their constituents*

Nucleic acids are constituents of the cell nucleus and form the genetic material. They are polymers of nucleotides which are, in turn, phosphate esters of nucleosides. The nucleosides are ribose and deoxyribose derivatives of the nitrogenous bases called purines (adenine, A; and guanine, G) and pyrimidines (uracil, U; thymine, T; and cytosine, C). These building blocks are important in the transmission of genetic information and in the transfer of chemical energy during cellular metabolism.

Certain precautions need to be taken during sample preparation for chromatography of nucleic acid constituents. Ribonucleotides, particularly the triphosphates, tend to break down in acid media. The extraction process should be performed at low temperature and as rapidly as possible; the tissue matrix is preferably cooled under liquid nitrogen prior to extraction. For the analysis of deoxyribonucleotides, special procedures such as periodate oxidation are needed to eliminate the ribonucleotides which are usually present in relatively large amounts and which may mask the deoxy analogues.

Being non-volatile compounds, these materials are generally analysed by LC; reversed-phase, ion-exchange and ion-pairing techniques are most frequently used for the smaller fragments and ion-exchange, size-exclusion and affinity methods are advantageous for the polymeric compounds. Both cation- and anion-exchange resins may be used in appropriate pH ranges. Polystyrene-based resins are more commonly used for the small molecules and, in view of their permeability, carbohydrate-derived ion exchangers may be used for the chromatography of the larger molecules. Protonation and charge can be controlled by pH variation. On strong cation-exchange resins, 2N hydrochloric acid elutes the bases in the order of their basicity (U, C, G, A). The glycosides, however, are unstable to acid and are therefore separated on anion exchangers of the Dowex 1 type. At low pH ($<4$), bases and nucleosides are selectively excluded by cation exclusion, but they may be separated at higher pH. Oligonucleotides are separated on DEAE-cellulose using salt and/or pH gradient, with or without urea, which suppresses the non-ionic interactions.

Dextran and polyacrylamide gels of different pore sizes have generally been used for size separation. However, traditional or high-performance size-exclusion methods have been of only limited success in the separation of the larger molecules. Anion-exchange chromatography allows resolution of nucleic acids according to the net negative charge which, in general, is directly related to the size of the molecule. Hydrophobic interaction chromatography, which is essentially reversed-phase chromatography (RPC), separates these molecules according to the nucleotide sequence and the size; elution is normally achieved by decreasing mobile phase polarity through addition of organic solvents (methanol, acetonitrile or isopropanol) or through decreasing the salt concentration. The so-called RPC5 columns (polytrifluorochloroethylene beads coated with a thin film of a tetra-alkylammonium halide which acts as a weak anion exchanger) offer a mixed bed approach wherein both ion-exchange and hydrophobic interactions could operate; in certain cases, the resolution is claimed to be better than either the ion-exchange or hydrophobic interaction methods. The mixed mode chromatography has been widely used for the isolation of transfer RNAs. Affinity methods, particularly the template chromatographic methods, are very effective for the nucleic acids. Most eukaryotic messenger RNAs have 3′-polyadenylate tails with 20–250 Å residues. At high ionic strengths (1.0M NaCl), they can form base pairs with small stretches of poly-T or poly-U, bound to a solid matrix (cellulose or Sepahrose). After washing off the accompanying materials, the mRNA can be eluted by decreasing the salt concentration to about 0.1M. The use of chromatographic techniques in the area of nucleic acid research has received considerable attention in recent years because of its importance in the field of recombinant DNA and biotechnology. There are several recent monographs and reviews on chromatography in nucleic acid research [41,64–67].

#### *9.2.3.4.8 Antibiotics*

By definition, antibiotics are produced by micro-organisms as secondary metabolites and are effective in inhibiting the growth of other micro-organisms. They are thus useful in curing diseases caused by micro-organisms in either humans, animals or plants. Following the discovery of penicillins more than four decades ago, the antibiotics have become the most widely used therapeutic agents. They may be broadly classified as antibacterial, antiviral, antialgal, antifungal, antiprotozoal, antiphage and cytotoxic. Chemically, they may belong to alicyclic (e.g. usnic acid), aromatic (chloramphenicol and tetracyclines), heterocycles (mitomycins and polyoxins), β-lactams (penicillins and cephalosporins), peptides (actinomycins and bleomycins), polyenes (nystatin), macrolides (erythromycin), carbohydrates (streptomycin), etc. In addition there are several synthetic analogues and semi-synthetic antibiotics which are used either for testing or as intermediates in the microbial synthesis of antibiotics. Over 50,000 semi-synthetic antibiotics are estimated to have been prepared, though only a tiny fraction could find some utility in clinical medicine.

As rapid methods of microscale qualitative and quantitative analysis, chromatographic techniques find extensive application in several aspects of antibiotics research, production and use: (1) screening of the culture broths for the presence of either known or new antibiotics; (2) selection of the strains for production of antibiotics; (3) monitoring of the production of the active compounds; (4) control of feed additives during fermentation; (5) control of isolation and purification; (6) preparative chromatography for the large-scale isolation; (7) identification, quantitation and quality control; (8) control of formulations; (9) study of biosynthesis and metabolism; (10) taxonomy of the micro-organisms (see below); and (11) determination of dosage by monitoring the drug levels in the serum, etc.

Almost all the chromatographic methods available have been used for the separation, purification and identification of antibiotics [68,69]. For rapid screening of a large number of cultures, the TLC methods are very convenient. The bioautographic methods are particularly advantageous in this field, when the activity of an antibiotic against other organisms is required to be tested. In this method, the dried thin-layer chromatogram is placed on an agar gel, seeded with the appropriate organism to be tested against and, after exposure to the plate for a certain period, incubated for a period of time (2–5 days). The organism will grow leaving clear areas representing the spots where the antibiotics are present [70]. Silica gel, cellulose, Sephadex, activated carbon and ion-exchange layers have been used. As in most other cases with non-volatile compounds, use of GC has been extensively examined during the 1960s and early 70s only to be replaced by the HPLC methods.

It is impossible to cover the chromatographic methods of analysis of the very large number of different types of antibiotics. β-Lactam antibiotics (penicillins and cephalosporins) are chosen for a brief discussion for illustration for they not only belong to a class of compounds not dealt elsewhere in the book, but also represent the largest single group of antibiotics; for more thorough discussion on a large number of antibiotic groups, see Ref. 69.

Except in the case of patients with hypersensitivity to penicillins, administration of somewhat larger than the required dose does not cause complications. Therefore, the main purpose of analysis of the penicillin antibiotics is not so much to quantify the serum levels but mainly to monitor the production of penicillins or to study the pharmacokinetics of the drug or to determine the stability of the antibiotic in pharmaceutical products. Because of their non-volatility and thermal instability, GC methods cannot be used for the analysis of these compounds; this is true of most other antibiotics also. Even for LC, the sample preparation could be quite demanding, primarily due to the narrow, near-neutral pH range over which they are stable. In practice, the culture filtrate is diluted 1:100 with phosphate buffer (pH 7.0) for direct injection or extraction into butyl acetate for subsequent precipitation as an alkali-metal salt.

RPLC systems are most frequently used; with various proportions of phosphate buffer and methanol (or occasionally acetonitrile). Alternatively, anion-exchange columns may be used. Penicillins are UV opaque and detection can be conveniently made using a 254 nm (or 220 nm) absorbance detector. The HPLC system has also been used to follow the tendency of ampicillin to polymerize; it has been shown that ampicillin, when allowed to stand in solution for three days, yields essentially a mixture of polymers with the degree of polymerization from 2 to 8; and with the dimer predominating.

Cephalosporins are closely related to the penicillins both structurally and in their antibacterial activity; several of them have an additional advantage of being resistant to penicillinase activity of some of the bacteria. These compounds may be analysed using similar systems as the penicillins.

Both GC and LC have been successfully used not only for antibiotics analysis but also to monitor the fermentation and other microbiological processes. Automatic sampling systems permit periodical withdrawal of a sample from the broth and suitable treatment for injection. The chromatogram is evaluated automatically and the signal actuates the appropriate valves in a microprocessor-controlled fermentor to adjust the pH, etc., and to add required substrates and nutrients; the system can also signal completion of fermentation for further processing. Identification of micro-organisms by HPLC is described later.

#### *9.2.3.5 Structure–retention–activity relationships*

One of the most exciting, though yet to be fully realized, potentials of chromatography is the relationship between the chromatographic behaviour (mobility or retention) of a substance and its chemical structure. It can be easily perceived that there must be a correlation between the two, since the structure of a compound determines its solubility and other physicochemical properties, which, in turn, determine the chromatographic properties. Chromatography being the result of a triangular interaction between the solute, stationary phase and the mobile phase, if the last two are held constant (or varied systematically within limits), it must be, at least in theory, possible to predict the chromatographic behaviour on the basis of the three-dimensional structure of the solute molecule or, conversely (and more interestingly), determine the structure of the compound from its chromatographic properties. If the nature of the molecular interactions between the three species can be quantified, the chromatographic behaviour of the solute can be predicted. Three of the features that characterize a molecule, i.e. the hydrophobicity, polarity and steric factors, also determine its chromatographic behaviour on one side and its physiological activity on the other. Considerable attention has been paid to this aspect which forms the subject matter of several reviews [71–74] and books [75].

Correlation of the chromatographic behaviour of a compound with its structure and its physiological activity is advantageous from several angles: (1) the chromatographic methods are simple to carry out and rapid; (2) they use very small quantities of the samples; (3) they can handle compounds with wide range of partition coefficients; (4) the material to be examined need not be pure; (5) many compounds in a complex mixture can be simultaneously studied; and (6) correlation of retention parameters in different solvents is possible. TLC may be particularly convenient as a large number of samples (about 25 on a plate) can be examined under exactly the same conditions. However, correlation between chromatographic data obtained from different sources may be unreliable owing to a number of factors such as variation in temperature, equilibration, length of the run or the flow-rate of the mobile phase from plate to plate. Column chromatographic techniques appear more reliable, though more time-consuming. It is preferable to use inert phases rather than 'active' phases and, to the extent possible, either the stationary phase or the mobile phase (or, preferably, both) should remain unchanged for meaningful correlation. Thus, GC appears more promising [76], since the same inert gas can be used as the mobile phase and a large number of compounds can be chromatographed on a given stationary phase. On LC, reversed-phase systems are preferable since the mobile phase variation can be limited to a large extent to that of the ratio of water and methanol or acetonitrile. Precise control of the chromatographic conditions by means of computers in the modern instruments should help reproducibility and reliability of the retention data. The computers also help use of pattern-recognition techniques and complicated quantum mechanical calculations (which have been used, though without much success, for the correlation of the molecular orbital models with the retention data). More accurate structure–retention correlations thus appear feasible.

Several numerical structural descriptors are needed to describe the molecular structure of the solute quantitatively. The number of such descriptors being unlimited, correlations should be made with caution. The main problem in quantification of the structural differences among solutes is to determine the right molecular or submolecular features. As noted in Chapter 2 and in section 9.2.2.1, a number of physicochemical parameters can be related to the chromatographic retention behaviour. These include the additive parameters of molecular features like the carbon number, molecular weight, parachor, polarizability, molecular volume, specific surface area, molar refractivity, etc. These easily determinable molecular parameters can be related to the chromatographic properties in the absence of specific interactions that could change the retention characteristics. Thus, the correlations are more dependable in cases where only dispersive (non-polar) interactions are expected; e.g. in GC and LC with non-polar stationary phases. The more complex polar interactions are more difficult to describe and quantify using a single property.

Among the better understood and easily quantifiable physical properties of substances is the solubility parameter. The intermolecular interactions that contribute to the solubility parameter may be divided into dispersive, dipolar, induction, ionic and hydrogen-bonding interactions. The dispersion forces do not differ significantly in a homologous series and the best correlations are to be found among homologues in a non-polar environment. Several topological indices (e.g. the Randic's connectivity index) have been examined for correlation of retention properties and so also for the quantum mechanical approaches. However, the correlations based essentially on solubility and partition have only been reasonably successful [74].

As early as 1944, Martin (see Chapter 1) had suggested that there must be a relation between the structure of the molecule and its $R_F$ value in partition chromatography. The partition coefficient $K$ of a substance $A$, for ideal solutions, is related to the free energy change or the chemical potential ($\mu$) required to transport one mole of $A$ from one phase to another:

$$\ln K = \frac{\mu_A}{RT} \tag{9.4}$$

This feature, i.e. one process (the free energy change) being a linear function of another (here, partition) is called the linear free energy relationship (LFER). The processes which are subject to LFER can be generally related to the molecular structure of the compounds involved. The chemical potential, thus, is an additive function in the sense that the addition of a group X to the molecule should change the partition coefficient by a factor depending on the nature of X and, if $m$ groups of X and $n$ groups of Y, etc., are added to the molecule A to give B:

$$RT \ln K = \mu_B = \mu_A + m\mu_X + n\mu_Y + \ \ldots \tag{9.5}$$

The most commonly used term in the structure–retention studies is $R_M$, which is related to the $R_F$ on TLC and the capacity factor $k$ in LC according to the equation:

$$R_M = \log\left(\frac{1}{R_F} - 1\right) = \log k \tag{9.6}$$

The $R_M$ value is directly related to the capacity factor and the partition coefficient and thus to the chemical potential, which is an additive function of the constituent groups. The $R_M$ values of homologous series of compounds have also been correlated to their biological activity. It must be mentioned that accurate determination of the retention data and correlation with those from other laboratories is dependent on a number of factors, even when the data is collected on identical chromatographic systems and the structure of the compound is the only variable considered. Nevertheless, useful correlations could be made under favourable conditions. Thus, a detailed study [71] of the retention parameters for a large number of simple phenols, hydroquinone monoethers, esters, vitamins E and K, alcohols, quinones and chromenols has revealed constancy of $R_M$ increments for the following groups and structural features: H, $CH_2$, ring-attached $CH_2$, double bond, branching, oxygen in ethers and isoprene units in long chains. The $R_M$ values in different classes of compounds have also been compiled for methyl, benzyl, phenyl, cycloalkyl, hydroxyl, methoxyl, amino, nitro, halogeno, formyl and carboxyl groups and linear relationships between the number of groups and retention parameters demonstrated [72]. If linearity in the change of retention data with changes in mobile-phase or stationary-phase composition is assumed, the data on different systems may also be considered. However, the additivity principle often breaks down when steric effects (e.g. *ortho* effects in aromatic compounds), intramolecular hydrogen-bonding, electronic effects (i.e. charge distribution or separation), intramolecular hydrophobicity, chain-branching, etc., operate. The anamolies may also be due to the solvent effects in the chromatographic system and are more pronounced in the normal-phase systems compared to the reversed-phase ones.

While complete structural analysis based on chromatographic behaviour is neither possible nor necessary, retention data for a given class of compounds can frequently lead to useful extrapolations and, thus, to structure determination. The Kovats retention index is the most frequently used parameter in this connection. Linearity of retention indices with the carbon number has been demonstrated for homologous series of hydrocarbons, alcohols, carbonyl compounds and esters. Structure-retention correlations among fatty acids [44], flavonoids [54], steroids [77], ketones [77], etc., are well documented.

The quantitative structure–retention relationships (QSRR) discussed above actually followed the earlier attempts at quantitative structure–activity relationships (QSAR) among medicinally important products. The QSAR studies on physiologically active compounds continue to be pursued in view of their obvious and potential utility. Naturally, the retention–activity relationships also received much attention. If reliable quantitative data on the physiological activity for a series of compounds is available, they can be correlated to the molecular descriptors representing the hydrophobic, polarity and steric properties, just as in the case of the QSRR studies.

On the premise that the permeability into, or interaction with, the living cell is a prerequisite for biological activity and that the orientation of the molecules of *n*-octanol could mimic the lipid bilayer of the cell wall, the majority of the QSAR studies have been made on the basis of the partition coefficient of substances in an *n*-octanol/water system. While the premise and the choice of the system may be questionable, the very large amount of data already available on this system made it convenient to adopt the same system for the retention–activity studies. The early attempts at such correlations employed liquid–liquid partition chromatography using *n*-octanol as the stationary phase and *n*-octanol–saturated water as the mobile phase. Though useful correlations could be made using this system, the octadecylsilyl (ODS)-bonded silica is now preferred almost universally. As already noted, close correlation between the structure and retention is observed in most cases on the reversed-phase layers.

The substituent parameter most frequently used in both QSAR and QSRR studies is the Hansch hydrophobic constant ($\pi$), which is defined as the logarithm of the ratio of *n*-octanol/water partition coefficient of a substituted to an unsubstituted derivative. Benzene derivatives are used to obtain the $\pi$ values, which are additive. When the separation mechanism is exclusively partition, the hydrophobic constants can be related to chromatographic data according to the equation:

$$\log K_a = m \log K_b + n \tag{9.7}$$

where $K_a$ and $K_b$ are the partition coefficients determined on organic/water systems (a and b) and $m$ and $n$ are constants that are characteristic of the particular solvent systems. The $R_M$ or the log $k$ values can be linearly related to log $K$ except that the relationship fails when the polarity difference between the two organic solvents is large. In other words, the partition data can be obtained from chromatographic data, using either RP–TLC or, better, RP–HPLC. Indeed, linear relationships were found for homologous or congeneric series of solutes between standard *n*-octanol/water partition coefficients and the capacity factors determined on ODS phases. As expected, the linearity between log $K$ and log $k$ is not found when the solutes are structurally diverse. Since the exact physicochemical nature of retention on these phases is not known, the quantitative relationship between the two may be much more complex than that predicted by eq. (9.7) (generally known as the Collander equation).

A number of strategies have been developed to unify the retention data (indices) for use in medicinal chemistry [78]. One such is the retention index system, analogous to the Kovats index in GC, using 2-ketoalkanes with 3–23 carbon atoms [79]; the index is obtained by substituting the adjusted retention time ($t'_R$) of eq. (9.1) by $k$, the capacity factor of the appropriate 2-alkanone. By definition, the retention index of the 2-ketoalkane is $100N$, where $N$ is the number of carbon atoms. Another index of hydrophobicity, $R_Q$, is given by the equation [80],

$$R_Q = \log\left(\frac{1 - R_o}{R_T}\right) \tag{9.8}$$

where $R_o$ and $R_T$ are the retention times of the unretained solute and the compound being studied. The $R_Q$ values are claimed to correlate excellently with the log $K$ values.

Applications of chromatography in QSAR studies have been reviewed [71]. The reported studies include: structure–retention–activity relationships of (1) potency of antibiotics like penicillins and cephalosporins against *E. coli, S. aureus* and *T. pallidum*, (2) minimum lethal dose of cardiac glycosides in cats, (3) fungicidal activity of some substituted phenols, (4) inhibitory potency of barbiturates on rat-brain oxygen uptake, (5) anabolic activity of androlone steroids, (6) serum albumin-binding constants, antihemolytic activity as well as the inhibition of $Na^+K^+$-activated adenosine triphosphate activity for a series of phenothiazine derivatives, and (7) bioactivity of steroids, acetophenones, phenols and benzodiazepines and several others. Most often, satisfactory correlations have been reported; in certain cases, however, silicone oil-impregnation and other non-polar stationary phases appeared to correlate better than the ODS packings. Hydrophobicity parameters of phenols have also been related to the human detection threshold of olefactory activity. For more details and references, see Ref. 74.

## 9.3 BIOLOGICAL AND BIOMEDICAL APPLICATIONS

The above discussion on structure–retention–activity is but one example of the way chromatography could bridge biology and chemistry. Chromatographic applications in biology are so varied and important that a predominant proportion of books now appearing on chromatography are fully devoted to the subject. While the subject is ramifying in different directions, the present treatment is limited to briefly mentioning just two areas for illustration.

### 9.3.1 Biochemical analysis

Analytical methodology is a very important aspect of biochemical research. The revolutionary changes that have occurred in biochemical research during the past four decades probably could not have come about but for the developments in chromatography. Martin's use of amino-acid derivatives in his first experiments on partition chromatography might have been coincidental but it signalled the beginning of the widespread use of chromatography as a biochemical technique. The 1940s also saw the development of ion-exchange chromatography for amino-acid analysis and determination of the amino-acid sequence of insulin by Sanger. While the discovery was indeed epoch-making, the analysis required large quantities of protein hydrolysate and almost 4 days for completion. With the subsequent developments in post- and pre-column derivatization and detection techniques, amino acids can now be analysed using materials in the femtogram ($10^{-15}$ g) range [81], so that automated microsequencing of peptides and proteins is possible using picomole quantities. The development of size-exclusion and affinity chromatographic techniques in the 1960s was again a landmark. In addition to the three classical methods noted above, the reversed-phase LC, using octadecylsilyl-bonded silica as stationary phase, is now widely used. Apart from the convenience of using an aqueous mobile phase (which is compatible with most of the common biological samples),the RPLC permits use of such techniques as ion-pair chromatography that have wide applicability in biochemical analysis.

Chromatographic methods for different classes of natural (biological!) products, including lipids, carbohydrates, amino acids, peptides, proteins, nucleic acids have been briefly described in the foregoing section. While these can be readily adapted to most of the situations that may prevail in biochemical laboratories, the current trend for the instrument companies has been to market dedicated instruments for different biochemical analyses. Thus, several columns and methods are designed specifically for certain types of analyses; e.g. a well-modulated aminopropylsilyl-bonded and diol-bonded silica column is convenient for carbohydrate analysis using pure water as the mobile phase. Amino-acid analysis systems, based on HPLC, post-column derivatization with *o*-phthaldehyde, and fluorimetric detection, are also available. Similarly, the lipid-analysis or the protein-purification columns require much less operator involvement in the optimization of the analysis. All the operating parameters like the mobile-phase composition, temperature, flow-rate and detector choice and control are normally furnished by the manufacturer of the HPLC systems. Most instrument suppliers also periodically publish application notes and newsletters describing the newer methods of interest to biochemists.

One such dedicated LC system for biochemical and biotechnological research is the FPLC system of Pharmacia, Sweden. Though originally meant to be an acronym for 'fast protein liquid chromatography', it is claimed to be useful for the analysis of a variety of biomolecules, including proteins (enzymes, monoclonal antibodies, membrane proteins), peptides, polynucleotides and even smaller molecules like vitamins and amino acids. Advantage is apparently taken of some common features in most biochemical separations. The separations are mostly based on size-exclusion, ion-exchange, affinity or reversed-phase chromatography and the mobile phase is invariably water-based, consisting of salts and buffers. The flow-rates may be high, but high pressures are not required if careful attention is given to uniformity of the stationary-phase particle size and shape. The FPLC thus uses Pharmacia's proprietary 'MonoBeads' technology for the stationary phases. As in other instrumental LC systems, the FPLC consists of solvent reservoirs, injection ports, valves, columns, pumps, gradient controllers, detectors, fraction collectors and recorders.

By choice of appropriate pumps (580 psi or 3500 psi) and compatible injection ports, plumbing and columns, the FPLC can be adapted to either medium-pressure or high-pressure operation, analytical or preparative applications. Typical applications include purification of enzymes, plasmids, restriction fragments and other compounds of interest in molecular biology, biotechnology, monoclonal antibody research, clinical research, food analysis and the like. Analysis of small molecules like the amino acids, peptides, nucleotides, oligonucleotides and vitamins is possible using the reversed-phase columns. Literature on the possible options and applications is available from the manufacturer on request [82]. Medium-pressure liquid chromatography (MPLC) systems are available from several other manufacturers.

Apart from the major classes of natural products, the compounds of general interest to biochemical investigators include nutrients (vitamins, minerals and other micronutrients), drugs, pharmaceuticals, xenobiotics and their metabolites, hormones, neurotransmitters (catecholamines or biogenic amines), constituents of the body fluids and the like. Some of these are briefly discussed here.

#### *9.3.1.1 Nutrition*

Nutritional research involves study of the dietary factors, their biological role, requirements for normal health and growth, and manifestations of the nutritional deficiencies as well as their treatment. The possibility of analysing and quantifying microquantities of several of these by modern HPLC techniques and instrumentation is vital for nutritional and metabolic studies [83]. The improved and convenient analytical procedures have certainly contributed to our understanding of the processes by which animals ingest, digest, absorb, transport, use and excrete the various constituents of food. Thus, it has long been known that dietary purines influence uric-acid metabolism and that excess uric acid in blood causes gout. HPLC has made it possible to determine individual purines in food and evaluate their role. Earlier studies relied on the spectroscopic determination of total purines. It is now well recognized that the different purines have their individual roles and that it is important to evaluate the components in foods. The information on the individual purines helps the understanding of the metabolism of purine nucleic acids, which may have an important role in medicine. Purines, being basic nitrogen heterocycles, are readily protonated and exist as positively charged species in acidic solution. The degree of protonation can be controlled by change of pH. Taking advantage of this feature, they can be readily separated and analysed on strong cation-exchange resins of the sulphonic acid type. Elution is effected by increasing the acid strength of the mobile phase. More recently, reversed-phase (ODS) columns and phosphate buffer–methanol gradient have been successfully used. The very high absorptivity of these compounds in the UV region permits detection in the nanomole range.

It is common knowledge that vitamins are important dietary ingredients. Their importance stems from the fact that, while they are essential for normal health and growth, they are not synthesized by the human body; food is, therefore, the source of the vitamins. Many of these compounds are present in minute quantities in foods, often as complex mixtures of closely related compounds. They are also generally unstable to heat, light and atmospheric oxygen, in other words, to most isolation procedures. HPLC, again, proves to be a very convenient and effective method of determination of the various vitamins and their co-factors in foods, tissues and body fluids [84].

The vitamins may be broadly classified as fat-soluble and water-soluble. To the former group belong the vitamins A, D, E and K, and to the latter, vitamins B complex, C and others. The fat-soluble vitamins can be readily separated either on silica gel (hexane–ether) or reversed-phase (ODS) packings using water–acetonitrile or water–methanol. The vitamin A activity is exhibited not only by retinol but also by several other carotenoids. These occur as esters of fatty acids and other lipids. They are therefore normally saponified prior to analysis to free them from the lipid portion. The carotenoids are insoluble in water and even on the reversed phases, the mobile phase is essentially entirely organic (e.g. dichloromethane–acetonitrile, 30:70). Most of the fat-soluble vitamins strongly absorb radiation in the UV region, owing to the presence of conjugated double bonds; many of them (e.g. vitamin E) also fluoresce in the UV region. A variable wavelength detector or a fluorescence detector is, therefore, commonly used.

Among the water-soluble vitamins, vitamin C (ascorbic acid) is widely distributed in fruits, particularly the citrus fruits. It is normally chromatographed on ODS packings (using methanol–water or methanol–aqueous buffers at pH between 4.5 and 5) or anion-exchange resins (using borate buffer, pH 8). The vitamin B complex is composed of different nitrogen heterocycles with different solubility and spectral properties. The LC systems are based mainly on RPLC packings and aqueous methanolic buffers, occasionally with ion-pairing agents.

#### *9.3.1.2 Endocrinology*

Endocrinology deals with the metabolism and function of hormones which control the growth and communication in the animal body. The endocrine system is a complex messenger network and includes the glands that produce the hormones, the blood stream that transports the hormones to the target organ or the sites of action, inaction and excretion, as well as the feedback mechanisms that control the production of the hormone. Though hormone research was actively pursued prior to the advent of the modern chromatographic techniques, the pace of progress was hampered by the slow nature of the biological methods and the minute quantities in which hormones are produced. With the introduction of HPLC, which is now being extensively used for the isolation, purification and analysis of the hormones, the understanding of the structure and function of these compounds increased enormously [85]. For the purpose of the present discussion, the hormones may be broadly divided into three chemically distinct groups; the peptides, steroids and biogenic amines (neurotransmitters).

The peptide and protein hormones form a fairly large group of compounds, produced by different glands and with diverse biological activity. Examples of this class include the corticotropin-releasing factor (CRF, produced by the hypothalamus), which controls the production of adrenocorticotropin (ACTH) by the anterior pituitary. The ACTH, in turn, regulates steroid-hormone synthesis in the adrenal cortex. Another well-known peptide hormone is insulin, which is produced in the pancreas and which regulates glucose metabolism. Incidentally, insulin was the first peptide hormone whose structure elucidation marked the extensive use of ion-exchange chromatography in biochemical separations.

All the problems generally associated with the separation of peptides and proteins are valid for peptide hormones also. Traditionally, they have been analysed using affinity, size-exclusion and electrophoretic techniques. While these techniques are still being widely used, HPLC, owing to its superior resolving power and speed, is gradually taking their place. Reversed-phase separation of these compounds is based essentially on hydrophobic interaction of the amino-acid residues with the stationary phase. Adsorption and elution is thus essentially controlled by the mobile phase, each peptide eluting at a specific composition. Gradient elution is, therefore, invariably used for chromatography. The selectivity can be modulated by changing the length of the hydrophobic carbon chain (chain lengths from 1 to 18 carbon atoms have been studied). Permeability to larger peptides may be enhanced by increasing the pore-size from the usual 6 nm to about 30 nm. Gradient elution with increasing proportion of 1-propanol in water (containing a trace of trifluoroacetic acid to improve the peak shape) is the most frequently used system. The retention may be optimized by addition of an organic modifier like acetonitrile and methanol. Since no reversible mechanisms are involved, the resolution is largely unaffected by the column length. Thus, longer columns may be used only if larger sample loadability is required, but at the cost of recovery.

UV absorbance detection is widely used in HPLC, but unless the peptide contains aromatic amino acids the absorbance is poor. In the absence of information on the amino acid composition, a photodiode-array detector (which scans the complete UV region and detects the aromatic amino-acid containing peptides) may be used. Alternatively, the high biological activity of the peptide hormones can be used in the detection by radio-immunoassay (RIA) methods. The RIA methods are, generally, very simple to perform, precise and sensitive but, in the absence of prior separation by one of the chromatographic techniques, can lead to erratic results owing to low specificity and cross-reactivity. Thus, the combination LC–RIA turns out to be an extremely powerful tool for biochemical research. For example, they have been used for characterization of neuropeptides like the endorphins and encephalins. Use of different mass spectrometric techniques for the analysis of neuropeptides in combination with HPLC has been reviewed [86].

Steroid hormones may be functionally classified into corticosteroids (mineralocorticoids and glucocorticoids) and sex hormones. Androsterones and other mineralocorticoids are involved mainly in salt and water metabolism, while the glucocorticoids (e.g. cortisol) have a very important role in carbohydrate, protein and other metabolisms. Androgens and oestrogens are the male and female sex hormones that control a variety of reproductive activities. Chromatography of steroids has been discussed earlier (section 9.2.3.4.3); many of these systems can be used for the analysis of steroid hormones. However, while both GC and LC may be used, the requirement of elaborate sample preparation and derivatization may be avoided in HPLC; RPLC using methanol–water systems is most commonly employed. Similar systems may also be used for analysing steroid conjugates (glucuronides and sulphates) which are the derivatives excreted by the body through urine. These may also be hydrolysed by acid or enzymes to free steroids prior to chromatography. Detection is normally by UV absorbance at a wavelength (240 or 254 nm) depending on the chromophore present, but when very high sensitivity is needed, post-column derivatization followed by laser-induced fluorescence detection could be used with sensitivities in the femtogram range. Sterols react with 7-[(chlorocarbonyl)-methoxy]-4-methylcoumarin (CMMC) and sterones with dansylhydrazine to give fluorescent derivatives.

The biogenic amines are formed in the body from amino acids. Tryptophan and tyrosine on hydroxylation, followed by decarboxylation, yield serotonin and dopamine. Dopamine, together with its metabolites, adrenaline (epinephrine) and noradrenaline (norepinephrine) belongs to the important class of catecholamines. The catecholamines, along with serotonin and acetylcholine, are an important class of neurotransmitters. The different catecholamines play a very vital role in different human diseases. Much of the importance attached to the development of analytical methodologies for the determination of the biogenic amines emanates from the large number of different types of disease states associated with the changes in their concentration levels. These include tumours in the neural crest and outside, endocrine disorders, hereditary and metabolic disorders, cardiovascular diseases and affective (psychiatric) disorders. By implication and experience, drugs that change the disease state also alter the catecholamine levels.

An exhaustive survey of chromatography of the neurotransmitters has been published recently [87]; there are also several other reviews and a book [88]. A conspicuous feature of HPLC determination of biogenic amines is the current dominance of electrochemical detection methods (particularly in the RPLC mode) over the fluorescence and other detectors, though these are also still being used. The popularity of the EC detector is due to its very high sensitivity which is required in view of the low concentration of the biogenic amines in the body fluids (pg/ml). The use of the ECD, however, precludes gradient-elution methods. The separation system that is most frequently used is based on RPLC and a buffer based on phosphate or phosphate-citrate, so that the ionic strength required for the efficient working of the EC detector is maintained. Selectivity is achieved by addition of organic modifiers (methanol) and ion-pairing reagents (octanesulphonic acid).

### 9.3.2 Biomedical applications

The largest impact of the two decades of development of HPLC has been in the area of biomedicine. As noted in Chapter 1, there are even two journals completely dealing with the subject (*Journal of Chromatography: Biomedical Applications* and *Journal of Biomedical Chromatography*). There can be no pretence that the very large number of different ways in which chromatography has been used in this area can even be listed here. However, as in the case of biochemical analysis, only two examples are given, just as an illustration of the range of biomedical applications of chromatography. These are the two major paths in which a disease state can be handled. The first deals with metabolic profiling of body fluid changes, which may be indicative of a disease state. The second deals with therapeutic drug monitoring which means determining the drug levels so that the drug dosage to be administered can be calculated.

#### *9.3.2.1 Profiling of body fluids*

The concept of 'metabolic pattern' was introduced in 1951 by Roger Williams [89]. In principle, every individual possesses a characteristic pattern of metabolism that is reflected in the body-fluid composition. There have also been reports (1950s) of fingerprinting of chromatograms. The term metabolic profile was introduced much later (1971) to describe gas chromatographic patterns of a group of metabolically and analytically related metabolites. However, the subject of metabolic profiling itself is still fairly new and is yet to be actually used in all areas where it is applicable. Two distinct types of profiling are possible. One is the profile of a set of closely related compounds (carbohydrates, lipids, etc.) that are present in the subject and the other is complete analytical data on a biological material (tissue, serum or excreta) from the subject.

The concept of profiling must be distinguished from the screening of clinical samples as is normally done by the modern autoanalysers where 24 (or even more) predetermined analyses for specific compounds of the body fluids are routinely analysed and the results for each component are independently compared with a set of standard values for clinical diagnosis. In addition to the absolute values, the profiling also takes into consideration the relative concentrations of the individual components. Since the profiling is normally carried out by a single analytical method using a particular tissue, the analytes are both metabolically and chemically related.

For example, the same analysis does not give both a lipid and carbohydrate profile. On the other hand, since profiling reveals the concentration pattern of all the components of a given type (even when the particular compound is not known and targeted for analysis), it may reveal changing patterns yet to be associated with the disease.

Using paper chromatography and other techniques, Williams had analysed several related physiological compounds for each subject under study and represented the profile graphically using vector angles to designate different analytes and the vector distances to represent the concentrations. Thus, the profile was a two-dimensional representation of a complex, multidimensional physiological state. These graphs revealed individuality of pattern for each subject. A genetic basis for the individuality was inferred when the differences were minimal for identical twins. While immediate changes in the profile may be indicative of a change in the state of the person's health, the profiles collected over a period of time may also reveal certain hidden diseases not immediately reflected by other symptoms. The profiles may also be compared for similarities or otherwise with a reference set of profiles for a certain disease (e.g. cancer or schizophrenia). Because of the ease of obtaining such data and the possibility of computerized storage, retrieval, pattern recognition and comparison of the profiles, the concept of metabolic profiling of body fluids has the potential to be a major contributor to the diagnosis of specific metabolic disorders.

Chromatography is the natural choice as the analytical method for the profiling studies because (1) it can analyse a large number of compounds of a certain class in a given sample, (2) a high degree of selectivity can be achieved when desired, (3) both qualitative and quantitative analysis is possible on the same sample, (4) the detection system does not normally introduce severe bias to the quantitation of the different analytes, (5) the large dynamic range of several chromatographic detectors permits determination of compounds that may be present in widely differing concentrations, (6) graphic representation of the peak identity and its concentration by the chromatogram (permitting immediate recognition of any changes in the pattern), and (7) the feasibility of automation in both sample processing and data analysis.

The GC–FID has provided the largest dynamic range of six to seven orders of magnitude, probably the largest available on any instrument of concentration-related measurement. The various choices of column dimensions, stationary phases and operating parameters allows optimization of the method to separate the various components of a complex mixture. Indeed, the separation potential of capillary or open-tubular GC is so high for even different classes of compounds that it had led to the development of the concept of GC-only profiling. The output of the GC detector is directly fed to an IBM-PC on which a software program, GC-MET (GC-metabolites) is available. For any given biological sample from a living species, the number of components and their concentration range is limited. Therefore, the principles of the closed loop (or set) may be used where the components present in the sample may be used as retention standards, without the need for external standards. The errors due to chemical differences between the standards and sample components are thus avoided. The software first locates the peaks that are recognizable and uses these as reference standards for calculation of retention-index data on all the compounds present in the sample. The peaks are identified as those present or absent in the library, irrespective of whether they are known or unknown. The output routine highlights those peaks that are not present so that these may be immediately recognized. The 'result of an analysis file' gives compound number, amount and the 'confidence factor' (which takes into account such variables to accuracy as the resolution, retention and concentration differences). The results from several samples are then statistically evaluated to form the reference sets with which the individual files are compared. The profiling softwares, in general, consist of three routines; namely, the library builder, data analysis and the statistical package [90]. The combination of GC–MS with a data system makes the profile analysis even more powerful and reliable, but certainly more complex.

Thin-layer chromatography offers certain special advantages in metabolic profiling. The low cost and high speed apart, the possibility of comparing the profiles of several samples at one glance makes it very convenient for the analyst. Standard profiles may be stored as photographs for direct comparison, either visually or instrumentally.

As with the other applications in biological sciences, HPLC has proved particularly convenient in handling the aqueous samples for metabolic profiling. Unlike in GC, several of the biologically important metabolic constituents (e.g. amino acids and carbohydrates) do not require derivatization prior to analysis. While the limited capability of universal detection in LC is still a drawback, the feasibility of adaptation of the LC–MS and LC–FTIR techniques is encouraging.

The utility of metabolic profiling is obviously in the area of clinical medicine. The methods for profiling of body fluids with respect to the individual classes of compounds have been recently reviewed [89]. These include carbohydrates, glycoproteins, glycolipids, lipids, hormonal and urinary steroids, polyamines, amino acids, neuropeptides, nucleic acids and constituents, ketone bodies and organic acids. Two examples are chosen here to illustrate the application of the profiling methods to the diagnosis of cancer and diabetes.

The polyamines, spermine and spermidine and their precursor putrescine (1,4-diaminobutane), play an important role in the normal growth and proliferation of eukaryotic cells. They are present in all body fluids and tissues. The levels of their excretion in urine (mainly as the monoacetyl derivatives) is very low but may increase significantly on intoxication, starvation and other pathophysiological changes. Abnormal levels of polyamines in the body tissues is indicative of abnormal cell growth. While it is still uncertain whether the polyamine metabolites could indeed be markers of tumours and other disease states, the data from individuals may give important information on the efficacy of certain therapies and the recurrence of the particular disease [91].

The polyamines, like the biogenic amines and certain amino acids, are present in the body fluids in the range of micromoles per litre. The structural feature common to all these compounds being the presence of an amino group, a high degree of specificity and sensitivity is required for the determination of the polyamines in the presence of the congeners. As the polyamines do not possess any other functional group to bring about selectivity in separation, specificity can be achieved only through efficient separation. Recently, over-pressured TLC on ion-exchange layers, followed by photo-densitometric evaluation of the ninhydrin-stained spots has been used for rapid determination of polyamines and other amines in the nanomole range. Ion exchange is also the principal separation mode for HPLC, to be followed by reaction with ninhydrin or *o*-phthalaldehyde; polyamines could be determined in the 25–100 pmol and 3–15 pmol range, respectively, with the two reagents. These methods, however, require the polyamine conjugates in the body fluids to be hydrolysed prior to chromatography. Ion-pair chromatography on reversed-phase columns using *n*-alkanesulphonic acids has also been used for the analysis of the complete range of polyamines. Since small molecules carry the readily derivatizable amino groups, GC analysis is quite convenient. Trimethylsilyl and polyhalogenated acyl and alkyl derivatives (section 3.6.2) have been used for their separation on non-polar (methyl polysiloxane) stationary phases. Depending on the derivatives, FID, ECD or NPD may be used for detection.

*Diabetes mellitus* (commonly known simply as diabetes), is a metabolic disease, characterized by a number of hormone-induced abnormalities and long-term complications [92]. Though primarily recognized as a defect in the glucose metabolism (mainly because of the ease of monitoring the glucose levels in the urine and serum), diabetic complications arise out of a number of abnormalities in lipid and amino-acid metabolism, leading to changes in the levels of ketones and organic acids in the body fluids. The major manifestations of the disease include increased total 4-heptanone (the sum of 2-ethyl-3-oxohexanoic acid and its decarboxylation product, 4-heptanone, which are elevated four-fold from the normal level of 200–250 $\mu$g/day), increased higher molecular weight ketone bodies (e.g. 2-pentanone, 3-penten-2-one and 2-heptanone, in addition to acetone), increased levels of dicarboxylic acids, hydroxy acids and oxomonocarboxylic acids (in addition to the fatty acids, over 60 of which need to be monitored). Determination of the serum and urinary glucose, pH, and ketones is routinely carried out in a clinical laboratory to monitor the disease. However, because of the large number of metabolic changes in diabetes, a profile analysis is particularly useful for following the intensity of the disease or the course of the treatment.

In view of the complexity of the ketone bodies (which requires a high-resolution technique for the profiling), and their volatility, open-tubular GC is generally preferred for the profiling. Acetone and other neutral ketones are isolated from the body fluids by gas-phase extraction and adsorption on a porous polymer like Tenax GC; the efficiency of adsorption is only 50–80% in the case of acetone, but 95–100% for the higher ketones. (Solvent extraction may be used for the isolation and determination of specific ketones.) The adsorbed ketones are released by purging the polymer with helium at 220–300°C and recondensed on a cooled pre-column, prior to GC analysis. Organic acids are isolated from the body fluids (after suitably precipitating the proteins) by either solvent extraction (ether, dichloromethane or ethyl acetate) or by retention on an anion-exchange resin. The acidic components in serum or urine being very complex mixtures, pre-fractionation may be necessary when the minor acids are to be determined. This is conveniently achieved by preparative TLC (see later) or by HPLC. Oxoacids and other carboxylic acids may be suitably derivatized for GC. Both non-polar (e.g. DB-1 or the cross-linked methyl polysiloxane) and polar (DB-5) may be used for the GC analysis. The profile analysis of the volatile substances showed that, in addition to the ketone bodies, aliphatic alcohol levels are also elevated in the serum and urine of diabetic patients. GC–MS techniques (mass fragmentography) are also available for the quantitation of ketones and alcohols.

#### *9.3.2.2 Therapeutic drug monitoring and toxicology*

When a foreign compound or xenobiotic (i.e. any compound which is not a nutrient and is not required for the normal growth and maintenance) is introduced into a living system, it may either pass unaffected or may undergo biotransformation by a variety of processes, depending on the structure of the compound and the organism. Finally, the compound or its metabolite may get conjugated with glucuronic acid or hippuric acid to a water-soluble compound and be excreted through urine. The several kinds of drugs introduced into the system to fight disease are also recognized by the body as potential toxicants and detoxification mechanisms set in to metabolize the drug and eliminate it from the system. The drug, or its metabolite, may act on the system or the disease-causing microbe. Two problems, however, are frequently experienced in chemotherapy. The drug may not act unless a certain concentration is attained in the blood, but, beyond a certain concentration, it may be toxic for the system. In other words, the drugs have a therapeutic range. To prevent acute toxicity, a drug is normally administered in repetitive doses, the frequency and dose being determined by the life of the active drug in the body. Both the time it takes for a drug to reach an effective concentration, its life and its toxic concentration are highly 'individualized' in the sense that these figures vary not only with the species, sex and age, but even within a group, from person to person. The dosage of the drug should therefore compensate for the individual variation in 'drug utilization' patterns. Even in a particular individual, the 'utilization' pattern varies widely with changing physiological state. For example, infants, adolescents, pregnant women and the aged have different rates of absorption and metabolism and the drug dosage must compensate for these differences. It is also known that the clinical response correlates much better with the plasma concentration of the drug than with the dosage of the drug.

The management of chemotherapy requires an interdisciplinary approach to (1) determine the identity and purity of the drug, (2) study its rate and mechanism of absorption by the body, (3) monitor its distribution in the various organs of the body and the time course, (4) determine its toxic effects and the mechanism and threshold levels, (5) determine the mechanism of detoxification of the organism, the metabolism of the drug in the body, its kinetics and mode of excretion, and (6) study the dose response of the disease, and so on. The specialities that are required for therapy management include pharmacology, pathology, toxicology, clinical chemistry, analytical chemistry, biochemistry and medicine. It must be noted that therapeutic drug monitoring (or TDM) is particularly effective when the drug has a narrow therapeutic range and is potentially toxic beyond a certain dose or when administered for extended periods of time (in case of chronic illnesses). The close relationship between TDM and toxicology is thus obvious.

It may be seen that the essential feature of all the above studies, is that they require an efficient analytical system that can accurately analyse the various samples in which the drug levels are to be monitored and determine the minute quantities of the drugs or their metabolites with a high degree of precision. The method must be unaffected by the matrix effects and be free of interference from the accompanying substances. Since the life of certain drugs in the body may be short, the analytical method must be fast. Undoubtedly, chromatography is the most effective method of analysis that satisfies all the above noted requirements. Here again, GC was extensively used in the 1960s, to be largely replaced in recent years by HPLC in view of the latter's less stringent demands on sample preparation and derivatization.

It must be noted, however, that there are several methods at the clinician's disposal to monitor the drug levels using microquantities of body fluids; e.g. radio-immuno-assay (RIA), enzyme-multiplied immuno-assay techniques (EMIT), enzyme-linked immunosorbent assay (ELISA), substrate-labelled fluorescent immuno-assay (SLFIA), fluorescence polarization immuno-assay (FPI), etc. As the names indicate, these methods are based on immunology and require the appropriate antibodies. However, a large number of drugs for which antibodies are not available (and for which, therefore, these methods cannot be used) do require to be monitored. Also, the immunochemical techniques are subject to interference by other substances present in the medium. Chromatographic methods are, again, needed to separate these components.

The decision in regard to the analytical method is normally based on the chemical structure and the physicochemical properties of the drug and the most probable pathways of biotransformation. However, the specificity and sensitivity required of the analysis is governed by the dosage, its frequency and the pharmacokinetics (i.e. the time course of drug and metabolite levels in the various body fluids, tissues and faeces).

As noted above, chromatography is now used extensively for TDM studies. The low cost, high specificity and sensitivity, speed, the feasibility of automation and compatibility with most biological samples naturally made the LC methods, particularly in the reversed-phase, very attractive for the purpose. The various methods of sample preparation, derivatization, method selection, optimization, detection and data processing have been discussed in general terms in the previous chapters and need not be elaborated here. It may, however, be noted that sample preparation has an important bearing on the precision of the analysis in TDM and toxicology in view of the very minute quantities of the analytes and the very diverse and numerous samples that may have to be handled. The very high-sensitivity detectors often make interference from contaminants a serious problem. The interfering compounds may originate from any source including the glass or plastic used for the manufacture of the sampling vials or even air. Evacuated blood collection tubes, disposable syringes, siliconized needles, which are especially designed for such high-sensitivity analyses are now commercially available.

The chromatographic method used depends on the nature of the drug. Systems for antiarrhythmics, antiasthmatics, antibiotics, anticonvulsants, antidepressants, antineoplastics, cardiovascular drugs, immunosuppressants and sedatives have been reviewed extensively [93,94]. Antidepressants and antihypertensives are invariably analysed by HPLC. The most frequently used stationary phase is octadecylsilyl silica. The mobile phase is either aqueous acetonitrile or aqueous methanol, suitably buffered and occasionally containing an ion-pairing agent as an additive. The three most frequently used detectors are the UV, fluorescence and electrochemical. Many of the drugs require post-column derivatization for adequate sensitivity with the detectors. LC–MS is now becoming more practicable but is still to be widely used. A fully automated HPLC system for drug monitoring consists of a sample processor, a programmable autosampler, connected through two alternating pre-columns (for sample enrichment) to a reversed-phase column, detectors and data processor.

### 9.3.3 Taxonomy

Taxonomy involves classification, identification and nomenclature of living organisms; i.e. plants, animals and microorganisms, with the aim of bringing together or dividing them on a logical basis. The classification of the order, family, genus or species is generally based on readily recognizable, visible characteristics and morphology. In recent years, biochemical, physiological and genetic features have assumed considerable importance. These features are reflected in the nature of the constituents or the secondary metabolites of the organism, by the analysis of which the specific chemical markers for the respective organisms can be established. The study of the variation of the chemical composition (i.e. the distribution of lipids, amino acids, carbohydrates, alkaloids, etc.) in plant, microbial or animal products, and the use of these characters for their classification and identification is called chemotaxonomy.

As long as two decades ago, Harborne [54], by extensive use of paper chromatography, studied the distribution of flavonoids (the colour principles of many flowers, leaves and woods) in a number of plants and established a clear taxonomic basis for their occurrence in different plants. Not only the skeletal variation (i.e. anthocyanins, flavones, isoflavones, chalcones, aurones, etc.) but also hydroxylation, glycosylation and other finer structural features could be used to identify the family and genus of the plant. Subsequently, other chemical classes (particularly the monoterpenoids, mainly due to the convenience of their analysis by GC) have also been used for the purpose [95]. More recently, animal and microbial chemotaxonomy has gained widespread attention.

A variety of techniques is available for studying the chemistry of the microbial cells and for establishing the chemical profiles from the biomass. Chemotaxonomy can possibly provide a better classification of the organisms, trace their evolution and identify the clinically important ones. Extracts, whole defatted cells or whole cells may be used for the analysis. The extracts contain capsular polysaccharides, pigments, nucleosides, nucleotides, flavins and amino acids. The whole defatted cells provide simpler gas chromatograms and facilitate classification and identification of relatively unknown bacteria. Single bacterial colonies could be analysed for their sugar and fatty-acid profile by capillary GC. Single-colony analysis is more advantageous than that of liquid-grown cultures owing to the speed and the small quantity of the material required. Pyrolysis of the bacterial colonies in an inert atmosphere yields a series of low molecular weight, volatile substances which can be readily identified by GC–MS techniques. Pyrolysis GC–MS has been used for characterization of mycobacteria and yeast species. Pyrolysis GC patterns have also been used for classification and recognition of fungi and bacteria.

Fatty acids are the most frequently used lipid constituents in the chemotaxonomic studies on bacteria. Both volatile and non-volatile fatty acids are useful in their characterization. Usually, they occur as free acids in archaebacteria and in the acylated form in eubacteria. The structural variation in bacterial fatty acids include the number of carbon atoms (both odd- and even-numbered carbon chains), branching, unsaturation and hydroxylation. The fatty-acid pattern can be used to differentiate several groups, genera and species of bacteria. The short-chain metabolic fatty acids are useful in the taxonomy of anaerobic bacteria. Cyclopropane fatty acids dominate in Gram-negative bacteria. As discussed earlier, the lipids are conveniently analysed by a variety of chromatographic techniques, particularly, GC. Certain commercial microbial identification systems (e.g. from Hewlett-Packard, U.S.A.) contain fingerprint libraries of fatty-acid profiles from several thousand bacterial strains for easy identification.

Metabolic products of anaerobic bacteria (e.g. lactic, acetic, propionic, butyric, isobutyric and isovaleric acids, ethanol, butanediol, acetone and certain amines) have proved to be of taxonomic value. Thus, the production of triethylamine from acetylchloline is a characteristic of *Proteus* species. The production of ethanol, methanethiol, dimethyl disulphide and trimethylamine is used as criteria for the detection of the enterobacteria such as *Escherichia, Klebsiella, Citrobacter* and *Proteus* species in urine samples. These can be conveniently detected in the fermentation broths by headspace GC techniques. Aliphatic dicarboxylic and phenolic acid metabolites may be directly analysed by HPLC and semi-quantitative profiles obtained. Identification of specific metabolites of bacteria in the human body has proved to be a vital diagnostic tool particularly because these bacteria cannot be cultured outside the body. A very elaborate review on the chemistry of the marker substances and their application to microbial characterization has appeared very recently [96].

More recently, chromatographic methods have also been used to distinguish between animal species. An interesting example is that of identification of strains of bees, one European and the other Africanized [97]. The Africanized bee is produced by interbreeding the European with the African species. The cuticle or the sheath covering the whole body of the bee contains high molecular weight hydrocarbons, characteristic of the species. To identify the species, 60 European bee specimens and 49 Africanized bee specimens were collected, their sheaths extracted with hexane and the hexane evaporated. From the gas chromatograms of the residue, ten peaks were selected and by use of the pattern-recognition techniques, the profile for each class could be made. The results showed that the profiles are distinctly different for the cuticular hydrocarbons and could be effective for taxonomic information concerning the bees.

## 9.4 INDUSTRIAL AND MISCELLANEOUS APPLICATIONS

Several manufacturing industries are closely linked to chemistry and biology. Some of these (like petroleum products and industrial polymers) may be considered to be essentially chemical (chemical technology), while some others, like fermentation and the more recent recombinant DNA products may be essentially biological (biotechnology); most others (like the pharmaceuticals and foods) are on the borderline. Analytical chemistry (particularly chromatography) forms an important component in the management of all these industries, particularly in two areas: (1) process control and (2) quality control. As an analytical method amenable to automation, chromatography has an important role in both these activities. The quality-control applications of chromatography are legion [98]. Use of laboratory information management systems in the quality-control laboratory has been briefly mentioned earlier in this chapter (section 9.2.1.3.4).

The availability of real-time qualitative and quantitative data on process streams would help not only to monitor and control the process, but also to optimize the raw-material consumption, energy and time, and effectively minimize effluent release [99]. The recent emphasis on 'quality cost' of production [100] and the developments in recombinant-DNA and other biotechnologies requires very close and efficient control of every aspect of the processes. The electronic and biotechnological industries also place a very high premium on the purity of the raw materials which must be ensured by efficient analytical methodologies. Many of the chemical process industries and fermentation industries are now essentially automated, but the feedback to the computers is a signal from the analytical instrument. Aliquots are periodically withdrawn from the reaction mixture, analysed and the process parameters altered depending on the results of the analysis.

The simplicity of instrumentation and the total inertness (safety) of the mobile phase make on-line GC preferable. The earlier drawbacks of packed-column GC (with respect to the stationary-phase instability to prolonged use) and the sample introduction problems in open-tubular GC are now largely eliminated with the advent of wide-bore capillary columns and high-temperature resistant bonded phases. Even so, the limited range of samples that can be directly introduced into GC is a handicap and LC is also frequently used. Use of small-particle, reversed-phase packings and small-bore columns largely reduces the risks and disposal problems associated with the use of solvents. More recently, the supercritical-fluid chromatography has become a realistic alternative that combines the favourable features of both GC and LC. Use of solvent modifiers and pressure-programming in SFC permits analysis of a variety of samples with very short cycle time and high analytical throughput.

Two other aspects of applied analytical research in manufacturing industry must be mentioned. One is the product analysis to explore the possible applications. Analysis of coal-derived products and flavours of foods and essential oils are briefly discussed here as examples of this type. The second aspect is analysis for pollution control as a consequence of the industrial growth. Here again, the applications are as diverse as the industrial products themselves and the treatment here is aimed at providing a brief overview of the subject. Some of the special applications of particular chromatographic methods to industrial products have been mentioned in the previous chapters; e.g. size-exclusion chromatography for determination of molecular weight and molecular-weight distribution and pyrolysis GC for characterization of polymers. Large- or process-scale chromatography (section 7.7), though shunned by the industry for a long time in view of the presumed high cost, also finds enormous application for some speciality products, particularly in the field of biotechnology [101].

### 9.4.1 Fuel sources

It is well known that the early boost to the development of GC was given by the petroleum industry and GC remains the most powerful tool for the analysis of the petroleum-derived fuels, i.e. the hydrocarbons [43]. The hydrocarbon fuels have dominated the energy scene for most of the post-war period and are likely to remain so, the depleting petroleum reserves notwithstanding. Chromatographic analysis of hydrocarbons has been discussed earlier (section 9.2.3.4.1). The present discussion is limited to the analysis of the raw materials; namely, the natural gas, petroleum and coal.

The natural gas contains a large number of low-molecular-weight hydrocarbons and permanent gases, including hydrogen, hydrogen sulphide, water vapour and $C_2$ to $C_4$ hydrocarbons. For analysis, the gas is first dried and then passed through absorbing oil (which is mainly the hexane fraction of petroleum ether, maintained at −20°C); The hydrocarbons, except methane are dissolved in the oil and the two fractions thus obtained are analysed. Gas–solid chromatography on Porapak Q (styrene-divinylbenzene co-polymer) or squalane-coated alumina may be used with a thermal conductivity detector, which responds to both organic and inorganic gases. Usually a multicolumn technique with both types of columns is employed.

Petroleum is a complex material with highly variable composition. The physical state may either be in the form of a light brown or black, thin fluid of low viscosity, to a thick paste. The crude oil is normally of low viscosity and dark colour and contains gases and dispersed solids. The major constituents of the oil are hydrocarbons (with 1 to 200 carbon atoms) along with small amounts of sulphur, oxygen and nitrogen compounds as impurities. Chemically these may be saturated aliphatic or alicyclic hydrocarbons, olefins, aromatics, thiols, sulphides, alcohols, phenols, organic acids, and nitrogen heterocycles. The elemental composition of petroleum is approximately as follows: carbon 83–87%, hydrogen 11–15%, sulphur 0.1–6%, nitrogen 0.1–1.5% and oxygen 0.3–1.2% [102].

Coal has a cross-linked and polymeric structure, but compositionally is very similar to that given above. (For this reason, much of the following discussion treats the analysis of the two materials together using the coal products as examples; differences are mentioned wherever applicable.) Coal was the most important source of organic chemicals until the petrochemical industry developed. Coal as a source of fuels again receives attention with increasing cost of petroleum. Several strategies are employed to liquefy coal to volatile fuels, mainly by high-temperature hydrogenolysis. Optimum use of coal would require the knowledge of exact composition. GC, which has been extensively used for the analysis of petroleum, however, cannot be used unless the coal is converted into products which are sufficiently volatile [103]. One such technique is pyrolysis GC, which has been widely employed in coal analysis; both filament- and furnace-type pyrolysers have been used. The polynuclear aromatic hydrocarbons present in coals are often extracted in a Soxhlet apparatus with suitable solvents (e.g. benzene–ethanol, 7:3) for analysis by either GC or LC; the extraction, even from pulverized coal, may require as long as 250 h. However, use of a flow-blending apparatus, originally designed for extraction of organics in rocks, could bring down the period to about 10 min for exhaustive extraction (solvent: chloroform, acetone and methanol; 47:30:23, wt.%). Supercritical extraction with toluene at 350–420°C has also been used for extraction of soluble organics from coal.

As noted above, liquefaction of coal for use as a fuel has considerable economic importance and has been extensively investigated. High-pressure hydrogenation of coal at 450°C converts about 95% of it into benzene-soluble products. Bituminous coals can be converted to about 80%-pyridine solubles in 1 min at 425°C. Extraction of the product with hexane and benzene gives three fractions in the molecular weight range 50–300 (hexane soluble), 300–1000 (benzene soluble, asphaltenes) and 400–2000 (benzene insoluble, pre-asphaltenes). The first fraction, consisting mainly of mono-, di- and tri-cyclic aromatics, is amenable to GC analysis. The other two are usually analysed by LC techniques, particularly, the size-exclusion methods.

Prefractionation using open-column methods is frequently used in coal analysis in view of the complexity of the products. Silica gel and alumina are the natural choice for the purpose. Often they are rendered more selective by impregnating agents like silver nitrate (olefins) and ferric chloride (nitrogenous compounds). Elution of de-asphalted coal liquid (adsorbed on an alumina column) with hexane yields aliphatic hydrocarbons; subsequent elution with benzene, 0.75% ethanol–chloroform and 10% ethanol–tetrahydrofuran furnishes polynuclear aromatics (PNAs), nitrogen-containing PNAs and hydroxylated PNAs, respectively. The benzene fraction, which may contain oxygen- and sulphur-containing PNAs may be refractionated on a silica column and eluted with benzene and benzene–methanol (1:1). The fraction yields PNAs and *O*-PNAs, while the second may contain *S*-PNAs. The *N*-PNA fraction is also refractionated on silica gel; elution with hexane–benzene (1:1), benzene and benzene-ether (1:1) would elute secondary, primary and tertiary nitrogenous PNAs in that order. Further fractionation of coal fuels requires 'utilization' of the subtle structural differences. Thus, separation of *n*-alkanes from branched-chain alkanes is possible by adsorption on molecular sieves (aluminosilicates) or by formation of inclusion complexes with urea or thiourea.

Analysis of alkanes and alkenes present in coal fluids is usually carried out by open-tubular GC on non-polar phases. The chromatograms are used to determine the carbon preference index (CPI), which indicates the predominance of odd over even carbon numbers; the CPI is said to be lower for high-ranking coals. Alkenes are usually present in pyrolysis tars. Isoprenoids are also detected in coals and are naturally derived from the phytyl chain of chlorophylls. Cyclic mono-, di-, tri- and penta-terpenes are most probably formed during thermal or 'bacterial diagenesis' of steroids, terpenoids and squalanes in plants and algae; their detection in the coal fluids may be used as biological markers to determine the evolutionary history of the coal. They are readily identifiable by GC and retention indices or by GC–MS.

Over 200 polynuclear aromatic hydrocarbons (PAHs) in coal tars have been separated using open-tubular non-polar and medium-polar columns and identified by the retention index and GC–MS. The major nuclei found in coals are benzene, naphthalene, phenanthrene, chrysene and picene. The aromatic distribution patterns also reveal the nature and origin of the coals.

Coal extracts, pyrolysates and hydrogenated fluids contain considerable amounts of oxygen, sulphur and nitrogen compounds, which all have an important bearing on the utility of the coal. Phenols are the predominant constituents of these products; they can be analysed by GC and GC–MS techniques after derivatization to acetyl, methyl or trimethylsilyl derivatives. Over 40 phenols have been identified using open-tubular columns coated with Superox 20M (equivalent to the more common Carbowax 20M). The range of carbon numbers of fatty acids present in coals is also characteristic of the coals; e.g. GC analysis of the extracts from sapropelic coals showed that they contained predominantly the $C_{10}$–$C_{36}$-carbon fatty acids, while higher than $C_{18}$ fatty acids are more common in the other coals.

### 9.4.2 Food analysis

The analysis of the important ingredients in food; namely, carbohydrates, proteins, lipids, colourants, vitamins and micronutrients, has been referred to in the previous sections of this chapter, though not specifically in the context of food analysis. The analysis of food assumed considerable importance with the growth of the food industry. Introduction of the various preservatives, colourants, sweeteners, taste modifiers, antioxidants, etc., and the possibility of spoilage and toxin formation during long storage of processed food necessitated a certain degree of regulation. Food analysis, declaration of the ingredients and testing by the regulatory agencies have become a statutory requirement in most countries, out of concern for food safety. There are also several toxins that are either present naturally or formed as metabolites of organisms that infest the food; e.g. mycotoxins and aflatoxins. These and the pesticide residues are harmful to humans even if present in concentrations as low as parts per billion (ppb, $10^{-9}$). Methods of analysis should, therefore, be sensitive enough to routinely and accurately determine the toxin at these levels.

Gas chromatography has been extensively used in food analysis for a variety of compounds, including non-volatile components like carbohydrates, proteins, lipids, organic acids and additives [104]. Most of the non-volatile materials are now routinely analysed by HPLC, mostly by the reversed-phase and ion-exchange methods [105,106]. However, there is one field of activity where GC remains the analytical technique *par excellence* and that is, flavour research.

#### *9.4.2.1 Food flavours and fragrances*

Broadly, the term flavour encompasses several features that contribute to the sensory appeal of food, including the aroma, taste, colour and the texture. The aroma of food is the combined effect of the various constituents present in the raw materials (grain, meat, vegetables, fruits, spices and condiments) and the changes brought about by processing. In addition to the essential food, large quantities of pleasure foods and stimulant foods are consumed mainly for their flavour characteristics. In this category may be included, ice creams, confectionary products, coffee, tea, cocoa, aerated soft drinks and alcoholic beverages. The commercial stake of the role of flavour in food acceptability is thus very significant. The flavour also helps recognition of food by animals and helps digestion by stimulating the production of the gastric juices. A closely related area of immense commercial importance is that of fragrances of flowers and other aromatic material that constitute perfumes.

Apart from the consumer acceptability, there are several reasons for identifying the aroma components in food. A study of the changes occurring during processing, storage and maturation of the food products helps to hasten the process on the one hand and prevent spoilage on the other. It also helps to develop rapid methods of screening for grading the raw materials and quality control on the basis of their flavour-component profile (see Fig. 9.2, p. 426).

The key to the understanding of the flavour of the whole food is to establish the chemical nature of the constituents of the food or its ingredients. The flavour compounds are generally small in size (MW, generally less than 250) and highly volatile. They include several different types of compounds including hydrocarbons, alcohols, esters, acids, carbonyl compounds, mercaptans, sulphides, disulphides, heterocyclic compounds containing oxygen, nitrogen and sulphur, etc. The complexity of the flavour composition may vary from essentially single components (e.g. eugenol in cloves) to over 500 in coffee aroma. A major limitation is that the size of the sample available for analysis is often very low, some of which may be present in parts per million or, as in the case of certain vegetables, parts per billion level. It may be of interest to note that the nose can readily detect 2-methoxy-3-isobutylpyrazine, the character-impact component of the green peppers, at a dilution level of one part in $10^{12}$, a sensitivity not easily attainable even on expensive instruments.

In view of their volatility, gas chromatography (particularly the open-tubular GC) is routinely used for the analysis of flavour compounds. Non-polar (dimethylpolysiloxane) and medium-polar (polyethylene glycol) type of stationary phases and the flame-ionization detector are the most frequently used systems for analysing these materials. Since the quantity available is very small (often a few $\mu$l) and the complexity also very high, it is seldom possible to isolate the compounds for structure determination; one needs to use an on-line detector, GC–MS being the most powerful among them. Several printed and computerized GC–MS libraries are commercially available [107,108]. Other flavour ingredients, like those responsible for the taste (sweeteners and bitter principles), colourants and additives are often analysed by LC methods.

While the chromatographic methods of analysis of flavours and fragrances are fairly straightforward, the main exercise that determines the success of the analysis is the method of sample preparation. As noted above, they may be extremely volatile and, unless proper care is taken, many of them may be lost during the process. They are often present in extremely diluted form, associated with a wide range of matrices, including flowers, fruits, animal glands (e.g. musk odours) and clay. In their natural habitat, they may be encapsulated in the matrix and associated with a variety of chemical entities like the lipids, carbohydrates, proteins and water. They may be widely differing in polarities, solubilities and volatilities. Several of the sample preparation methods discussed in the beginning of this chapter may be used, depending on the sample and the purpose of the analysis. Steam distillation has been a popular method for isolation of the essential oils of flowers, fruits and spices on a larger scale. However, the possibility of artefact formation and loss of the more volatile constituents during the prolonged heat treatment would prevent its use as a general method. Steam distillation with simultaneous solvent extraction (e.g. in a Likens–Nickerson type apparatus), sweep co-distillation with a volatile solvent, headspace collection and supercritical-fluid extraction are the most frequently used methods for isolation and concentration of the flavours for analysis [109,110]. Two typical examples of the methods of flavour analysis are exemplified below.

#### *9.4.2.1.1 Flavour of distilled beverages*

The distilled beverages are traditionally obtained by fermentation of sugar-cane molasses (for rum), fruits (brandies), and barley malt (whiskey) followed by distillation and maturation in oak casks for several years. Some of the volatile congeners formed during the yeast fermentation are lost during the process. A number of constituents (mainly lignins and sugars) are extracted from the wood and, over the period, interact among themselves and with the congeners, as well as with the water and ethanol, to yield a complex blend of flavours that characterize the liquor. Three problems characterize the analysis of these flavours: (1) the very large number of compounds (over 500 identified in rum), (2) the wide variety of compounds (alcohols, esters, acetals, phenols, sulphur compounds, heterocycles and acids) and (3) very low concentration. The matrix also may cause problems in the sense that most of the components are insoluble in water but volatile with it and are very soluble in ethanol which constitutes about 40% of the product. The high concentration of ethanol prevents direct use of solvent extraction, steam distillation or sweep co-distillation methods and necessitates several-fold dilution of an already dilute sample.

The flavour analysis of rum and correlation of the separated components with their sensory characteristics is a typical example of the methodologies used in this field [111]. The liquor (rum) was diluted so that it contained less than 10% alcohol and about 2000 l of it was extracted in 75 l batches with a mixture of *n*-pentane and ether (2:1) for 24 h. Each batch of the organic extract was carefully distilled on a 2 m packed distillation column to remove about 80% of the solvent, which was reused for the next batch. The combined concentrate was reconcentrated to about 1 l. Acids and phenols were removed from the concentrate by extraction with dilute aqueous sodium hydroxide. Phenols were liberated from the basic extract by adjusting the pH to 9.5 and acids subsequently at pH 1. These were etherified or esterified using diazomethane prior to GC analysis. When the pentane and ether were removed from the neutral portion, the residue contained a large excess of the so-called 'fusel' oil that is essentially isoamyl alcohol. The flavour components were isolated first by room-temperature distillation at 10 mm Hg and then by molecular distillation to yield about 50 g of oil with the characteristic odour of the liquor. This was separated into 22 fractions on an open-column of silica gel using gradient elution techniques with pentane–isopropyl chloride–ether mixtures. Of these, two non-polar and 11 polar fractions could be selected on the basis of sensory evaluation for further analysis. Further fractionation of these selected portions was carried out on preparative HPLC using a 50 cm × 9.4 mm (o.d.) column packed with microparticulate silica (10 $\mu$m) and gradient elution with the same solvents as mentioned above; 57 fractions could be collected. From these, organoleptically selected fractions were further separated by preparative GC on open-tubular SCOT columns (50 m × 0.75 mm i.d.) using different stationary phases (including among others, the more common Carbowax 20M and Silicone OV-17). The individual compounds were then identified by IR, NMR and, occasionally, synthesis.

The characteristic rum aroma could be isolated by the headspace collection technique, using Chromosorb 102 for trapping the volatiles swept out by a draft of helium. The trapped volatiles were desorbed by helium at 120°C and collected in a cooled (−196°C) 1 mm capillary Y-tube for GC and GC–MS analysis. Over 400 compounds could be identified by the above method. Using an effluent splitter and sniffing the components as they elute, the characteristic aromas could be recorded.

*9.4.2.1.2 Essential oils*

The analysis of the essential oil from the Bulgarian rose by Kovats remains one of the masterpieces in flavour and fragrance research. Though the work was actually carried out over twenty-five years ago, the economic importance of the work prevented its publication. Details of the investigation, describing the isolation and identification of 127 compounds (hydrocarbons, alcohols, aldehydes, ketones, ethers, esters, acids and phenols) from the Bulgarian rose oil appeared only recently [112]. The number of compounds subsequently identified increased to 275 but the importance of the original work, which amply demonstrates the complementary nature of the different analytical methods, remains undiminished. The methodology is an interesting blend of classical and relatively more recent chromatographic methods.

The Bulgarian rose oil (1200 g) was distilled (using a Vigreux column at a reduced pressure of 2 mm; traps maintained at −70°C with solid $CO_2$ and acetone) to yield 62.1 g of volatile oil which separated into two fractions. Saturation of the aqueous layer with potassium carbonate left an oily neutral fraction (45.2 g). The oil was redistilled yielding two main fractions, one boiling 74–77°C (720 mm; 21.3 g, fraction A1) and the other at 30–60°C (12 mm; 11.8 g). The second fraction (hydrocarbons) was chromatographed on silica gel and using pentane (2.9 g hydrocarbons) and then with methanol (4.3 g oil, A2).

The main (undistillable) fraction of 1140 g was diluted with pentane (400 ml) and extracted with aqueous sodium carbonate (2.5 g acids, fraction B), sodium hydroxide (6.5 g phenols, C) and 2N hydrochloric acid (0.25 g bases, D), leaving 1130.8 g neutral fraction. 140 g of this fraction (+80 ml pentane) was subjected displacement chromatography on silica gel (600 g) using pentane, 1-chloropropane and methanol as displacers. After loading the sample first in an upward direction and irrigating with pentane, the column was turned 180° for washing with 1-chloropropane; it was again brought to the original position and washed with methanol. The remainder of the oil was similarly fractionated in eight batches to yield totally 236.1 g non-polar (with pentane, E), 74.6 g slightly polar (with 1-chloropropane, F) and 815.6 g of polar (G) fractions.

Fractions B and C (acids and phenols) were treated with diazomethane to yield the corresponding methyl esters and ethers and then subjected to GC analysis on a 100 m capillary column (1 mm i.d.) using Apiezon L as the stationary phase. Identification of the individual peaks was by comparison of the retention indices. By a judicious combination of distillation, chemical fractionation, derivatization and preparative gas chromatography, all the other compounds were isolated and identified by techniques such as GC, UV, IR, NMR, and MS. Quantitation was by TCD detection and peak-area measurement.

***9.4.2.2 Pesticide-residue analysis***

The rise in population and the consequent increase in the demand for food requires the use of a variety of chemicals in agriculture. The use of the so-called agrochemicals is aimed at directly or indirectly aiding food output; they include fertilizers, plant-growth regulators and pesticides. Among these, pesticides form the largest group, with thousands of them in use. They may be classified as insecticides, fungicides, herbicides, nematocides, miticides, molluscicides, etc. Chemically, they may be grouped into carbamates, triazines, organochlorine compounds, organophosphorus and sulphur compounds, etc. The effectiveness of these chemicals in controlling the pests that are harmful to the crops has resulted in almost unlimited, and somewhat indiscriminate, application. While many of them, except the organochlorine compounds, undergo rapid natural degradation, their continued use has resulted in the accumulation of the compounds in food and the environment. Some of these compounds (e.g. polychlorinated biphenyls, or PCBs) could be highly hazardous to health and their occurrence as residues in food needs to be monitored.

A majority of the pesticides and their residues (after the natural, often partial, degradation) are non-polar to medium polar and volatile and can be readily analysed by GC, TLC or HPLC [41,113–115]. The most commonly used stationary phases in GC are the polysiloxanes, based on methyl-, methyl phenyl-, cyanoethylmethyl-, and trifluoropropylmethyl polysiloxanes; the optimum temperature is generally 200°C. Both packed and open-tubular columns are used for routine work. When packed columns are used, care must be taken to avoid on-column decomposition by silanization (trimethylsilylation) of all active sites on the solid support. Apart from their sufficient volatility, the most important factor responsible for the effective use of GC is the sensitivity of some of the specific detectors, particularly the electron-capture detector. The ECD, whose sensitivity increases with the number of halogen atoms present in the molecule, can detect chlorinated pesticide residues in grain and vegetables, serum of exposed individuals, water, soil or air at less than 1 part per trillion ($10^{-12}$). With optimum configuration and operating conditions, the linear dynamic range is $10^4$. Similarly, the thermionic or the alkali-flame ionization detector is especially useful for the compounds containing phosphorus and sulphur. By use of appropriate filters (at two specific wavelengths), both phosphorus and sulphur can be detected and determined simultaneously, with a sensitivity of $10^{-8}$ g for P and $4 \times 10^{-8}$ g for S. GC–MS is very useful for trace detection and simultaneous identification.

The HPLC retention data for several pesticides have been compiled [116]. HPLC and TLC is carried out on normal or, more frequently on reversed phases, using UV or fluorescence detection, depending on the nature of the samples; often, post-column derivatization techniques need to be used to form chromogenic or fluorogenic compounds, suitable for sensitive detection. Similar approaches on high-performance TLC yield adequate sensitivity for a number of pesticides, but accurate determination would require extensive sample clean-up (e.g. using a small column of Florisil or some of the commercially available disposable extraction columns). When a large number of samples are to be analysed for residue detection, this method may be preferable.

Sample preparation for pesticide analysis by HPLC and TLC usually involves extraction with a solvent, evaporation and partitioning of the residue between hexane and acetonitrile to remove the fat. The acetonitrile fraction is evaporated and fractionated on a Florisil column; elution with hexane yields polychlorinated biphenyls, while the more polar pesticides may be eluted with increasing proportion of ether in hexane. The continuing serial publication, edited by Zweig [117] is a good source of information in the field; the work describes detailed analytical methodology for the determination of several individual pesticides in a variety of samples.

### 9.4.3 Environmental applications

The industrial revolution has brought an enormous improvement in life-styles and comforts, but concomitantly has polluted the environment. Large-scale application of pesticides severely contaminates (and denatures) the soil, industrial waste disposal renders river waters undrinkable and automobiles probably make fresh air a rare commodity in urban areas. The workers in many industrial sites are constantly exposed to much more serious occupational health hazards. A variety of organic chemicals, though they may be present in trace quantities in the environment, are carcinogenic or otherwise harmful to health.

The concern for occupational safety and public health has made several countries impose stringent measures to control, regulate and monitor industrial effluent management and waste disposal. In the United States of America, for example, three agencies are responsible for determining and controlling the levels of pollutants in the environment; these are the U. S. Environmental Protection Agency (USEPA), Occupational Safety and Health Administration (OSHA) and the National Institute for Occupational Safety and Health (NIOSH). The USEPA has defined 129 specific toxic (priority) pollutants that need to be analysed in environmental samples. These include acrolein, acrylonitrile, benzene, benzidine, various chloroalkanes, chlorobenzenes, chloronaphthalenes, chlorinated ethers, halomethanes, chlorofluoromethanes, nitroaromatics, nitrophenols, *N*-nitrosoamines, chlorophenols, polynuclear aromatic hydrocarbons, phthalate esters, various pesticides and their metabolites. The OSHA has identified 400 compounds whose presence in the working environment must be controlled. The NIOSH is responsible for providing the appropriate, reliable methods for sampling and analysis.

Chromatographic methods are extensively used in environmental analysis. Application of GC [118], LC [119] and TLC [120] methods in environmental analysis have been extensively reviewed. The sample preparation methods (section 9.2.1.1), particularly the methods of enrichment of the trace quantities of pollutants in air and water (section 9.2.1.4) have been described earlier in this chapter. The present account is, therefore, meant to be a brief overview of the subject.

#### *9.4.3.1 Air pollutants*

Industrial chimneys and automobile exhausts are the most visible sources of air pollution. Air is also in constant contact with water and soil and many of the volatiles in these two systems can partition into it. Air may carry as many as one hundred thousand organic compounds, hundreds of which can be carcinogenic. The most serious among these are the polynuclear aromatic hydrocarbons (PNAs or PAHs) and the polychlorinated hydrocarbons (PCHs), particularly the polychlorinated biphenyls (PCBs). They may be present in parts per billion or, often, even parts per trillion level and this fact, coupled with the large numbers of the compounds, poses a formidable challenge to the analytical chemist. Determinations at such low levels are valid only when carried out in relation to some standards. Solid or liquid standards are more stable than the gaseous standards because the latter is more susceptible to loss by adsorption or diffusion through the container wall and to photochemical decomposition. Standards at very low levels (ppb range) must be prepared as and when needed. A number of standardization procedures are available [118]. Data on air pollutants may be expressed as volume ratio (ppm or ppb) or by weight/volume as, for example, $mg/m^3$ or $\mu g/m^3$.

As already mentioned, sampling and concentration are crucial manipulations in trace analysis of air. The moisture content, even in not-so-humid air, can be several orders of magnitude larger compared to the organics and can interfere with the trapping procedures. Plastic bags or evacuated containers may be used for sample collection. Preferably, on-site concentration using cryogenic (liquid air or liquid nitrogen) traps or adsorbents (carbon, active metals or organic porous polymers) may be employed, in a manner similar to headspace trapping techniques. Both packed and open-tubular GC have been used in combination with FID, ECD and MS for the analysis of air pollutants. The stationary phases include porous polymers, non-polar phases (dimethyl polysiloxane) and medium-polar phases (polyethylene glycol).

#### *9.4.3.2 Water pollutants*

Because of its high dissolving power, water naturally contains millions of compounds even when unpolluted. The composition varies depending on the source, which may be ground water or water from lakes, rivers or seas. Added to these are the waters used for the various irrigation, cleansing and scavenging purposes in agriculture, industries and sewage, respectively.

Any of the several methods discussed earlier may be used for preparing samples for chromatography. Direct injection and injection on pre-columns containing water absorbers are sometimes used but a pre-concentration step is often required. Solvent extraction or adsorption of organics on charcoal or on Amberlite XAD-2 is common. The organics adsorbed on the porous polymers can be fractionated into acidic, basic and neutral pollutants by washing the polymer with very dilute (0.05 M) sodium hydroxide, 1 M HCl and solvent ether, respectively.

Only about 5% of the organics normally present in water are volatile enough for analysis by GC. RPLC may be used for both pre-concentration and subsequent chromatographic analysis. The methods of analysis and detection naturally depend on whether complete analysis, or detection, or determination of specific compounds is required.

#### *9.4.3.3 Soil chemistry*

Many of the variety of chemicals (e.g. pesticides) used in agriculture leave residues in the soil and these eventually find their way into foods; the methods of their analysis have been discussed earlier. Even without the added chemicals, the soil chemistry is extremely complex, being composed of water, minerals, organic matter, air and a large number of insects, worms, mites, etc., in addition to billions of bacteria, fungi, algae and protozoa. The micro-organisms convert plant and animal biomass into a variety of compounds; they are also capable of degrading several of the pesticides used. Chemicals in the soil thus vary widely from place to place and, at the same place, from time to time.

The quality of the soil is reflected by the gases and other compounds present in the soil, which include carbon dioxide from the respiration of organisms, ammonia and oxides of nitrogen from microbial and chemical reactions in the soil, organosulphur compounds, such as carbon disulphide, carbonyl sulphide and hydrogen sulphide, formed by decomposition of the sulphur-amino acids, hydrocarbons of either organic or mineral origin, volatile fatty acids, alcohols, carbonyl compounds, etc.

While there are no specific guidelines for soil analysis, any of the common sample preparation methods can be used including headspace analysis and solvent extraction. Pesticides and PCBs can be extracted with iso-octane, while semivolatiles like aromatic hydrocarbons are extractable with dichloromethane. The soil is preferably ground to about 1 mm sized particles prior to extraction.

## REFERENCES

[1] C. F. Poole and S. A. Schuette, *Contemporary Practice of Chromatography*, Elsevier, Amsterdam, 1984.
[2] T. L. Peters, *Anal. Chem.*, **52**, 211 (1980).
[3] M. Godefroot, P. Sandra, and M. Verzele, *J. Chromatogr.*, **203**, 325 (1981).
[4] G. Hawk and J. Strimaitis (eds.), *Advances in Laboratory Automation Robotics*, Zymark Corp., Marlboro, Massachusetts, 1984,1985.
[5] C. H. Lochmueller, K. R. Lung, and M. R. Kushman, *J. Chromatogr. Sci.*, **23**, 429 (1985).
[6] M. V. Budahegyi, E. R. Lombosi, T. S. Lombosi, S. Y. Meszaros, Sz. Nyiredy, G. Tarjan, I. Timar, and J. M. Takacs, *J. Chromatogr.*, **271**, 213 (1983).
[7] G. Vernin and M. Chanon (eds.), *Computer Aids to Chemistry*, Ellis Horwood, Chichester, 1986.
[8] T. Wang and Y. Sun, *J. Chromatogr.*, **407**, 79 (1987).
[9] J. K. Haken, *Adv. Chromatogr.*, **14**, 367 (1976).
[10] J. E. Picker and R. E. Sievers, *J. Chromatogr.*, **203**, 29 (1981).
[11] D. A. Cronin, *J. Chromatogr.*, **64**, 25 (1972).
[12] M. Beroza and M. N. Inscoe, in *Ancillary Techniques of Gas Chromatography*, L. S. Ettre and W. H. McFadden, (eds.), Wiley, New York, 1969.
[13] J. Novak, *Adv. Chromatogr.*, **11**, 1 (1974).
[14] F. J. Debbrecht, in *Modern Practice of Gas Chromatography* R. L. Grob (ed.), Wiley, New York, 1985.
[15] E. Katz (ed.), *Quantitative Analysis Using Chromatographic Methods,* Wiley, Chichester, 1987.
[16] C. E. Reese, *J. Chromatogr. Sci.*, **18**, 201 (1980).
[17] H. M. Richard, G. F. Russel, and W. G. Jennings, *J. Chromatogr. Sci.*, **9**, 560 (1971).
[18] K. Beyermann, *Organic Trace Analysis*, Ellis Horwood, Chichester, 1984.
[19] G. R. Umbreit, in *Modern Practice of Gas Chromatography* R. L. Grob (ed.), Wiley, New York, 1985.
[20] S. Ahuja (ed.), *Ultratrace Analysis of Pharmaceutical and Other Compounds of Interest*, Wiley, New York, 1986.
[21] R. J. Laub and R. L. Pecsok, *Physicochemical Applications of Gas Chromatography,* Wiley, New York, 1978.
[22] J. R. Conder and C. L. Young, *Physicochemical Measurement by Gas Chromatography,* Wiley, New York, 1979.
[23] M. A. Kaiser and C. Dybowski, in *Modern Practice of Gas Chromatography,* R. L. Grob (ed.), Wiley, New York, 1985.
[24] T. L. Hafkenscheid and E. Tomlinson, *Adv. Chromatogr.*, **25**, 1 (1986).
[25] G. Schwedt, *Chromatographic Methods in Inorganic Analysis,* Huethig, Heidelberg, 1981.
[26] T. R. Crompton, *Gas Chromatography of Organometallic Compounds,* Plenum Press, New York, 1982.
[27] M. Lederer, in *Chromatography* E. Heftmann (ed.), Part B, Elsevier, Amsterdam, 1983.
[28] P. C. Uden, in *Inorganic Chromatographic Analysis,* J. C. Macdonald (ed.), Wiley, New York, 1985.
[29] F. C. Smith, Jr. and R. C. Chang, *The Practice of Ion Chromatography,* Wiley, New York, 1983.
[30] D. T. Gjerde and J. S. Fritz, *Ion Chromatography,* Huethig, Heidelberg, 1987.
[31] L. Ebdon, S. Hill, and R. W. Ward, *Analyst,* **111**, 1113 (1986); **112**, 1 (1987).
[32] P. C. Uden, *J. Chromatogr.*, **313**, 3 (1984).
[33] H. Veening and B. R. Willeford, *Adv. Chromatogr.*, **22**, 117 (1983).

[34] M. Quereshi (ed.), *Handbook of Chromatography: Inorganics,* CRC Press, Boca Raton, Florida, 1986.
[35] T. S. Ma and R. C. Rittner, *Modern Organic Elemental Analysis,* Marcel Dekker, New York, 1979.
[36] V. Rezl, *J. Chromatogr.,* **251**, 35 (1982).
[37] R. H. Liu and W. W. Ku, *J. Chromatogr.,* **271**, 309 (1983).
[38] R. Dappen, H. Arm, and V. R. Meyer, *J. Chromatogr.,* **373**, 3 (1986).
[39] D. W. Armstrong, *Anal. Chem.,* **59**, 84A, (1987).
[40] W. H. Pirkle and T. C. Pochapsky, *Adv. Chromatogr.,* **27**, 73 (1987).
[41] E. Heftmann (ed.), *Chromatography: Fundamentals and Applications of Chromatographic and Electrophoretic Methods,* Elsevier, Amsterdam, 1983.
[42] E. R. Adlard, A. W. Bowen, and D. G. Salmon, *J. Chromatogr.,* **186**, 207 (1980).
[43] K. H. Algelt and T. H. Gouw (eds.), *Chromatography in Petroleum Analysis,* Marcel Dekker, New York, 1979.
[44] M. S. F. Lie Ken Jie, *Adv. Chromatogr.,* **18**, 1 (1980).
[45] H. K. Mangold, *CRC Handbook of Chromatography: Lipids* , CRC Press, Boca Raton, Florida, 1984.
[46] A. Kuksis and J. J. Myher, *J. Chromatogr.,* **379**, 57 (1986).
[47] Y. Masada, *Analysis of Essential Oils by Gas Chromatography and Mass Spectrometry,* Wiley, New York, 1976.
[48] D. R. Reeve and A. Crozier, in *Isolation of Plant Growth Substances,* J. R. Hillman (ed.), Cambridge University Press, Cambridge, 1978.
[49] R. F. Taylor, *Adv. Chromatogr.,* **22**, 157 (1983).
[50] J. C. Touchstone, *Steroids,* CRC Press, Boca Raton, Florida, 1986.
[51] S. Gorog, *Quantitative Analysis of Steroids,* Elsevier, Amsterdam, 1983.
[52] Z. J. Petryka, *Adv. Chromatogr.,* **22**, 215 (1983).
[53] S. Roy, *J. Chromatogr.,* **391**, 19 (1987).
[54] J. B. Harborne, *Comparative Biochemistry of Flavonoids,* Academic Press, New York, 1967.
[55] S. S. Churms, *Handbook of Chromatography: Carbohydrates,* CRC Press, Boca Raton, Florida, 1982.
[56] K. Kakehi and S. Honda, *J. Chromatogr.,* **379**, 27 (1986).
[57] T. Hanai, *Adv. Chromatogr.,* **25**, 279 (1986).
[58] K. Robards and M. Whitelaw, *J. Chromatogr.,* **373**, 81 (1986).
[59] Z. Deyl, J. Hyanek, and M. Horakova, *J. Chromatogr.,* **379**, 177 (1986).
[60] R. F. Pfeifer and D. W. Hill, *Adv. Chromatogr.,* **22**, 37 (1983).
[61] J. L. Meek, *Proc. Natl. Acad. Sci. U.S.A.,* **77**, 1632 (1980).
[62] I. Kerese, *Methods of Protein Analysis,* Ellis Horwood, Chichester, 1984.
[63] W. S. Hancock, *CRC Handbook of HPLC Separation of Amino Acids, Peptides and Proteins,* CRC Press, Boca Raton, Florida, 1984.
[64] H. Schott, *Affinity Chromatography: Template Chromatography of Nucleic Acids and Proteins,* Marcel Dekker, New York, 1984.
[65] P. R. Brown (ed.), *HPLC in Nucleic Acid Research: Methods and Applications,* Marcel Dekker, New York, 1985.
[66] P. A. Perrone and P. R. Brown, in *Ion-Pair Chromatography: Theory and Biological and Pharmaceutical Applications,* M. T. W. Hearn (ed.), Marcel Dekker, New York, 1985.
[67] L. W. McLaughlin, *Trends Anal. Chem.,* **5**, 215 (1986).
[68] M. J. Weinstein and G. H. Wagman, *Antibiotics: Isolation, Separation and Purification,* Elsevier, Amsterdam, 1978.
[69] G. H. Wagman and M. J. Weinstein, *Chromatography of Antibiotics,* Elsevier, Amsterdam, 1984.
[70] V. Betina, *Meth. Enzymol.,* **43**, 100 (1975).
[71] J. Green and S. Marcinkiewicz, *Chromatogr. Rev.,* **5**, 58 (1963) and accompanying papers (Parts I–VII).
[72] E. Tomlinson, *J. Chromatogr.,* **113**, 1 (1975).
[73] J. K. Haken, *Adv. Chromatogr.,* **14**, 367 (1976).
[74] R. Kaliszan, *CRC Crit. Rev. Anal. Chem.,* **16**, 323 (1986).
[75] R. Kaliszan, *Quantitative Structure–Chromatographic Retention Relationships,* Wiley, New York, 1987.
[76] F. A. Vandelheuvel and A. S. Court, *J. Chromatogr.,* **39**, 1 (1969).
[77] J. Raymer, D. Wiesler, and M. Novotny, *J. Chromatogr.,* **325**, 13 (1985).
[78] R. M. Smith, *Adv. Chromatogr.,* **26**, 277 (1987).
[79] J. K. Baker and C.-Y. Ma, *J. Chromatogr.,* **169**, 107 (1979).
[80] S. Toon, J. Mayer, and M. Rowland, *J. Pharm. Sci.,* **73**, 625 (1984).
[81] F. Lottspeich, *J. Chromatogr.,* **326**, 321 (1985).

[82] *Pharmacia FPLC System: High Performance Purification of Biomolecules,* Promotional Literature, Pharmacia AB, Uppsala, Sweden, 1986.
[83] A. J. Clifford, *Adv. Chromatogr.,* **14**, 1 (1976).
[84] A. DeLeenheer and W. Lambert (eds.), *Modern Chromatographic Analysis of the Vitamins,* Marcel Dekker, New York, 1985.
[85] R. L. Patience and E. S. Penny, *Adv. Chromatogr.,* **27**, 37 (1987).
[86] D. M. Desiderio, *Adv. Chromatogr.,* **22**, 1 (1983).
[87] E. Gelpi, *Adv. Chromatogr.,* **26**, 321 (1987).
[88] A. M. Krstulovic (ed.), *Quantitative Analysis of Catecholamines and Related Compounds,* Ellis Horwood, Chichester, 1986.
[89] Z. Deyl and C. Sweeley (eds.), Profiling of Body Fluids and Tissues, (*J. Chromatogr.,* Special Issue, **379**), Elsevier, Amsterdam, 1986.
[90] J. F, Holland, J. J. Leary, and C. C. Sweely, *J. Chromatogr.,* **379**, 3 (1986).
[91] N. Seiler, *J. Chromatogr.,* **379**, 157 (1986).
[92] H. M. Liebich, *J. Chromatogr.,* **379**, 347 (1986).
[93] S. H. Y. Wong (ed.), *Therapeutic Drug Monitoring and Toxicology by Liquid Chromatography,* Marcel Dekker, New York, 1985.
[94] I. D. Watson, *Adv. Chromatogr.,* **26**, 117 (1987).
[95] J. B. Harborne and B. L. Turner, *Plant Chemosystematics,* Academic Press, New York, 1984.
[96] I. Brondz and I. Olsen, *J. Chromatogr.,* **379**, 367 (1986).
[97] B. Lavine and D. Carlson, *Anal. Chem.,* **59**,468A (1987).
[98] K. Tsuji and W. Morozowich (eds.), *GLC and HPLC Determination of Therapeutic Agents,* Marcel Dekker, New York, 1978.
[99] K. J. Clevett, *Process Analyzer Technology,* Wiley, New York, 1986.
[100] J. B. Callis, D. L. Illman, and B. R. Kowalski, *Anal. Chem.* , **59**, 624A (1987).
[101] J. A. Asenjo and J. Hong (eds.), *Separation, Recovery and Purification in Biotechnology,* American Chemical Society, Washington, DC, 1986.
[102] J. A. Kent (ed.), *Riegel's Handbook of Industrial Chemistry*, 7th edn., Reinhold, New York, 1974.
[103] J. M. Charlesworth, *Fuel Proc. Technol.,* **16**, 99 (1987).
[104] R. A. Barford and P. Magidman, in *Modern Practice of Gas Chromatography,* R. L. Grob (ed.), Wiley, New York, 1985.
[105] R. Macrae (ed.), *HPLC in Food Analysis,* Academic Press, New York, 1982.
[106] D. W. Gruenwedel and J. R. Whitaker (eds.), *Food Analysis: Principles and Techniques,* Vol. 4, Marcel Dekker, New York, 1986.
[107] Y. Masada, *Analysis of Essential Oils by Gas Chromatography and Mass Spectrometry,* Wiley, New York, 1976.
[108] M. C. ten Noever de Brauw, J. Bouman, A. C. Tas, and G. F. LaVos, *Compilation of Mass Spectra of Volatile Compounds in Foods, Vol. 1–10, Central Institute of Nutrition and Food Research-TNO, The Netherlands, 1980.*
[109] R. Teranishi, R. A. Flath, and H. Suginawa (eds.), *Flavor Research,* Marcel Dekker, New York, 1981
[110] D. A. Cronin, in *Food Flavours,* I. D. Morton and A. J. McLeod (eds.), Elsevier, Amsterdam, 1982.
[111] R. ter Heide, H. Sclaap, H. J. Wobben, P. J. deValois, and R. Timmer, in *The Quality of Foods and Beverages,* G. Charalambous and G. Inglett, (eds.), Vol. 1, Academic Press, New York, 1981.
[112] E. sz. Kovats, *J. Chromatogr.,* **406**, 185 (1987).
[113] K. G. Das (ed.), *Pesticide Analysis,* Marcel Dekker, New York, 1981.
[114] S. Blackburn, *CRC Handbook of Chromatography: Pesticides,* CRC Press, Boca Raton, Florida, 1986.
[115] G. Zweig, in *Modern Methods of Food Analysis,* K. K. Stewart and J. R. Whitaker (eds.), AVI, Westport, Connecticut, 1984.
[116] J. F. Lawrence, *High-Performance Liquid Chromatography of Pesticides,* Academic Press, New York, 1982.
[117] G. Zweig (ed.). *Analytical Methods for Pesticides and Plant Growth Regulators,* Vol. 1–13, Academic Press, New York, 1963–84.
[118] R. L. Grob (ed.), *Modern Practice of Gas Chromatography,* Wiley, New York, 1985.
[119] J. F. Lawrence (ed.), *Liquid Chromatography in Environmental Analysis,* Humana, Clifton, New Jersey, 1984.
[120] J. Touchstone and J. Sherma (eds.), *Techniques and Applications of Thin-Layer Chromatography,* Wiley, New York, 1985.

# Chemical Index

Trade names are printed in bold-face type. Bold page numbers indicate chemicals used as stationary phases.

# General index

Trade names are printed in bold-face type. Bold page numbers indicate whole chapters or sections on the subject.